Living Shorelines
The Science and Management
of Nature-Based Coastal Protection

Marine Science Series

The CRC Marine Science Series is dedicated to providing state-of-the-art coverage of important topics in marine biology, marine chemistry, marine geology, and physical oceanography. The series includes volumes that focus on the synthesis of recent advances in marine science.

CRC MARINE SCIENCE SERIES

SERIES EDITORS

Michael J. Kennish, Ph.D. and Judith S. Weis

PUBLISHED TITLES

Acoustic Fish Reconnaissance, I.L. Kalikhman and K.I. Yudanov

Artificial Reef Evaluation with Application to Natural Marine Habitats, William Seaman, Jr.

The Biology of Sea Turtles, Volume I, Peter L. Lutz and John A. Musick

Chemical Oceanography, Third Edition, Frank J. Millero

Coastal Ecosystem Processes, Daniel M. Alongi

Coastal Lagoons: Critical Habitats of Environmental Change, Michael J. Kennish and Hans W. Paerl

Coastal Pollution: Effects on Living Resources and Humans, Carl J. Sindermann

Climate Change and Coastal Ecosystems: Long-Term Effects of Climate and Nutrient Loading on Trophic Organization, Robert J. Livingston

Ecology of Estuaries: Anthropogenic Effects, Michael J. Kennish

Ecology of Marine Bivalves: An Ecosystem Approach, Second Edition, Richard F. Dame

Ecology of Marine Invertebrate Larvae, Larry McEdward

Ecology of Seashores, George A. Knox

Environmental Oceanography, Second Edition, Tom Beer

Estuarine Indicators, Stephen A. Bortone

Estuarine Research, Monitoring, and Resource Protection, Michael J. Kennish

Estuary Restoration and Maintenance: The National Estuary Program, Michael J. Kennish

Eutrophication Processes in Coastal Systems: Origin and Succession of Plankton Blooms and Effects on Secondary Production in Gulf Coast Estuaries, Robert J. Livingston

Habitat, Population Dynamics, and Metal Levels in Colonial Waterbirds: A Food Chain Approach, Joanna Burger, Michael Gochfeld

Handbook of Marine Mineral Deposits, David S. Cronan

Handbook for Restoring Tidal Wetlands, Joy B. Zedler

Intertidal Deposits: River Mouths, Tidal Flats, and Coastal Lagoons, Doeke Eisma

Living Shorelines: The Science and Management of Nature-Based Coastal Protection, Donna Marie Bilkovic, Molly M. Mitchell, Megan K. La Peyre, and Jason D. Toft

Marine Chemical Ecology, James B. McClintock and Bill J. Baker

Ocean Pollution: Effects on Living Resources and Humans, Carl J. Sindermann

Physical Oceanographic Processes of the Great Barrier Reef, Eric Wolanski

Pollution Impacts on Marine Biotic Communities, Michael J. Kennish

Practical Handbook of Estuarine and Marine Pollution, Michael J. Kennish

Practical Handbook of Marine Science, Third Edition, Michael J. Kennish

Restoration of Aquatic Systems, Robert J. Livingston

Seagrasses: Monitoring, Ecology, Physiology, and Management, Stephen A. Bortone

Trophic Organization in Coastal Systems, Robert J. Livingston

Living Shorelines
The Science and Management
of Nature-Based Coastal Protection

Edited by
Donna Marie Bilkovic
Molly M. Mitchell
Megan K. La Peyre
Jason D. Toft

CRC Press
Taylor & Francis Group
Boca Raton London New York

CRC Press is an imprint of the
Taylor & Francis Group, an **informa** business

CRC Press
Taylor & Francis Group
6000 Broken Sound Parkway NW, Suite 300
Boca Raton, FL 33487-2742

First issued in paperback 2020

ISBN 13: 978-0-367-57383-6 (pbk)
ISBN 13: 978-1-4987-4002-9 (hbk)

Library of Congress Cataloging-in-Publication Data

Names: Bilkovic, Donna Marie, editor.
Title: Living shorelines : the science and management of nature-based coastal
protection / [edited by] Donna Marie Bilkovic, Molly M. Mitchell, Megan K.
La Peyre, and Jason D. Toft.
Description: Boca Raton, FL : Taylor & Francis, 2017. | Series: Marine
science series | Includes bibliographical references and index.
Identifiers: LCCN 2016039319| ISBN 9781498740029 (hardback : alk. paper) |
ISBN 9781498740036 (ebook)
Subjects: LCSH: Coastal zone management. | Shore protection. | Coastal
ecology.
Classification: LCC HT391 .L58 2017 | DDC 333.91/7--dc23
LC record available at https://lccn.loc.gov/2016039319

Visit the Taylor & Francis Web site at
http://www.taylorandfrancis.com

and the CRC Press Web site at
http://www.crcpress.com

Contents

Part I
Background

Chapter 1
Donna Marie Bilkovic, Molly M. Mitchell, Jason D. Toft, and Megan K. La Peyre

Chapter 2
Katie K. Arkema, Steven B. Scyphers, and Christine Shepard

Part II
Management, Policy, Design

Chapter 3
Niki L. Pace

Chapter 4
Kateryna M. Wowk and David Yoskowitz

Chapter 5
Andrew Rella, Jon Miller, and Emilie Hauser

Chapter 6
Kevin R. Du Bois

Part IV
Synthesis of Living Shoreline Science: Biological Aspects

Mark Anthony Browne and M.G. Chapman

**Part V
Summary and Future Guidance**

Jana Davis

Jason D. Toft, Donna Marie Bilkovic, Molly M. Mitchell, and Megan K. La Peyre

Foreword

It is likely that only the few who have never been exposed to the shores of oceans or large lakes are unfamiliar with the human struggle to supplant nature's dominance over shoreline stability. The so-called hardening of shorelines is a historic and pervasive alteration of coastal environments to counteract change, understandably to counter the devastative effects of major storms and sea level rise but occasionally simply for cosmetic purposes (Charlier et al. 2005; Nordstrom 2000). It has virtually been incorporated into the DNA of those who have occupied or managed shorelines since the first human millennium. Consider the persistent remnants of the seawall still standing on the shores of Batroun Bay, Lebanon, built by the Phoenicians in ca. 1st century BC, or the oak castrum seawall-equivalent that protected the 10th-century St. Donatian's church in Bruges, Belgium, in an era when the medieval town once fronted the sea and Vikings rode the waves. The long history of human desire to dominate nature is now manifest in vast coastal infrastructures of sea walls, groins, revetments, gabions, breakwaters, and other static engineered structures for coastal protection.

This irony is, of course, that the ecosystem goods and services provided by natural shorelines are the consequence of their naturally dynamic character. What we increasingly recognize as the functions and values of erodible shorelines more often than not depend on energetic erosion and accretion processes that maintain a dynamic equilibrium, but not necessarily a spatially static land-form. As Dean (1999) observed, "Shoreline hardening to thwart nature's ebb and flow is therefore the antithesis of beach conservation."

Increasing recognition of the physical–ecological processes that account for resilient, sustain-able shorelines has necessitated reassessment of the static shoreline model. With accelerated sea level rise, as well as cumulative development along coasts, we have begun to recognize coastal zones as linked social–ecological systems, where human effects and natural processes complicate system dynamics (Kittinger and Ayers 2010). However, reinstituting natural ecosystem processes to promote a full suite of natural ecosystem goods and services is generally unfeasible under all but the most reversible conditions. While "ecohydrology" and other physicochemical principles can be employed to shift shoreline ecosystems more toward their remaining natural potential, "ecoengi-neering" approaches are often necessary to adapt to the persistent effects of shoreline degradation, climate change, and socioeconomic and societal constraints that limit or target delivery of specific goods and services, such as public safety (Elliott et al. 2016). The result is more often than not a "novel" ecosystem state, where rehabilitation or reallocation are the only options to restoration (Aronson and Le Floc'h 1996; Bullock et al. 2011; Hobbs et al. 2013). While not restoration *per se*, such "hybrid" nature-based approaches to living shorelines may be intended or even designed to provide shoreline protection and ecological function as a "win–win" for both society and ecology, albeit with acknowledged trade-offs (Elliott et al. 2016; Rosenzweig 2003).

This volume is likely the first consolidation of the science and application of living shorelines that encapsulates diffuse approaches to and lessons learned from such "win–win" ecoengineering. Although the authors' context of living shorelines is broad—constrained only by the degree to which the connection between aquatic and terrestrial habitats is maintained and engineered structures dominate—they capture the common purpose of protecting shorelines and infrastructure as well as conserving, creating, or restoring natural shoreline functions in estuarine, marine, and aquatic systems (Bilkovic et al., Chapter 1). That the impetus to pursue living shorelines is accelerating, perhaps commensurate with coastal squeeze, argues for synthetic critique of available nature-based tools, documented ecosystem goods and services, social or economic metrics, legal and policy con-siderations, and approaches to community engagement that this volume offers. The regions, ecosys-tems, scales, and perspectives represented across the 24 chapters capture much of the variability in approaches to and results from living shorelines around the world. Various "beach to reach" scale investigations are represented from estuaries around North America, particularly from Chesapeake

Bay, the Louisiana–Mississippi Gulf of Mexico, and Australia, to broader, programmatic-scale examples provided from the Netherlands, United Kingdom, and France. Diverse ecosystems are also well represented, from confined estuaries to estuarine complexes, such as Chesapeake Bay, San Francisco Bay, and Puget Sound, to coastal shorelines of Europe. As imagined from the plethora of approaches to shoreline armoring, the applications are as divergent, from removal/modification of coastal levees and other extensively engineered features of open shores to seawalls of urban and port settings. Perhaps most attractive to the manager and practitioners of living shorelines, the perspectives span the spectrum of factors they will need to evaluate, including social and regulatory considerations they will need to build a supporting constituency, to detailed scientific and technical information that will be required to justify and design living shoreline projects. Perhaps the intrinsic value available in these chapters, and particularly in Davis' Chapter 23 on knowledge gaps, may be the lessons learned that authors have sought to synthesize and extrapolate into what is required in moving forward to advance the state of knowledge. In many respects, a thorough reading of this volume should provide the essential experience for adaptive learning to the next era of shoreline armoring. This is particularly the case for many of the examples and recommendations for metrics to assess the need (e.g., wave power, Chapter 11), structural effectiveness (e.g., structure–current interactions, Chapter 12), or ecosystem goods and services responses (e.g., faunal biodiversity and populations, Chapters 17, 19, 20, and 22) of alternative living shorelines.

Perhaps one of the most notable sources of living shorelines rationale represented in these chapters are clear measures of ecosystem goods and services that can derive from living shorelines elements. Diminution of wave and tidal surge effects on shoreline erosion are the most intuitive, especially as presented as guidance based on technical information on responses of tidal marshes to wave power (e.g., Chapters 11 and 13). However, nutrient reduction (e.g., Chapter 14) and particularly fauna colonization and diversity (e.g., Chapters 15, 17, 19, 20, 21, and 22) substantiate the potential contributions of different living shoreline approaches. Strayer and Findlay (Chapter 16) measurably advance this assessment further by providing an analysis across metrics of ecosystem structure (biodiversity), functions (decomposition), and services (recreation).

As with any multifaceted volume of this breadth, where most contributions are reviews of very different perspectives on living shorelines in specific regions and ecosystems, the level of detail and generality vary. Accordingly, the reader should recognize that much of the real value is the background cited studies that the authors draw on and relate to. Note that there remain considerable uncertainties about the approach and benefit of living shorelines recognized both implicitly and explicitly in these chapters, and most comprehensively by Davis in Chapter 23. For instance, it is still a struggle to find in this volume and the supporting literature examples of rigorously scientific (e.g., BACI, randomized control) comparisons of the ecosystem outputs, goods, services, or functions among typically armored shorelines, completely natural shorelines, and living shoreline constructs that are propositioned as alternatives (see Gap #9, Chapter 23). Similarly, quite often the application of idealized living shoreline features, particularly oyster reefs and seagrasses, to the construction of living shoreline projects for shoreline protection is not explicitly transferable (e.g., often the findings are from regions of estuaries and coasts not particularly vulnerable to shoreline erosion). An analysis that may have to be addressed in the next iteration of this volume is the cumulative and interactive effects of living shoreline elements, as it seems this approach or issue has yet to be addressed opportunistically or experimentally.

If there is any perspective that still dominates living shorelines, it is that "natural elements" are broadly recognized as the primary tool of living shorelines. Except for the large, coastal-scale approaches (e.g., managed realignment), ecosystem process-based approaches are less often considered as viable alternatives, either in socioeconomic analyses of trade-offs or in presenting long-term prognoses of shoreline change with stakeholders. For a vast array of shoreline protection scenarios, novel ecosystems are the only feasible outcome of such hybrid approaches that involve implanting specific features, either for the purpose of enhanced biodiversity and ecological function or for specific ecosystem goods and services. Enhancing the ecological and other functions in socio-economically constrained settings such as seawalls is a given win–win. However, in the predicted future of rising seas and intensifying climate events, the sustainability of living shorelines will need to be assessed much more meticulously with nature-based approaches scaled from the long-term synthetic plans to incremental, site-specific solutions that take advantage of natural processes rather than just unmaintainable features. This volume provides critical insights into the science and technical, sociocultural, and practical factors that will ultimately be required for decisions about how to move in that direction.

C. A. Simenstad
School of Aquatic and Fishery Sciences
University of Washington
Seattle, Washington

REFERENCES

Aronson, J., and E. Le Floc'h. 1996. Vital landscape attributes: Missing tools for restoration ecology. *Restoration Ecology* 4: 377–387.

Bullock, J.M., J. Aronson, A.C. Newton, R.F. Pywell, and J.M. Rey-Benayas. 2011. Restoration of ecosystem services and biodiversity: Conflicts and opportunities. *TREE* 26: 541–549.

Charlier R.H., M.C.P. Chaineux, and S. Morcos. 2005. Panorama of the history of coastal protection. *Journal of Coastal Research* 21: 79–111.

Dean, C. 1999. *Against the Tide: The Battle for America's Beaches.* Columbia University Press, New York. 279 pp.

Elliott, M., L. Mander, K. Mazik, C. Simenstad, F. Valesini, E. Wolanski, and A. Whitfield. 2016. Ecoengineering with ecohydrology: Successes and failures in estuarine restoration. *Estuarine, Coastal Shelf Science* 176: 12–35.

Hobbs, R.J., E.S. Higgs, and C.A. Hall. 2013. *Novel Ecosystems: Intervening in the New Ecological World Order.* Wiley-Blackwell, Oxford, UK.

Kittinger, J.N., and A.L. Ayers. 2010. Shoreline armoring, risk management, and coastal resilience under rising seas. *Coastal Management* 38: 634–653.

Nordstrom, K.F. 2000. *Beaches and Dunes on Developed Coasts.* Cambridge University Press, Cambridge, UK. 338 pp.

Rosenzweig, M.L. 2003. *Win–Win Ecology: How the Earth's Species Can Survive in the Midst of Human Enterprise.* Oxford Univ. Press, Oxford.

Acknowledgments

We are exceedingly grateful to the authors of the chapters in this book who expertly shared their wealth of knowledge on living shorelines and were patient and responsive throughout the lengthy process of editing and publishing this volume.

The chapters in the book have benefited by insightful peer review and thoughtful discussions with numerous colleagues. We extend our gratitude to the following: Michael Piehler, University of North Carolina; Christine Shepard, The Nature Conservancy; Lesley Baggett, University of Mobile; Katherine Dafforn, University of New South Wales; Karen Dyson, University of Washington; Pamela Mason, Virginia Institute of Marine Sciences; Pam Morgan, University of New England; Judy Haner, The Nature Conservancy; Judith Weis, Rutgers University; Chris Boyd, Troy University; Greg Tolley, Florida Gulf Coast University; Ariana Sutton-Grier, University of Maryland; Rachel Gittman, Northeastern University; Steve Jacobus, New Jersey Department of Environmental Protection; Kate Boicourt, NY–NJ Harbor & Estuary Program; Colleen Mercer Clarke, Partnership for Canada–Caribbean Climate Change Adaptation; Niki Pace, Mississippi–Alabama Sea Grant Legal Program; Mike Vasey, San Francisco Bay National Estuarine Research Reserve; Jennifer Ruesink, University of Washington; Eliza Heery, University of Washington; Stuart Munsch, University of Washington; Carl Hershner, Virginia Institute of Marine Science; Karl Nordstrom, Rutgers University; Sid Narayan, University of California; Erik Van Slobbe, Wageningen University and Research; Joanna Rosman, University of North Carolina; Karinna Nunez, Virginia Institute of Marine Science; Joost Stronkhorst, Deltares; Rob Francis, King's College London; Michael Chadwick, King's College London; Louise Wallendorf, United States Naval Academy; Stephen Scyphers, Northeastern University; Jon Miller, Stevens Institute of Technology; Julie Bradshaw, Virginia Institute of Marine Science; Neville Reynolds, VHB, Inc.; Kirk Havens, Virginia Institute of Marine Science; Jenny Davis, NOAA; Scott Hardaway, Virginia Institute of Marine Science; Karen Duhring, Virginia Institute of Marine Science; Amy Smith Kyle, The Nature Conservancy; and Bryan Piazza, The Nature Conservancy.

Special thanks are extended to John Sulzycki and Jennifer Blaise with CRC Press/Taylor & Francis Group, and Judy Weis with Rutgers University, who helped guide us through the process and were a pleasure to work with from start to finish.

List of Contributors

Robert Abbott
ENVIRON International Corporation
Emeryville, California

Katie K. Arkema
The Natural Capital Project
Stanford University
c/o School of Environmental and Forest
    Sciences
University of Washington
Seattle, Washington

Geana Ayala
San Francisco State University
Romberg Tiburon Center for Environmental
    Studies
Tiburon, California

and

University of California, Davis
Department of Environmental Science
    and Policy
Davis, California

Aaron J. Beck
Virginia Institute of Marine Science
College of William & Mary
Gloucester Point, Virginia

Robert Beine
Biological and Agricultural Engineering
Louisiana State University Agricultural Center
Baton Rouge, Louisiana

Donna Marie Bilkovic
Virginia Institute of Marine Science
College of William & Mary
Gloucester Point, Virginia

D.G. Blair
Stewardship Centre for BC
Canada

Bas W. Borsje
University of Twente
Department of Water Engineering
    and Management
Enschede, The Netherlands

Katharyn Boyer
San Francisco State University
Romberg Tiburon Center for Environmental
    Studies
Tiburon, California

Mark Anthony Browne
Evolution and Ecology Research Centre
School of Biological, Earth and Environmental
    Sciences
University of New South Wales
Sydney, New South Wales, Australia

Dorothy Byron
Dauphin Island Sea Lab
Dauphin Island, Alabama

Donna Campbell
Department of Biology
University of Central Florida
Orlando, Florida

Matthew Campbell
North Carolina State University
Raleigh, North Carolina

Just Cebrian
Dauphin Island Sea Lab
Dauphin Island, Alabama

and

Department of Marine Sciences
University of South Alabama
Mobile, Alabama

Randy M. Chambers
College of William & Mary
Williamsburg, Virginia

M.G. Chapman
Centre for Research on Ecological Impacts
    of Coastal Cities
School of Life and Environmental Science
University of Sydney
Sydney, New South Wales, Australia

Jeffery R. Cordell
School of Aquatic and Fishery Sciences
University of Washington
Seattle, Washington

Carolyn A. Currin
NOAA
NCCOS Center for Coastal Fisheries and
    Habitat Research
Beaufort, North Carolina

Jana Davis
Chesapeake Bay Trust
Annapolis, Maryland

Jenny Davis
CSS
NOAA
NCCOS Center for Coastal Fisheries and
    Habitat Research
Beaufort, North Carolina

Susan De La Cruz
U.S. Geological Survey
Western Ecological Research Center
San Francisco Bay Estuary Field Station
Vallejo, California

Sierd de Vries
Delft University of Technology
Department of Hydraulic Engineering
Delft, The Netherlands

Kevin S. Dillon
Department of Coastal Sciences
The University of Southern Mississippi
Ocean Springs, Mississippi

Melinda Donnelly
Department of Biology
University of Central Florida
Orlando, Florida

Kevin R. Du Bois
Former staff to the Norfolk Wetlands Board
and
Living Shoreline Practitioner
Norfolk, Virginia

Brian Emmett
Archipelago Marine Research
Victoria, British Columbia, Canada

Luciana S. Esteves
Faculty of Science and Technology
Bournemouth University
Talbot Campus
Poole, Dorset, United Kingdom

Nicole Faghin
Coastal Management Specialist
Washington Sea Grant
Seattle, Washington

Stuart E.G. Findlay
Cary Institute of Ecosystem Studies
Millbrook, New York

Nate Geraldi
King Abdullah University of Science
    and Technology
Thuwal, Saudi Arabia

Maureen Goff
School of Aquatic and Fishery Sciences
University of Washington
Seattle, Washington

Edwin Grosholz
University of California, Davis
Department of Environmental Science
    and Policy
Davis, California

Steven G. Hall
Marine Aquaculture Research Center
Biological and Agricultural Engineering
North Carolina State University
Raleigh, North Carolina

Emilie Hauser
Hudson River National Estuarine Research
    Reserve
NEIWPCC New York State Department
    of Environmental Conservation
Staatsburg, New York

Kenneth L. Heck, Jr.
Dauphin Island Sea Lab
Dauphin Island, Alabama

and

Department of Marine Sciences
University of South Alabama
Mobile, Alabama

Stephanie K.H. Janssen
Deltares
and
Delft University of Technology
Department Multi Actors Systems
Delft, The Netherlands

Stephanie Kiriakopolos
San Francisco State University
Romberg Tiburon Center for Environmental
    Studies
Tiburon, California

and

University of California, Davis
Department of Environmental Science and
    Policy
Davis, California

Damien Kunz
Environmental Science Associates
San Francisco, California

Megan K. La Peyre
U.S. Geological Survey
Louisiana Fish and Wildlife Cooperative
    Research Unit
School of Renewable Natural Resources
Louisiana State University Agricultural Center
Baton Rouge, Louisiana

Marilyn Latta
State Coastal Conservancy
Oakland, California

Jeremy Lowe
Environmental Science Associates
and
San Francisco Estuary Institute
San Francisco, California

Arjen P. Luijendijk
Deltares
and
Delft University of Technology
Department of Hydraulic Engineering
Delft, The Netherlands

Kelly Major
Department of Biology
University of South Alabama
Mobile, Alabama

Amit Malhotra
JHT
NOAA
NCCOS Center for Coastal Fisheries
    and Habitat Research
Beaufort, North Carolina

Christopher A. May
The Nature Conservancy in Michigan
Lansing, Michigan

Earl Melancon
Department of Biological Sciences
Nicholls State University
Thibodaux, Louisiana

Jen Miller
San Francisco State University
Romberg Tiburon Center for Environmental
    Studies
Tiburon, California

Jon Miller
Stevens Institute of Technology
Hoboken, New Jersey

Lindsay Schwarting Miller
School of Renewable Natural Resources
Louisiana State University Agricultural Center
Baton Rouge, Louisiana

Shea Miller
School of Renewable Natural Resources
Louisiana State University Agricultural Center
Baton Rouge, Louisiana

Molly M. Mitchell
Virginia Institute of Marine Science
College of William & Mary
Gloucester Point, Virginia

Julien Moderan
San Francisco State University
Romberg Tiburon Center for Environmental
    Studies
Tiburon, California

Stuart H. Munsch
School of Aquatic and Fishery Sciences
University of Washington
Seattle, Washington

Rena Obernolte
Isla Arena Consulting
Emeryville, California

Michelle Orr
Environmental Science Associates
San Francisco, California

Tyler Ortego
OraEstuaries
Metairie, Louisiana

Niki L. Pace
Louisiana Sea Grant Law & Policy Program
Louisiana State University
Baton Rouge, Louisiana

Mark S. Peterson
Department of Coastal Sciences
The University of Southern Mississippi
Ocean Springs, Mississippi

Cassie Pinnell
San Francisco State University
Romberg Tiburon Center for Environmental
    Studies
Tiburon, California

Rochelle Plutchak
Department of Marine Sciences
University of South Alabama
Mobile, Alabama

and

Dauphin Island Sea Lab
Dauphin Island, Alabama

and

National Oceanic and Atmospheric
    Administration
Silver Spring, Maryland

Sean P. Powers
Department of Marine Sciences
University of South Alabama
Mobile, Alabama

and

Dauphin Island Sea Lab
Dauphin Island, Alabama

Walter I. Priest III
Wetland Design and Restoration
Bena, Virginia

Andrew Rella
Stevens Institute of Technology
Hoboken, New Jersey

Jon D. Risinger
Biological and Agricultural Engineering
Louisiana State Agricultural Center
Baton Rouge, Louisiana

Paul Sacks
Science Department
Winter Springs High School
Winter Springs, Florida

Steven B. Scyphers
Department of Marine and Environmental
  Sciences
Marine Center
Northeastern University
Nahant, Massachusetts

Christine Shepard
The Gulf of Mexico Program
The Nature Conservancy
Punta Gorda, Florida

Kevin Stockmann
San Francisco State University
Romberg Tiburon Center for Environmental
  Studies
Tiburon, California

David L. Strayer
Cary Institute of Ecosystem Studies
Millbrook, New York

Jason D. Toft
School of Aquatic and Fishery Sciences
University of Washington
Seattle, Washington

Vincent Vuik
Delft University of Technology
Department of Hydraulic Engineering
Delft, The Netherlands

and

HKV Consultancy
Lelystad, The Netherlands

Linda Walters
Department of Biology
University of Central Florida
Orlando, Florida

Jon J. Williams
Ports, Coastal & Offshore
Mott MacDonald
Croydon, United Kingdom

Kateryna M. Wowk
Harte Research Institute
Texas A&M University
Corpus Christi, Texas

David Yoskowitz
Harte Research Institute
Texas A&M University
Corpus Christi, Texas

Chela Zabin
University of California, Davis
Department of Environmental Science and
  Policy
Davis, California

PART I

Background

CHAPTER 1

A Primer to Living Shorelines

Donna Marie Bilkovic, Molly M. Mitchell, Jason D. Toft, and Megan K. La Peyre

CONTENTS

1.1 THE CHALLENGE: A HISTORY OF SHORELINE ARMORING

For centuries, estuarine and coastal shorelines have been dramatically modified by humans for far-ranging and, at times, conflicting purposes, such as water access, commerce, aquaculture, and property protection. Legal principles governing these uses often tend to favor the interests of coastal property owners or societal rights to access and exploit natural resources. For example, in most coastal communities, shoreline armoring is allowed and accepted if property is deemed to be at risk. As a result, shorelines have been extensively armored globally and ecosystem function has diminished. The amount of shoreline armoring along a given coast varies depending on surrounding land use, with major coastal cities often having more than 50% of their shores hardened (e.g., Chapman and Bulleri 2003). However, shoreline armoring is not restricted to dense urban areas, as coastal areas in the process of change (e.g., agriculture to suburban) may experience the fastest rates of shoreline hardening (Isdell 2014). In sum, the United States has roughly 14% (22,000 km) of its extensive coastline armored (Gittman et al. 2015). In Europe, more than half of the >15,000 km of coastline that is eroding is artificially stabilized (EC 2004). Likewise, more than half of Mediterranean coastlines are armored and developed (EEA 1999). In Japan, approximately half of its coastline is reportedly eroding (15,900 km of the 34,500 km of total coast) and approximately 27% of that coastline has been hardened (Koike 1993). In Australia, the densely populated coastal cities typically have more than 50% of their coastlines armored (Chapman 2003). Pressures to abate erosion and secure shorelines in place will only continue, and likely increase, as the proportion of the global population living within 100 km of the coasts grows from one-third to an expected one-half by 2030 (Small and Nicholls 2003) and sea level continues to rise. This may be particularly problematic in areas where heavily urbanized landscapes intersect with higher-than-average rates of sea level rise; for example, the North American Gulf and mid-Atlantic coasts have the highest rates of rise in the United States (Boon and Mitchell 2015) and respectively have the third and fifth fastest-growing coastal populations in the continental United States (Crossett et al. 2004).

Closely associated with shoreline modification is the loss or alteration of intertidal and shallow subtidal habitats (e.g., wetlands, seagrasses) and ecosystem function (e.g., Bilkovic and Roggero 2008; Chapman and Bulleri 2003; Dethier et al. 2016; Dugan et al. 2011; Peterson and Lowe 2009 and references within). This has implications for ecosystem service provision to coastal communities including shore protection, fisheries production, and water quality benefits (e.g., Arkema et al. 2013; Bilkovic et al. 2016; Gedan et al. 2011; Scyphers et al. 2015). Growing concern about the cumulative effects of piecemeal alterations to the coastlines has reinforced the need for alternative shoreline management strategies.

While wetlands have been long recognized as providing some level of protection to coastal communities from wave-induced erosion, the intentional use of natural habitat elements to reduce shoreline erosion was first reported in the early 1970s (Garbisch and Garbisch 1994). Since that time, the understanding and practical application of nature-based techniques have grown tremendously. In recent years, nature-based approaches are being extensively promoted and practiced globally primarily because of (1) growing acknowledgment of the value of ecosystem services provided by coastal habitats (Barbier et al. 2011; Costanza et al. 1997) and the adverse effects of traditional armoring to coastal systems, (2) the extensive ongoing loss of many threatened coastal habitats (marsh, seagrasses) (Duarte 2009; Halpern et al. 2008; Waycott et al. 2009), and (3) the realization that dynamic erosion protection approaches that incorporate natural ecosystem elements (e.g., marsh, beach) may be more responsive and resilient in some settings to storm events than traditional armoring (Gedan et al. 2011; Gittman et al. 2014). However, the field of shoreline restoration in human systems is in its nascent stages. In particular, challenges remain to reconcile conflicting uses and establish a path-forward to effectively manage shorelines for both ecological and human protections, leading to our topic of living shorelines.

1.2 WHAT'S IN A NAME?

With any new discipline, there tends to be a growing period when terminology has not been formally defined and accepted by all participants, which can lead to a certain amount of confusion, setbacks, and misinformation. This is especially true for shoreline management when goals can be at cross-purposes and depend on the viewpoint. The countless terms that exist to describe shoreline protection approaches that integrate nature elements illustrate the difficulties that remain for the field. These terms include living shorelines, nature-based shoreline protection, green shorelines, geomorphic engineering, soft stabilization, building with nature, and variances thereof. For simplicity, we will refer to these shoreline approaches as *Living Shorelines* henceforth. A near consensus among practitioners is that these approaches are intended to not only protect shorelines and infrastructure but also conserve, create, or restore natural shoreline functions in estuarine, marine, and aquatic systems. In practice, living shorelines should predominantly consist of organic techniques and materials that are characteristic of the local system, such as wetland, riparian, and dune plantings; beach nourishment; shellfish reefs; and emplaced large woody debris. The connection between aquatic and terrestrial habitats should be maintained, and dynamic movement of habitat features should be allowed in response to storm events and to promote sediment capture. Last, in higher-energy sites that require hybrid approaches with engineered structures to provide the requisite wave attenuation to sustain planted vegetation and habitats, engineered structures should be minimized to not overwhelm the living habitat features they are supporting (Figure 1.1).

Public policy actions to promote the use of living shorelines have been implemented by regional and national governmental entities in the United States and Europe. In the United States, many coastal states have opted for either legislative requirements or the more popular option of

(a)   (b)

(c)   (d)

Figure 1.1   Living shorelines: (a) planted marsh with supporting stone sill (from Bilkovic and Mitchell, Chapter 15, this book); (b) bio-engineered oyster reef (gabion mats) with marsh (From La Peyre et al., Chapter 18, this book); (c) oyster reef bases designed to encourage oyster recruitment (From Hall et al., Chapter 13, this book); (d) created pocket beach habitat for juvenile salmon excavated from a stretch of riprap-armored shoreline (From Cordell et al., Chapter 21, this book).

employing incentives (expedited permit process, fee waivers) or fiscal aid (e.g., low interest loans) for the implementation of living shorelines as a preferred shoreline management approach (Bilkovic and Mitchell, Chapter 15, this book). Likewise, federal agencies including the National Oceanic and Atmospheric Administration (NOAA), United States Environmental Protection Agency, and United States Army Corps of Engineers (USACE) have developed funding and planning initiatives to support living shoreline implementation. Moreover, the USACE and NOAA helped initiate a Community of Practice to enhance collaboration among academic institutes, nongovernmental organizations, and state and federal agencies called Systems Approach to Geomorphic Engineering (SAGE) to use living shorelines as a tool to achieve larger-scale community resilience (Bilkovic et al. 2016). In Europe (e.g., United Kingdom, Belgium, the Netherlands, and France), "building with nature" initiatives that involve the implementation of nature-based solutions as flood risk reduction measures are often driven by national and European legislation and are a growing trend (Borsje et al., Chapter 8, this book; Esteves et al., Chapter 9, this book).

These efforts are not without criticism. Many of the criticisms voiced are related to the (often unintentional) misidentification of a practice that does not meet standard criteria as a living shoreline (e.g., Pilkney et al. 2012). There are many practices, particularly in urban settings, designed to enhance or mitigate adverse effects of engineered shorelines (e.g., seawalls) in high-energy or high-risk settings where living shorelines are not suitable. For instance, a common practice in Europe is the inland realignment of a coastal protection line (seawall). Elsewhere, armored shorelines are being reengineered to increase their habitat value, such as adding complexity to the surface of a seawall to encourage invertebrate recruitment (Browne and Chapman, Chapter 22, this book) or incorporating beach and cobble areas along coastlines where seawalls maintain coastal integrity of highly urbanized areas (Cordell et al., Chapter 21, this book) (Figure 1.2). However, care should be taken to ensure that these efforts are termed living shorelines only if they fully meet the definition requirements in that the connection between aquatic and terrestrial habitats is maintained and engineered structures do not dominate. When engineered structures dominate the landscape, and shoreline protection is dependent solely on the maintenance of these engineered structures, the litmus test for a living shoreline has failed. For example, in some oyster reef restoration projects, a failure to recruit oysters that survive, grow, and build a reef over time may result in some level of shoreline protection, but this fails to be a living shoreline. In a final summary chapter of this book, we draw from the contributed works of the edited volume to more fully discuss the terminology of shoreline restoration and engineering and place living shorelines into a better context. Our hope is that this book will serve to better define living shorelines in a manner that can help refine the use of the term living shorelines as well as identify common concerns and the means to remedy or lessen some of these concerns.

(a)

(b)

(c)

Figure 1.2 Urban shoreline enhancement practices: (a) Managed realignment strategies practiced in Europe such as managed retreat and breaching of seawalls to support intertidal habitat creation most clearly fall within the definition of a living shoreline. Other practices developed for higher energy or at-risk urban settings such as inland realignment of the coastal protection line may not fulfill the definition of a living shoreline. (From Esteves, L.S. and Williams, J.J., Chapter 9, this book.) (b) Reengineered seawall practices to increase complexity and enhance habitat use and diversity. (From Browne, M.A. and Chapman, M.G., Chapter 22, this book.) (c) Habitat enhancements in the new construction of Seattle's seawall including light penetrating surfaces in the sidewalk, intertidal benches, and seawall relief and texturing. (From Cordell et al. Chapter 21, this book.)

1.3 ROLE FOR LIVING SHORELINES IN CONSERVATION AND CLIMATE ADAPTATION STRATEGIES

While living shorelines are grounded in well-established restoration science and practices, these projects are designed in different ways to accommodate both human use and ecological goals. Because of this, there is often a risk that ecological goals may be sacrificed for perceived shoreline or infrastructure protection. Therefore, living shorelines deviate from the traditional perspective of restoration or conservation, but these efforts can be complementary. The advantage that living shorelines confer is new feasible opportunities, particularly in urban or ex-urban settings, to offset previous or inevitable coastal habitat losses. The added benefit of protecting a shoreline can help overcome a "restoration" hurdle by providing an economically attractive rationale for the individual or community shouldering the cost. Further, the use of a living shoreline in place of armoring will act to curb additional loss to globally threatened habitats.

These shoreline management approaches represent a paradigm shift as they work with rather than against natural nearshore processes. Such "process-based" planning and design is a worthy goal that can be difficult to achieve, but not impossible. The potential added value of a dynamic living shoreline serving as a climate adaptation strategy to conserve coastal resilience makes that goal particularly meaningful. While in many settings where landward migration is restricted there may be a foreseeable end date for a living shoreline, the benefits of maximizing ecosystem service provision in the near term may still be substantial. Moreover, there are techniques and certain natural elements that may be incorporated to allow living shorelines to keep pace with sea level rise and extend their lifespan. For example, in the right location, an oyster reef used for shoreline protection can respond to changing conditions including subsidence and sea level rise (Casas et al. 2015; La Peyre et al., Chapter 18, this book; Mann and Powell 2007; Walles et al. 2015) and structural support can be designed to encourage the capture of sediment so that accretion rates are sufficient for the persistence of created marshes in place (Currin et al., Chapter 11, this book).

1.4 PURPOSE AND BOOK ORGANIZATION

To help guide future shoreline management and restoration, this edited volume assembles, synthesizes, and interprets the current state of the knowledge on the science and practice of living shorelines, as well as some outside the realm of living shorelines that still attempt to incorporate nature-based approaches, in order to provide context. Researchers, managers, and practitioners have gained valuable knowledge to inform the science of living shorelines, but that information has not been summarized in a single source, nor is it readily accessible to the broad spectrum of users in different disciplines (e.g., engineering, ecology, geology, and social sciences) that shoreline management and restoration require.

To cover the spectrum of topics, the book is divided into five major sections: (1) Background: History and Evolution; (2) Management, Policy, and Design; (3, 4) Synthesis of Living Shoreline Science: Physical and Biological Aspects; and (5) Summary and Future Guidance. After this introductory chapter, Arkema and others adeptly detail how ecosystem service concepts can inform living shoreline science and implementation. The second section of the book provides insights on significant management, policy, and other social factors that influence the success of living shoreline projects, as well as effective living shoreline designs. The third and fourth section of the book synthesizes the literature that is available with data from new studies on the physical aspects of living shorelines including information on trajectories of living shoreline ecosystem development. The fifth section of the book summarizes information presented throughout the book to help guide future research and management strategies. The first summary chapter details the gaps in our understanding of living shorelines. In the second summary chapter, the editors synthesize author perspectives on three focal areas to examine commonalities and differences among the contributed works. The three focal areas are as follows:

1. Lessons learned from the practice of shoreline restoration/conservation
2. Longevity and stability of projects in the near and long term with considerations for climate change and human development
3. What is the path forward? Research needs, strategies for working across different disciplines, training options, and future opportunities

This book will serve as a valuable reference to guide scientists, students, managers, planners, regulators, environmental and engineering consultants, and others engaged in the design and implementation of living shorelines. Our intent is to provide a background and history of living shorelines; understandings on management, policy, and project designs; technical synthesis of the science related to living shorelines including insights from new studies; and the identification of research needs, lessons learned, and perspectives on future guidance. To capture international efforts, the book includes perspectives from leading researchers and managers in Europe; the East, West, and Gulf coasts of the United States; and Australia who are working on natural approaches to shoreline management. The broad geographic scope and interdisciplinary nature of contributing authors will help facilitate dialogue and transfer of knowledge among different disciplines and across different regions. It is our hope that this book will provide coastal communities with the scientific foundation and practical guidance necessary to implement effective shoreline management that enhances ecosystem services and coastal resilience now and into the future.

REFERENCES

Arkema, K.K., G. Guannel, G. Verutes, S.A. Wood, A. Guerry, M. Ruckelshaus, P. Kareiva, M. Lacayo, and J.M. Silver. 2013. Coastal habitats shield people and property from sea-level rise and storms. *Nature Climate Change* 3: 913–918.

Barbier, E.B., S.D. Hacker, C. Kennedy, E.W. Koch, A.C. Stier, and B.R. Silliman. 2011. The value of estuarine and coastal ecosystem services. *Ecological Monographs* 81: 169–193.

Bilkovic, D.M. and M. Roggero. 2008. Effects of coastal development on nearshore estuarine nekton communities. *Marine Ecology Progress Series* 358: 27–39.

Bilkovic, D.M., M. Mitchell, P. Mason, and K. Duhring. 2016. The role of living shorelines as estuarine habitat conservation strategies. *Coastal Management* 44: 161–174.

Boon, J.D. and M. Mitchell. 2015. Nonlinear change in sea level observed at North American tide stations. *Journal of Coastal Research* 31: 1295–1305.

Casas, S.M., J.F. La Peyre, and M.K. La Peyre. 2015. Restoration of oyster reefs in an estuarine lake: Population dynamics and shell accretion. *Marine Ecology Progress Series* 524: 171–184.

Chapman, M.G. 2003. Paucity of mobile species on constructed seawalls: Effects of urbanization on biodiversity. *Marine Ecology Progress Series* 264: 21–29.

Chapman, M.G. and F. Bulleri. 2003. Intertidal seawalls—New features of landscape in intertidal environments. *Landscape and Urban Planning* 62: 159–172.

Costanza, R., R. d'Arge, R. De Groot, S. Faber, M. Grasso, B. Hannon, K. Limburg, S. Naeem, R.V. O'Neill, J. Paruelo, and R.G. Raskin. 1997. The value of the world's ecosystem services and natural capital. *Nature* 387: 253.

Crossett, K.M., T.J. Culliton, P.C. Wiley, and T.R. Goodspeed. 2004. Population Trends along the Coastal United States: 1980–2008. NOAA Coastal Trends Report Series. Table 2.

Dethier, M.N., W.W. Raymond, A.N. McBride, J.D. Toft, J.R. Cordell, A.S. Ogston, S.M. Heerhartz, and H.D. Berry. 2016. Multiscale impacts of armoring on Salish Sea shorelines: Evidence for cumulative and threshold effects. *Estuarine, Coastal and Shelf Science* 175: 106–117.

Duarte, C.M. 2009. Global loss of coastal habitats: Rates, causes and consequences. Fundación BBVA, Bilbao, Spain.

Dugan, J.E., L. Airoldi, M.G. Chapman, S.J. Walker, and T. Schlacher. 2011. Estuarine and coastal structures: Environmental effects, a focus on shore and nearshore structures. In: Wolanski, E. and D. McLusky (Eds.), *Treatise on Estuarine and Coastal Science*. Vol. 8, pp. 17–41. Waltham, MA: Academic Press.

EC. 2004. Living with Coastal Erosion in Europe—Sediment and Space for Sustainability. OPOCE, Luxembourg. http://www.eurosion.org/project/eurosion_en.pdf (accessed May 2016).

EEA. 1999. State and Pressures of the Marine and Coastal Mediterranean Environment. Environmental Issues Series 5. Luxembourg: OPOCE. Online. http://www.eea.europa.eu/publications/ENVSERIES05 (accessed July 14, 2016).

Garbisch, E.W. and J.L. Garbisch. 1994. Control of upland bank erosion through tidal marsh construction on restored shores: Application in the Maryland portion of Chesapeake Bay. *Environmental Management* 18(5): 677–691.

Gedan, K.B., M.L. Kirwan, E. Wolanski, E.B. Barbier, and B.R. Silliman. 2011. The present and future role of coastal wetland vegetation in protecting shorelines: Answering recent challenges to the paradigm. *Climatic Change* 106(1): 7–29.

Gittman, R.K., F.J. Fodrie, A.M. Popowich, D.A. Keller, J.F. Bruno, C.A. Currin, C.H. Peterson, and M.F. Piehler. 2015. Engineering away our natural defenses: An analysis of shoreline hardening in the US. *Frontiers in Ecology and the Environment* 13(6): 301–307.

Gittman, R.K., A.M. Popowich, J.F. Bruno, and C.H. Peterson. 2014. Marshes with and without sills protect estuarine shorelines from erosion better than bulkheads during a Category 1 hurricane. *Ocean & Coastal Management* 102: 94–102.

Halpern, B.S., K.L. McLeod, A.A. Rosenberg, and L.B. Crowder. 2008. Managing for cumulative impacts in ecosystem-based management through ocean zoning. *Ocean & Coastal Management* 51(3): 203–211.

Isdell, R.E. 2014. Anthropogenic modifications of connectivity at the aquatic-terrestrial ecotone in the Chesapeake Bay. Master's Thesis. College of William and Mary, Williamsburg, Virginia.

Koike, K. 1993. The countermeasures against coastal hazards in Japan. *Geojournal* 38: 301–312.

Mann, R. and E.N. Powell. 2007. Why oyster restoration goals in the Chesapeake Bay are not and probably cannot be achieved. *Journal of Shellfish Research* 26(4): 905–917.

Peterson, M.S. and M.R. Lowe. 2009. Implications of cumulative impacts to estuarine and marine habitat quality for fish and invertebrate resources. *Reviews in Fisheries Science* 17: 505–523.

Pilkney, O.H., N. Longo, R. Young, and A. Coburn. 2012. Rethinking Living Shorelines, Program for the Study of Developed Shorelines, Western Carolina University, http://www.wcu.edu/WebFiles/PDFs /PSDS_Living_Shorelines_White_Paper.pdf, accessed May 17, 2016.

Scyphers, S.B., T.C. Gouhier, J.H. Grabowski, M.W. Beck, J. Mareska, and S.P. Powers. 2015. Natural shorelines promote the stability of fish communities in an urbanized coastal system. *PLoS One 10*(6): p.e0118580.

Small, C. and R.J. Nicholls. 2003. A global analysis of human settlement in coastal zones. *Journal of Coastal Research* 19: 584–599.

Walles, B., J. Salvador de Paiva, B. van Prooijen, T. Ysebaert, and A. Smaal. 2015. The ecosystem engineer *Crassostrea gigas* affects tidal flat morphology beyond the boundary of their reef structures. *Estuaries and Coasts* 1: 1–10.

Waycott, M., C.M. Duarte, T.J. Carruthers, R.J. Orth, W.C. Dennison, S. Olyarnik, A. Calladine, J.W. Fourqurean, K.L. Heck, A.R. Hughes, and G.A. Kendrick. 2009. Accelerating loss of seagrasses across the globe threatens coastal ecosystems. *Proceedings of the National Academy of Sciences* 106(30): 12377–12381.

CHAPTER 2

Living Shorelines for People and Nature

Katie K. Arkema, Steven B. Scyphers, and Christine Shepard

CONTENTS

2.1 INTRODUCTION

Rising seas, expanding coastal development, and increases in the frequency of extreme weather catastrophes are putting shoreline communities around the world at risk from erosion and flooding (Day et al. 2007; Nicholls et al. 1999; US Army Corps of Engineers [USACE] 2015). Spurred by a widened awareness surrounding the loss of coastal habitats and the deficiencies of traditional erosion control structures, much progress has been made to advance the science and implementation of nature-based approaches to coastal protection (Bilkovic et al. 2016; Currin et al. 2008, 2009; Jones et al. 2012; National Oceanographic and Atmospheric Administration [NOAA] 2015; National Research Council [NRC] 2014). Researchers, practitioners, and the private sector have developed a suite of alternative techniques for stabilizing shorelines, such as replanting saltmarsh or restoring oyster reefs (Berman et al. 2005; Currin et al. 2008; Gittman et al. 2016; NOAA 2015; Piazza et al. 2005; Scyphers et al. 2011). Recent studies suggest that the design of living shorelines (e.g., width of marsh, presence of sill) influences various outcomes, including the abundance of ecologically and economically important fish and invertebrates, water quality, and erosion control (Bilkovic and Mitchell 2013; La Peyre et al. 2013; Scyphers et al. 2011; Toft et al. 2013).

In the United States, policy interest in nature-based approaches to coastal protection is also growing. A new memorandum from the Executive Office of the President directs federal agencies to incorporate the value of "green infrastructure" and ecosystem services into planning and decision-making (OMB 2015). Recent state and federal policies highlight living shorelines as the preferred alternative for erosion control, especially along protected coasts (Maryland Living Shoreline Protection Act of 2008; NOAA 2015; NRC 2007; Virginia 2011). A report released by the Obama Administration in August of last year recommends prioritized federal research into ecosystem services and coastal green infrastructure to inform risk reduction, resilience planning, and decision-making (National Science and Technology Council [NSTC] 2015). The hope is that infusing natural features into shoreline stabilization practices will afford protection for communities while maintaining or restoring the multiple benefits of coastal habitats for people and ecosystems now and in the future (Jones et al. 2012).

Despite this technical innovation and political support, several challenges preclude broad uptake of living shorelines into coastal management. One challenge is the lack of tools, guidance, and capacity for identifying appropriate protection measures for a particular setting (e.g., Berman and Rudnicky 2008; NSTC 2015; Restore America's Estuaries [RAE] 2015) and forecasting which ecosystem services are likely to be gained or lost with habitat conversion (Bilkovic et al. 2016; La Peyre et al. 2015). Few designers and engineers are familiar with nature-based techniques for coastal protection, which further hampers implementation (RAE 2015). A second challenge is harnessing the support of coastal communities and key stakeholders for effective project implementation and sustainability (Olsson et al. 2004; Scyphers et al. 2014, 2015). Poor community engagement can lead to a mismatch between project outcomes and community values and expectations (Schultz 2011). These challenges point to the need for science that will help practitioners anticipate trade-offs and potential impacts to target resources, encourage a common language through which multiple agencies and other stakeholders can define goals, and support the development of performance standards that capture social as well as ecological outcomes from living shoreline projects (NSTC 2015; Olander et al. 2015; RAE 2015). Incorporating ecosystem service approaches and tools into living shoreline science and practice provides an opportunity to address these challenges.

Ecosystem services are the benefits that nature provides to people (Daily 1997; Tallis and Polasky 2009). With over a third of the world's population concentrated near the shore (IPCC 2007), humans rely heavily on benefits delivered from coastal and marine ecosystems (Barbier et al. 2011; Tallis et al. 2011). Wetlands, oyster reefs, coral reefs, subtidal vegetation, dunes, and coastal forests provide a suite of ecosystem services. These services include diverse natural resources, protection for public and private property from coastal hazards, coastal and maritime jobs, and opportunities for recreation, tourism, and aesthetic enjoyment (Barbier et al. 2011; Grabowski et al. 2012). Living shorelines have the potential to contribute to the resilience and sustainability of this diversity of important services. Yet, nature-based stabilization projects infrequently define or monitor objectives using an ecosystem services framework that explicitly communicates and quantifies the benefits of coastal restoration to people (Tallis and Polasky 2009; Tallis et al. 2011).

A fundamental assumption of living shoreline projects is that natural features provide a set of benefits that hard infrastructure may not provide (Bilkovic and Mitchell 2013; Chesapeake Bay Foundation 2007; Center for Coastal Resources Management 2006; Davis et al. 2015; Gittman et al. 2016). However, a review of the scientific and gray literature suggests a disconnect between living shorelines, ecosystem services, and human well-being metrics (Figure 2.1). The number of peer-review papers, reports, proposals, and other documents referring to *living shorelines* increases

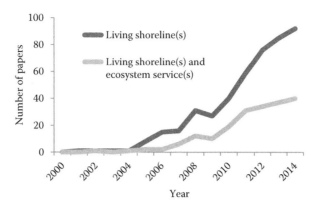

Figure 2.1 Trends in the number of peer-reviewed papers, documents, reports, and other types of gray literature with the phrases "living shoreline(s)" versus "living shoreline(s) and ecosystem service(s)" from 2000 to 2014 based annual searches in Google Scholar.

steadily from 2000 to 2014. However, in each year, only approximately half of these documents also discuss *ecosystem services*. Moreover, of those papers that are about both living shorelines and ecosystems services across all years, just a little more than 50% articulate benefits from living shorelines in terms that directly matter to people (e.g., food from fisheries [e.g., Kalinowski and Baker 2014; La Peyre et al. 2013] or recreational opportunities [e.g., Hoatson 2010; Pendleton 2010]). Even fewer papers (12.5%) actually provide quantitative information using social or economic metrics (e.g., Grabowski et al. 2012; Humphries and La Peyre 2015).

Part of this disconnect likely stems from the breadth of living shoreline definitions. Some definitions describe living shorelines as alternatives to hardening that employ natural habitat elements to protect shorelines from erosion while also providing critical habitat for wildlife and water quality benefits (NOAA 2015; NRC 2007; Virginia 2011), while others refer to specific techniques that baffle wave energy and reduce chronic erosion (Berman et al. 2005). A new NOAA report addresses this issue by clearly describing a continuum of shoreline stabilization strategies as a gradient from natural to built with hybrid structures in the middle and living shorelines falling left of center (NOAA 2015). In general, however, many of the definitions provided by state and federal agencies, nongovernmental organizations (NGOs), and various researchers focus more on ecological and physical outcomes than social or economic ones. Even those definitions that suggest benefits to people, such as "improved water quality" or "shoreline protection," rarely state it explicitly (e.g., improved water quality for waterfront residents or tourists) or define them using social or economic endpoints such as increased visitation to coastal waters with higher water quality (for lakes recreation example, see Keeler et al. 2015).

Endpoints that resonate with people are core elements in ecosystem services science (Tallis et al. 2011) and implementation (Olander et al. 2015) and could provide explicit goals for designing living shoreline projects and monitoring outcomes. An important step to operationalizing the ecosystem service concept is a production function model that relates change in ecosystems to change in the production of services that matter to people. For example, in a living shoreline context, change in ecosystem structure (e.g., oyster restoration; Figure 2.2 (i)) can lead to change in ecosystem function (i.e., reduction in wave height; Figure 2.2 (ii)) and benefits to people (e.g., reduction in erosion of coastal property and associated property damage; Figure 2.2 (iii) and (iv)). Of course, not all assessments of living shoreline projects need to have a quantified benefit for people to be worth implementing. Instead, some projects can remain limited to restoring ecosystem structure and function to provide habitat for species. However, when public or stakeholder participation is involved,

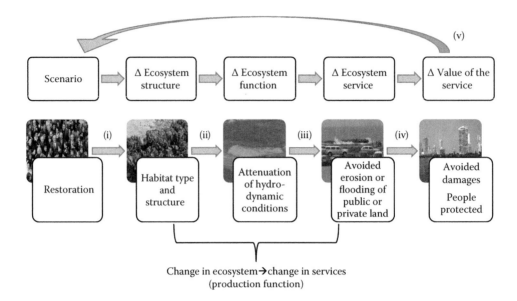

Figure 2.2 General framework for modeling ecosystem services applied to coastal protection provided by living shorelines. The framework links changes in habitat structure (e.g., width and height of oyster reef) as a result of restoration to changes in biophysical conditions (e.g., wave attenuation) to changes in the service provided (e.g., avoided erosion) and ultimately to changes in the value of the service (e.g., avoided damages), which could, in turn, influence the propensity for public and private decision-makers to restore habitat in the future. (Photo credits: Katie Arkema [Natural Capital Project] and Bo Lusk [The Nature Conservancy].)

an important component of the ecosystem services framework is the feedback between change in service value and peoples' decisions (Figure 2.2 (v)). If waterfront communities understand and, ideally, value the suite of benefits of living shorelines, they may be more likely to take actions that further increase the flow of benefits in the future (Scyphers et al. 2015, in review).

Here we argue that drawing upon ecosystem service concepts and tools to inform living shoreline science and implementation has the potential to advance the effectiveness and uptake of nature-based shoreline stabilization techniques. In Section 2.2, we review a suite of benefits provided by natural and enhanced living shorelines, explicitly focusing on endpoints that resonate with people. In Section 2.3, we review emerging research that links geophysical, biological, and climate science with social and economic information to estimate services and understand the factors that are likely to influence their delivery over the short and long term. We also review results from projects that engaged communities to assess how people's perceptions and values are likely to influence the implementation and longevity of living shoreline projects. In Section 2.4, we describe several recent cases that involved heightened calls for ecosystem services information and discuss how a services framework can facilitate better implementation pathways for living shoreline approaches. We end by identifying key ecological and social science research questions and opportunities for advancing ecosystem services science to inform development and implementation of living shorelines.

2.2 ECOSYSTEM SERVICES PROVIDED BY LIVING SHORELINES FOR PEOPLE

People desire numerous services from shoreline ecosystems, including fisheries benefits and wave attenuation. Some of these services are substitutable with built infrastructure. However, hardening a shoreline may be less cost-effective than restoring coastlines, especially after accounting for the influence of armoring and bulkheads on a full suite of ecosystem services (Jones et al. 2012). Several of the subsequent chapters in this book highlight important ecosystem services provided by living shorelines from an ecological perspective. To complement these chapters, we focus this section on the societal relevance of living shorelines. We discuss the evidence for five ecosystem services that living shorelines provide to coastal communities: food and livelihoods from fisheries, protection from coastal hazards for people and property, opportunities for recreation and tourism, carbon storage and sequestration, and human health and well-being. Our goal with this section is to illustrate an ecosystem services approach to framing potential benefits of living shorelines to people.

2.2.1 Habitat to Support Fish, Fisheries, and Livelihoods

People around the world rely on marine and coastal fisheries to provide food resources and support livelihoods. Numerous species of crabs, shrimp, and finfishes that are particularly valued by human communities rely on natural habitats such as saltmarsh, seagrass, and oyster reef for essential habitat and feeding grounds (Beck et al. 2001; Jordan et al. 2012; Peterson et al. 2003). In contrast, some of the most common traditional armoring techniques, such as bulkheads and sea-walls, provide a poor habitat alternative and often result in biophysical changes that further degrade natural shorelines (e.g., Bilkovic and Roggero 2008; Bozek and Burdick 2005). A guiding principle that has emerged from recent living shorelines research (e.g., Gittman et al. 2016; Humphries and La Peyre 2015; Scyphers et al. 2011) is the necessity of complex habitat (i.e., high rugosity) for bio-diversity and fisheries production. Fisheries benefits are also projected as among the first benefits to appear after implementation (La Peyre et al. 2014) and thus may provide a valuable tool to com-municate outcomes from living shorelines to a wide audience of stakeholders and decision-makers. For instance, experimental studies in the Gulf of Mexico found that compared to degraded natural shorelines, oyster reef breakwaters strongly enhanced abundances of blue crabs (*Callinectes sapidus*) and several highly desirable finfish species such as red drum (*Sciaenops ocellatus*) and spotted seatrout (*Cynoscion nebulosus*) (Scyphers et al. 2011, 2015). Conservative assessments of the eco-nomic value of enhanced fish production have estimated a hectare of oyster reef to yield approxi-mately $4000 in commercial landings (Grabowski et al. 2012). In addition to being highly valued by recreational and commercial fishing industries, many of these fisheries species are key components of identity and culture in coastal communities (Dyer and Leard 1994).

2.2.2 Coastal Protection for People and Property

Coastal communities persistently face a wide variety of hazards including erosion, ecological degradation, flooding, sea level rise, and storms, and we now know that many traditional shoreline structures can often exacerbate rather than reduce these threats. Oyster and coral reefs, seagrass, mangroves, dunes, and other natural and enhanced coastal and marine ecosystems have the poten-tial to attenuate waves, stabilize shorelines, and buffer storm surge (Ferrario et al. 2014; La Peyre et al. 2015; Scyphers et al. 2011; Shepard et al. 2011), reducing the effects of hazards from which coastal communities suffer (e.g., Barbier et al. 2013; Spalding et al. 2014). Until very recently, little work existed comparing the influence of nature-based options to armored shorelines. However, Gittman et al. (2014) show that natural and enhanced shorelines may be more resilient to storms in some settings. Nearly all of the studies we reference and others have described the nonlinear and inherently context-dependent nature of erosion and storm buffering (e.g., Koch et al. 2009) and highlighted the need for more empirical data across a range of biophysical conditions and settings (NSTC 2015; Pinsky et al. 2013; Sutton-Grier et al. 2015). The context dependency of coastal pro-tection services also extends to humans in the system. A study mapping coastal risk reduction owing to habitats fringing US shorelines (Arkema et al. 2013) (Figure 2.3) suggests that coastal protection

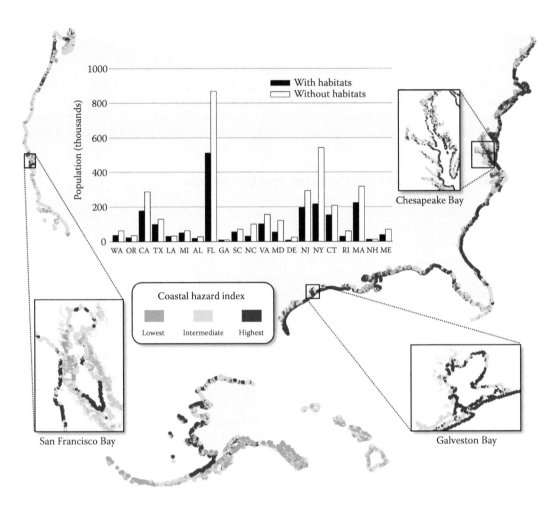

Figure 2.3   Vulnerability of the United States coastline and population to sea level rise in 2100 and storms. (a) Coastal exposure to hazards from the A2 sea level rise scenario and storms. Warmer colors in map and insets indicate regions with the greatest exposure to hazards (index values > 3.36). (b) Black bars show the total population in each coastal state living in areas most exposed to hazards (red in map) with protection provided by habitats, and the increase in vulnerability that would occur if habitats were lost because of climate change or human impacts (white bars). Data depicted in the inset maps are zoomed-in views of the nationwide analysis. This figure first appeared in Nature Climate Change in Arkema et al. (2013).

benefits of living shorelines depend on spatial variation in social and economic factors such as property values and demographics of coastal communities. Taken together, these papers indicate that living shorelines can play an important role in sheltering waterfront residents and communities from coastal hazards, but the ecological, physical, and social context in which these benefits occur needs to be better understood.

2.2.3 Coastal and Marine Recreation and Tourism

Healthy coastal and marine ecosystems provide tourism opportunities and support livelihoods. In the United States alone, ocean-related tourism and recreation generated 9.1 million jobs and $89 billion in economic output in 2010 (Kildow et al. 2014). Fishing, swimming, beach-going, and boating are among the top recreational uses of coastlines for both residents and tourists (Pendleton et al. 2001, 2006). In addition to the commercial values described above, the fisheries enhancement of living shorelines could translate to increased recreational satisfaction among residents and opportunities for visitors (Fedler 1984). Recent studies also suggest that living shorelines play an important role in promoting favorable water quality by filtering out excess nutrients such as nitrogen and phosphorus, as well as stabilizing sediments (Piehler and Smyth 2011). The closure of beaches to swimming as a result of poor water quality is not uncommon can be detrimental for coastal economies (Brown and Clarke 2007; Parsons et al. 2009). Reef-building bivalves such as oysters also contribute to water quality services through filter-feeding and benthic–pelagic coupling; nitrogen removal from one hectare of oyster reef was valued at more than $6700 annually (Grabowski et al. 2012).

2.2.4 Blue Carbon Storage and Sequestration

Globally, terrestrial and ocean ecosystems remove large amounts of carbon dioxide from the atmosphere, helping to regulate the Earth's climate. Seagrasses and mangroves, for example, store and sequester carbon, such that degradation and conversion of these systems globally release 0.15–1.02 Pg (billion tons) of carbon dioxide annually (Pendleton et al. 2012). These emissions are equivalent to 3%–19% of those from deforestation globally and result in economic damages of $US 6–42 billion annually (Pendleton et al. 2012). The degradation of vegetated shoreline can contribute to both the export of stored carbon and diminished sequestration potential (Theuerkauf et al. 2015). By incorporating vegetative elements such as saltmarshes and seagrass in their design, living shorelines have the potential to contribute to the carbon captured by the world's coastal and ocean ecosystems, frequently termed "blue carbon" (Davis et al. 2015). Several studies documenting carbon storage and sequestration rates find that restored seagrass beds and marshes, like their natural counterparts, accumulate carbon through in situ production and in their sediments (Davis et al. 2015; Donato et al. 2011; Mcleod et al. 2011). Rates vary depending on age of the restored plots and density of vegetation. For instance, a recent study of fringing marsh shorelines in coastal North Carolina revealed that older and more mature marshes are much better at storing carbon than younger marsh sites (Davis et al. 2015). Like many of the aforementioned services, restored ecosystems will take time to accumulate carbon at a rate that is comparable to natural systems (Bilkovic and Mitchell 2013; Craft et al. 2003; Davis et al. 2015; Greiner et al. 2013). Unlike other services, however, increases in carbon storage and sequestration resulting from habitat restoration benefit all coastal communities equally, regardless of their location because the atmosphere is well mixed (Mandle et al. 2015).

2.2.5 Human Health and Well-Being

A growing number of studies find that nature positively influences human well-being, including mental health (Bratman et al. 2015; Cracknell et al. 2015; Sandifer et al. 2015). For instance, spending 90 minutes walking in the woods was recently found to lower people's levels of rumination (repetitive thoughts focused on negative aspects of self), which is a known risk factor for mental illness (Bratman et al. 2015). In another study, watching marine life at the National Aquarium in Plymouth, UK, reduced both heart rate and blood pressure. In fact, people were more captivated and their moods more positive with more fish in the tank (Cracknell et al. 2015). Water features appear to be particularly restorative (White et al. 2010) with access to coastal areas reducing stress, increasing physical activity, and resulting in stronger communities (Depledge and Bird 2009). In Britain, the Universities of Exeter and Plymouth are deploying the "Blue Gym" program to examine health benefits from time spent in ocean environments and to create a national network of ocean and coastal activities to promote mental and physical well-being (Depledge and Bird 2009). The aforementioned research suggests the powerful service that coastlines restored with natural materials could provide to waterfront residents and coastal communities. This is likely to be particularly important in urban settings where new studies suggest that city dwellers have a higher risk for anxiety, depression, and other mental illnesses than people living outside urban centers (Lederbogen et al. 2011; Peen et al. 2010).

2.3 ADVANCEMENTS IN ECOSYSTEM SERVICE SCIENCE AND PRACTICE THAT SUPPORT THE DESIGN AND IMPLEMENTATION OF LIVING SHORELINES

Our growing understanding of the ways in which nature benefits people has the potential to inform the science and implementation of living shorelines. Recent research has focused on how change in the structure and function of coastal ecosystems, as a result of restoration, development, and other actions in coastal zones, leads to change in ecosystem services and outcomes for particular groups of people that benefit from those services (Barbier 2013; Olander et al. 2015; Tallis et al. 2011). Quantifying relationships between nature and people is critical for identifying trade-offs among services and effectively engaging stakeholders about their values, beliefs, and expectations for living shoreline projects in the short and long term.

2.3.1 Modeling and Measuring Ecosystem Service Outcomes to Inform Living Shoreline Projects

Advancements in modeling ecosystem services have the potential to provide insights and tools that can be used to forecast outcomes of living shoreline projects in terms that matter to people (Arkema et al. 2015; White et al. 2012). These models take advantage of established approaches in fisheries (Jordan et al. 2012 and references within), coastal engineering (Guannel et al. 2015 and references within), and other disciplines and couple these with new insights about coastal ecosystems and novel data sources (Wood et al. 2013) to model ecosystem function and delivery of services to people. The geographic, ecological, physical, and social context often influence the outcome of management actions, suggesting that the specific location where a living shoreline project takes place matters (Ruckelshaus et al. in press). Working with diverse data sources and across disciplines is integral for predicting where living shoreline approaches are likely to be most effective for achieving multiple goals for nature and people.

Empirical evidence suggests that multiple factors influence the effect of living shorelines on provisioning of services, including the size, shape, and maturity of the project, as well as local environmental conditions. Though the number of field projects that monitor living shorelines and assess ecosystem services over multiple years is growing (Figure 2.1), overall, the empirical data quantifying these relationships are limited. For instance, scientists and practitioners commonly assume that ecosystem service provisioning by living shorelines scales with size (Gedan et al. 2011), yet this relationship has not been verified across multiple living shoreline projects. Another major factor contributing to variability in ecosystem services provided by living shorelines is project age and maturity of nearby habitats (Bilkovic and Mitchell 2013; Gittman et al. in 2016; La Peyre et al. 2014; NOAA 2015). However, in general, very little is known about the rate of provisioning over time as monitoring data rarely extend beyond 1–2 years after construction (if monitoring is even conducted at all). The lack of consistent and long-term monitoring data is a major impediment, making it difficult to test how environmental characteristics, such as salinity, water depth, and wave energy, influence project performance. Adequate funding for pre- and post-project monitoring is essential for elucidating the conditions under which living shorelines are able to achieve desired goals, ultimately leading to the development of well-tested tools and approaches to inform when, where, and how to implement living shoreline projects (Bilkovic et al. 2016; Currin et al. 2009; NRC 2007).

2.3.2 Understanding and Quantifying the Beneficiaries of Living Shoreline Projects

Our Google Scholar search suggests that living shoreline projects that explicitly consider outcomes for people often measure these in economic units (e.g., Grabowski et al. 2012; Humphries and La Peyre 2015). Economic metrics for ecosystem services are compelling for policy- and decision-makers and can serve as a common unit of measurement across several objectives (Arkema et al. 2015; White et al. 2012). Indeed, for a long time, scientists and practitioners assumed that quantifying ecosystem services meant putting a dollar value on nature. However, this assumption is changing. New research into the benefits of nature for human well-being offers insights for living shoreline projects.

Ecosystem services are delivered in a variety of ways with many different values to people, including nutritional benefits from pollinators (Eilers et al. 2011), human health benefits (Myers et al. 2013), and safety from hazards (Arkema et al. 2013; Beck et al. 2013) (Figure 2.3). The magnitude of these values varies tremendously and depends not only on biophysical factors but also on the scarcity of services, their utility to specific communities, and the availability of built and technological substitutes. Recent studies find that quantifying services using a variety of social metrics, such as poverty, literacy, and age can reveal important patterns that strictly economic results may mask (Arkema et al. 2013; Mandle et al. 2015; USACE 2015). For example, Arkema et al. (2013) showed that coastal habitats were important for reducing the number of poor families at high risk in southern Texas and the total value of property at high risk in Florida. Valuing coastal habitats in terms of just the latter economic metric—reduction in total value of property at risk—would have suggested incorrectly that the habitats in Florida were more important than those in Texas for coastal protection services (see p. 32 of NSTC 2015 for more discussion of socioeconomic considerations). Further, quantifying ecosystem services with multiple values tends to be useful for community engagement, as different metrics resonate with different groups of people (Arkema et al. 2015; Barbier 2013).

For living shorelines to continue to grow in popularity as alternatives or complements to traditional engineering approaches (Erdle et al. 2006; NRC 2007), project goals and outcomes must align with stakeholder values. In systems with a high proportion of residential shoreline, understanding how waterfront homeowners make decisions is critical, considering how the management of private yards and shorelines scales up to influence the health and resilience of local ecosystems (Cook et al. 2012). Surveys of homeowner decision-making reveal that biophysical consequences of neighboring armored shorelines, exacerbated by misperceptions of maintenance-related costs, can overshadow aesthetic preferences of natural shorelines and environmental awareness (Figure 2.4) (Scyphers et al. 2015). However, ecosystem service-related social values and economic incentives may provide powerful tools for communicating the benefits of living shorelines and encouraging implementation (Scyphers et al. in review). Along public property and at larger spatial scales, living shorelines planning and implementation hinge on leveraging a broader society informed of the social, ecological, and economic costs; benefits; and trade-offs. Interactive mapping and decision support systems, such as The Nature Conservancy's (TNC) Coastal Resilience decision support tool and the Natural Capital Project's InVEST models, provide unprecedented capabilities to integrate data on both ecosystem services and societal dimensions (Beck et al. 2013).

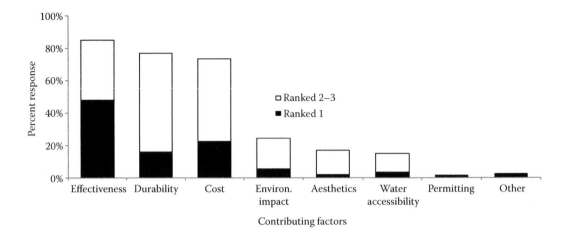

Figure 2.4   Waterfront homeowners are key actors in the decision to implement living shorelines or traditional armoring structures. Scyphers et al. (2015) reported the findings of a survey of waterfront residents in coastal Alabama that showed that economics-centric attributes of effectiveness, durability, and cost far outweighed potential other influences such as concern for environmental impacts, aesthetics, water access, and permitting. Notably, most homeowners perceived armored structures as superior to natural shorelines in terms of the three most influential criteria; however, reported maintenance costs revealed this to be an inaccurate perception as natural shorelines required approximately half the investment of armored structures.

2.3.3 Community Engagement and Capacity Building to Foster Implementation of Living Shorelines

Lack of awareness of nature-based approaches is another limiting factor in the use of living shorelines. Historically, many decision-makers were unfamiliar with options other than seawalls and bulkheads to combat shoreline erosion, and living shorelines were absent from the menu of protection measures offered by contractors and engineering firms. However, community outreach and education efforts, such as those led by NGOs (e.g., TNC, Sea Grant) and federal and state agencies (e.g., NOAA, Maryland Department of Natural Resources, North Carolina Division of Coastal Management), are increasing awareness about the potential for using living shorelines to reduce erosion and restore habitats. Collaborative relationships between conservation NGOs and private engineering forms, such as those recently initiated by TNC and a suite of major engineering firms including CH2M, are increasing the use of nature-based approaches and technologies. Despite these advances, further capacity building and education are necessary to address barriers to living shoreline uptake and implementation (RAE 2015). These can leverage an ecosystem services framework to make clear how living shorelines contribute to a variety of public values beyond those enjoyed exclusively by the landowner.

2.4 LINKING SCIENCE AND PRACTICE

In Sections 2.4.1 through 2.4.3, we describe how ecosystem service concepts and science are being used to inform living shoreline implementation in three case studies. The case studies include ecosystem-based adaptation (EbA) in the Caribbean, Rebuilding after Hurricane Sandy in the northeastern United States, and Restoration of the Gulf of Mexico. While this book focuses on the US context, we have the potential to learn from ecosystem service science and practice occurring around the world; thus, we include an example from the Caribbean. In the case of both EbA and postdisaster restoration, a suite of national and regional policies are calling for management decisions informed by ecosystem service considerations. The following initiatives are examples of how ecosystem services framing, science, and tools can help facilitate implementation pathways for nature-based shoreline stabilization.

2.4.1 Ecosystem-Based Adaptation to Climate Change—Caribbean

Global climate change is one of the most serious threats to sustainable development around the world. This is particularly true for small island developing countries such as those in the Caribbean. To confront this threat, the CARICOM Heads of State commissioned a framework to guide regional climate adaptation (Caribbean Community Climate Change Centre 2009). One component of the overall strategy is EbA, an approach that offers the opportunity to increase resilience to climate change while maintaining benefits from the natural environment that support the region's economy. With so many Caribbean countries heavily reliant on fisheries, tourism, forestry, and agriculture, governments in the region are looking to mainstream EbA into development planning. An ecosystem services framework has the potential to facilitate implementation of living shoreline projects for climate adaptation and economic development.

As an example, Placencia is a small city in southern Belize that specializes in high-end ecotourism resorts where visitors pay more to experience the intact natural environment. Over the last several years, World Wildlife Fund has been engaging stakeholders to identify feasible climate adaptation strategies (e.g., mangrove restoration; Figure 2.5a) on the local scale. For example, despite the threat of storm damage, many stakeholders are opposed to seawalls because of a tourism-driven economy that relies on beautiful beaches and cabanas nestled within mangrove forests to draw visitors. Local residents and resort owners worry about negative effects of seawalls on beaches and would prefer to repair infrastructure after a storm than suffer the irreparable loss of coastal ecosystems to the tourism industry (Nadia Bood, personal communication). Policy-makers and resource managers pursuing integrated management in Belize (Arkema et al. 2015) are asking for evaluations of alternative adaptation options to understand more explicitly these kinds of trade-offs among coastal defense measures, tourism revenue, fisheries, and other services (Figure 2.5b). They foresee using this information to identify where coastal habitat restoration can be most useful for adapting to climate change and anticipating unintended consequences of management decisions.

2.4.2 Rebuilding after Hurricane Sandy—United States

In the northeastern United States, the dense human populations and historical fishing communities highlight the complex trade-offs of how we develop coastlines. In 2012, Hurricane Sandy brought that conversation front and center as many northeast and Atlantic communities suffered from unprecedented flooding and devastation. In the aftermath, several planning initiatives and funding opportunities that emphasized nature-based defenses as central to the rebuilding strategy and future coastal resilience were launched (e.g., Mayor Bloomberg's "A Stronger More Resilient New York," the US federal government's "Hurricane Sandy Rebuilding Strategy," and the National Fish and Wildlife Foundation's "Hurricane Sandy Coastal Resiliency Competitive Grant Program").

As an example of one of the post-Sandy programs, a relatively small number of private organizations and federal agencies launched the "Rebuild by Design" initiative aimed at promoting a paradigm shift in how planners and governments approach disaster preparedness and response. Nature-based approaches to coastal protection and living shorelines in particular have garnered a large role in this initiative with more than half of the 10 finalists incorporating habitats or ecosystem services in their designs. A winning design that focused on Long Island, New York, integrates

(a)

| | | Adaptation scenarios | |
	No action	Integrated	Reactive
NPV of total benefits	$0.790	$1.300	$0.650
NPV lobster fishing	$0.008	$0.009	$0.006
NPV tourism and recreation	$0.782	$1.273	$0.702
NPV carbon storage and sequestration	–	$0.013	–$0.061
NPV of total implementation costs	–$0.005	–$0.015	–$0.191
NPV of erosion damages from sea level rise and storms	–$2.517	–$2.556	–$2.005
Total NPV of all benefits, costs and damages	–$1.731	–$1.275	-$1.550
NPV compared to no action scenario	–	$0.456 billion	$0.181 billion

(b)

Figure 2.5   (a) Mangrove restoration in Placencia, Belize, informed in part by a (b) cost–benefit analysis comparing the net present value (NPV) of Integrated, Reactive, and No Action climate adaptation scenarios. The Integrated scenario that emphasizes nature-based approaches to coastal protection results in more than twice the NPV of the Reactive scenario, owing to higher returns from a suite of ecosystem services provided by storm-buffering mangroves and coral reefs and lower implementation costs. Seawalls were more common in the Reactive scenario and assumed to completely withstand hazards, thus leading to lower damages from storms. Higher damages in the Integrated scenario relative to the No Action scenario are a result of increased coastal development for tourism and corresponding higher property values. (Adapted from Rosenthal, A., K. Arkema, G. Verutes et al. 2013. Identification and Valuation of Adaptation Options in Coastal-Marine Ecosystems: Test Case from Placencia, Belize, Inter-American Development Bank Report. http://community.eldis.org/.59c095ef/Identification%20and%20Valuation%20of%20Adaptation%20Options%20in%20Coastal%20Belize%20FINAL.pdf; and Arkema, K.K., G.M. Verutes, S.A. Wood et al. 2015. Embedding ecosystem services in coastal planning leads to better outcomes for people and nature. *PNAS* 112(24): 7390–7395.)

"eco-edges" of saltmarsh to reduce wave action, improve ecological functioning, and generate new recreational opportunities (Rebuild by Design 2014).

2.4.3 Restoring the Gulf of Mexico—United States

The US Gulf Coast exemplifies a region where the widespread decline of natural shoreline habitats has prompted investments in restoration and living shorelines (e.g., Coastal Protection and Restoration Authority 2012). Vast expanses of saltmarshes, fringing oyster reefs, and sandy shorelines contribute to highly productive commercial and recreational fisheries, as well as vital tourism economies. Historically, shoreline and watershed development, coupled with excessive and destructive harvest of oysters, have been primary drivers in the decline of these habitats. However, this region has also benefited from a long history of coastal restoration. In recent years, multimillion dollar investments in living shorelines were facilitated by the American Recovery and Reinvestment Act, which aimed to restore ecosystems and economies (Edwards et al. 2013). Hurricane Katrina was a particularly powerful focusing event for conversations on the importance of coastal wetlands for storm buffering (Barbier et al. 2013). Now, there is a tremendous need for the ecosystem services and living shoreline research discussed in this chapter and the rest of this book in the Gulf of Mexico where large-scale constructed living shoreline and breakwater projects are being planned across the region in response to the approximate $12.6 billion of restoration funding provided by legal settlements from the Deepwater Horizon Oil Spill (Figure 2.6, Shepard et al. 2015).

In coastal Alabama, Mobile Bay has provided an exemplary system for implementing and evaluating living shorelines (Roland and Douglass 2005; Scyphers et al. 2011, 2015). Importantly, as the science of living shorelines has progressed, implementation has followed suit. For nearly 20 years, living shorelines projects have been implemented at spatial scales ranging from waterfront residential properties to km-long stretches along public properties. Living shoreline initiatives have prospered in coastal Alabama, at least in part, because of strong partnerships among local and national NGOs, state and federal research and regulatory agencies, and local academic institutions working collaboratively. In addition to advancing the science and practice of living shorelines, these partnerships have ensured a broader focus, yielding social and economic benefits for coastal communities through restoration-related jobs and enhanced fisheries-related ecosystem services.

2.5 CONCLUSIONS AND FUTURE DIRECTIONS

Empirical studies of living shorelines have consistently documented enhanced ecological functions compared to armored and degraded shorelines. In this chapter, we extend our understanding of ecological outcomes to incorporate social considerations and metrics using an ecosystem services framework. We discuss the potential for ecosystem service science and practice to inform several challenges facing living shorelines. In particular, framing possible outcomes of living shoreline projects in terms that resonate with people can facilitate stakeholder engagement, enhance project implementation, and influence success. Ecosystem service models that link change in the ecosystem structure and function of natural and enhanced shorelines to services that matter to people have the potential to inform selection of living shoreline sites and provide tools and approaches for engaging community members from the start of a project through to postimplementation monitoring. While empirical evidence and models can help explore potential trade-offs, the reality of on-the-ground engagement is that decisions are rarely the direct result of the science. Rather than suggesting a "best choice" scenario for living shoreline projects, ecosystem service

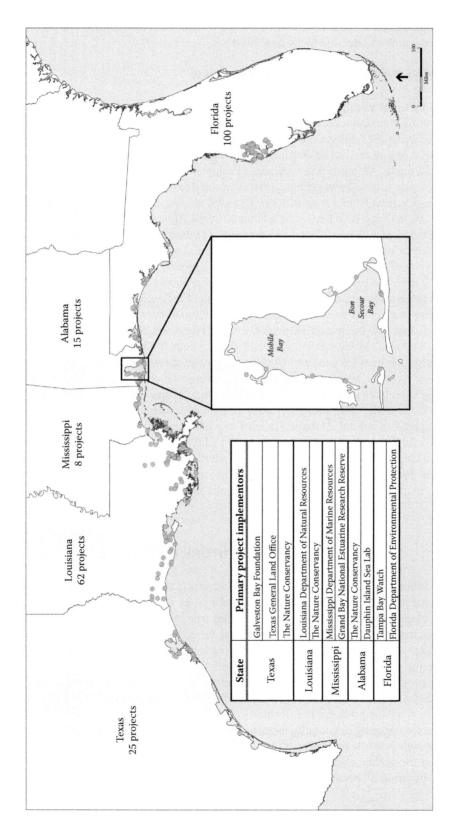

Figure 2.6   A total of 253 living shoreline projects have been identified in The Nature Conservancy's preliminary Gulf of Mexico Living Shoreline Inventory.

tools and approaches provide visualization and measures through which policy-makers, natural resource managers, and waterfront residents can gain awareness about alternative approaches for shoreline stabilization and revise and refine interventions to meet multiple objectives for a diversity of beneficiaries.

The rapid popularization of living shoreline approaches points to the need for science to keep pace with implementation. This need bridges basic and applied ecology by calling for studies on the relative efficacy of various living shoreline types and technologies for people (i.e., through the provisioning of ecosystem services) and wildlife across varying scales of space and time (NSTC 2015). The following are some potential research questions that remain:

- How does the structural longevity of various living shoreline strategies compare to traditional shoreline armoring structures?
- How do vertical growth rates for living shorelines (e.g., for oyster reefs, marsh, intertidal flats) compare to control sites and to local rates of sea level rise (see Rodriguez et al. 2014 for an assessment of oyster reefs)?
- How can living shoreline projects be optimally designed to attenuate waves and reduce rates of erosion/increase rates of deposition shoreward of the treatments?
- How do the ecosystem services provided by living shorelines composed of multiple habitats compare to single-habitat projects (Guannel et al. 2016)?

Likewise, there is a need for understanding the social dynamics of coastal human communities where living shorelines are typically implemented (e.g., with societal and stakeholder surveys, census data, public/private lands information, etc.) and how human outcomes and decisions relate to the ecological and site-specific characteristics of the projects (RAE 2015). Some potential key research questions are as follows:

- How do the environmental values, beliefs, and social norms of waterfront homeowners and other key decision-makers align with the characteristics of living shorelines and traditional armoring?
- How important are ecosystem services in the shoreline management decisions of waterfront homeowners?
- How does the economic value of natural and restored shorelines compare to armored structures in real estate pricing and property appraisal?
- What is the role of homeowner associations and similar institutions for establishing policies and social norms regarding shoreline management?

In addition to living shoreline research needs, several new frontiers in ecosystem service modeling may soon be useful for living shoreline science and practice. These include quantifying the marginal value of each new unit of area of habitat (Ricketts and Lonsdorf 2013), incorporating interactions between different habitat types in providing services (Guannel et al. 2016), modeling temporal trajectories for habitat restoration (La Peyre et al. 2014), and, last, confronting models with empirical data. For instance, understanding the relative importance of patch size could help prioritize waterfront parcels for maximizing ecological gains, and improved abilities to predict ecosystem service trajectories could benefit both planning for postimplementation monitoring and managing stakeholder expectations.

As the numbers and types of living shoreline projects implemented in the United States and around the world continue to increase, we have an opportunity to compare across projects. This learning, however, will only occur if project funders make project monitoring a priority. A meta-analysis of project performance (based in part on ecosystem service provisioning and beneficiaries) will advance restoration science and living shoreline implementation by documenting the design elements, site characteristics, and environmental factors that lead to the greatest ecological and societal benefits.

REFERENCES

Arkema, K.K., G. Guannel, G. Verutes et al. 2013. Coastal habitats shield people and property from sea-level rise and storms. *Nature Climate Change* 3(10): 913–918.

Arkema, K.K., G.M. Verutes, S.A. Wood et al. 2015. Embedding ecosystem services in coastal planning leads to better outcomes for people and nature. *PNAS* 112(24): 7390–7395.

Barbier, E.B. 2013. Valuing ecosystem services for coastal wetland protection and restoration: Progress and challenges. *Resources* 2(3): 213–230. DOI:10.3390/resources2030213.

Barbier, E.B., I.Y. Georgiou, and B. Enchelmeyer. 2013. The value of wetlands in protecting Southeast Louisiana from hurricane storm surges. *PLoS ONE* 8, e58715.

Barbier, E.B., S.D. Hacker, C. Kennedy et al. 2011. The value of estuarine and coastal ecosystem services. *Ecological Monographs* 81(2): 169–193.

Beck, M.W., B. Gilmer, G.T.Z. Ferdana et al. 2013. Using interactive decision support to integrate coast hazard mitigation and ecosystem services in Long Island Sound, New York and Connecticut USA, in Renaud, F.G., K. Sudmeier-Rieux, and M. Estrella (Eds.), *The Role of Ecosystems in Disaster Risk Reduction* 140–163. UNU Press.

Beck, M.W., K.L. Heck Jr., K.W. Able et al. 2001. The identification, conservation, and management of estuarine and marine nurseries for fish and invertebrates: A better understanding of the habitats that serve as nurseries for marine species and the factors that create site-specific variability in nursery quality will improve conservation and management of these areas. *Bioscience* 51(8): 633–641.

Berman, M., H. Berquist, and P. Mason. 2005. Building arguments for living shorelines. Proceedings of the 14th Biennial Coastal Zone Conference, New Orleans, Louisiana.

Berman, M. and Rudnicky, T. 2008. *The Living Shoreline Suitability Model, Worcester County, Maryland. Coastal Zone Management Program.* Maryland Department of Natural Resources, Annapolis, Maryland.

Bilkovic, D.M. and M. Mitchell. 2013. Ecological trade-offs of stabilized salt marshes as a shoreline protection strategy. *Ecological Engineering* 61: 469–481.

Bilkovic, D.M., M. Mitchell, P. Mason, and K. Duhring. 2016. The role of living shorelines as estuarine habitat conservation strategies. *Coastal Management* 44(3): 161–174.

Bilkovic, D.M. and M.M. Roggero. 2008. Effects of coastal development on nearshore estuarine nekton communities. *Marine Ecology Progress Series* 358:27.

Bozek, C.M. and D.M. Burdick. 2005. Impacts of seawalls on saltmarsh plant communities in the Great Bay Estuary, New Hampshire USA. *Wetlands Ecology and Management* 13:553–568.

Bratman, G.N., J.P. Hamilton, K.S. Hahn et al. 2015. Nature experience reduces rumination and subgenual prefrontal cortex activation. *PNAS* 112(28): 8567–8572. DOI:10.1073/pnas.1510459112.

Brown, R.R. and J.M. Clarke. 2007. *Transition to Water Sensitive Urban Design: The Story of Melbourne, Australia.* Facility for Advancing Water Biofiltration, Monash University, Melbourne, Australia.

Caribbean Community Climate Change Centre (CCCCC). 2009. Climate Change and the Caribbean: Regional Framework for Achieving Development Resilient to Climate Change (2009–2015). http://www.caribbean climate.bz/ongoing-projects/2009-2021-regional-planning-for-climate-compatible-development-in-the -region.html.

Center for Coastal Resources Management. 2006. Living Shorelines. Rivers and Coast 1(2).

Chesapeake Bay Foundation. 2007. Living Shorelines for the Chesapeake Bay Watershed.

Coastal Protection and Restoration Authority (CPRA) of Louisiana. 2012. *Louisiana's Comprehensive Master Plan for a Sustainable Coast*; Office of Coastal Protection and Restoration, Baton Rouge, LA, USA.

Cook, E.M., S.J. Hall, and K.L. Larson. 2012. Residential landscapes as social–ecological systems: A synthesis of multi-scalar interactions between people and their home environment. *Urban Ecosystems* 15: 19–52.

Cracknell, D., M.P. White, S. Pahl et al. 2015. Marine biota and psychological well-being: A preliminary examination of dose–response effects in an aquarium setting. *Environment and Behavior*, July, 0013916515597512. DOI:10.1177/0013916515597512.

Craft, C., P. Megonigal, S. Broome et al. 2003. The pace of ecosystem development of constructed *Spartina alterniflora* marshes. *Ecological Applications* 13(5): 1417–1432.

Currin, C.A., P.C. Delano, and L.M. Valdes-Weaver. 2008. Utilization of a citizen monitoring to assess the structure and function of natural and stabilized fringing salt marshes in North Carolina. *Wetlands Ecology Management* 16: 97–118.

Currin, C.A., W.S. Chappell, and A. Deaton. 2009. Developing alternative shoreline armoring strategies: The living shoreline approach in North Carolina. Puget Sound shorelines and the impacts of armoring. *Proceedings of the State of the Science Workshop* 91–102.

Daily, G. 1997. *Nature's Services: Societal Dependence on Natural Ecosystems.* Island Press.

Davis, J.L., C.A. Currin, C. O'Brien et al. 2015. Living shorelines: Coastal resilience with a blue carbon benefit. *PLoS ONE* 10: e0142595.

Day, J.W., D.F. Boesch, E.J. Clairain et al. 2007. Restoration of the Mississippi Delta: Lessons from Hurricanes Katrina and Rita. *Science* 315: 1679–1684.

Depledge, M.H. and W.J. Bird. 2009. The Blue Gym: Health and wellbeing from our coasts. *Marine Pollution Bulletin* 58(7): 947–948.

Donato, D.C., J.B. Kauffman, D. Murdiyarso et al. 2011. Mangroves among the most carbon-rich forests in the tropics. *Nature Geoscience* 4(5): 293–297.

Dyer, C.L. and R.L. Leard. 1994. Folk management in the oyster fishery of the U.S. Gulf of Mexico, in Dyer, C.L., and J.R. McGoodwin (Eds.), *Folk Management in the World's Fisheries: Lessons for Modern Fisheries Management.* University Press of Colorado, Niwot, Colorado.

Edwards, P.E.T., A.E. Sutton-Grier, and G.E. Coyle. 2013. Investing in nature: Restoring coastal habitat blue infrastructure and green job creation. *Marine Policy* 38: 65–71.

Eilers, E.J., C. Kremen, S. Greenleaf et al. 2011. Contribution of pollinator-mediated crops to nutrients in the human food supply. *PLoS ONE* 6(6): e21363. doi:10.1371/journal.pone.0021363.

Erdle, S.Y., J.L. Davis, and K.G. Sellner. 2006. Management, policy, science, and engineering of nonstructural erosion control in the Chesapeake Bay. *Proceedings of the 2006 Living Shoreline Summit.* CRC Publication No. 08-164.

Fedler, A.J. 1984. Elements of motivation and satisfaction in the marine recreational fishing experience. *Marine Recreational Fisheries* 9: 75–83.

Ferrario, F., M.W. Beck, C.D. Storlazzi et al. 2014. The effectiveness of coral reefs for coastal hazard risk reduction and adaptation. *Nature Communications* 5: 3794.

Gedan, K., M. Kirwan, E. Wolanski et al. 2011. The present and future role of coastal wetland vegetation in protecting shorelines: Answering recent challenges to the paradigm. *Climatic Change* 106(1): 7–29.

Gittman, R.K., C.H. Peterson, C.A. Currin et al. 2016. Living shorelines can enhance the nursery role of threatened estuarine habitats. *Ecological Applications* 26(1): 249–263. http://dx.doi.org/10.1890/14-0716.1.

Gittman, R.K., A.M. Popowich, J.F. Bruno et al. 2014. Marshes with and without sills protect estuarine shorelines from erosion better than bulkheads during a category 1 hurricane. *Ocean & Coastal Management* 102(December): 94–102. DOI:10.1016/j.ocecoaman.2014.09.016.

Grabowski, J.H., R.D. Brumbaugh, R. Conrad et al. 2012. Economic valuation of ecosystem services provided by oyster reefs. *BioScience* 62(10): 900–909.

Greiner, J.T., K.J. McGlathery, J. Gunnell et al. 2013. Seagrass restoration enhances 'blue carbon' sequestration in coastal waters. *PLoS ONE* 8(8): e72469. DOI:10.1371/journal.pone.0072469.

Guannel, G., K. Arkema, P. Ruggiero et al. 2016. The power of three: Coral reefs, seagrasses and mangroves protect coastal regions and increase their resilience. *PLoS ONE* 11: e0158094.

Guannel, G., P. Ruggiero, J. Faries et al. 2015. Integrated modeling framework to quantify the coastal protection services supplied by vegetation. *Journal of Geophysical Research: Oceans* 120(1): 324–345. DOI:10.1002/2014JC009821.

Hoatson, S. 2010. The Effectiveness of Ecotourism as an Ecological Restoration Tool: Exploring Function, Proximity and Feasibility in the Chesapeake Bay Watershed. The Evergreen State College. http://archives.evergreen.edu/masterstheses/Accession86-10MES/hoatson_sMES2010.pdf.

Humphries, A.T. and M.K. La Peyre. 2015. Oyster reef restoration supports increased nekton biomass and potential commercial fishery value. *PeerJ* 3(August): e1111. DOI:10.7717/peerj.1111.

IPCC. 2007. Fourth Assessment Report: Climate Change. 6.2.2 Increasing human utilisation of the coastal zone.

Jones, H.P., D.G. Hole, and E.S. Zavaleta. 2012. Harnessing nature to help people adapt to climate change. *Nature Climate Change* 2: 504–509.

Jordan, S.J., T. O'Higgins, and J.A. Dittmar. 2012. Ecosystem services of coastal habitats and fisheries: Multiscale ecological and economic models in support of ecosystem-based management. *Marine and Coastal Fisheries* 4: 573–586.

Kalinowski, P. and Y. Baker. 2014. Tidal Wetlands Protection in Virginia. Virginia Coastal Policy Clinic. http://law.wm.edu/academics/programs/jd/electives/clinics/vacoastal/reports/VCPC%20Tidal%20 Wetlands%20Report%20Web.pdf.

Keeler, B.L., S.A. Wood, S. Polasky et al. 2015. Recreational demand for clean water: Evidence from geo-tagged photographs by visitors to lakes. *Frontiers in Ecology and the Environment* 13(2): 76–81. DOI:10.1890/140124.

Kildow, J.T., C.S. Colgan, J.D. Scorse et al. 2014. State of the US Ocean and Coastal Economies 2014. *Publications* Paper 1. http://cbe.miis.edu/noep_publications/1/.

Koch, E.W., E.B., Barbier, and B. Silliman. 2009. Non-linearity in ecosystem services: Temporal and spatial variability in coastal protection. *Frontiers in Ecology and the Environment* 7: 29–37.

La Peyre, M.K., A.T. Humphries, and S.M. Casas. 2014. Temporal variation in development of ecosystem services from oyster reef restoration. *Ecological Engineering* 63: 34–44.

La Peyre, M.K., L. Schwarting, and S. Miller. 2013. Baseline data for evaluating the development trajectory and provision of ecosystem services by created fringing oyster reefs in Vermilion Bay, Louisiana. US Geological Survey. OFR 2013-1053.

La Peyre, M.K., K. Serra, T.A. Joyner et al. 2015. Assessing shoreline exposure and oyster habitat suitability maximizes potential success for sustainable shoreline protection using restored oyster reefs. *PeerJ* 3(October): e1317. DOI:10.7717/peerj.1317.

Lederbogen, F., P. Kirsch, L. Haddad et al. 2011. City living and urban upbringing affect neural social stress processing in humans. *Nature* 474(7352): 498–501. DOI:10.1038/nature10190.

Mandle, L., H. Tallis, L. Sotomayor et al. 2015. Who loses? Tracking ecosystem service redistribution from road development and mitigation in the Peruvian Amazon. *Frontiers in Ecology and the Environment* 13(6): 309–315. DOI:10.1890/140337.

Mcleod, E., G.L. Chmura, S. Bouillon et al. 2011. A blueprint for blue carbon: Toward an improved understanding of the role of vegetated coastal habitats in sequestering CO_2. *Frontiers in Ecology and the Environment* 9(10): 552–560.

MDGA. Maryland General Assembly, 2008. HB973 Living Shoreline Protection Act of 2008. Available from: http://www.dnr.state.md.us/ccs/pdfs/ls/dnr/scm/2008_LSPA.pdf.

Myers, S., L. Gaffikin, C.D. Golden et al. 2013. Human health impacts of ecosystem alteration. *PNAS* 110: 18753–18760.

National Oceanographic and Atmospheric Administration (NOAA). 2015. Guidance for considering the use of living shorelines. Living Shorelines Workgroup Synthesis Report. National Research Council. 2007. *Mitigating Shoreline Erosion along Sheltered Coasts*. The National Academies Press, Washington, D.C.

National Research Council. 2014. *Reducing Coastal Risks on the East and Gulf Coasts*. Committee on U.S. Corps of Engineers Water Resources Science, Engineering, and Planning: Coastal Risk Reduction. Water Science and Technology Board and Ocean Studies Board. The National Academies Press, Washington, D.C.

National Science and Technology Council (NSTC). 2015. Ecosystem service assessment: Research needs for coastal green infrastructure. Report published by the White House Office of Science and Technology Policy. Accessed at: https://www.whitehouse.gov/sites/default/files/microsites/ostp/NSTC/cgies_research _agenda_final_082515.pdf.

Nicholls, R.J., F.M.J. Hoozemans, and M. Marchand. 1999. Increasing flood risk and wetland losses due to global sea-level rise: Regional and global analyses. *Global Environmental Change* 9(Supplement 1): S69–S87.

Olander, L., R.J. Johnston, H. Tallis et al. 2015. Best Practices for Integrating Ecosystem Services into Decision Making. Nicholas Institute Working Paper. https://nicholasinstitute.duke.edu/sites/default/files /publications/es_best_practices_fullpdf_0.pdf.

Olsson, P., C. Folke, and T. Hahn. 2004. Social–ecological transformation for ecosystem management: The development of adaptive co-management of a wetland landscape in southern Sweden. *Ecology and Society* 9.4, 2.

OMB Management Memorandum 16-01. Incorporating Ecosystem Services into Federal Decision Making, October 7, 2015.

Parsons, G.R., A.K. Kang, C.G. Leggett et al. 2009. Valuing beach closures on the Padre Island National Seashore. *Marine Resource Economics* 213–235.

Peen, J., R.A. Schoevers, A.T. Beekman et al. 2010. The current status of urban–rural differences in psychiatric disorders. *Acta Psychiatrica Scandinavica* 121(2): 84–93.

Pendleton, L. 2010. Measuring and Monitoring the Economic Effects of Habitat Restoration: A Summary of a NOAA Blue Ribbon Panel. Nicholas Institute for Environmental Policy Solutions, Duke University and Restore America's Estuaries.

Pendleton, L., D.C. Donato, B.C. Murray et al. 2012. Estimating global "blue carbon" emissions from conversion and degradation of vegetated coastal ecosystems. *PLoS ONE* 7(9): e43542.

Pendleton, L., J. Kildow, and J.W. Rote. 2006. The non-market value of beach recreation in California. Shore and Beach 74(2): 34.

Pendleton, L., N. Martin, and D.G. Webster. 2001. Public perceptions of environmental quality: A survey study of beach use and perceptions in Los Angeles County. *Marine Pollution Bulletin* 42(11): 1155–1160.

Peterson, C.H., J.H. Grabowski, and S.P. Powers. 2003. Estimated enhancement of fish production resulting from restoring oyster reef habitat: Quantitative valuation. *Marine Ecology Progress Series* 264: 249–264.

Piazza, B.P., P.D. Banks, and M.K. La Peyre. 2005. The potential for created oyster shell reefs as a sustainable shoreline protection strategy in Louisiana. *Restoration Ecology* 13: 499–506.

Piehler, M.F. and A.R. Smyth. 2011. Habitat-specific distinctions in estuarine denitrification affect both ecosystem function and services. *Ecosphere* 2(1): art12.

Pinsky, M.L., G. Guannel, and K. Arkema. 2013. Quantifying wave attenuation to inform coastal habitat conservation. *Ecosphere* 4(8): art95.

Rebuild by Design. 2014. Living with the Bay: A Comprehensive Regional Resilience Plan for Nassau County's South Shore. Interboro Team, Long Island, New York. http://www.rebuildbydesign.org/project/interboro-team-final-proposal/.

Restore America's Estuaries. 2015. *Living Shorelines: From Barriers to Opportunities*. Arlington, VA.

Ricketts T. and E. Lonsdorf. 2013. Mapping the margin: Comparing marginal values of tropical forest remnants for pollination services. *Ecological Applications* 23: 1113–1123.

Rodriguez, A.B., F.J. Fodrie, J.T. Ridge et al. 2014. Oyster reefs can outpace sea-level rise. *Nature Climate Change* 4(6): 493–497.

Roland, R.M. and S.L. Douglass. 2005. Estimating wave tolerance of Spartina alterniflora in coastal Alabama. *Journal of Coastal Research* 21: 453–463.

Rosenthal, A., K. Arkema, G. Verutes et al. 2013. Identification and Valuation of Adaptation Options in Coastal-Marine Ecosystems: Test Case from Placencia, Belize, Inter-American Development Bank Report. http://community.eldis.org/.59c095ef/Identification%20and%20Valuation%20of%20Adaptation%20Options%20in%20Coastal%20Belize%20FINAL.pdf.

Ruckelshaus, M., G. Guannel, K. Arkema et al. In press. Evaluating the benefits of green infrastructure for coastal areas: Location, location, location. *Ocean and Coastal Management*.

Sandifer, P., A.E. Sutton-Grier, and B.P. Ward. 2015. Exploring connections among nature, biodiversity, ecosystem services, and human health and well-being: Opportunities to enhance health and biodiversity conservation. *Ecosystem Services* 12: 1–15. http://dx.doi.org/10.1016/j.ecoser.2014.12.007

Schultz, P. 2011. Conservation means behavior. Conservation Biology 25: 1080–1083.

Scyphers, S.B., M.W. Beck, J. Haner et al. In review. Social values and economic incentives enhance habitat conservation along residential shorelines.

Scyphers, S.B., J.S. Picou, R.D. Brumbaugh et al. 2014. Integrating societal perspectives and values for improved stewardship of a coastal ecosystem engineer. *Ecology and Society* 19.

Scyphers, S.B., J.S. Picou, and S.P. Powers. 2015. Participatory conservation of coastal habitats: The importance of understanding homeowner decision making to mitigate cascading shoreline degradation. *Conservation Letters* 8: 41–49.

Scyphers, S.B., S.P. Powers, K.L. Heck Jr. et al. 2011. Oyster reefs as natural breakwaters mitigate shoreline loss and facilitate fisheries. *PLoS ONE* 6(8): e22396.

Shepard, C., C. Crain, and M.W. Beck. 2011. The protective role of coastal marshes: A systematic review and metaanalysis. *PLoS ONE* 6(11): e27374. http://bit.ly/vfAHvT.

Shepard, C., B. Gilmer, J. DeQuattro et al. 2015. *Charting Restoration: Gulf Restoration Priorities and Funded Projects Five Years after Deepwater Horizon*. The Nature Conservancy, Washington, DC, 32 pp.

Spalding, M.D., A.L. McIvor, M.W. Beck et al. 2014. Coastal ecosystems: A critical element of risk reduction. *Conservation Letters* 7: 293–301.

Sutton-Grier, A.E., K. Wowk, and H. Bamford. 2015. Future of our coasts: The potential for natural and hybrid infrastructure to enhance the resilience of our coastal communities, economies and ecosystems. *Environmental Science and Policy* 51: 137–148.

Tallis, H., S.E. Lester, M. Ruckelshaus et al. 2011. New metrics for managing and sustaining the ocean's bounty. *Marine Policy* 36(1): 303–306. DOI:10.1016/j.marpol.2011.03.013.

Tallis, H. and S. Polasky. 2009. Mapping and valuing ecosystem services as an approach for conservation and natural-resource management. *Annals of the New York Academy of Sciences* 1162: 265–283.

Theuerkauf, E.J., J.D. Stephens, J.T. Ridge et al. 2015. Carbon export from fringing saltmarsh shoreline erosion overwhelms carbon storage across a critical width threshold. *Estuarine, Coastal and Shelf Science* 164: 367–378.

Toft, J.D., A.S. Ogston, S.M. Heerhartz et al. 2013. Ecological response and physical stability of habitat enhancements along an urban armored shoreline. *Ecological Engineering* 57: 97–108. DOI:10.1016/j.ecoleng.2013.04.022.

US Army Corps of Engineers (USACE). 2015. Resilience adaptation to increasing risk. Main report, North Atlantic Comprehensive Coastal Study. Accessed at: http://www.nad.usace.army.mil/Portals/40/docs/NACCS/NACCS_main_report.pdf.

Virginia. 2011. Senate Bill No. 964 § 28.2-104.1. Living shorelines; development of general permit; guidance (under general powers and duties).

White, C., B.S. Halpern, and C.V. Kappel. 2012. Ecosystem service tradeoff analysis reveals the value of marine spatial planning for multiple ocean uses. *Proceedings of the National Academy of Sciences* 109(12): 4696–4701.

White, M., A. Smith, K. Humphryes et al. 2010. Blue space: The importance of water for preference, affect, and restorativeness ratings of natural and built scenes. *Journal of Environmental Psychology* 30: 482–493.

Wood, S.A., A.D. Guerry, J.M. Silver et al. 2013. Using social media to quantify nature-based tourism and recreation. *Scientific Reports* 3: 2976.

PART II

Management, Policy, Design

CHAPTER 3

Permitting a Living Shoreline

A Look at the Legal Framework Governing Living Shoreline Projects at the Federal, State, and Local Level

Niki L. Pace

CONTENTS

3.1 INTRODUCTION

Now that you have an understanding of the importance of various types of shoreline methods to control coastal erosion, perhaps you are wondering how to permit your project. If not, you should be. Relatively new shoreline management technologies and techniques can be challenging to permit when they do not easily fit within existing permitting frameworks. It is easy to overlook the various levels of regulations that can affect your project. Project managers may not realize, for instance, that projects require approval from federal, state, and possibly local government agencies.

Regulation of living shorelines occurs at all levels of government. Installation of shoreline stabilization systems is subject to federal and state regulation and in some geographic areas by local regulation as well. For instance, projects will require permits from the US Army Corps of Engineers (USACE) under the Clean Water Act, state coastal and wetland permitting programs under state coastal programs, and possibly zoning permits from local planning departments. These overlapping authorities have resulted in a legal framework that varies widely from jurisdiction to jurisdiction. For this reason, you should always consult with permitting agencies early in the process to ensure permit approval or to determine potential obstacles that might be encountered throughout the process.

To put this discussion in context, this chapter will use a hypothetical waterfront property owner—Anne. Anne is interested in installing a living shoreline along her property. Anne has a next-door neighbor, Bob, who is not familiar with living shorelines or how Anne's living shoreline may affect his property rights. As we go through this chapter, we will return to Anne's living shoreline at the end of each section and discuss how the project will be affected.

The key objective of this chapter is to expose the reader to the various laws and levels of government involved in regulating living shoreline projects. The first discussion will set the stage by reviewing jurisdictional boundaries, the public trust doctrine, and shoreline property law. Then, the next few sections will examine project permitting from the federal, state, and local level, respectively. Last, we will put it all together in the context of a hypothetical living shoreline project.

3.2 THE PUBLIC TRUST, SHORELINE PROPERTY
LINES, AND JURISDICTIONAL BOUNDARIES

Managing coastlines for any reason is a complex undertaking, as are the legal concepts surrounding coastal properties. To fully make sense of the myriad of laws and regulations in place along our shores, it is helpful to understand some basic tenets of property law that are unique to coastlines. These legal principles affect how and why we regulate the shoreline as we do. For that reason, a brief review of the public trust doctrine, shoreline property lines, and jurisdictional boundaries will set the stage for the discussions that follow.

3.2.1 The Public Trust Doctrine

Before delving into the regulatory hurdles of shoreline projects, let's take a step back and examine why our shorelines are managed for public benefit. In other words, what is the public trust doctrine and why does it matter? The public trust doctrine can be traced back to Roman law under

Emperor Justinian who declared that the shorelines and waterways were public goods, to be used by all. Under the Justinian Code, the sea and its shorelines were deemed property intended for the use and benefit of the public and were thus incapable of being privately owned. This principle was adopted by English common law and was later incorporated into American law (Wilkins and Wascom 1992). Upon independence, each of the original 13 states became the trustee of the submerged lands within its borders for the use by the people. As additional states were admitted into the Union, they too received the title to these lands under the equal footing doctrine, which provided that newly recognized states entered the Union with the same sovereign powers and rights as the original colonies (Pace 2011).

The specifics of the public trust doctrine may vary among states but there are some common general principles that apply across the country. Under the public trust doctrine, the state holds title to submerged lands underlying navigable waters in trust for the public to protect the traditionally public nature of these lands (Martin v. Waddell, 41 U.S. 367, 409–411 (1842)). As characterized by the US Supreme Court in *Illinois Central Railroad Co. v. Illinois*, "[i]t is a title held in trust for the people of the state, that they may enjoy the navigation of the waters, carry on commerce over them, and have liberty of fishing therein, freed from the obstruction or interference of private parties" (Illinois Central Railroad Co v. Illinois, 146 U.S. 387, 452 (1892)). In other words, the public trust doctrine recognizes and protects the public's right to use waters for navigation, commerce, and fishing.

Although the public trust generally embodies these common elements, actual implementation of the public trust varies by state. For instance, some states have expanded the public trust doctrine to encompass recreational use and environmental preservation. The public trust doctrine also operates as a limitation on states' ability to transfer or sell submerged lands "unless conveyed for uses promoting the interest of the public" (Craig 2007; Klass 2006). Accordingly, the public may access and use state-owned submerged lands provided that such use does not interfere with the rights of other members of the public.

With respect to shoreline management, the public trust doctrine means that there are two property owners to take into account when considering any shoreline stabilization approach—the waterfront property owner who holds title to the uplands (the area above and landward of the mean high tide line or low high tide line, depending on state) and the government ownership of the foreshore and submerged waterbottoms held in trust for the public benefit. State property boundaries are usually tied to the water line, often either the mean high tide line or the mean low tide line. This is of particular importance in the context of permitting living shorelines that require the placement of materials on the waterbottoms, as will be discussed in more detail under state regulation in Section 3.4 (Figure 3.1).

3.2.2 Ambulatory Property Lines

In addition to public ownership, another curious aspect of coastal property is that the property lines can move. Shorelines move naturally through accretion and erosion, causing the shore to either gain or lose dry land. Those migrations can have legal consequences for the waterfront property owners. In many states, traditionally scientific terms like "accretion," "erosion," and "avulsion" have legal definitions outlining how these processes will affect property lines. These legal definitions, discussed in more detail below, often further distinguish between shoreline changes that occur naturally and those that are caused by man, and some may even distinguish those caused by the upland property owner for his benefit versus those caused by the actions of a neighbor or third party. The result is a variety of rules courts use to interpret who owns newly accreted or submerged land.

To address this wide-ranging mix of circumstances, common law developed rules to address accretion, erosion, and avulsion (Kalo 2005). Common law refers to laws developed by courts through legal decisions as opposed to statutory law that is created through legislation enacted

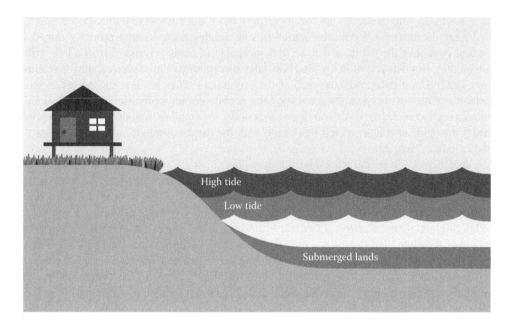

Figure 3.1   Shoreline property boundaries.

by state or federal legislative bodies (such as Congress). Under common law, shoreline migration resulting from gradual changes is treated distinctly from sudden, or avulsive, events (Christie 2010). In the first instance, the waterfront boundary shifts either as a result of accretion (where sediment is added to the shoreline) or erosion (where waterfront owners lose land). As the US Supreme Court noted in *Stop the Beach Renourishment v. Florida Department of Environmental Protection*, whether by accretion or erosion, the property line continues to track the water line (130 S. Ct. 2592, 2599 (2010)). In other words, the small and slowly occurring changes will not alter the property line because the changes are so small as to go unnoticed on a day-to-day basis.

However, property lines remain static after an avulsion. Avulsion literally means "a tearing away" and refers to the sudden removal of soil and sand after a hurricane or flood. As characterized by the US Supreme Court, "regardless of whether an avulsive event exposes land previously submerged or submerges land previously exposed, the boundary between littoral [waterfront] property and sovereign land does not change; it remains (ordinarily) what was the mean high-water line before the event" (*Stop the Beach Renourishment*, 130 S. Ct. at 2599 [summarizing the avulsion doctrine in the context of Florida law]). In other words, an avulsive event, by holding the property line immovable, severs the previous migration of the property line from the shoreline and fixes the boundary at a static location. It is important to note, however, that these rules can vary by state. Some states may allow a property owner to "reclaim" land lost through avulsion, though this may require actual physical reclamation through filling (Boyd and Pace 2013).

Like an avulsive event, construction of hardened shorelines such as bulkheads, even if placed above the waterline, can also sever the migration of the property line as a practical matter. This happens over time, as waves lap against the shoreline slowly eroding any remaining land below the bulkhead, moving the shoreline inland. Eventually, the waterline comes into direct contact with the bulkhead, often referred to as the bathtub effect (Douglass and Pickel 1999). When this happens, the so-called wet beach essentially disappears. As a result, this upland erosion defense results in permanently fixing the property line at the bulkhead, at least until a future storm or flood washes out the bulkhead.

Successful living shoreline projects, unlike hardened structures, can have completely different impacts on a waterfront property line. Living shorelines are unique from upland bulkheads in that they often require placement of fill on the waterbottoms, and in certain cases, the living shoreline may be so successful at erosion prevention that the property actual accretes land. When accretion occurs, the issue then becomes determining who owns this newly created land. As the land is accreted as a result of man-made efforts, it is generally considered "artificially accreted" land. Though state rules vary, states often continue to hold title over these newly created lands because the law does not allow a property owner to fill in state public trust lands for personal benefit.

This legal distinction between natural accretion and artificial accretion has in some cases led to the misconception that a public beach may crop up at the edge of a waterfront property. This is an improbable scenario. Living shorelines are unlikely to result in large deposition of sand significant enough to support public recreation. Furthermore, waterfront property owners retain their riparian rights over the land. Riparian rights refer to a collection of common law principles that give waterfront property owners special legal rights of access to the water. This means that the property owner will be allowed to traverse the land to access the water as well as construct piers and boathouses along the water's edge, within the confines of existing law. In other words, a waterfront property owner who installs a living shoreline is unlikely to notice any change in his legal or practical access to the water.

3.2.3 Jurisdictional Boundaries

Although shoreline stabilization projects generally take place in nearshore (state) waters, a brief overview of the water-ward jurisdictional boundaries of state and federal governments provides context for the layers of regulatory oversight. Sixty years ago, Congress enacted the Submerged Lands Act, aimed at resolving state and federal jurisdictional disputes over coastal waters and waterbottoms. The Submerged Lands Act gives coastal states jurisdiction over submerged lands and resources extending three geographic miles from shore while allocating the remainder of the territorial sea to federal jurisdiction.* The territorial sea extends from 3 to 12 nautical miles from the coast and is regulated by the federal government. Notwithstanding some initial litigation (which led to extended state boundaries of three marine leagues in Texas and the Gulf side of Florida),† these boundaries have been the operational norm for sometime. Therefore, pursuant to the Submerged Lands Act, state regulations will apply to projects within 3 miles of the shore. This state regulation, however, remains overlaid by additional federal regulation.

3.2.4 Hypothetical

Returning to our hypothetical landowner Anne, how do these background rules of law affect her installation of a living shoreline project? First, the land below the waterline is owned by the state for the public benefit. If Anne seeks to place materials on that submerged land, she may need permission from the state (as will be discussed in greater detail under state regulation).

Second, Anne needs to be aware that the living shoreline, like any erosion control project, may alter her property line. To the extent that her property accretes new land after the installation of the project, this artificially created new land will generally remain the property of the state. However,

---

* The Submerged Lands Act set the state seaward boundary at three geographic miles, unless a state could show a greater seaward boundary at the time of entering the Union or by prior Congressional approval. 43 U.S.C.A. § 1312. For qualifying states, seaward boundaries were capped at three marine leagues in the Gulf of Mexico and three geographic miles in the Atlantic and Pacific (43 U.S.C.A. § 1301(b)).

† The US Supreme Court resolved the matter as to the five gulf states in 1960 when it ruled that only Texas and Florida were eligible for extended boundaries based on the historical record. Similar claims raised by Alabama, Louisiana, and Mississippi were rejected by the Court (*United States v. Louisiana, Texas, Mississippi, Alabama & Florida*, 363 U.S. 1 (1960)). Of note, North Carolina has never claimed more than a one-marine league boundary. See N.C. Gen. Stat. § 141-6.

Anne will retain rights to cross the land to reach the water and to construct piers across this land to access the water.

Last, Anne should consider how the project might affect her neighbors to the left and right. At least one state has required that property owners installing living shoreline projects develop an erosion monitoring program to ensure that the living shoreline does not cause increased erosion to neighboring property. Interestingly, the state does not require this same monitoring of armored shorelines even though hardened shorelines can also cause erosion to neighboring properties. Another consideration is the proximity of the project to the neighbor's property line. Many states will restrict projects located within 10 ft of a neighbor's property line, again to protect neighboring property rights. These so-called setback requirements are often applied equally to any construction taking place water-ward of the shoreline, such as a pier or dock.

3.3 FEDERAL SHORELINE REGULATION

Every state manages its coastline uniquely. However, certain federal laws overlay state coastal management. Such laws include the Clean Water Act, the Coastal Zone Management Act, the Magnuson–Stevens Act, the Endangered Species Act, and others. This section provides a brief overview of provisions of federal laws that may affect shoreline stabilization and restoration projects.

3.3.1 Clean Water Act

The Federal Water Pollution Control Act, commonly known as the Clean Water Act, plays the greatest role in federal oversight of shoreline restoration and stabilization projects. The Clean Water Act regulates activities that affect US waters.* This jurisdiction extends to all coastal waters within the territorial sea (from the shoreline extending out 12 nautical miles). Clean Water Act regulations may affect a living shoreline project in two ways. First, the Clean Water Act regulates the dredge and fill of US waters under § 404. Section 404 provisions are generally referred to as the wetland regulations and naturally come into play with shoreline projects. Second, the Clean Water Act prohibits the discharge of materials into US waters without a permit under § 402. These provisions generally regulate water quality and may be combined with the regulation of stormwater and runoff. Discharge regulations can affect living shoreline projects if the project causes sediment disturbance to a degree that it affects water quality standards.

The wetland and water quality provisions of the law are overseen by two different agencies—the US USACE and the Environmental Protection Agency (EPA), respectively. EPA permitting of water quality has been largely delegated to state agencies. As such, that program will be discussed within the state permitting section. The remainder of the Clean Water Act permitting discussion will focus on wetland permitting by USACE.

3.3.1.1 Wetland Permitting

The USACE oversees wetland permitting throughout the country, with permitting issued on a regional level through local district offices. All activities in waters of the United States will require a permit from the USACE. There are three different types of wetland permits that may come into play, depending on project design and regional jurisdiction: (1) nationwide general permits, (2) regional general permits, or (3) individual permits. Nationwide and regional general permits allow a project,

---

* The definition of "waters of the United States" has been the subject of much litigation and is currently the subject of a new final rule from the EPA, clarifying the definition. 80 Fed. Reg. 37054 (June 29, 2015). The new definition is currently being litigated in the courts. There is no question, however, that coastal waters fall within the scope of the CWA protections.

falling within certain parameters, to proceed without an individualized assessment of the project. Use of a general permit may still require that certain notifications and assurances be provided to the USACE, and, in some cases, official USACE confirmation that work may proceed. An individual permit will be required if the project does not qualify for a nationwide or regional general permit. The individual permit can require a lengthy and complex application process that includes a particularized assessment of the project and the opportunity for public notice and comment.

In the context of living shorelines, two nationwide permits are of particular relevance. First, Nationwide Permit 13 is used to permit "bank stabilization" projects designed to prevent erosion. To qualify, the project must be less than 500 linear feet and use less than 1 m^3 of fill per running foot. Nationwide Permit 13 is most frequently used to permit bulkheads and hardened shoreline stabilization projects. For this reason, the permit often faces criticism from living shorelines proponents, who point to the availability of a general permit for bulkheads without a similar mechanism for living shorelines. Although living shorelines often require an individual permit, at least one jurisdiction uses Nationwide Permit 13 to permit living shoreline projects as well as other erosion control efforts.*

The second relevant nationwide permit is Nationwide Permit 27, which is used for Aquatic Habitat Enhancement and Restoration Activities. This permit is frequently used as part of wetland mitigation projects, including shoreline restoration projects. Nationwide Permit 27 has been used in some jurisdictions for permitting natural shoreline stabilization projects such as oyster reef restoration and living shoreline installations. Use of Nationwide Permit 27 often requires preconstruction notification to the USACE.

Nationwide permit criteria are reauthorized every 5 years, with the next revision scheduled for 2017. During reauthorization, the USACE evaluates the impacts of the permit to determine if those impacts are consistent with guiding regulations (§ 404(b)(1) Guidelines). Under these guidelines, the USACE is required to take into account the cumulative impacts of the permit and select a practicable alternative that is the least damaging to the environment. Many in the living shoreline community maintain that the current standards under Nationwide Permit 13 fail to do this. Rather, they argue for revised permit conditions that would require more careful consideration of the permit's cumulative and secondary impacts to the environment (Restore America's Estuaries [RAE] 2015). Advocates believe that this approach will lead to the "adoption of more hierarchical criteria for the evaluation of bank stabilization projects" that will take into account the environmental benefits of living shorelines (RAE 2015). If the USACE incorporates these suggestions, Nationwide Permit 13 may gain use in the living shoreline permitting context. Alternatively, the USACE may choose to develop a new nationwide permit specifically tailored to living shoreline projects. These changes may arrive as early as 2017. In June 2016, the USACE released a new proposed living shoreline permit that, if approved, will take effect in spring 2017. The proposed nationwide permit provides a more efficient regulatory approach to living shoreline projects. The permit encompasses placement of native vegetation coupled with natural hard structures, like oyster reefs or rock sills. The proposed permit limits structures and fill to a maximum of 30 feet from the mean high water mark. Any project that extends beyond the maximum requires a waiver from the district engineer and must show that the project will present minimal adverse environmental impacts (81 Fed. Reg. 35207). The proposed permit also requires preconstruction notification for all living shoreline projects prior to receiving a nationwide permit. In contrast, hardened structures, like bulkheads, only require preconstruction notification for structures greater than 500 feet in length. At the time of publication, the USACE is considering public comments about the new permit, and some changes may be made before a final release.

If the project does not qualify for a nationwide permit, there may still be a regional general permit available. Regional general permits are similar to nationwide permits but are designed for

* USACE Jacksonville Florida District Office.

Table 3.1 Wetland Permitting Quick Reference

- **Nationwide Permit (NWP)**

  A general permit that allows the USACE to authorize activities across the country that cause minimal adverse impacts.

  Permitted activity must satisfy all of the permit conditions, which include compliance with state or local laws and regulations.

  NWPs relevant to restoration projects include NWP 13 and NWP 27.

- **Proposed Nationwide Permit for Living Shorelines**

  Allows placement of native vegetation and natural hard structures, like oyster reefs or rock sills.

  Requires pre-construction notification.

- **Nationwide Permit 13: Bank Stabilization**

  Authorizes activities necessary to prevent erosion and stabilize shorelines.

  Limited to projects no more than 500 ft in length, unless waived by a district engineer citing minimal adverse effects.

  Permitted activity must also comply with any regional or state laws and regulations.

- **Nationwide Permit 27: Aquatic Habitat Restoration, Establishment, and Enhancement Activities**

  Authorizes activities associated with the restoration, enhancement, and establishment of tidal and nontidal wetlands.

  Specifically authorizes the construction of oyster habitat in tidal waters.

  Permitted activity must also comply with any regional or state laws and regulations.

- **Regional General Permit (RGP)**

  Typically required for projects that fall somewhere between an IP and NWP in terms of their proposed impacts.

  Usually includes provisions intended to protect the environment and resources of a specific region that shares similar interests.

- **Individual Permit (IP)**

  Issued for projects that propose extensive impacts, or impacts to rare or fragile aquatic environments.

  Generally required for projects whose proposed impacts will be greater than 1 acre of wetland or stream, but USACE can choose to review any project under an individual permit, regardless of its impact or size.

  Most detailed and time-consuming wetland permitting process.

a specific geographic area. Regional general permits can be used to permit localized projects with minimal environmental impacts that are not already allowed under an existing nationwide permit. For example, a regional general permit for living shoreline activities has been adopted for the Alabama and Mississippi coasts by the USACE Mobile District Office. The type of activity allowed and conditions required for regional general permit will depend on the district.

Last, it is important to keep in mind that states have the authority to object to, or condition, the use of nationwide permits within their jurisdictions to the extent that the permits are not consistent with federally approved state coastal programs created pursuant to the Coastal Zone Management Act, discussed in Section 3.3.2. For instance, the New England District has suspended use of all nationwide permits and instead uses state-specific general permits. In other words, just because Nationwide Permit 13 exists, that does not mean it will be available to use in every jurisdiction. States may also place regional conditions on the use of nationwide permits within their state. These conditions afford states the opportunity to tailor nationwide permits to meet the needs and objectives of the state's coastal management policy. Project managers should check in early with their local USACE district office to determine what type of wetland permits are applicable to their project. For a quick review of the wetland permits discussed, see Table 3.1.

3.3.2 Coastal Zone Management Act

As previously mentioned, participating states that maintain a federally approved coastal program have authority to object to the use of wetland general permits within their boundaries.

This authority is derived from the Coastal Zone Management Act (16 U.S.C. §§ 1451–1466 (2012)). The Coastal Zone Management Act is a federal law overseen by the National Oceanic and Atmospheric Administration. Unlike the Clean Water Act, the Coastal Zone Management Act is a voluntary federal law that does not require states to take action, but encourages and rewards the states that do. The law encourages states to adopt legally enforceable coastal management programs. The programs will vary by state but almost always include state coastal wetland permitting programs. In return, the federal government agrees to be consistent with state coastal programs, meaning that federal permits may not be issued unless the permitted activity is consistent with the state's coastal program. This trade-off is commonly referred to as the "consistency requirement." This consistency review is usually incorporated into the state permitting process.

3.3.3 Other Federal Laws

In the context of shoreline projects, a wide variety of additional federal laws may come into play. Often the application will depend on the geographic location of the project, the physical characteristics of the site, and the habitat or species located in that area. During the permitting process with the USACE, additional federal agencies will be consulted so that any potential issues with these other laws can be identified and addressed. Again, this reinforces the importance of early consultation with permitting agencies. If minor problems with a project's design or location can be identified early on, then time and money can be saved by working with the agency for solutions or considering new site locations to avoid problem areas, such as an area that contains endangered species habitat. The following is a brief overview of a few of these other laws (Table 3.2).

Table 3.2   Legal Framework for Living Shorelines Permitting: Quick Reference

Law	Responsible Agency	Scope
Federal		
Clean Water Act § 404	USACE	Fill placed in US waters, wetlands
Rivers and Harbors Act	USACE US Coast Guard	Navigational hazards
Endangered Species Act	US Fish and Wildlife NMFS	Listed species, critical habitat
Magnuson–Stevens Act	NMFS	Essential fish habitat Areas of particular concern
Coastal Barrier Resources Act	US Fish and Wildlife	Federal funding for projects located in the CBRS
State		
State Wetland Permitting	State Coastal Program or Environmental Agency	Fill of wetlands
Water Quality Regulations	State Environmental Quality Agency	Discharges into waters, stormwater, runoff
State Consistency Review	State Coastal Program or Environmental Agency	State oversight of USACE permits
Submerged Lands	Varies	Permission to use state-owned waterbottoms
Local		
Land Use Regulations	Community Planning Department	Varies by locality

3.3.3.1 *Rivers and Harbors Act*

The Rivers and Harbors Act is a much older law that predates the Clean Water Act. The USACE administers the Rivers and Harbors Act, like the Clean Water Act. The aim of the Rivers and Harbors Act is to protect navigation on US waters from obstruction (33 U.S.C. §§ 401–467n (2012)). To accomplish this goal, the act regulates where structures can be placed along navigable waters. The construction of piers, breakwaters, bulkheads, jetties, and any other structures that might impede navigation will require permitting approval by the USACE. These requirements are usually dealt with during the Clean Water Act permitting process, so a project manager may not notice an additional permitting layer.

There is one noticeable difference, however. As part of the Rivers and Harbors Act permitting process, consultation with the US Coast Guard may be required to ensure that the shoreline project does not impose any navigation dangers to public recreational users of the water. A project can be prohibited if it is located too close to a navigational channel. In practice, this generally affects shoreline stabilization hybrid structures, more than natural stabilization efforts (Boyd 2007). For instance, a project that includes the use of submerged concrete wave attenuation devices may require signage alerting boaters to the hidden navigational hazard.

3.3.3.2 *Endangered Species Act*

The Endangered Species Act provides protection for endangered and threatened species by prohibiting harm to those species as well as preventing any significant habitat modification or degradation (16 U.S.C. § 1538(a)(1)(B)). Under the act, no person can destroy or harass a protected species. In the context of shoreline projects, the biggest areas of concern are the prevention of harassment and harm. In the context of the Endangered Species Act, harassment is defined as actions that will likely injure "wildlife by annoying it to such an extent as to significantly disrupt normal behavioral patterns which include, but are not limited to, breeding, feeding, or sheltering" (50 C.F.R. § 17.3(c)). Harm is defined to include "significant habitat modification or degradation where it actually kills or injures wildlife by significantly impairing essential behavioral patterns, including breeding, feeding or sheltering" (50 C.F.R. § 17.3(c)). Engaging in any of these behaviors can result in criminal penalties.

The Endangered Species Act also designates critical habitat. When a species becomes protected under the act, the implementing agency will also designate critical habitat for that species' recovery. Critical habitat refers to specific geographically designed areas that the agency determines are essential for the conservation of the species. Critical habitat can include areas that are not currently being used by the species but are necessary for the species' recovery.

For this reason, the USACE will consult with the administering agencies during the wetland permitting process. Two agencies share enforcement of the Endangered Species Act: the US Fish and Wildlife Service and the National Marine Fisheries Service (NMFS). In limited circumstances, the agencies may choose to issue an Incidental Take Permit that allows a project to move forward, but only if the project cannot be redesigned to avoid harm to the species. In this instance, the permit applicant will also be required to develop a habitat conservation plan in order to receive a permit. A habitat conservation plan is a plan to minimize and mitigate harm to the affected species during the project and must be implemented by the project manager.

The practical takeaway for small-scale shoreline stabilization projects is that the presence of protected species or critical habitat has the potential to severely affect the project's feasibility and design. If you have reasons to believe that protected species may be affected, consult with the applicable agencies early on to determine if the project is feasible. Depending on the site, the impacts may be minimized by slight project redesigns. On the other hand, the site may be so critical to a protected species that the project is simply unfeasible and will not be approved.

3.3.3.3 Magnuson–Stevens Act

The Magnuson–Stevens Fishery Conservation and Management Act is a law aimed at fishery conservation and management (16 U.S.C. §§ 1801–1891d (2012)). The National Marine Fisheries Service manages fisheries through regional councils throughout the United States. While the law does not carry the same level of species protections found in the Endangered Species Act, the law does regulate harvests and protects essential fish habitat. Essential fish habitat refers to "those waters and substrate necessary to fish for spawning, breeding, feeding, or growth to maturity" as identified by the regional councils (16 U.S.C. § 1802(10) (2012)). Actions that may adversely affect essential fish habitat are broad and include physical, chemical, or biological changes to the water, injury to the species or habitat, and reduction of quality or quantity of essential fish habitat. Essential fish habitat can often include nearshore areas and grassy, marshy areas that may be affected by shoreline stabilization efforts. For instance, most of the nearshore waters in Washington State are considered essential fish habitat, thus affecting all shoreline projects in the region.

Another component of the Magnuson–Stevens Act that may affect shoreline restoration projects is whether or not the project is located in "habitat areas of particular concern." These special areas are designated by the regional councils and represent high-priority areas for conservation and protection. Depending on the nature of the project, these areas might be ideal locations for long-term shoreline restoration projects, so long as the project meets the objectives of the regional council by improving habitat while restoring the shoreline.

If the proposed project site involves either essential fish habitat or habitat areas of particular concern, the USACE will contact the National Marine Fisheries Service for consultation during the permitting project. The National Marine Fisheries Service will have the opportunity to comment on the proposed shoreline stabilization project and may make recommendations to minimize any adverse impacts to fisheries.

3.3.3.4 Coastal Barrier Resources Act

The Coastal Barrier Resources (CBRS) Act is a federal law that restricts the expenditure of federal money for development on coastal barrier islands (16 U.S.C. §§ 3501–3510 (2012)). The goal of the Coastal Barrier Resources Act is to minimize loss to human life from hazardous coastal development, preserve natural resources of barrier islands, and restrict federal assistance for coastal development on barrier islands. The US Fish and Wildlife Service manages the law and maintains periodically updated maps that display all areas within the Coastal Barrier Resources System. The Coastal Barrier Resources System encompasses undeveloped coastal barriers along the Gulf of Mexico, Atlantic, and Great Lakes coasts.

The act does not limit development of property within the System. However, it does restrict "future Federal expenditures and financial assistance which have the effect of encouraging development of coastal barriers" (16 U.S.C. § 3501). This includes federal flood insurance coverage, nondisaster emergency relief, government loans, new infrastructure such as roads and bridges, and other federal assistance.*

Although this law places limitations on using federal dollars for hardening a shoreline within the System, it actually encourages the use of living shoreline techniques in these areas. Specifically, the act exempts "nonstructural shoreline stabilization projects that mimic, enhance, or restore a natural shoreline system" so long as the project is consistent with the law. Exemptions also apply to projects that protect fish and wildlife resources and habitats, including acquisition of related lands.

---

* Financial assistance "includes any contract, loan, grant, cooperative agreement, or other form of assistance, including the insurance or guarantee of a loan, mortgage, or pool of mortgages" (16 U.S.C. § 3505).

For these reasons, the Coastal Barrier Resources Act is unlikely to impact construction of natural shoreline stabilization projects. As mentioned, the law only applies to projects funded with federal dollars. The act may be triggered if, for instance, your project is funded through a federal grant or assistance program. However, the project is still likely to qualify for one of the exemptions so long as the project is designed to mimic a natural shoreline. If a proposed project is located within the mapped area of the System, the project manager should contact the local US Fish and Wildlife Service office to discuss any potential concerns.

3.3.4 Hypothetical

Returning to the hypothetical, Anne and her project manager should contact the district USACE to set up a preconstruction meeting. During this time, the project design and location can be discussed. Anne can also find out information about the type of wetland permit that will likely be needed. In addition, this preapplication meeting gives the USACE the opportunity to advise Anne of any known or suspected challenges related to her project location, such as navigational issues, presence of protected species, or essential fish habitat. Depending on these variables, Anne should get a better idea of how long the permitting process may take and whether or not her location or design should be modified to ease permitting requirements. As previously mentioned, in many jurisdictions, Anne's living shoreline will require an individual permit from the USACE. Anne should be prepared for a more lengthy permitting process in this instance.

To recap, the following federal laws may affect your living shoreline project:

- Clean Water Act
- Coastal Zone Management Act
- Rivers and Harbors Act
- Endangered Species Act
- Magnuson–Stevens Act
- Coastal Barrier Resources Act

3.4 STATE REGULATION OF SHORELINE

State regulation of shorelines adds to the complexity of permitting living shorelines. Again, early consultation with permitting agencies is essential. There are 30 coastal states including the Great Lakes, and each state has its own way of doing things. Since it is beyond the scope of this chapter to explore 30 different coastal management regimes, this section will speak in generalities of state law. Most states regulate within common themes, even if the mechanics are different. Those themes include state coastal programs (including coastal wetlands), state water quality programs, and leasing of state-owned submerged waterbottoms. This section will include these three themes and then examine a few states that have adopted laws specific to living shorelines.

3.4.1 Coastal Programs and Wetland Permitting

Just as all living shoreline projects are likely to require a wetland permit from the USACE, these same projects will require state permitting as well. However, the project may not require two separate permitting applications. Some state agencies and the local USACE district work together through a joint permit application process. As this is not always the case, project managers should work closely with state permitting authorities and be aware of local policies.

States generally regulate shoreline management projects through state coastal programs, which include wetland permitting in coastal waters. As previously discussed, these programs may place state-specific conditions on the use of USACE nationwide permits or eliminate the use of the permit altogether. State policies can be more restrictive than federal policies, meaning that a project may qualify for a federal permit from the USACE but not satisfy the state permitting requirements. Many permittees find this scenario confusing, not realizing that permission from both levels of government is required (RAE 2015).

3.4.2 Water Quality Regulation

A living shoreline project may also require water quality permitting. Water quality permitting requirements are derived from the Clean Water Act, as discussed in Section 3.3.1. The EPA oversees the water quality regulations but has largely delegated that role to state agencies, such as a state Department of Environmental Quality or similarly named agency. Depending on design, a living shoreline project may cause the disturbance of sediments or runoff into the water during construction. These discharges could trigger state water quality permitting.

Water quality standards are set by the state for each body of water. The standard used will depend on the designated use of the water body. For instance, a drinking reservoir will have a much higher water quality standard than a river used primarily for shipping. States enforce these standards by regulating how much pollution (including sediment) can be released into a particular body of water through the issuance of water quality permits (sometimes referred to as stormwater permitting). Even small construction projects can trigger water quality oversight. When consulting with state agencies about wetland permitting, project managers should inquire as to the local water quality regulations and contact the permitting agency (if permitted separately). In some states, impacts to water quality will be considered during the coastal permitting process.

3.4.3 State-Owned Submerged Lands

Projects like living shorelines, which require placement of materials on the waterbottoms, may require additional state permissions for the use of those waterbottoms. Referring back to the public trust doctrine, states hold title to submerged waterbottoms and manage those lands for the benefit of the public. Therefore, many states require special licenses to place materials on state waterbottoms from the state agency managing these submerged lands. In some states, this will be the same agency that oversees coastal permitting. In other states, it will be an entirely different agency.

The actual permission, terminology and requirements vary by state. It may be called a lease, easement, or license (RAE 2015). Some states require a fee be paid, while others will waive the fee if the project is restoring or maintaining a natural shoreline. In rare instances, the state agency may require a deed restriction be placed on the waterfront property owner's land, acknowledging that any accreted land will still belong to the state.

In some states, this review may be incorporated into the state coastal permitting process, without requiring a separate permission. For instance, Connecticut incorporates this review into the state environmental permitting process rather than having a separate leasing program. For a survey of state waterbottom leasing programs in 21 coastal states, see Inventory of Shellfish Restoration Permitting and Programs in the Coastal States (Pace et al. 2014).

3.4.4 Examples of State Living Shoreline Policies

As the science supporting living shorelines has become increasingly robust, more states are considering incorporating living shorelines into their coastal management plans. Maryland has become the oft-cited poster child of this movement, after adopting a statewide policy in 2008. Since that time, other states have modified policies to promote the use of natural shoreline stabilization methods. This section will briefly look at some of these policies.

3.4.4.1 Maryland

Maryland's Living Shoreline Protection Act, passed in 2008, was the first statewide effort to promote living shorelines over hardened erosion control methods. In the Act's preamble, the legislature stated that living shorelines were the preferred shoreline protection method because they trap sediment, filter pollution, and provide important aquatic and terrestrial habitat (2008 Md. Laws Ch. 304 (H.B. 973)). The preambles goes on to establish that state public policy favors the preservation of natural habitat; shoreline protection practices should consist of nonstructural "living shoreline" erosion control measures wherever technologically and ecologically appropriate (Boyd and Pace 2013).

In addition, the law requires that erosion control projects consist of nonstructural shoreline stabilization measures that preserve the natural environment, such as marsh creation, except in areas designated by the Maryland Department of Natural Resources (MDNR) Mapping Department as appropriate for structural shoreline stabilization measures (Md. Code Ann., Envir. § 16-201). Property owners may seek a variance from this requirement if they can demonstrate that nonstructural stabilization is not feasible. A variance is a mechanism that allows the permittee to request a deviation from the rule. Reasons for the variance may include excessive erosion, areas subject to heavy tides, or areas too narrow for effective use of nonstructural shoreline stabilization measures (Boyd and Pace 2013).

3.4.4.2 Virginia

While Virginia has not gone as far as Maryland, the state is making strides to incorporate living shorelines into its coastal management plan. In 2011, the state passed a law requiring the Marine Resources Commission and the Department of Conservation and Recreation to work with the Virginia Institute of Marine Science to develop a state general permit for the use of living shorelines (Va. Code Ann. § 28.2–104.1 (West 2011)). That law was amended in 2014 to include the Department of Environmental Quality. The 2014 amendment also included a requirement for developing an expedited permit review process for qualifying living shoreline projects (Va. Code Ann. § 28.2–104.1 (West 2014)). In September 2015, Virginia adopted regulations authorizing its first statewide general permit for living shorelines (4 Va. Admin. Code § 20-1300-10 et seq. (2015)). The new regulations aim to encourage property owners to use living shorelines and to promote the restoration of natural tidal wetland vegetation.

3.4.4.3 Alabama

In Alabama, the state has begun to encourage the use of living shorelines, though the state policies are not as forceful as those of Maryland. In 2014, Alabama amended its regulations that govern shoreline erosion control structures to specifically include a living shoreline option. The regulation describes living shoreline techniques as including "the planting of native vegetation, the placement of wave attenuation structures, the placement of fill materials, and/or other techniques" (Ala. Admin. Code r. 220-4-.09(4)(b)(7) (2014)). With respect to property lines, the regulation clearly points out that any newly accreted land below the mean high tide line that results from the project will not change the original property lines. In other words, that newly accreted land will remain state-owned lands.

3.4.5 Hypothetical

Going back to the hypothetical, Anne and her project manager, again, should contact state regulators early in the process. If Anne happens to be in a state with a joint permitting process with the USACE, Anne may be able to proceed with either the USACE district or the state permitting agency, depending on which agency (state or USACE) is administering the permitting process. If Anne is in a state that does not have a joint permitting application process, she should proceed independently but simultaneously with the state agency. She should also inquire about any special requirements related to using the waterbottoms for her living shoreline project.

To recap, the following state law issues may affect your living shoreline project:

- State Coastal Programs and Wetland Permitting
- Water Quality Regulations
- State-Owned Submerged Lands

3.5 LOCAL GOVERNMENT REGULATION

In addition to state and federal permits, some cities or counties may have localized regulations that affect shoreline restoration and stabilization projects. In practical terms, this means that the project may require additional permitting from the local government, usually through the planning department. While local government regulation is still relatively rare, an increasing number of communities are using their planning and zoning authority to restrict the use of bulkheads and hardened shorelines in favor of softer, natural approaches. To that extent, these local governments are actually increasing and promoting the use of living shorelines in their communities. In this section, local government regulatory authority will be discussed as well as a few examples of local government regulations that promote soft and natural shorelines.

3.5.1 Local Government Regulatory Authority

Local governments derive their regulatory authority from the state. State regulatory authority originates with the Tenth Amendment of the US Constitution, which provides: "The powers not delegated to the United States by the Constitution, nor prohibited by it to the States, are reserved to the States respectively, or to the people" (U.S. Const. am. 10). This is commonly referred to as state police power, and allows for state laws designed to protect the public health, safety, morals, and general welfare.

States, in turn, have passed along much of this authority to local governments through enabling legislation that allows local governments to implement laws aimed at planning and zoning within their communities. This is the authority local governments use to regulate construction along shorelines, including placing restrictions on the type of shoreline erosion control structures that can be used within their community.

3.5.2 Local Examples

Using this planning authority, several local governments throughout the United States have implemented local laws aimed at promoting the use of alternative shoreline stabilization techniques within their jurisdictions. Some communities refer to these techniques as living shorelines, while others more commonly refer to practices that are more natural. Regardless of the terminology, the end goal is the same. These proactive local governments are advancing more environmentally sensitive shoreline erosion controls on the local level. The advantages of local approaches are quite simple. The local government can act now, rather than waiting on a statewide policy to be adopted. This also gives local governments the freedom to tailor the regulations to fit the needs of each individual community. The only major restrictions are that the local laws not conflict with state or federal law.

3.5.2.1 *Kent County, Maryland*

Kent County, Maryland, adopted a countywide policy that allows the county to be more proactive in requiring the use of living shorelines. As previously discussed, Maryland has a statewide policy favoring living shorelines, adopted in 2008. However, Kent County acted before the state, adopting a policy aimed at requiring living shorelines 6 years earlier, in 2002. Under this local law, Kent County requires property owners considering installation of hardened shoreline armoring to demonstrate that a living shoreline would be ineffective for that site. The ordinance clearly sets out the law's purpose as encouraging "the protection of rapidly eroding portions of the shoreline in the County by public and private landowners" through the use of "nonstructural shore protection measures" where effective and practical, in order to "conserve and protect plant, fish and wildlife habitat" (Kent County, Md., Code § 6-3.10).

The ordinance goes on to set out specific criteria that will be taken into account during the permitting review process:

- Nonstructural practices shall be used whenever possible.
- Structural measures shall be used only in areas where nonstructural practices are impractical or ineffective.
- Where structural measures are required, the measure that best provides for the conservation of fish and plant habitat and which is practical and effective shall be used.
- If significant alteration of the characteristics of a shoreline occurs, the measure that best fits the change may be used for sites in that area.

This clear definition of conditions makes implementation and enforcement of the ordinance easier to understand for both the permitting agency and the property owner.

In addition to adopting the new ordinance, the local planning department in Kent County undertook extensive education and outreach efforts to ensure that the community as a whole understood the importance of these policy changes (Thomas-Blate 2010). Another important component of this law's successful implementation is the availability of shoreline erosion data. The MDNR created the Maryland Coastal Atlas to make these data accessible by the public (Boyd and Pace 2013). In addition, Prince George's County (Prince George's County, Md., Code § 5B-124 (2011)) and Salisbury, Maryland (Salisbury, Md., Code § 12.20.120 (1996)) have also included living shorelines in their local land use laws.

3.5.2.2 *Brevard County, Florida*

In Brevard County, Florida, local laws prohibit the construction of new bulkheads in certain areas. Living shorelines, therefore, become the preferred shoreline stabilization method in areas where bulkheads and reinforced rock revetment habitats are prohibited (Brevard County, Fla., Code § 62-3666(9) (2011)). Under the ordinance, the definition of shoreline stabilization specifically includes living shorelines: "Shoreline stabilization means alteration of the shoreline or the surface water protection buffer from its natural state for the purpose of minimizing erosion utilizing riprap material, interlocking brick systems, rock revetments, vegetation, living shorelines, retaining structures located in uplands, or other allowable methods."

To reduce confusion, the ordinance also includes a definition of living shorelines. In this case, Brevard County adopted the following living shoreline definition: "erosion management techniques, such as the strategic placement of plants, stone, sand, and other structural and organic materials, that are used primarily in areas with low to moderate wave energy, and are designed to mimic natural coastal processes" (Brevard County, Fla., Code § 62-3661 (2011)). Clearly defining the term gives greater clarity to the local permitting agency, as well as property owners and contractors.

3.5.2.3 *Hawaii Counties*

Two Hawaiian counties have adopted shoreline management approaches that limit the use of hardened shorelines. In Honolulu County, Hawaii, permitting authorities can deny a property owner's request to build a shoreline hardening structure unless the structure is minor and does not significantly interfere with natural processes (Honolulu County, Haw., Code § 23-1.8 (2010)). Honolulu County's goal is to protect and preserve the natural shoreline, especially sandy beaches (Codiga and Wager 2011). In Kaua'i County, Hawaii, local regulations place restrictions on construction of new erosion control structures. Those structures cannot adversely affect beach processes, artificially fix the shoreline, interfere with public access or public views along the shoreline, impede natural processes and movement of the shoreline and sand dunes, or alter the grade of the shoreline setback area (Kaua'i County, Haw., Code § 8-27.7(b)(4) (2013)).

3.5.2.4 *Others*

In addition to the localities discussed above, several other communities have adopted local laws that include specific reference to living shorelines. For instance, Wilmington, North Carolina, restricts the placement of shoreline stabilization structures to above the upland buffer except for living shorelines, which can be placed within the buffer or conservation area (Wilmington, N.C., Code § 18-341 (2005)). In Virginia, several counties and cities allow for the use of living shorelines within wetland areas so long as they comply with state permitting requirements (Suffolk, Va., Code § 34-298 (1998)).

3.5.3 Hypothetical

At this stage in the process, Anne should have held preliminary meetings with state and federal agencies and be proceeding with the permitting application process. The only additional step that is added is to inquire whether the applicable city or county government has further regulations that apply to her living shoreline project. This can be accomplished by contacting her local planning department. As can be seen from the examples above, where local regulations apply to shoreline erosion control, they often favor natural and living shoreline approaches.

3.6 CONCLUSION

As can be seen from this chapter's discussion, permitting a living shoreline project can be complex and time-consuming, though strides are being made to improve the process. Yet, the general criticism remains. Bulkheads and hardened structures are far easier to permit than more environmentally sound natural approaches (RAE 2015). However, efforts are being made to shift mindsets away from bulkheads in favor of living shorelines on many different levels. For instance, outreach on the benefits of living shorelines to the public, property owners, contractors, and regulators raise awareness. More requests for living shoreline permits can lead to regulator willingness to develop general permits for this type of shoreline stabilization. As demonstrated by state and local efforts highlighted above, regulators are becoming increasingly aware of the environmental damage caused by hardened shorelines and are incrementally moving toward requiring more natural approaches.

REFERENCES

Boyd, C.A. 2007. Shoreline protection alternatives. Mississippi–Alabama Sea Grant Consortium MASGP-07-26.

Boyd, C.A. and N. Pace. 2013. Coastal Alabama Living Shorelines Policies, Rules, and Model Ordinance Manual. Mississippi–Alabama Sea Grant Consortium MASGP-13-023. http://masglp.olemiss.edu/Advisory/livingshorelines/Coastal-Alabama-Living-Shorelines-Policies-Manual.pdf.

Christie, D. 2010. Of beaches, boundaries and SOBs. *Journal of Land Use & Environmental Law* 25: 39.

Codiga, D. and K. Wager. 2011. Sea-Level Rise and Coastal Land Use in Hawai'i: A Policy Tool Kit for State and Local Governments. Center for Island Climate Adaptation and Policy.

Craig, R.K. 2007. A comparative guide to the eastern public trust doctrines: Classifications of states, property rights, and state summaries. *Penn State Environmental Law Review* 16: 1–114.

Douglass, S.L. and B.H. Pickel. 1999. The tide doesn't go out any more—The effect of bulkheads on urban bay shorelines. *Shore & Beach* 67: 19–25.

Kalo, J. 2005. North Carolina oceanfront property and public waters and beaches: The rights of littoral owners in the twenty-first century. *North Carolina Law Review* 83: 1434–1440.

Klass, A. 2006. Modern public trust principles: Recognizing rights and integrating standards. *Notre Dame Law Review* 82: 704.

Pace, N. 2011. Wetlands or seawall? Adapting shoreline regulation to address sea level rise and wetland preservation in the Gulf of Mexico. *Journal of Land Use & Environmental Law* 26: 327–364.

Pace, N., S. Otts, T. Bowling, C. Janasie, and C.A. Boyd. 2014. Inventory of shellfish restoration permitting and programs in the coastal states. Mississippi–Alabama Sea Grant Consortium. http://masglp.olemiss.edu/projects/files/tnc-report.pdf.

Restore America's Estuaries. 2015. Living shorelines: From barriers to opportunities. https://www.estuaries.org/images/stories/RAEReports/RAE_LS_Barriers_report_final.pdf.

Thomas-Blate, J.C. 2010. Living shorelines: Impact of erosion control strategies on coastal habitats. Atlantic State Marine Fisheries Commission, Habitat Management Series #10. http://www.fws.gov/northeast/marylandfisheries/reports/hms10livingshorelines.pdf.

Wilkins, J. and M. Wascom. 1992. The public trust doctrine in Louisiana. *Louisiana Law Review* 52: 861–906.

CHAPTER 4

Socioeconomic and Policy Considerations of Living Shorelines—US Context

Kateryna M. Wowk* and David Yoskowitz

CONTENTS

4.1 INTRODUCTION

Around the world, interest in using living shorelines (or natural infrastructure) to strengthen coastal resilience is rising, and evidence that such approaches can protect communities from coastal hazards, while providing a multitude of additional services, is becoming clearer. In the United States, a spotlight was shone on living shorelines after Hurricane Sandy, which wreaked havoc along a vast stretch of densely populated coastline,[†] exposing the vulnerabilities of an area with enormous investments in critical infrastructure and lifelines for the region and nation—an area that is also home to key global institutions. As the winds died, the water subsided, and the Northeast and Mid-Atlantic regions began to heal, a question about the future of our coasts emerged. How can we best reduce the impacts of weather and climate-related events, enhance our resilience, and secure the safety and well-being of the people and places we cherish? Living shoreline approaches, which had been previously investigated but never gained significant traction across broader geographic scales, reemerged in the national spotlight as an increasingly desirable option.

---

* The chapter was drafted when Dr. Wowk was consultant and senior social scientist to the US National Oceanic and Atmospheric Administration (NOAA), though she wrote in her personal capacity. Any views expressed do not necessarily reflect those of NOAA.

[†] Hurricane Sandy cumulatively caused 377 deaths and more than $110 billion in damages (NOAA National Climatic Data Center [NCDC] 2013).

However, implicit in that question—though perhaps somewhat hidden—are a number of value judgments. Once strategies and methodologies for strengthening coastal resilience are identified, who says which are best? By which criteria shall they be judged? If resilience entails an ability to "bounce back" from a shock or disturbance,* should we be aiming to return to that same exact state, or something better? What does "better" mean? Further, by that token, whose concept of "well-being" are we striving to achieve? Ask 10 people these questions and one may very well get that exact number of distinct, potentially conflicting, responses.

In an era of increasingly intense and frequent extreme weather events, with sea level rise encroaching on our coasts bit by constant bit, we must act to reduce the rising risk of damaging impacts. But to find the best way forward, a way that is both effective and will endure, we must recognize that we are operating in a social system where there are multiple values at play, and that the costs and benefits of any approach are in large part a function of such values. Envisioning the future of our coasts must recognize this by pursuing a systems approach to management—a system that includes many different objectives, infrastructure types, economic components, ecosystems, and, ultimately, humans.

4.2 SETTING THE STAGE: INCREASING SUPPORT FOR LIVING SHORELINE APPROACHES

US coastal counties are home to nearly 4 in 10 Americans, are centers of commerce, and are vital to the national economy (National Oceanic and Atmospheric Administration [NOAA] 2014). In 2011, coastal shoreline counties contributed $6.6 trillion to the US Gross Domestic Product (GDP)—just under half of GDP that year (NOAA 2012). Particularly in densely populated coastal areas, resilience strategies to protect vital assets against weather and climate hazards have typically relied on hardened, built infrastructure including seawalls, levees, groins, and bulkheads. Yet these approaches may limit commercial and recreational opportunities, have limited purpose and life-span, and can be expensive through planning and engineering, implementation, and maintenance phases (Jonkman et al. 2013). Such approaches can also exacerbate flooding conditions if breached, and they are inherently static—without intervention, they cannot adapt to shifting environmental parameters imposed by climate change or by dynamic coastal processes.

In recent years, a substantial and growing body of work has been established that looks to living shorelines to strengthen coastal resilience to weather and climate impacts, address some of the aforementioned challenges, and meet additional societal objectives. The preservation and restoration of natural coastal features—alone or in conjunction with hardened structures—can be attractive alternatives to strategies based solely on "gray" infrastructure. A core group of natural coastal features, including wetlands, mangroves, coral and oyster reefs, beaches, and dunes, has been shown to buffer the impacts of coastal storms and sea level rise while also delivering a host of desirable co-benefits to surrounding communities (Axley 2013; Barbier et al. 2011; Koch et al. 2009; Taylor et al. 2015).

In the United States, the ability of natural features to provide services that enhance our daily lives has received high levels of support and promotion. At the national level, this topic gained significant attention with the release of a 2011 report from the President's Council of Advisors on Science and Technology (PCAST) on "Sustaining Environmental Capital: Protecting Society and the Economy." The report found that "[h]uman well-being depends on capital of three kinds and the goods and services that flow from these: [economic capital, sociopolitical capital, and environmental capital]," with environmental capital being ecosystems and supporting geophysical conditions

---

* A capability to anticipate, prepare for, respond to, and recover from significant multihazard threats with minimum damage to social well-being, the economy, and the environment (Melillo et al. 2014).

that provide a range of processes, resulting in ecosystem services that confer benefits on human society (PCAST 2011).

High-level and coordinated attention also has specifically increased within the government in using natural or restored features along the coasts to provide protective and other services. Both Hurricanes Katrina in 2005 and Sandy in 2012 strengthened federal interest in the use of living shorelines for coastal protection. These approaches became further elevated in 2013 when President Obama created the Hurricane Sandy Rebuilding Task Force, which focused significant attention on building resilient infrastructure as a key component of the $50 billion disaster relief package to communities affected by Sandy (113th Congress of the United States of America 2013). Though the funding was primarily focused on restoring damaged built infrastructure, to help meet this goal, the strategy and its guidelines also emphasized the need for environmentally sustainable and innovative solutions that consider natural infrastructure options in all federal Sandy infrastructure investments. In addition, the President's Climate Action Plan, released in June 2013, highlighted the need for resilient infrastructure approaches and called for using natural ecosystems, including forests and wetlands, to help sequester carbon and mitigate the effects of climate change (Executive Office of the President 2013). The recognition of the climate mitigation benefits of natural ecosystems is timely, given that one of the top vulnerabilities and risks to society as a result of climate change in the latest Intergovernmental Panel on Climate Change (IPCC) Report is injury or death from sea level rise, storm surge, and coastal flooding (IPCC Working Group II 2014).

Support from the administration as well as state, local, nongovernmental organization (NGO), and private sector partners also led to significant efforts to both promote the state of the science regarding the efficacy of living shoreline approaches and begin to fill some substantial knowledge gaps that currently impede their more widespread adoption. Although such approaches have been demonstrably effective at enhancing climate resilience under certain circumstances, it is clear that their potential varies among regions, scenarios, and decision contexts. Optimal use of living shorelines requires a thorough understanding of the benefits and trade-offs associated with different infrastructure approaches under different conditions, as well as greater consistency of methods used to identify, quantify, and value these outcomes.

To target those knowledge gaps in most need of attention, the Obama Administration released an *Ecosystem Service Assessment: Research Needs for Coastal Green Infrastructure*. The report identifies key information needed by federal planners and decision-makers to advance the broad integration of natural infrastructure along the coast and prioritizes research areas related to the use of natural infrastructure to reduce vulnerability and enhance resilience to climate-related threats in coastal areas (some of the key information gaps are discussed further in Section 4.6). The report further builds on, and is aligned with, recent and ongoing efforts to advance the integration of ecosystem services into federal decision-making (OSTP 2015). It is hoped that federal agencies will collaborate in focused efforts to fill these research gaps, in concert with state, local, nongovernmental, academic, and private sector partners.

However, while it is critical to continue to fill research gaps to facilitate broader implementation of living shoreline approaches, it is also important to recognize that, in select cases, we have enough information to act now. For some approaches and under certain conditions, we already have the key data and information needed to guide management efforts on adopting living shorelines for coastal protection and other benefits (see, e.g., OSTP 2015; Sutton-Grier et al. 2015).

4.3 FROM SUPPORT TO VIABILITY: ECOSYSTEMS INCLUDE PEOPLE (*AND PEOPLE HAVE VALUES*)

Living shoreline approaches can provide significant benefits to people. This includes not only important protection and risk reduction benefits but also the direct harvesting of natural goods (e.g., fish, timber), improvements to water quality, carbon sequestration, and cultural, recreational, and aesthetic opportunities (Arkema et al. 2013; Barbier et al. 2011; Beseres-Pollack et al. 2013; Ferrario et al. 2014; Gedan et al. 2011; Rodriguez et al. 2014; Shepard et al. 2011; Zhang et al. 2012). In mitigating inundation, for example, a recent study examined wave reduction benefits and erosion protection benefits of salt marshes under storm surge conditions. The authors found that vegetation was responsible for 60% of wave attenuation during storm events and that even when waves were large enough to break salt marsh vegetation stems, the plants protected the soil from eroding during major storm events (Möller et al. 2014). In looking at trends across the literature, Sutton-Grier et al. (2015) also found that for smaller hurricanes and larger storm events, living shorelines have effectively protected some coastal areas from erosion at a level that surpasses many built approaches. There are further opportunities for using living shorelines to protect areas that have limited defenses, as well as in designing shorelines that include a combination of natural and built infrastructure (known as "hybrid" infrastructure). These approaches may be more cost-effective in the long run in comparison to built infrastructure on its own, can maintain the provisioning of coastal ecosystem services, and can be used to effectively prevent loss of life and property.

There is also significant potential for living shorelines to play an important role in providing added resilience for communities in the face of the slower onset impacts of climate change (Barbier 2014), such as an increase in chronic flooding. This is key, as most costs to society from natural hazards are actually incurred from localized, smaller events (Axley 2013). The ability of natural infrastructure to absorb chronic impacts may become even more important in the face of rising sea level, which is already increasing flooding in some areas. In Boston Harbor, for example, incidents with chronic flooding (high tide plus 2.5 ft) jumped from just 21 occurrences between 1920 and 1990 to 15 occurrences between January 2012 to April 2014 (Aiken et al. 2014; NOAA et al. 1920–1990, 2012–2014).

Yet, though evidence is mounting on the effectiveness of these approaches, when decision-makers are considering alternatives, the efficacy of living shorelines in providing resilience and other benefits is only part of the equation. Natural, built, and hybrid solutions all have one exact thing in common—they are all place based. They all, eventually, must be implemented in a place, within a local community that is linked to that place, and within the context of a value system that community has across the capital they held, hold, or desire. Understanding the values of a community—and priorities across those values—is no easy task. As the PCAST report elaborated, the "three kinds of capital—economic, sociopolitical, and environmental—cannot sensibly be ranked in importance because all three are absolutely indispensable, and they are completely interdependent" (PCAST 2011). However, at a scaled-down level, local circumstances can highlight dominant values, and stakeholder engagement processes can be used to determine whether certain approaches are more likely to succeed in a local context than others. For example, in Manasquan, New Jersey, evidence exists that a series of dunes provided protection and lessened the damaging impacts from Hurricane Sandy (Barone et al. 2014). But at a 2015 town hall meeting, the majority of residents came out strongly against rebuilding the dunes, citing that the dunes do not provide much additional protection and believing that in some cases they can make the situation worse. Many comments also echoed in the room that dunes blocked the ocean view (FoxNews 2015). This raises a question, outside of considering community members located further inland that may benefit from additional protection at the shore—if the view is highly valued by some key community members, might other natural infrastructure or hybrid approaches, those that do not obstruct the ocean view but afford similar levels of protection, be accepted?

The success of living shoreline approaches and overall enduring resilience of our coasts necessarily depends on understanding *who* (and what) is behind the shorelines we seek to protect, including a local community's level of understanding of their vulnerabilities and options to reduce risk, the short- and long-term trade-offs across those options (including individual and stakeholder incentives), and a community's capacity and interest to engage in a process to map the way forward. Further, land-use planning occurs at local levels; thus, targeting smaller communities in advancing resilience approaches is a sensible strategy. It is only with the interest and support of local decision-makers and stakeholders that we will be able to break away from business as usual and secure a more resilient future for our coasts and societies that abut their shores. We need to apply frameworks that can help us engage communities in understanding threats, vulnerabilities, potential solutions, and the trade-offs across those solutions that affect all types of capital, involving the entire system.

4.4 A SYSTEMS APPROACH FOR EFFECTIVE MANAGEMENT

Critical to the effective uptake of living shorelines is the recognition that a particular habitat or biophysical feature (e.g., oyster reef, marsh) does not exist in isolation. It is part of a system, and its connection to the rest of what makes up the system is key to the success of any one effort to enhance ecological, economic, and social resilience through the use of living shorelines. This "systems" approach is also captured in the ideal of ecosystem-based management (EBM), a broad holistic approach for the management of multiple species, natural resources, and humans, which are all components of a larger ecosystem. There are typically three general criteria that characterize EBM: sustainability, ecological health, and inclusion of humans (Arkema et al. 2006). Yet the application of EBM principles still seems to be elusive, especially as it relates to the inclusion of humans. A recent study on the use of EBM in US federal agency programs overseeing coastal and marine management found that consideration of the "human dimension" is lacking (Dell'Apa et al. 2015). While the vision and intention of EBM is in the right place, the uptake and impact of the approach still struggles.

The concept of EBM, therefore, has evolved into the realm of *system resilience*. Whereas EBM was disconnected from social and economic systems, a holistic resilience approach includes these components along with ecological systems. The final report from the National Research Council Committee on the effects of the Deepwater Horizon oil spill on ecosystem services (NRC 2013) found that "Communities with highly diversified economies and social structures are better placed to withstand disturbances to ecosystems that affect the provision of ecosystem services." At the same time, the enhancement of natural systems and the ecosystem services they support can provide significant benefits to communities and economies.

For living shorelines, it is critical that the particular biophysical feature (marsh, oyster or coral reef, seagrass, mangrove) be considered as part of a "system" and that it and the system are explicitly connected to human well-being. The system must include the human dimension. A *social–ecological systems* (SES) approach, especially around issues of resilience, recognizes this connection and the positive feedback loop effect that human action can have on natural environments and on lessening the likelihood of natural hazards turning into disasters (Adger 2000; Folke et al. 2002). Yoskowitz and Russell (2015) expand on that concept through the development of the *social–ecological cycle* for coastal systems, where components (management and policies, drivers and stressors, biophysical features, and societal needs) are linked together through defined relationships (see Figure 4.1).

Living shorelines play an important role in acting as a management tool that meets certain societal needs such as shoreline erosion reduction, storm surge buffering, and nutrient removal. Being responsive to these needs and diminishing the impact of stressors can lead to increased ecological and *human* community resilience. Yet, there is still a need to explicitly account for this connection. This can be achieved through the lens of *ecosystem services*.

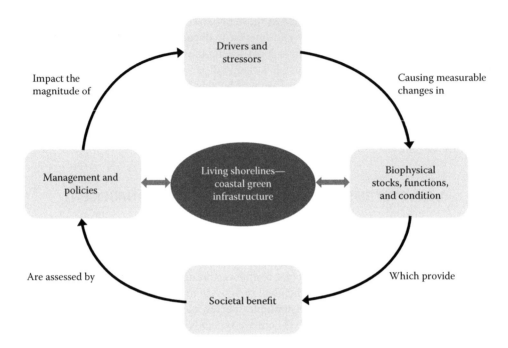

Figure 4.1   Social–ecological cycle with living shorelines.

4.5 LINKING COMMUNITY, ECONOMIC, AND ECOLOGICAL RESILIENCE

A significant corpus of knowledge has been established over the years, which deepens our understanding of community and economic resilience with respect to nature. For example, Berkes (2007) points to four "clusters of factors" that are relevant to building community resilience: living with change and uncertainty; nurturing diversity across ecological, social, and political spectra; enhancing learning and problem-solving knowledge; and creating opportunities for self-organization. These factors, the author offers, are even more critical considering the irreducibility of uncertainty with respect to natural hazards and that ecosystems are constantly changing. In interactions between humans and natural systems, the ability to learn and adapt is a significant factor. Cultivating shared knowledge and enhancing understanding of the values of different stakeholders—through the establishment of collaborative platforms and partnerships for sharing, dialogue, innovation, and learning—is highlighted throughout the literature on social and economic resilience (Cumming et al. 2005; Rose 2004, 2007; Simmie and Martin 2010).

Linking well-developed concepts of social and economic resilience specifically to ecological resilience is not a stretch, and indeed these and many other authors establish or attempt to elaborate such interconnections throughout the SES spectrum. Gunderson (2010) pointedly conducts a comparison across ecological and human community systems, finding important management implications including that both systems have multiple meanings of resilience and highlight the need for diversity across resilience concepts; that it is important to consider the different types of capital across each as well as cross-scale interactions; and that, as others have pointed out, there is a need to foster experimentation and learning to build adaptive capacities. Adger et al. (2005) emphasize that we need a better understanding of the linkages between ecosystems and human societies to help reduce vulnerabilities, and that socioecological resilience must be considered at broader scales, including by generating multilevel social networks that can implement and support governance frameworks, that is, legal, political, and financial aspects. A deeper level of cooperation is needed across management authorities that span society and the environment, perhaps supported by a strong leader or by changing social norms within a management agency, again with established mechanisms for learning, building shared narratives, and enabling growth of adaptive capacity (Carpenter et al. 2001; Goldstein et al. 2012).

In so much of the literature on SES resilience, it is clear that pathways are needed to deepen collaboration and stakeholder involvement. Walker et al. (2002) provide a useful approach for thinking about how such connections might be established, detailing a framework of four steps, with careful initial consideration of necessary stakeholders to be involved as well as definition of the problem. This process-oriented guidance focuses on the following: conceptual model development of the system, led by stakeholders and including identification of priority ecosystem goods and services; identification of factors constituting uncertainty in the system as well as visions for future scenarios; resulting from these initial steps, consideration of attributes that would affect resilience through the development of simple models; and, ultimately, evaluation of the process and outcomes in light of implications for management and policy. The framework is predicated on the fact that human actions in response to predictions are reflexive, and that, given new information that is seriously considered and potentially dangerous, people will react to chart an alternative course for their future. We inhabit a complex and dynamic system wherein sustainability is "fundamentally dependent on the active, positive involvement of all stakeholders." Interestingly, the authors also note that options from such optimization processes are rarely implemented, given the long list of "other" factors that go into a decision-making process, for example, interactions among voters, industry, monetary policies, election cycles, and so on. They suggest that one way to gain necessary buy-in is by bringing enhanced understanding of resilience across an SES in front of voters, interest groups, and politicians. In addition to information on exposure to vulnerabilities and how resilient attributes of a system can be enhanced, further detail and focus on key ecosystem services are of increasing interest to many stakeholders and decision-makers.

4.6 CHALLENGES IN BROADER APPLICATIONS OF ECOSYSTEM SERVICES ASSESSMENT FOR LIVING SHORELINES

The concept of humans receiving benefits from nature is inherently anthropocentric by definition. However, making conservation and restoration decisions based solely on human needs would not be appropriate (Yoskowitz and Russell 2015). An ecosystem services approach helps transcend the natural and social science divide. It provides the mechanism and language to merge those disciplines into a cohesive field of research. Specific to living shorelines, it can help elucidate the impact of investment for these projects, which is necessary in order to garner short- and longer-term support. However, the approach is not without its limitations, the foremost challenge being that there still are significant gaps in our knowledge regarding specific benefits of living shoreline approaches. Additional information, including on performance and market/nonmarket values, is needed in order to incorporate natural and hybrid infrastructure solutions into policy, investment trade-offs, and decision-making. In a helpful analysis, the White House Ecosystem Services Assessment details the most pressing remaining challenges to economic assessment of ecosystem services provided by natural infrastructure (OSTP 2015):

- Limited data coverage and quality—high-quality scientific data are only available for a limited subset of ecosystem services (including co-benefits).
- Incomplete knowledge and responses—integrating services into planning and decision-making often includes stated-preference studies, the results of which can be highly dependent on study design (e.g., population surveyed, question format, characterization of resource under consideration, etc.).
- Changes in supply and demand—ecosystem-service valuations generate results for a set point in time—they can quickly become obsolete as conditions evolve.

In addition, to avoid the need to conduct costly, time-consuming, and expert primary studies on a local ecosystem service in question, benefit transfer studies are offered as acceptable alternatives, in that they apply monetary values estimated in a set of previously conducted empirical studies to assess the value of a service under question (EPA 1999). However, such techniques rely on accessibility of multiple robust primary studies, including considering ecosystems under similar conditions. To date, such a foundation of knowledge is typically elusive. Add to all of this that, though we know how living shorelines systems perform under some conditions, as alluded to above, there still remains uncertainty in the storm protection services provided under some scenarios, for example, in response to large, slow-moving storms or with storm surge greater than 1 m (Sutton-Grier et al. 2015). We need better data and information on how the structure and dynamics of different coastal ecosystems connected to living shoreline approaches result in outputs that are directly relevant and useful to decision-makers (OSTP 2015). Particularly with hybrid structures, the performance of these approaches under varying conditions is not well studied (Sutton-Grier et al. 2015). Equally as important, we also are missing information on the specific unit costs of separate approaches. Analysis on an empirical basis across different types of approaches or different mixes of approaches thus remains limited, resulting in an incomplete understanding of trade-offs across various potential solutions. This is even true for built infrastructure (Jonkman et al. 2013), which, as previously discussed, has been under analysis for much longer than living shoreline approaches. We need actual project data on unit costs of various approaches if we are truly to understand the costs and benefits born by one method over another. Considering these factors, it is not difficult to see why, on broader levels, living shorelines are not as trusted as quantifiable measures for risk reduction.

Yet, we also know many circumstances under which living shorelines do provide protection, and that they also offer multiple co-benefits that are of value. We may not have multiple similar studies from which we can extrapolate results to new project areas, but we do have a corpus of singular studies proving their effectiveness and value. In new project areas, there are methods to help decision-makers and stakeholders consider alternative, potentially more resilient strategies as they envision their future coasts. The Nature Conservancy (TNC) recently released a highly useful framework to this end. This "Guide for Incorporating Ecosystem Service Valuation into Coastal Restoration Projects" provides yet another detailed process to engage stakeholders in mapping a way forward, with the useful distinction that it is specifically focused on the coast and living shoreline alternatives. A second significant utility is that, while the authors emphasize the desire to detail and consider information on ecosystem services, the guidebook can be applied even without taking the final step of fully valuing benefits, which again can be a significant roadblock. The authors note that "applying the process still can lead to greater success for restoration projects and greater likelihood of increasing ecosystem service benefits to stakeholders, which can lead to increased stakeholder support for restoration projects" (Schuster and Doerr 2015). These are incredibly useful considerations as we consider operationalizing frameworks in decision-making processes toward resilient community action.

4.7 THE DECISION-MAKING CONTEXT

Ultimately, using a systems approach for the resilience of our coastlines must be done in a specific time, place, and decision-making context. Again, the White House Ecosystem Services Assessment (OSTP 2015) provides a useful framing of key factors any decision-making process should include when considering one or multiple approaches (Box 4.1).

As demonstrated in the literature, whether for short-term preparedness or longer-term resilience, the involvement of key stakeholders that resilient strategies will protect and affect is critical

BOX 4.1 ADAPTED FROM THE ECOSYSTEM SERVICE ASSESSMENT: RESEARCH NEEDS FOR COASTAL GREEN INFRASTRUCTURE

✓ Geographic and spatial context—*is the living shoreline/hybrid approach feasible?*

✓ Management and societal objectives—*is the approach desirable and why? Would key stakeholders accept the approach?*

✓ Social justice—*which communities might the approach help or harm?*

✓ Policy directives—*are there regulatory, political, or other barriers that would bar use of the approach?*

✓ Potential uncertainty—*how will the benefits received from the approach change through time/space?*

✓ Cost-effectiveness—*is the approach the most cost effective? Does cost-effectiveness differ over the short and longer term? What are the trade-offs?*

✓ Financing—*could financing for the approach be obtained?*

to any decision-making process. Especially with living shorelines, the intent is generally to implement and maintain the approaches for a very long time—multiple decades, if not longer. If the local community has not bought into the approach, the required resources for the approach to remain effective over the longer term are not likely to flow. It is further worth noting that this "buy-in" is not static. As conditions change and different priorities emerge, community memory can be short, and thus, especially for new approaches, iterative prioritization is required. Without longitudinal stakeholder buy-in, we are undercutting the ability of living shorelines to showcase their abilities, as over the years they grow stronger, their protection becomes more robust, and the services they provide become more and more valuable.

Fortunately, over the last few years, a wealth of useful tools, information, case studies, and lessons learned on community and stakeholder engagement, in specifically working on coastal management issues, has become fuller and fuller. Some of these are summarized in Table 4.1.

Table 4.1 Local Engagement in Coastal Management: A Selection of Case Studies, Tools, and Information

Resource	Summary
NOAA's Digital Coast online platform	Includes a host of resources for state and local planners, including a geospatial data needs matrix, mapping tools, planning guides, funding tips and training opportunities. NOAA also recently funded a policy analysis that details State and Local Decision-Making: potential questions and gaps state and local decision-makers may face in considering living shoreline approaches (ERG 2014).
Systems Approach to Geomorphic Engineering (SAGE)	A new governance approach modeled as a community of practice that engages federal agencies, state and local agencies, NGOs, academia and the financial sector. Operational efforts focus on regional demonstration pilots that engage community partners in producing tailored projects and usable products (SAGE 2015).
US Climate Resilience Toolkit	Online Federal resource launched in November 2014, provides scientific tools, information, and expertise to help people manage their climate-related risks and opportunities and improve their resilience to extreme events.

(Continued)

Table 4.1 (Continued)   Local Engagement in Coastal Management: A Selection of Case Studies, Tools, and Information

Resource	Summary
US Army Corps of Engineers North Atlantic Coast Comprehensive Study (NCCAS)	Provides a step-by-step approach, with advancements in the state of the science and tools to conduct three levels of analysis at the regional, state and local levels, and includes information on natural and built approaches (USACE 2015).
How Resilient Is Your Coastal Community? *A guide for evaluating coastal community resilience*	Released by USAID (U.S. Indian Ocean Tsunami Warning System Program 2007), this guide to evaluating coastal community resilience to tsunamis and other hazards provides a framework for collaboration in building community resilience across sectors involved in planning and implementing community development, coastal management and disaster management programs. It assists communities, step by step, in identifying weaknesses and gaps in resilience, establishing mechanisms for community participation and input, and developing and implementing plans to enhance resilience.
United Nations Environment Programme (UNEP) et al. Green Infrastructure Guide for Water Management	UNEP–DHI Partnership—Centre on Water and Environment, International Union for Conservation of Nature (IUCN), TNC, and the World Resources Institute (WRI) provide information on living shoreline solutions and associated cost benefits, including illustrative case studies that provide examples of these options in addressing water management challenges (Bertule et al. 2014).
The Nature Conservancy's (TNC) Coastal Resilience Network and Guide for Ecosystem Service Valuation in Coastal Restoration	Coastal Resilience is a global network of practitioners who are applying an approach and web-based mapping tool designed to help communities understand their vulnerability from coastal hazards, reduce their risk, and determine the value of nature-based solutions. Recently, a guide was also released by TNC that can help coastal communities understand how they can measure socioeconomic benefits of living shorelines, even in the absence of actually conducing valuation studies (Schuster and Doerr 2015).
Designing with Water: Creative Solutions from around the Globe	Predicated on Boston's "Designing with Water" efforts, this report provides 12 case studies describing how cities around the world are using Designing with Water strategies to decrease their potential flood damage without losing the vibrancy and livability of their communities (Aiken et al. 2014).
Rebuild by Design	Funded by Hurricane Sandy relief and led by the Department of Housing and Urban Development (HUD), this competition relied on an intensely collaborative process to create proposals to build resilient infrastructure projects in the Sandy-affected area by using both public and private resources. Each of the winning six proposals is featured online and details how experts engaged local governments and communities to ensure integrity and innovation in resilient designs (Rebuild by Design 2014).
Dutch Dialogues	These workshops resulted from extended interactions between Dutch engineers, urban designers, landscape architects, city planners and soil/hydrology experts, and their Louisiana counterparts. The process was "unique in that it brought together community members, international, national, and local academics and professionals from a variety of disciplines, and high-level government officials to discuss best practices, tour sites, and work across disciplines and scales to develop concepts" (Meyer et al. 2009; Reinhardt 2009).
Examining Adaptation Action in Florida (forthcoming)	William Butler, assistant professor in the Department of Urban and Regional Planning at Florida State University (FSU), and colleagues looked at 42 counties and municipalities in Florida pursuing adaptation strategies. The study resulted in a useful framework for considering how communities move along the continuum from problem understanding to taking bold action, and the factors behind those shifts (Butler et al., submitted for publication).

Clearly, much progress is being made in involving stakeholders to map toward a future they want by using informative and collaborative community engagement and participatory processes.

4.8 THE WAY FORWARD

Stakeholder engagement is the key to resilient coasts and communities that are able to sustain and grow critical sources of capital. Broader application of multi-stakeholder frameworks that provide guidance on meaningfully engaging communities at regional, state, and local scales, while also advancing progress on known challenges to the implementation of living shorelines, is needed. This chapter points to a selection of especially useful frameworks that are enjoying success at regional and local levels, and are fostering a sense of ownership toward collective, resilient outcomes. Not every framework will work well in each circumstance, but whichever is applied, frameworks should ultimately help decision-makers to:

- *Identify key stakeholders* that need to be at the table (including by thinking outside of the box on key partners that have interest in and could help sustain the longer-term success of living shorelines, e.g., private sector partners).
- *Iteratively engage key stakeholders in mapping their future through proven methods* and social processes that enable participants to devise shared visions, consider multiple objectives and options (including the trade-offs therein), resolve conflicts, and achieve multiple goals across all three types of capital.
- *Consider known challenges early on*, such as permitting issues,* justification requirements for new projects (as in cost–benefit analysis or requirements to apply certain discount rates to new projects), or incomplete cost information at the project level. By considering these challenges early on in the decision-making process, it is more likely that solutions can be derived through consultation with local or external experts.

The resources included in this chapter and book can be used by communities in determining which resilient coastal strategies best meet their needs. Ultimately, we hope to see uptake and application of the targeted, useful guidance that exists. Place-based decision-makers can rely on these proven sources when considering the short- and long-term trade-offs they face and the outcomes they hope to secure. Further, as on-the-ground actions advance, the broader community should also focus efforts on filling key knowledge gaps by collating information through one centralized access point. Today, much of the useful data and information that could meet decision-makers' needs are scattered and are not interoperable, and thus the knowledge base that is needed to extrapolate results in considering possible futures remains elusive.

As we work to gather more evidence and empirical research, we also must remain cognizant that we will never have perfect information. We will always operate under some amount of uncertainty, especially when implementing new approaches and technologies. This is where a systems approach becomes even more critical. A systems approach allows us to take a holistic view—to look at the known and unknown, the vulnerabilities, risks, strengths, and the capabilities of any given community and area, at any scale. At smaller scales, at local and county levels, it becomes possible not only to examine the various types of natural capital that exist and are at risk but also to use existing information to derive common understanding and goals through social engagement. It is a concept that relies on actual processes to build trust and shared meaning

---

* As explained in Sutton-Grier et al. (2015), "permitting for [living shoreline] projects can take much longer than a permit for built infrastructure … because living shorelines projects often have to apply for an individual Clean Water Act 404 permit, while bulkheads can often be covered under an Army Corps Nation Wide Permit (which are generally granted more quickly)."

through collective thinking, reminiscent of David Bohm's work *On Dialogue*. As the theoretical physicist stated, "[T]hought is a system. That system not only includes thought and feelings, but it includes the state of the body; it includes the whole of society—as thought is passing back and forth between people in a process by which thought evolved from ancient times." We must move away from fragmented management of thoughts and of capital types that we know are intricately linked. Issue-by-issue management, sector-by-sector engagement, and even individualism will not stand up to the challenges we face along our coasts. We must consider and engage the collective whole through a systems approach—a system that includes humans. As imperfect as our information is, as imperfect as we may be, we can collectively define, design, implement, and sustain strategies to secure our resilient future.

REFERENCES

113th Congress of the United States of America. 2013. Disaster Relief Appropriations Act. 113th Congress of the United States of America, p. 47.

Adger, W.N., T.P. Hughes, C. Folke, S.R. Carpenter, and J. Rockstrom. 2005. Social–ecological resilience to coastal disasters. *Science* 309.

Adger, W.N. 2000. Social and ecological resilience: Are they related? *Progress in Human Geography* 24: 347–364.

Aiken, C., N. Chase, J. Hellendrung, and J. Wormser. 2014. *Designing with Water: Creative Solutions from Around the Globe*. Preparing for the Rising Tide Series. Volume 2.

Arkema, K.K., S.C. Abramson, and B.M. Dewsbury. 2006. Marine ecosystem-based management: From characterization to implementation. *Frontiers in Ecology and the Environment* 4: 525–532. http://dx.doi.org/10.1890/1540-9295(2006)4[525:MEMFCT]2.0.CO;2.

Arkema, K.K., G. Guannel, G. Verutes, S.A. Wood, A. Guerry, M. Ruckelshaus, P. Kareiva, M. Lacayo, and J.M. Silver. 2013. Coastal habitats shield people and property from sea-level rise and storms. *Nature Climate Change* 3: 913–918.

Axley, J. 2013. Green Infrastructure to Enhance Resistance to Natural Hazards. URS FEM TARC 12-J-0009:White Paper. Yale University.

Barbier, E.B. 2014. A global strategy for protecting vulnerable coastal populations. *Science* 345: 1250–1251.

Barbier, E.B., S.D. Hacker, C. Kennedy, E. Koch, A.C. Stier, and B.R. Silliman. 2011. The value of estuarine and coastal ecosystem services. *Ecological Monographs* 81(2): 169–193.

Barone, D., K. McKenna, and S. Farrell. 2014. Hurricane Sandy: Beach-dune performance at New Jersey Beach Profile Network sites. *Shore & Beach* 82(4).

Berkes, F. 2007. Understanding uncertainty and reducing vulnerability: Lessons from resilience thinking. *Natural Hazards* 41: 283–295.

Bertule, M., G.J. Lloyd, L. Korsgaard, J. Dalton, R. Welling, S. Barchiesi, M. Smith, J. Opperman, E. Gray, T. Gartner, J. Mulligan, and R. Cole. 2014. Green Infrastructure Guide for Water Management: Ecosystem-based management approaches for water-related infrastructure projects. UNEP.

Beseres-Pollack, J., D.W. Yoskowitz, H.C. Kim, and P. Montagna. 2013. Role and value of nitrogen regulation provided by oysters (*Crassostrea virginica*) in the Mission-Aransas Estuary, Texas, USA. *PLoS ONE* 8(6): e65314. DOI: 10.1371/journal.pone.0065314.

Butler, W., R. Deyle, and C. Mutnansky. Submitted for publication. Low-regrets incrementalism: Land use planning adaptation to accelerating sea level rise in Florida's coastal communities. *Journal of Planning Education and Research*.

Carpenter, S., B. Walker, J.M. Anderies, and N. Abel. 2001. From metaphor to measurement: Resilience of what to what? *Ecosystems* 4: 765–781.

Committee on the Effects of the Deepwater Horizon Mississippi Canyon-252 Oil Spill on Ecosystem Services in the Gulf of Mexico, National Research Council (NRC). 2013. *An Ecosystem Services Approach to Assessing the Impacts of the Deepwater Horizon Oil Spill in the Gulf of Mexico*. Washington, DC: National Academies Press (US).

Cumming, G.S., G. Barnes, S. Perz, M. Schmink, K.E. Sieving, J. Southworth, M. Binford, R.D. Holt, C. Stickler, and T. Van Holt. 2005. An exploratory framework for the empirical measurement of resilience. *Ecosystems* 8: 975–987.

Dell'Apa, A., A. Fullerton, F. Schwing, and M.M Brady. 2015. The status of marine and coastal ecosystembased management among the network of U.S. federal programs. *Marine Policy* 60: 249–258.

Environmental Protection Agency (EPA). 1999. Section 7.3: Benefits Transfer. In Economic Analysis Resource Document. Innovative Strategies and Economics Group, Office of Air Quality Planning & Standards.

Executive Office of the President. 2013. The President's Climate Action Plan. The White House, Washington, D.C.

Ferrario, F., M.W. Beck, C.D. Storlazzi, M. Fiorenza, C.C. Shepard, and L. Airoldi. 2014. The effectiveness of coral reefs for coastal hazard risk reduction and adapation. *Nature Communications*.

Folke, C., S. Carpenter, T. Elmqvist, L. Gunderson, C.S. Holling, and B. Walker. 2002. Resilience and sustainable development: Building adaptive capacity in a world of transformations. *AMBIO: A Journal of the Human Environment* 31(5): 437–440.

FoxNews. 2015. Jersey shore town refuses to rebuild dunes ravaged by Superstorm Sandy. *FoxNews*.

Gedan, K.B., M.L. Kirwan, E. Wolanski, E.B. Barbier, and B.R. Silliman. 2011. The present and future role of coastal wetland vegetation in protecting shorelines: Answering recent challenges to the paradigm. *Climatic Change* 106: 7–29.

Goldstein, B.E., A.T. Wessells, R. Lejano, and W. Butler. 2012. Narrating resilience: Transforming urban systems through collective storytelling. *Urban Studies* 52(7): 1285–1303.

Gunderson, L. 2010. Ecological and human community resilience in response to natural disasters. *Ecology and Society* 15(2): 18.

IPCC Working Group II. 2014. Climate Change 2014: Impacts, Adaptation, and Vulnerability. Intergovernmental Panel on Climate Change.

Jonkman, S.N., M.M. Hillen, R.J. Nicholls, W. Kanning, and M. van Ledden. 2013. Costs of adapting coastal defences to sea-level rise—New estimates and their implications. *Journal of Coastal Research* 29(5): 1212–1226.

Koch, E.W., E.B. Barbiers, B.R. Silliman, D.J. Reed, G.M. Perillo, S.D. Hacker et al. 2009. Non-linearity in ecosystem services: Temporal and spatial variability in coastal protection. *Frontiers in Ecology* 7(1): 29–37.

Melillo, J.M., T.C. Richmond, and G.W. Yohe, Eds. 2014. Climate Change Impacts in the United States: The Third National Climate Assessment. U.S. Global Change Research Program, 841 pp. DOI:10.7930/J0Z31WJ2.

Meyer, H., D. Waggonner, and D. Morris. 2009. Dutch Dialogues—New Orleans Netherlands. Sun Publishing.

Möller, I., M. Kudella, F. Rupprecht, T. Spencer, M. Paul, B.K. van Wesenbeeck, G. Wolters, K. Jensen, T.J. Bouma, M. Miranda-Lange, and S. Schimmels, S. 2014. Wave attenuation over coastal salt marshes under storm surge conditions. *Nature Geoscience* 7: 727–731.

National Climatic Data Center (NCDC). 2013. Billion-Dollar U.S. Weather/Climate Disasters 1980–2012. National Oceanic and Atmospheric Administration.

National Oceanic and Atmospheric Administration (NOAA) et al. 1920–1990. Water Level Reports—8443970 Boston. Retrieved from NOAA Tides and Currents: http://tidesandcurrents .noaa.gov/reports.html?type=maxmin&bdate=19200101&edate=19900101&units=standard&id =8443970&retrieve=Retrieve.

National Oceanic and Atmospheric Administration (NOAA). 2012. Spatial Trends in Coastal Socioeconomics Demographic Trends Database: 1970–2010.

National Oceanic and Atmospheric Administration (NOAA) et al. 2012–2014. Water Level Reports—8443970 Boston. Retrieved from NOAA Tides and Currents: http://tidesandcurrents .noaa.gov/reports.html?type=maxmin&bdate=20120101&edate=20140401&units=standard &id=8443970&retrieve=Retrieve.

National Oceanic and Atmospheric Administration (NOAA). 2014. NOAA's State of the Coast.

OSTP. 2015. Ecosystem Service Assessment: Research Needs for Coastal Green Infrastructure. National Science and Technology Council.

PCAST Report. 2011. Sustaining Environmental Capital: Protecting Society and the Economy, 2011: Prepared by the President's Council of Advisors on Science and Technology.

Rebuild by Design. 2014. Finalists.

Reinhardt, J. 2009. Changing Paradigms through Education, Technical Assistance, and Dialogue: The Role of the Planning Profession and Its Professional Organization in the Recovery of the Gulf Coast following Hurricane Katrina. American Planning Association.

Rodriguez, A.B., F.J. Fodrie, J.T. Ridge, N.L. Lindquist, E.J. Theuerkauf, S.E. Coleman, J.H. Grabowski, M.C. Brodeur, R.K. Gittman, D.A. Keller, M.D. and Kenworthy. 2014. Oyster reefs can outpace sealevel rise. *Nature Climate Change*.

Rose, A. 2004. Defining and measureing economic resilience to disasters. *Disaster Prevention and Management* 13(4): 307–314.

Rose, A. 2007. Economic resilience to natural and man-made disasters: Multidisciplinary origins and contextual dimensions. *Environmental Hazards* 7: 383–398.

Schuster, E. and P. Doerr. 2015. A Guide for Incorporating Ecosystem Service Valuation into Coastal Restoration Projects. The Nature Conservancy, New Jersey Chapter. Delmont, NJ.

Shepard, C.C., C.M. Crain, and M.W. Beck. 2011. The protective role of coastal marshes: A systematic review and meta-analysis. *PLoS ONE* 6.

Simmie, J. and R. Martin. 2010. The economic resilience of regions; towards an evolutionary approach. *Cambridge Journal of Regions, Economy and Society* 3: 27–43.

Sutton-Grier, A., H. Bamford, and K. Wowk, K. 2015. Future of Our Coasts: The Potential for Natural and Hybrid Infrastructure to Protect Our Coastal Communities and Economies.

Systems Approach to Geomorphic Engineering (SAGE). 2015. "About SAGE." Accessed December 2015 from: http://sagecoast.org/index.html.

Taylor, E.B., J.C. Gibeaut, D.W. Yoskowitz, and M.J. Starek. 2015. Assessment and monetary valuation of the storm protection function of beaches and foredunes on the Texas coast. *Journal of Coastal Research* 31(5): 1205–1216.

USACE. 2015. North Atlantic Coast Comprehensive Study: Resilient Adaptation to Increasing Risk—Main Report.

US Indian Ocean Tsunami Warning System Program. 2007. How Resilient Is Your Coastal Community? A Guide for Evaluating Coastal Community Resilience to Tsunamis and Other Coastal Hazards. US Indian Ocean Tsunami Warning System Program supported by the United States Agency for International Development and partners, Bangkok, Thailand. 144 pp.

Walker, B., S. Carpenter, J. Anderies, N. Abel, G. Cumming, M. Janssen, L. Lebel, G. Peterson, and R. Pritchard. 2002. Resilience management in social–ecological systems: A working hypothesis for a participatory approach. *Conservation Ecology* 6(1): 14.

Yoskowitz, D. and M. Russell. 2015. Human dimensions of our estuaries and coasts. *Estuaries and Coasts* 38(S1): 1–8. DOI: 10.1007/s12237-014-9926-y.

Zhang, K., H. Liu, Y. Li, H. Xu, J. Shen, J. Rhome, and T.J. Smith. 2012. Role of mangroves in attenuating storm surges. *Estuarine Coastal and Shelf Science* 102: 11–23.

CHAPTER 5

An Overview of the Living Shorelines Initiative in New York and New Jersey

Andrew Rella, Jon Miller, and Emilie Hauser

CONTENTS

5.1 BACKGROUND

Compared to the long-established and mature living shorelines programs in states such as Virginia, Maryland, and North Carolina, the living shorelines programs in New York and New Jersey are still in a critical developmental stage. Both states have displayed a strong commitment to the implementation of living shorelines, acknowledging that living shorelines can play an important role in helping communities and ecosystems become more resilient. Rather than beginning from scratch, both New York and New Jersey have wisely chosen to look outward and learn from the experiences of others; however, because of each state's combination of unique landscape, climate, and socioeconomic factors, it is not possible to simply translate successful approaches from the

Mid-Atlantic and Southeast. Unique factors that differentiate New York and New Jersey from other areas include the presence of ice, heavy vessel traffic, and a highly urbanized coast, which leads to "coastal squeeze." Because of these differences and the varying shoreline types throughout the two states, New York and New Jersey typically view the implementation of living shorelines as a suite of ecologically grounded principles that can be adapted to a diverse range of environments, rather than a specific technique or techniques.

5.2 NEW YORK

The diverse geography and sheer size of New York State make the adoption of a consistent state-wide approach to living shorelines difficult. The shoreline types in New York State range from the open ocean shores of Long Island, to the heavily urbanized shorelines of New York City, to the shores along the Great Lakes and Lake Champlain, to large rivers including the Mohawk, Hudson, Niagara, and Saint Lawrence. The water along these shores ranges from fresh to saline, and ranges from heavily polluted to near pure. Similarly, the energy along these coasts varies from fully exposed along the ocean, Long Island Sound, and Great Lakes coasts to the quiescent shorelines found along inland streams, rivers, and lakeshores. These shorelines are important not only for their human use but also for the critical habitats they support. Unfortunately, these habitats are at risk from impending shoreline changes associated with development, sea level rise, and human intervention intended to prevent erosion and flooding of shoreline properties. Dramatic changes are imminent as aging structures need to be replaced, public and private agencies make investments in waterfront revitalization, and climate change imposes new stresses on the shoreline. Sea level rise projections for the end of the century are as much as 3 to 6 ft, with flooding from more intense rain storms and ocean-driven storm surges making shorelines a "ground zero" for policy-makers and the public dealing with erosion and inundation challenges (Blair et al. 2015b). Out of necessity, approaches to managing these diverse shorelines and critical habitats vary greatly.

The promotion of living shoreline stabilization approaches in New York State originated in 2003 for Lake Champlain (Northeast Regional Planning Commission 2003) and in 2005 for the Hudson Estuary (Allen et al. 2006). Alternative approaches were recommended in the NY State Sea Level Rise Task Force Report (2010) and in the Urban Waterfront Adaptive Strategies (2013) for New York City. The passage of the NYS Community Risk and Resiliency Act in September 2014 calls on the NYS Department of Environmental Conservation (DEC) and NYS Department of State to develop guidance on "the use of resiliency measures that utilize natural resources and natural processes to reduce risk" by January 2017. This chapter highlights three major living shoreline initiatives that attempt to balance the ecological and engineering needs of New York's shorelines—the Hudson River Sustainable Shorelines Project (HRSSP) will be discussed in full detail, and the Coastal Green Infrastructure Research Plan for New York City and Waterfront Edge Design Guidelines (WEDG) will be briefly discussed.

5.2.1 Hudson River Estuary

The Hudson River Estuary extends 152 miles from the river's mouth in New York City, north past Albany to the Troy Federal Lock and Dam in Troy, New York. The tidal influence of the Atlantic Ocean is only minimally dampened over the course of the river, causing semi-diurnal tidal variations on the order of 4.5 ft, with Spring tides ranging up to 6 ft. Water levels in the estuary are predominantly determined by the tides, but are also influenced by strong flows from the Hudson's many tributaries, as well as by storm surges during extreme storms such as Sandy. During periods of ebb flow, the freshwater flows north to south, but the strong tidal flows (often greater than 2 ft/s) reverse the direction of water flow every 6 h throughout the estuary and are roughly 10 times as large as downriver flow of freshwater (Rella and Miller 2012a). The transition from freshwater to saltwater typically occurs in the lower half of the river, near the Tappan Zee Bridge (approximately 25 miles north of Midtown Manhattan) and is dependent on seasonal and daily variations in freshwater flows and tides. The maximum width of the river is 3 miles at the Tappan Zee; however, most of the river is 0.5–1 mile wide, and the upper section near Albany is less than 0.5 miles wide. The average channel depth of the river is approximately 50 to 60 ft, with a maximum of 175 ft (Blair et al. 2015b). The upper third of the estuary has a 32-ft-deep navigation channel maintained by regular dredging. However, the estuary also contains extensive shallow-water areas that are less than 5 ft deep at low tide, many of which support wetlands or beds of submerged aquatic vegetation. Much of the river bottom is sand or mud, although there are patches of gravel and cobble (Rella and Miller 2012b).

Relict oyster reefs also exist in the southern portion of the estuary (Rella and Miller 2012a) and provide one telling example of how human activities have affected the estuary and dramatically affected the ecology of the estuary. From the 17th to 19th centuries, oysters occupied more than 350 square miles of the Hudson–Raritan estuary from as far south as Sandy Hook to as far north as Ossining, New York (NY/NJ Baykeeper 2006). The Dutch colonists were so impressed by the abundance of oysters in New York Harbor that they named what is now called Liberty Island "Great Oyster Island" (NY/NJ Baykeeper 2005). In the 20th century, severe pollution caused New York State to ban the harvesting of all oysters for human consumption, and the vital resource that once played a central role in the estuary's web of life went virtually extinct. Oysters are now functionally extinct, with a loss greater than 99% in the Hudson–Raritan Estuary (Harbor & Estuary Program [HEP] 2012).

5.2.2 Hudson River Sustainable Shorelines Project

The HRSSP seeks to protect, conserve, and enhance Hudson River Estuary shorelines, their resilience to climate change and sea level rise, and their ability to serve as vital habitats and corridors (Blair et al. 2015b). This Project is designed to provide science-based information about the best shoreline management options for preserving important natural functions of the Hudson River Estuary's shore zone, especially as sea level rise accelerates and storms increase in intensity (Blair et al. 2015a). The Project began in 2005, and has received significant support from the Cooperative Institute for Coastal and Estuarine Environmental Technology and the NERRS Science Collaborative, through the following funded phases of the Project:

- Mitigating Shoreline Erosion along the Hudson River Estuary's Sheltered Coasts (Phase 1)
- Sustainable Shorelines along the Hudson River Estuary: Promoting Resilient Shorelines and Ecosystem Services in an Era of Rapid Climate Change (Phase 2)
- What Made Shorelines Resilient: A Forensic Analysis of Shoreline Structures on the Hudson River Following Three Historic Storms (Phase 3)
- Assessing Ecological and Physical Performance of Sustainable Shoreline Structures (Phase 4)

The primary partners in this ongoing project are staff of the Consensus Building Institute, Stevens Institute of Technology, Cary Institute of Ecosystem Studies, as well as the groups previously mentioned (Blair et al. 2015b). Leadership is provided by the DEC Hudson River National Estuarine Research Reserve (HRNERR) and the New York State Department of Environmental Conservation (NYSDEC) in cooperation with the Greenway Conservancy for the Hudson River Valley. The HRSSP has four overarching objectives:

1. Characterize the present and future estuary and shoreline conditions
2. Determine the ecological, engineering, and economic trade-offs of different shoreline management options

3. Demonstrate innovative shorelines and best management practices
4. Characterize shoreline decision-making arenas so that useful decision-making tools and effective communication products can be generated (Blair et al. 2015a)

In reflecting on the diversity of the environments and shoreline types along the Hudson River Estuary, a decision was made early on in the HRSSP to refer to the suite of ecologically responsible shoreline stabilization alternatives as ecologically enhanced engineered shoreline protection defined in (Hauser 2012) as:

A subset of shore protection methods that incorporate measures to attract and support both terrestrial and aquatic biota and desirable ecological functions. These can be either modifications to existing structures through the addition of plantings and other ecological measures or the design of new structures incorporating ecologically-friendly materials, geometry, or placement. If correctly designed, ecologically-engineered structures serve to prevent or reduce shore erosion while emulating the physical and biological conditions of naturally occurring, stable shorelines. Valuable ecosystem services are enhanced or restored; including provision of habitat for terrestrial and aquatic species, maintenance of water quality, aesthetic, resilience and sustainability.

There is now a trend in New York State and nationally (Executive Order 11988) to use the term *nature-based shoreline protection*. The Guidelines for Implementing Executive Order 11988 (FEMA 2015) *defines nature-based approaches as features (sometimes referred to as "green infrastructure") designed to mimic natural processes and provide specific services such as reducing flood risks and/or improving water quality. Nature-based approaches are created by human design (in concert with and to accommodate natural processes) and generally, but not always, must be maintained in order to reliably provide the intended level of service.*

The multidisciplinary, regional research team responsible for carrying out the work uses a collaborative, user-driven process to identify the information and tools needed by regulators, engineers, resource managers, and other experts to understand the appropriate settings, to develop the tools and best approaches, and to address the key questions about ecological benefits and trade-offs, and physical constraints for living shorelines projects (Blair et al. 2015b). An Advisory Committee made up of key users of the information provided advice throughout the project. Their diverse perspectives helped identify key challenges, barriers, and opportunities. One of the key pieces of information derived from this collaborative process was a detailed understanding of several of the impediments to the implementation of living shorelines projects. These barriers included the following:

• Evidence that these alternative approaches will perform satisfactorily under the conditions in the Hudson River Estuary, even during severe storms
• A lack of information on design conditions (currents, wind-driven waves, ice scour, and vessel wakes)
• The need for more robust information about sustainable shoreline management options and guidance materials for permit applicants
• The need for detailed guidance about specific structural refinements that can enhance the ecological function of shoreline treatments or increase structural stability under future conditions (Blair et al. 2015b)

Since its inception, the HRSSP has generated new information to help overcome these barriers and answer the questions that decision makers and advocates have about shorelines, from ecological, engineering, economic, social, legal, and other perspectives. All HRSSP findings are vetted with users and communicated through project reports, summaries, meetings and training workshops, and a centralized project website (https://www.hrnerr.org/hudson-river-sustainable-shorelines/). These efforts are supporting more widespread adoption of nature-based, sustainable shorelines on the tidal Hudson River, and are adding to the body of knowledge on sustainable, nature-based shorelines, to the benefit of other states and regions. The net result is that communities and regulators are better able to consider the full benefits, trade-offs, and consequences of particular decisions as shorelines are modified to adapt to climate change, protect infrastructure, revitalize waterfronts, and support multiple human uses of the river (Blair et al. 2015b). The brief description of several of the products developed through the HRSSP that follows is organized according to the four overarching objectives of the HRSSP described above.

5.2.2.1 Characterize Present and Future Estuary and Shoreline Conditions

The first step in making good decisions about shorelines and how they might change in the future is to understand the characteristics of the varying shoreline types and the physical forces affecting them (Blair et al. 2015a). The forces impinging on the Hudson's shores include wind-driven waves, wakes from commercial and recreational vessels, currents from tides and downriver flow, and floating debris and ice driven onshore by these forces. Depending on their exposure to wind, currents, wakes, and ice, and their position relative to the navigation channel and protective shallows, different parts of the Hudson's shores receive very different inputs of physical energy. Likewise, land uses on the landward side of the shore and water-dependent uses on the riverward side of the shore are highly variable along the Hudson. As a result, different parts of the Hudson place very different demands on engineered structures along the shore (Rella and Miller 2012a).

In order to better understand the habitats of the Hudson River Estuary, the NYSDEC Hudson River Estuary Program and HRNERR, in collaboration with numerous partners, supported the mapping of biological, chemical, and physical characteristics of the estuary. This effort included mapping of the shoreline, benthic conditions, contaminates, unique freshwater tidal wetlands and submerged aquatic vegetation. Data from these projects are located on the New York State Geographic Information System Clearinghouse and include remote sensing images, point sample data, and interpretive maps (ny.gis.gov). In the summer and fall of 2005, Miller, Bowser, and Eckerlin completed an inventory of shoreline types along the 127-mile stretch of Hudson River from the Tappan Zee Bridge to Troy. A three-level classification system was used based on (1) the character of the shoreline segment (hard engineered, soft engineered, or natural); (2) the type of "structure" (revetment, gabion, cribbing, bulkhead, broadleaf vegetation, woody vegetation, unvegetated, woody debris, other); and (3) the primary structure material or substrate (timber, sheet pile, concrete, mixed mud/sand, unconsolidated rock, solid bedrock, sand with brick, other). The project also collected information on land use and feature geometry (HRNERR 2012). Details of the methodology can be found at http://gis.ny.gov/gisdata/metadata/nysdec.hr shoreline_type.xml. Overall, the results showed that of the 250 miles of shorelines inventoried, 42% were hard engineered, 47% were natural, and 11% were natural with remnants of engineering structures. The most common shoreline structure was riprap (32%), followed by woody (29%) and unvegetated (16%) slopes. The dominant substrate found within the region was unconsolidated rock (52%), mud/sand (16%), and mixed soil/rock (12%) (Miller et al. 2006).

Phase 2 of the HRSSP was focused on understanding the physical forces acting on the shoreline. Before the HRSSP study, currents, wind-driven waves, ice, and vessel wakes had not been systematically measured or characterized (Blair et al. 2015a). The approach taken combined numerical simulations of the hydrodynamics of the estuary with field observations of wakes and ice (Miller et al. 2015).

In 2011, Miller and Georgas compiled water circulation statistics from a high-resolution hydrodynamic model. The model was used to simulate a typical year worth (2010) of flows in the estuary including contributions from tributaries, storm surges, and wind waves. The model was validated against 6-min total water level (tide and surge) observation records at five real-time gage locations. The data set characterizes the energy regimes in the Hudson and can be used to assist in the identification of suitable shoreline stabilization alternatives. Details of the study and methodology are provided in Miller and Georgas (2015) and the results of the modeling study are available on the

Table 5.1   Summary of Site Characteristics

Parameter	Climatology
WL_{max} (ft NAVD88)	4.90
H_{max} (ft)	1.77
H_{med} (ft)	0.13

NYS GIS Clearinghouse. An example summary of the site characteristics for Esopus Meadows is presented in Table 5.1. The maximum simulated water level in 2010 at the site (WL_{max}) was 4.90 ft NAVD88. The maximum (H_{max}) and median (H_{med}) modeled wind wave heights were 1.77 and 0.13 ft, respectively.

The ice climatology (Georgas et al. 2015) is a compilation of statistics on ice characteristics (ice thickness, percent ice cover, and ice types) based on United States Coast Guard (USCG) daily ice reports from 2004 to 2011. The ice thickness and ice cover area climatology was produced by fitting Generalized Extreme Value Cumulative Distribution Functions to the empirical mean daily data for each ice region, when ice was present and reported (Figures 5.1 and 5.2). The ice reports have a limited history and scope but empirically provide a probabilistic picture of the regional ice climatology in the Hudson intended to address the paucity of ice information and to inform engineers of ice conditions. The uncertainty around the data is within a factor of 2 of the reported regional ice thickness percentiles (Georgas et al. 2015). An example of the potential usefulness of this unique data set is described in a way in "Large Seasonal Modulation of Tides Due to Ice Cover Friction in a Mid-Latitude Estuary."

In some areas of the Hudson, vessel wakes can be several times the size of the typical wind waves and therefore represent a critical design parameter. Previous work has shown that wakes from commuter ferry traffic in the lower Hudson River produce a continual source of waves even

Figure 5.1   Photo of Ice at Norrie Point Environmental Center.

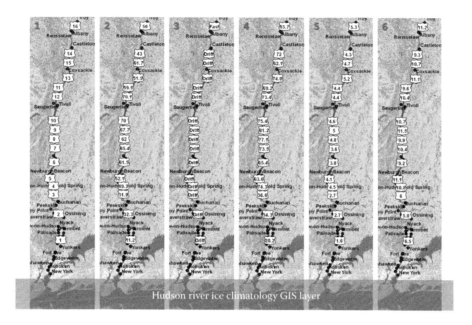

Figure 5.2 Ice climatology GIS layer.

when wind waves are not present, and can have a significant impact on the shoreline (Bruno 2002). Stevens Institute of Technology began a study in 2012 to determine the wake heights produced from recreational and commercial vessels as they travel along the Hudson River. The results of the wake study are described in LaPann-Johannessen et al. (2015) and represent a preliminary analysis of the wake energy area along the 127-mile stretch from the Tappan Zee Bridge to the Federal Dam at Troy. Based on the extent of the study area, and the available labor and funding, a simple, visual wake measurement approach was utilized at 32 locations. The study took place over the course of 4 days in 2012 (June 26–28 and July 19) and 3 days in 2013 (June 28 and 29 and July 1). The total number of recorded wakes at each site ranged from a minimum of 3 at Coxsackie to 59 at Highland Falls. The average number of observed wakes at each site was 25. The size of the observed wakes ranged from a minimum of 0 inches (imperceptible) at several sites to 42 inches at Kingston. The average observed wake was 4.2 inches. The observed data only represent a "snapshot" of the wake activity in the Hudson River; however, based on the magnitude of the observed wakes relative to the modeled wind waves, the results suggest additional resources need to be dedicated to understanding the influence of wakes in the Hudson (LaPann-Johannessen et al. 2015).

5.2.2.2 Determine the Ecological, Engineering, and Economic Trade-Offs of Shoreline Management Options

The second overarching objective of the HRSSP is to improve the understanding of the ecological, engineering, and economic trade-offs of different shoreline management options. These management options critically affect the shore zone. People use shore zones for transportation (roads, railroads, and ports), waste disposal, resource extraction, and other industrial uses. People also live in or like to visit shore zones to swim, hike, watch birds, boat, or just relax and enjoy their beauty. Humans have affected freshwater shore zones by laterally compressing and hardening shorelines for stabilization, changing hydrologic regimes, tidying shore zones, increasing inputs of physical energy and pollution from resource extraction, and introducing alien species (Strayer 2011).

5.2.2.2.1 Ecological Trade-Offs

Shore zones are complexes of habitats that support high biodiversity and vital habitat for multiple life stages of many fish, birds, reptiles, amphibians, and invertebrates. Different shore zones provide different kinds and scales of habitat and, when aggregated, can significantly influence life in the Hudson River ecosystem (Blair et al. 2015a). Healthy shore zones can protect the water by capturing pollution running off from the land, protect the adjacent land by absorbing energy from waves and currents, and protect the land downstream from floods. Additional information on the ecological research conducted through the HRSSP is covered in Chapter 16 of this book ("Ecological Performance of Hudson River Shore Zones: What We Know and What We Need to Know") by Strayer and Findlay.

5.2.2.2.2 Engineering Trade-Offs

In 2006, ASA Analysis & Communication and Alden Research Laboratory, Inc. were contracted to investigate options for restoring ecological functions and enhancing shoreline habitats. The final report (Allen et al. 2006) presents a literature review; a synthesis of the field survey to qualitatively assess the current types, condition, and function of natural and engineered shoreline habitats; an evaluation of five potential shoreline restoration sites; and a summary of the regulatory process (HRNERR 2012). The project report was used as a basis for a training program in 2006 that engaged engineers and natural resource managers for the first time on alternative techniques. The stabilization techniques that were deemed suitable for the conditions of the Hudson were those that used hard features to stabilize soil and vegetation to enhance habitat: joint planting, vegetated geogrid, vegetated rock gabions, live crib walls, brush mattresses, and subtidal rock structures for habitat refuge. These techniques can withstand shear stresses greater than 2.5 lb/ft^2 (Allen et al. 2006).

In 2012, Rella and Miller (2012a) produced *Engineered Approaches for Limiting Erosion along Sheltered Shorelines: A Review of Existing Methods*, which contained a comprehensive summary of published literature on 28 engineering approaches for limiting erosion along sheltered shorelines. The document is intended to inform decision makers of the variety of different alternatives that have been utilized elsewhere, and is not specific to the Hudson River Estuary. The descriptions are broken down into the following six categories: *Description, Design and Construction, Adaptability, Advantages, Disadvantages,* and *Similar Techniques.* The *Description* section provides a short discussion of the specified stabilization technique. Wherever possible, pictures and figures showing cross sections or typical installations are provided. The *Design and Construction* section contains information on some of the basic design and construction considerations associated with each approach. When available, cost information and information about operation and maintenance considerations are also presented. The *Adaptability* section contains information related to the ability of the selected treatment to adapt to changing conditions either naturally or through anthropogenic intervention. Factors such as expected life span, durability, and ease of modification are considered. In the *Advantages* and *Disadvantages* sections, bulleted lists summarizing the positive and negative attributes of each method are provided. *Similar Techniques* lists the alternative approaches that are most similar to the one being discussed in terms of their ability to reduce erosion at the shoreline. A table summarizing these qualitative evaluations and glossary conclude the document (Figure 5.3) (Rella and Miller 2012a).

A series of historic storms in 2011 and 2012 provided a unique opportunity to address one of the questions identified by the HRSSP advisory committee as a barrier to the more widespread acceptance of living shorelines projects. In 2014, Stevens Institute of Technology conducted an assessment of the performance of six shoreline stabilization techniques in the wake of three historic storms—Tropical Storms Irene and Lee in 2011 and Post-Tropical Storm Sandy in 2012. The objective was not only to show that living shorelines projects could be successful but to also

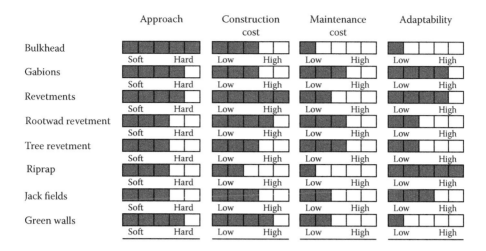

Figure 5.3 Comparison of stabilization techniques.

identify the factors critical to their performance (Rella and Miller 2015). On the basis of the advice of a technical advisory panel, a combination of six traditional and nontraditional shorelines with varying degrees of damage was selected for the analysis. Key questions addressed by the analysis were as follows:

- What factors were critical in the success or failure of each stabilization method?
- What aspects of structural design were pivotal in determining survival or failure?
- What were the impacts of water levels and currents?
- Did vegetation factor into the success or failure of the stabilization method?
- Were there patterns among structures that survived or failed?

Separate forensic analyses describing the impact of the three storms at each site, as well as two additional reports that describe the common project performance factors and methodology are available on the HRSSP website. Each analysis considered the following:

- A review of historic photographs
- A review of available design drawings/calculations
- A review of the physical forces climatology developed in Phase 2 of the HRSSP
- Interviews with project managers and designers
- Field data collection (bathymetry, water mark observation, photographic documentation, etc.)
- A review of FEMA flood maps
- A review of FEMA post-storm inundation modeling and watermark collection
- Modeling of the hydrodynamic conditions during each storm

Collectively, this information was used to create a holistic picture of each site, from which the critical project performance factors could be determined. A number of site-specific performance factors were identified in addition to the following six cross-cutting factors:

1. Maturity of vegetation—sites with mature vegetation performed better
2. Slope compatibility—sites where appropriate slopes were utilized performed better
3. Debris impact—debris flows during Sandy caused significant damage at several sites
4. Leeside erosion and impacts—erosion attributed to overtopping during Sandy damaged several sites
5. Maintenance and adaptive management—sites that were well maintained performed better
6. Adequate stone sizing—inappropriate stone sizes were identified as a cause of failure at several sites

On the basis of the findings of the forensic analyses, six recommendations were made:

1. More research needs to be done on the performance of ecologically enhanced stabilization approaches during heavy ice and debris conditions.
2. More research needs to be done in the area of plant material selection.
3. Proper monitoring and maintenance are important to the long-term performance of all projects; however, it is critical for ecologically enhanced shoreline projects.
4. Temporary stabilization measures should be provided to allow vegetation to mature.
5. Terracing or other measures should be used to avoid unnatural slopes.
6. Leeside forces should be addressed in design/construction of coastal structures (Rella and Miller 2015).

5.2.2.2.3 Economic Trade-Offs among Shoreline Options

The costs of managing shorelines can range in construction and long-term operation and maintenance. Different shoreline treatments are associated with different thresholds of risk, and some are more easily adapted over time to sea level rise and other changing physical conditions, with implications for long-term cost. The natural services associated with different shorelines also have values that are part of the equation. Both policymakers and individual shoreline owners and managers must weigh this complex set of factors in the context of other factors, such as regulations and incentives, to select shoreline treatments. This part of the project sought to better understand the costs involved with different shoreline stabilization techniques and provide useful information for comparing options (Blair et al. 2015a).

Van Luven (2011) developed a white paper for the HRSSP with recommendations for valuing the ecosystem services of different shoreline techniques. A full cost accounting should include not only the engineering costs of installation, maintenance, replacement, and the like but also ecosystem service values that currently are not incorporated into most cost–benefit analyses. The importance of these services is increasingly being recognized by policymakers and regulators, but estimating their values remains difficult. This report sets forth a framework for an economic assessment of nine shoreline treatment techniques: bulkhead, timber cribbing, live stakes/joint planting, revetment, live crib wall, vegetated geogrid, riprap, green (bio) wall, and living shorelines (HRNERR 2012).

Rella and Miller (2012a,b) performed a comparative cost assessment of 10 different shoreline stabilization approaches at three sites, under two sea level rise scenarios. The approaches, sites, and sea level rise scenarios were selected in consultation with the HRSSP advisory committee. A 70-year time frame was selected for the analysis. The basic life cycle costs were separated into four main categories.

1. **Initial construction (IC)**—the costs described above associated with constructing each of the alternatives as designed.
2. **Maintenance and repair (M&R)**—refer to the costs associated with inspecting and performing basic maintenance for each approach. All M&R costs are formulated in terms of a percentage of the initial cost. It is assumed that M&R costs for several of the approaches will increase under the rapid sea level rise scenario.
3. **Damage costs (DC)**—include costs outside of the typical M&R costs created by storm impacts below the design level (50 years). These storms may have specific impacts that require significant modifications to restore the original function of the shoreline stabilization approach
4. **Replacement Costs (RC)**—some structures such as bulkheads have a finite life span and, regardless of the storm conditions, will need to be replaced once or even twice with the 70-year period.

Within the limitations of the analysis, the results showed that, at most sites, there is a suite of alternatives for which the life cycle costs are relatively similar. Given the uncertainties associated with many aspects of the economic valuations, the error bands on the results are such that many of the costs are functionally equivalent. Within this grouping are generally one or more alternatives that incorporate some sort of ecological enhancement (Rella and Miller 2012b). This finding is consistent with a National Oceanic and Atmospheric Administration (NOAA) report entitled *Weighing Your Options* that determined that the costs of many "living shorelines" stabilization approaches were on par with bulkheads.

5.2.2.3 Demonstrating Innovative Shorelines and Best Management Practices

Property owners, engineers, regulators, and other decision makers need evidence that innovative, ecologically beneficial shoreline techniques will perform satisfactorily under the conditions in the Hudson. To respond to this need, the HRSSP team designed and constructed innovative shoreline projects, provided technical assistance to other groups designing shorelines, and created a Demonstration Site Network that uses case studies to describe innovative projects that met certain criteria. Shoreline engineers or project managers submitted information on their projects including design, costs, physical attributes of the site, and other planning considerations. If the project illustrated a standardized set of best management practices and innovation, then HRNERR staff wrote detailed case studies in collaboration with the engineer, property owner, and other involved parties. The case studies describe the planning, design, implementation, costs, and lessons learned and are available on the website (https://www.hrnerr.org/hudson-river-sustainable-shorelines/demonstration -site-network/). Some of the demonstration sites were used in the forensic analysis study described above and will be used in an ongoing study to assess their ecological and physical performance (Figures 5.4 and 5.5).

In addition, outside agencies have funded designs for sustainable shoreline projects at state parks. The sites in the demonstration site network are instrumental in advancing the use of sustainable shoreline principles, understanding that each site requires unique design solutions. By working with a variety of design experts, a community of practice has been established among and between engineers, ecologists, regulators, nongovernmental organizations (NGOs), resource management

Figure 5.4 Coxsackie boat launch demonstration site (before).

Figure 5.5   Coxsackie boat launch demonstration site (after).

staff, and community members who have shared understandings and are collaborating on new projects. From this effort, there is much more awareness of, interest in, and comfort with the idea of living/sustainable shorelines in the Hudson River Estuary.

5.2.2.4 *Characterizing Shoreline Decision-Making Arenas to Create Shoreline Decision Tools and Effectively Communicate Results*

Understanding the legal and decision-making arenas framing shoreline decisions is necessary to influence decision making. Under this objective, project leaders aimed to learn more about who is making shoreline decisions along the Hudson and how they are making them, as well as identifying opportunities for influencing these decisions (Blair et al. 2015a). The engagement of users and collaboration were also important components of the project. Several studies were carried out to understand the decision-making arenas around shoreline management including regulations, shoreline decision makers, and users. This process included several tiers of engagement to ensure that work was coordinated across topic areas and that it was thoroughly responsive to end users' needs. Focus groups and key informant interviews, surveys, and other social science methodology were used to understand the human dimensions of the project, to understand the decision-making arena, and to shape final products.

5.2.3 New York City Green Infrastructure Research Plan

The HRSSP has been successful at building momentum behind the use of living shorelines in New York. After Sandy, it became obvious that the status quo was insufficient and that to build truly resilient coastal communities, combinations of green and gray infrastructure would be required. In New York City, the potential benefits of green infrastructure were acknowledged early on; however, several critical knowledge gaps were also recognized. In an effort to maximize the efficiency of future research expenditures, the Hudson River Estuary Program and the New York City Mayor's

Office of Recovery and Resiliency and Department of City Planning commissioned the development of a research plan on coastal green infrastructure. The plan (Zhao et al. 2014) first assessed the current state of knowledge about six potential green infrastructure strategies (constructed wetlands and maritime forests, constructed reefs, constructed breakwater islands, channel shallowing, ecologically enhanced bulkheads and revetments, and living shorelines) and then outlined a prioritized research plan. Research objectives were separated into meta-research strategies that focused on topics of relevance to several of the proposed green infrastructure approaches and strategy-specific research plans. A total of 21 research agendas were proposed, with each agenda containing one or more research task. Each task was ranked across four categories: fundamental principles, chronology, regional applicability, and affordability. Summing the scores across all categories results in a rating from 1 to 10, where 10 signifies a highly recommended project (Zhao et al. 2014).

5.2.4 Waterfront Edge Design Guidelines

WEDG is a tool and scorecard to guide and improve waterfront projects, offering best practices to enhance the construction and management of waterfronts with approaches that are understandable, feasible, beneficial, and permit-friendly. WEDG has an extensive menu of engineering methods, ecological enhancements, and other designs that promote public access, offer ecologically smart options for healthy shorelines, promote state-of-the-art amenities, and encourage resiliency. The guidelines are practical and intended for use by maritime businesses, park administrators, property owners and developers, design practitioners, concerned citizens, civic groups, community boards, and elected officials. The Waterfront Alliance developed WEDG and its incentive-based ratings system with input from hundreds of waterfront experts, in order to achieve necessary change at the waterfront. The guidelines are intended to be used for properties directly touching a water body, but the program does include recommendations for water-dependent infrastructure and measures for improved resiliency for any type of structure one would build at the edge (WEDG 2014).

5.3 NEW JERSEY

New Jersey's living shorelines program has made remarkable strides over the last decade. New Jersey currently has a living shorelines working group within the New Jersey Department of Environmental Protection (NJDEP) to facilitate living shorelines projects and in July of 2013, as part of an emergency rule adoption to the Coastal Permit Program Rules N.J.A.C. 7:7 and Coastal Zone Management Rules N.J.A.C. 7:7E, adopted provisions to facilitate living shorelines. This included a general permit intended to simplify the permitting of habitat restoration and enhancement projects, which includes living shorelines projects. The two regulations have subsequently been combined and were adopted as the Coastal Zone Management Rules N.J.A.C. 7:7 on July 6, 2015. In the state of New Jersey, a living shoreline is defined as

> ...a shoreline management practice that addresses the loss of vegetated shorelines, beaches, and habitat in the littoral zone by providing for the protection, restoration or enhancement of these habitats. This is accomplished through the strategic placement of plants, stone, sand, or other structural and organic materials. The three types of living shorelines include natural, hybrid, and structural. Natural living shorelines incorporate natural vegetation, submerged aquatic vegetation, fill, and biodegradable organic materials. Hybrid living shorelines incorporate natural vegetation, submerged aquatic vegetation, fill, biodegradable organic materials, and low-profile rock structures such as segmented sills, stone containment, and living breakwaters seeded with native shellfish. Structural living shorelines include, but are not limited to, revetments, breakwaters, and groins (http://www.nj.gov/dep/landuse/activity/livingshore.html).

The definition is significant in that it includes a broad spectrum of projects applicable in diverse coastal environments. This recognition of living shorelines projects as distinct from traditional shoreline stabilization was an important first step in being able to regulate them in a more appropriate way.

In 2009, the NJDEP commissioned a white paper that highlighted some of the difficulties encountered in permitting one of the state's first living shorelines projects. The white paper entitled "Overcoming Regulatory Obstacles to Allow for Living Shorelines" (Frizerra 2011) presented an overview of the benefits of living shorelines along sheltered coasts as compared to hard structures, and included information on the different State and Federal agencies that support living shorelines alternatives. In the white paper, the authors put forth suggestions for streamlining the planning and permitting of living shorelines projects in New Jersey. Some of these suggestions included the creation of a shoreline management plan, a general permit from the Army Corps of Engineers (ACOE), a joint application and improved coordination between New Jersey and ACOE, and specification for exemptions based on the size of projects. Alterations were suggested for three existing permits in New Jersey's Rules and General Permits (7:7-7:18 Coastal General Permit for bulkhead construction and placement of associated fill, 7:7-7.1 Coastal General Permit for stabilization of eroded shorelines, and 7:7-7.29 Coastal General Permit for habitat creation and enhancement activities). Future recommendations to enhance projects and planning included collecting site-specific data, monitoring of projects, and possible incentives to encourage living shorelines projects instead of hard structures.

5.3.1 Overcoming Regulatory Obstacles

Recognizing that some of the obstacles identified by Frizerra (2011) originated from within the NJDEP itself, the agency began to take steps to address some of the internal impediments. In 2010, the State held a "regulatory rule writing summit" and invited experts from both inside and outside the state to discuss the elements of a general permit that would facilitate habitat enhancement projects, including living shorelines projects. After several years of modification and review, the State ultimately adopted Coastal General Permit 29, in July 2013. The general permit was subsequently recodified to General Permit 24 in July 2015. The General Permit (N.J.A.C. 7:7-6.24), commonly referred to as Coastal GP 24, authorizes living shorelines projects designed to protect, restore, or enhance habitat, provided certain criteria are met (NJDEP 2015).

Under Coastal GP 24,

- Projects must comply with all applicable coastal statutes and Coastal Zone Management rules including the provision of public access.
- Projects must maintain or improve the value and function of the local ecosystem and may disturb only the minimum amount of NJDEP-defined special areas (e.g., shellfish habitat, SAV, intertidal and subtidal shallows, and wetlands) as defined in N.J.A.C. 7:7-9.
- Projects constructed seaward of mean high water are limited in size to 1 acre or less, unless the applicant is a federal or state agency that can demonstrate the need for a larger project.

Associated with this is the requirement that restoration activities must take place within an area bounded by the shoreline as indicated on the 1977 state tidelands map. An exception is provided for structural components designed to reduce the wave energy that can be placed outside the 1977 tidelands boundary. An innovative provision that was included in the GP was that all projects must be sponsored by either a federal or state agency, NGO, or academic institution with experience designing living shorelines projects. This was done to ensure that proposed projects would have the greatest chance of success and be constructed as a research project from which significant future benefits could be gained. Before the 2013 revisions project, sponsorship required a monetary commitment. In the 2013 and subsequent 2015 adoptions, the monetary contribution requirement was removed to facilitate habitat restorations, including living shorelines.

5.3.2 Site Characterization and Design

One of the overriding concerns with the adoption of the Coastal GP 24 was that there was a lack of knowledge within the state on how to design, regulate, and ultimately build successful living shorelines projects. In order to ensure that living shorelines projects in New Jersey were well thought out and well engineered, the state of New Jersey commissioned Stevens Institute of Technology to develop a set of engineering guidelines for living shorelines projects (Miller et al. 2015). The document essentially provides a common starting point from which the developer, shoreline designer, regulators, and other relevant parties can build innovative solutions to what are often challenging and complex problems. A tiered approach is suggested in which a basic analysis of the relevant parameters is used to help select between alternatives and develop one or more conceptual designs. Once an alternative has been selected, the scale of the project and the complexity of the site are considered to determine whether a higher-level analysis is required for any of the critical parameters.

While there are many parameters that are relevant to the design of living shorelines projects, the guidelines focus on the 17 outlined in Table 5.2. While many of these parameters are familiar to engineers, some are not, and some are applied in a slightly different way when designing living shorelines projects.

In addition to the parameters listed in Table 5.2, there are a number of other considerations that play a significant role in the selection and design of an appropriate living shorelines project. Some of the more important factors include permitting, constructability, monitoring, and the consideration of end effects, debris impact, and invasive species. The guidelines provide two tables that can be used to help project designers select between five common living shorelines alternatives. The first (Table 5.3) qualitatively describes the most appropriate site characteristics for different project types, while the second (Table 5.4) provides approximate quantitative bounds for the ranges described in Table 5.3. The boundaries described in the table have been adopted from those used in guidance for other geographical areas and there is an intent to update the tables based on the results of monitoring (discussed below) of newly constructed projects. In addition to providing information on how to assess each of the 17 parameter types, the guidelines also include a section describing the role that the parameters play in the design of the five selected living shorelines alternatives.

The guidelines have become an integral part of an online decision-making tool being developed by The Nature Conservancy (TNC). In 2013, the State of New Jersey partnered with TNC, Stevens Institute of Technology, Rutgers University, and others on a resilient coastlines initiative designed to help communities make informed decisions about shoreline management. The initiative, which is supported by post-Sandy funding, through the NOAA's Coastal Resilience Networks

Table 5.2   Parameters Typically Used in the Design of Living Shorelines Projects

System Parameters	Ecological Parameters
Erosion history	Water quality
Sea level rise	Soil type
Tidal range	Sunlight exposure
Hydrodynamic Parameters	**Terrestrial Parameters**
Wind waves	Upland slope
Wakes	Shoreline slope
Currents	Width
Ice	Nearshore slope
Storm surge	Offshore depth
	Soil bearing capacity

Table 5.3 Appropriate Conditions for Various Living Shorelines Approaches

	Marsh Sill	Breakwater	Revetment	Living Reef	Reef Balls
System Parameters					
Erosion history	*Low–med*	*Med–high*	*Med–high*	*Low–med*	*Low–med*
Relative sea level	Low–mod	Low–high	*Low–high*	Low–mod	Low–mod
Tidal range	*Low–mod*	Low–high	Low–high	*Low–mod*	Low–mod
Hydrodynamic Parameters					
Wind waves	*Low–mod*	*High*	*Mod–high*	*Low–mod*	*Low–mod*
Wakes	*Low–mod*	*High*	*Mod–high*	*Low–mod*	*Low–mod*
Currents	Low–mod	Low–mod	Low–high	Low–mod	Low–mod
Ice	*Low*	*Low–mod*	*Low–high*	*Low*	Low–mod
Storm surge	Low–high	*Low–high*	*Low–high*	Low–high	Low–high
Terrestrial Parameters					
Upland slope	Mild–steep	Mild–steep	*Mild–steep*	Mild–steep	Mild–steep
Shoreline slope	*Mild–mod*	*Mild–steep*	*Mild–steep*	*Mild–mod*	Mild–steep
Width	*Mod–high*	*Mod–high*	Low–high	*Mod–high*	*Mod–high*
Nearshore slope	*Mild–mod*	*Mild–mod*	Mild–steep	*Mild–mod*	*Mild–mod*
Offshore depth	Shallow–mod	Mod–deep	Shallow–deep	Shallow–mod	Shallow–mod
Soil bearing	Mod–high	*High*	*Mod–high*	Mod–high	*Mod–high*
Ecological Parameters					
Water quality	Poor–good	Poor–good	Poor–good	*Good*	*Poor–good*
Soil type	*Any*	Any	Any	Any	Any
Sunlight exposure	*Mod–high*	Low–high	Low–high	Mod–high	Low–high

(CRest) Program, is designed to help communities identify restoration opportunities and develop conceptual designs based on the best available engineering and ecological data. The backbone of the project is the development of an online decision support tool called the "Restoration Explorer," which integrates the criteria presented in the *Engineering Guidelines*, with information on regulatory requirements, design considerations, and potential funding sources. The project includes a significant outreach component in which the tool has been piloted in several communities. Feedback from the meetings was used to refine the tool that went live in 2015.

5.3.3 Demonstration Projects

The lack of successful demonstration projects is nearly always cited as one of the most significant impediments to the acceptance of living shorelines projects. Unfortunately, successful projects constructed outside of the state have not resonated with local landowners, engineers, and even regulators. Ultimately, this attitude has led to a circular argument whereby demonstration projects do not exist because there are no demonstration projects to show they might work. Despite self-defeating arguments like this, several organizations have been persistent and have found ways to construct small demonstration projects. Many more are in the planning stages bolstered in part by Sandy recovery funding.

The Partnership for the Delaware Estuary (PDE) was one of the first organizations to overcome the impediments to living shorelines projects when they launched the Delaware Estuary Living Shoreline Initiative (DELSI) in partnership with Rutgers University. Some of the struggles encountered by PDE in permitting their early projects inspired the white paper, which ultimately led to a

Table 5.4 Criteria Ranges

Parameter	Criterion		
	Low/Mild	Moderate	High/Steep
System Parameters			
Erosion history	<2 ft/year	2 to 4 ft/year	>4 ft/year
Sea level rise	<0.2 in/year	0.2 to 0.4 in/year	>0.4 in/year
Tidal range	<1.5 ft	1.5 to 4 ft	>4 ft
Hydrodynamic Parameters			
Waves	<1 ft	1 to 3 ft	>3 ft
Wakes	<1 ft	1 to 3 ft	>3 ft
Currents	<1.25 kts	1.25 to 4.75 kts	>4.75 kts
Ice	<2 in	2 to 6 in	> 6 in
Storm surge	<1 ft	1 to 3 ft	>3 ft
Terrestrial Parameters			
Upland slope	<1 on 30	1 on 30 to 1 on 10	>1 on 10
Shoreline slope	<1 on 15	1 on 15 to 1 on 5	> 1 on 5
Width	<30 ft	30 to 60 ft	>60 ft
Nearshore slope	<1 on 30	1 on 30 to 1 on 10	>1 on 10
Offshore depth	<2 ft	2 to 5 ft	>5 ft
Soil bearing capacity	<500 psf	500 to 1500 psf	>1500 psf
Ecological Parameters			
Water quality	–	–	–
Soil type	–	–	–
Sunlight exposure	<2 h/day	2 to 10 h/day	>10 h/day

change in the way in which living shorelines projects are handled in New Jersey. The DELSI project was initiated to help address the loss of tidal salt marshes in the Delaware Bay but has led to numerous collaborations and projects elsewhere. One of the by-products of the DELSI project has been the development and documentation of the DELSI tactic for restoring impaired shorelines. The DELSI tactic incorporates the use of ribbed mussels, coir logs, bagged oyster clutches, and cordgrass communities in a way that mimics natural formations of dense wetland plant and mussel colonies in low–wave energy areas. These wetland and mussel communities protect the shoreline and provide ecological uplift to the surrounding area.

In 2011, the PDE, in association with Rutgers, published the "Practitioner's Guide: Shellfish-Based Living Shorelines for Salt Marsh Erosion Control and Environmental Enhancement in the Mid-Atlantic" (Whalen et al. 2011). The report provides information on the benefits and purpose of living shorelines projects, estimated construction costs, resources for external material, an overview of DELSI, and installation guidance for DELSI Tactic living shorelines projects. A step-by-step installation procedure is provided in addition to information on site selection, project timing, permit considerations, materials, proper installation, planting, and mussel application. The report breaks down the inventory of approaches into "bio-based" material, inert material, and hybrid options, and provides a set of physical criteria relevant for the selection of an appropriate living shorelines approach. A GIS-based methodology is also presented for determining the suitability of living shorelines based on the physical and biological characteristics of a shoreline.

As a complement to the Practitioner's Guide, a six-page brochure was developed for the general public. "Living Shorelines: Healthy Shores, Healthy Communities" (PDE 2012) was produced in

2012 and contains a clear explanation of what living shorelines are, and how they are beneficial to the environment and local communities. The brochure offers basic, general information on three examples of living shorelines techniques—shellfish and plant tactics for low-energy sites, marsh sills for medium-energy sites, and nearshore or offshore breakwaters for high-energy sites. The brochure also briefly covers the costs associated with different living shorelines techniques and provides a list of resources. Additional information is provided in *Final Report: Delaware Living Shoreline Possibilities* (PDE; Rutgers University Haskin Shellfish Research Lab 2012). The focus of the report is on identifying potential living shorelines solutions for locations along the Delaware Estuary. It contains information on a wide variety of living shorelines techniques and methods, including riparian vegetation management, sediment nourishment and dune restoration, tidal marsh enhancement and creation, bank grading, fiber logs, bivalve shellfish restoration with fiber logs, rock, prefabricated concrete materials, marsh toe revetment, marsh sill, marsh with groins, and nearshore or offshore breakwater system. The report contains a table that compares the benefits, design constraints, and costs of different living shorelines tactics for either low-energy or medium- to high-energy environments. It also provides a detailed analysis of potential project sites within the local area and offers conceptual plans, along with costs comparisons and recommended next steps.

More recently, the NJDEP has partnered with several NGOs, federal and state agencies, and academic institutions after Superstorm Sandy to provide technical assistance to communities as they plan, design, and construct living shorelines projects aimed at enhancing coastal resiliency. The project was funded by the Department of the Interior through a competitive grants program administered by the National Fish and Wildlife Foundation (NFWF). The technical team is charged with systemically identifying ecologically based resiliency strategies, developing them into successful, ready-to-use local actions, assessing large numbers of communities for the applicability of these strategies, and working with pilot communities to create quick wins and success models that will drive further green infrastructure adoption. The project team includes the NJDEP, Sustainable Jersey, National Wildlife Federation, PDE, and the NJ Sea Grant Consortium. Additional partners include nine communities that will complete specific ecologically based resiliency projects with project team assistance. The primary objectives of the project include the following:

- Developing vetted natural hazard mitigation Best Management Practices (BMPs) for New Jersey coastal municipalities that are ready to be implemented. These BMPs will make known strategies more accessible, and will break new ground on small-scale turnkey strategies that can be implemented by low-capacity local governments, such as ordinances, policies, outreach, and small-scale projects.
- Moving ecosystem-based strategies into the mainstream by educating relevant parties about the benefits and applicability of ecological systems and habitats as viable and effective options for communities to reduce their risk to natural hazards.
- Providing technical assistance on the feasibility and effectiveness of site-specific approaches to reduce natural hazards and build climate resilience.
- Establishing a continuum of technical support for municipalities at all stages of development, from assessment to design, to create a large and fast-moving pipeline of projects that provides communities with the technical expertise needed to determine the appropriate way to incorporate green infrastructure approaches into planning activities, including hazard mitigation, climate resiliency, and other efforts.

The deliverables for the project range from assistance in identifying potentially suitable living shorelines project sites, all the way up to the design, permitting, and construction of actual projects. This project began in 2015 and is funded through 2017. Several of the communities involved in the project are currently going through the permitting process and are scheduled to have completed projects by the spring of 2017. When the project is complete, New Jersey will have tripled its number of living shorelines demonstration sites and strengthened the internal infrastructure required to continue to build successful projects.

5.3.4 Monitoring

With several projects on the ground and the promise of many more in the near future, the topic of monitoring has taken on an increased sense of urgency in the Northeast. Monitoring protocols have been discussed in the states of Delaware, New York, and New Jersey, with the three states having made a varying degree of progress. At the same time, there are several initiatives underway nationally that are being used to frame these efforts. At the national level, these efforts are being led by NFWF and the US Army Corps of Engineers and are focused on developing standardized approaches to evaluate and compare federally funded projects. At the local level, efforts have been focused on developing locally relevant protocols that can be applied by a wide range of user groups. One of the realizations is that monitoring is often unfunded and that to be effective, simple cost-efficient approaches need to be developed that can provide representative and consistent results.

New Jersey's effort is predominantly based on methods developed by the PDE for monitoring of their living shorelines projects (CITE). The PDE approach is flexible; however, it emphasizes consistency and reliability. The flexibility is built into the approach through the recognition that each project has a set of specific objectives and that effective monitoring must be focused on whether a project meets those objectives. Flexibility also comes in through the realization that monitoring is likely to be carried out by a diverse set of stakeholders with vastly different budgets and technical capabilities. In order for long-term monitoring to truly be effective, there must be some means of integrating and comparing data collected by different approaches.

In the PDE approach, a metric is defined as a specific parameter used to assess project success and gauge attainment of project goals, while methods are the actual techniques that are used to assess the metrics. Metrics are segregated into two categories, core and supplemental, and can be either project type or goal based. Core metrics include a small number of common parameters that should be collected for all projects of a specific type, while supplemental metrics are those that apply only to certain project designs and in situ physical, biological, and chemical conditions, as well as local economic/social concerns. Project-type metrics are associated with each project type and serve to assess the general effectiveness of a restoration technique. Goal-based metrics are associated with project-specific goals and are evaluated to determine if the specific goals of a project are being met. Specific examples of the different types of metrics are provided in the original document.

Once the metrics are defined for a specific project, a monitoring plan can be developed to assess its effectiveness. PDE advocates utilizing a step-wise process in the development of a monitoring plan in which

1. Relevant metrics based on the project type and the specific goals of the project are identified.
2. User constraints including expertise, budget, timetable, permitting, scale, and intended analysis are considered.
3. Appropriate methods based on these constraints are selected.
4. A monitoring plan grounded in the evaluation of the effectiveness of the restoration/enhancement method and the ability of the project to meet its goals is developed.

The specific methods employed in the monitoring are selected from a set of acceptable techniques based on the user constraints. Two recently initiated projects have working groups focused on refining the information contained in the PDE report. The intent is to develop a monitoring approach that can be applied to an even more diverse set of projects than originally envisioned by PDE.

5.4 SUMMARY

Both New York and New Jersey have taken significant strides in advancing their states living shorelines programs over the last decade; however, the approaches have been quite different. The approach taken in New York has been to identify opportunities for influencing shoreline management decisions within the existing regulatory framework by providing the latest science-based information to the right people at the right time, which was highlighted here. The HRSSP has brought together engineers, ecologists, resource managers, property owners, regulators, and a variety of other stakeholders to help define the path forward for living shorelines. Together, this group has helped set a research agenda that has guided the development of several products aimed at improving the understanding of the function of living shorelines. Integral to this work has been the development of several communication tools and products intended to get the relevant information to the critical stakeholders with influence over shoreline development. Combined with the research agendas offered in the New York City Green Infrastructure Research Plan and the WEDG, the impact from these local initiatives will drive the implementation of living shorelines in New York State for years to come.

New Jersey's path has been different from New York's. Instead of trying to work within the confines of the existing regulatory system, New Jersey pursued a path that ultimately led to amendments to an existing general permit that incorporates living shorelines activities. The adoption of GP 24 represents a paradigm shift and has opened the door to a range of projects that would not have been possible under the old regulations. To ensure a level of consistency, New Jersey has funded the development of a set of engineering guidelines that are intended to provide a framework to guide project designers and regulators in the design and implementation of these unfamiliar projects. At the community level, the NJDEP is working with several groups including TNC and NJ Sea Grant to increase community awareness of the benefits of living shorelines projects. Perhaps most importantly, monitoring protocols are being developed such that the effectiveness of the projects can be evaluated, and future designs can be improved.

Living shorelines programs in New York and New Jersey are in a much better place today than they were just a decade ago. While several historic storms in the past decade, namely, Irene, Lee, and Sandy, have helped accelerate the pace at which living shorelines initiatives in each state have grown, the groundwork for the programs was laid well in advance of the storms. The projects discussed here are a testament to the commitment each state has made to advance the living shorelines approach. While much progress has been made, there is still much work to be done. The recently released New York City Green Infrastructure Research Plan highlights some of the local research needs. It could be said that the living shorelines programs in both states are entering the critical adolescent phase of their development, where the decisions made today will have a significant impact on the long-term outlook for living shorelines in the region.

ACKNOWLEDGMENT

We thank HRNERR staff including Betsy Blair, Lisa Graichen, and Dan Miller, as well as Steve Jacobus from NJDEP.

REFERENCES

Allen, G., T. Cook, E. Taft, J. Young, and D. Mosier. 2006. Hudson River Shoreline Restoration Alternatives Analysis. Prepared by Alden Research Laboratory, Inc. and ASA Analysis and Communications, Inc. for the Hudson River National Estuarine Research Reserve.

Blair, E., L. Graichen, and E. Hauser. 2015a. Project Overview. In association with and published by the Hudson River Sustainable Shorelines Project, Staatsburg, NY 12580, https://www.hrnerr.org/doc/?doc=260857495.

Blair, E., and E. Hauser. 2015b. Sustainable Shorelines along the Hudson River Estuary: Phase 2, Promoting Resilient Shorelines and Ecosystem Services in an Era of Rapid Climate Change. In association with the Hudson River Sustainable Shorelines Project, Staatsburg, NY 12580.

Bruno, M.S. 2002. *Field and Laboratory Investigation of High-Speed Ferry Wake Impacts in New York Harbor.* Center for Maritime Systems, Davidson Lab, Stevens Inst. of Technology, Hoboken, NJ 07030.

FEMA. 2015. The Guidelines for Implementing Executive Order 11988, Floodplain Management, and Executive Order 13690, Establishing a Federal Flood Risk Management Standard and a Process for Further Soliciting and Considering Stakeholder Input. https://www.fema.gov/media-library-data/1422653213069 -9af488f43e1cf4a0a76ae870b2dcede9/DRAFT-FFRMS-Implementating-Guidelines-1-29-2015r2.pdf.

Frizerra, D. 2011. Mitigating Shoreline Erosion along New Jersey's Sheltered Coast: Overcoming Regulatory Obstacles to Allow for Living Shorelines, Trenton, NJ: New Jersey Coastal Management Office.

Georgas, N., J.K. Miller, Y. Wang, and Y. Jian. 2015. Tidal Hudson River Ice Cover Climatology. Davidson Laboratory, Stevens Institute of Technology Prepared for the Hudson River Sustainable Shorelines Project. NYSDEC Hudson River National Estuarine Research Reserve.

Hauser, E. 2012. Terminology for the Hudson River Sustainable Shorelines Project. In association with and published by the Hudson River Sustainable Shorelines Project, Staatsburg, NY 12580.

HEP. 2012. The State of the Estuary 2012. New York–New Jersey Harbor and Estuary Program.

HRNERR. 2012. Publication and Resources. Hudson River National Estuarine Research Reserve New York State Department of Environmental Conservation, Hudson River Sustainable Shorelines Project, https:// www.hrnerr.org/hudson-river-sustainable-shorelines/publications-resources/.

LaPann-Johannessen, C., J.K. Miller, A. Rella, and E. Rodriquez. 2015. Hudson River Wake Study. In association with and published by the Hudson River Sustainable Shorelines Project, Staatsburg, NY 12580.

Miller, D., C. Bowser, and J. Eckerlin. 2006. Shoreline Classification in the Hudson River Estuary, unpublished, NYSDEC Hudson River National Estuarine Research Reserve. Geospatial data available at NYSGIS Clearinghouse: Shoreline Type, http://gis.ny.gov/gisdata/inventories/details.cfm?DSID=1136.

Miller, J.K., and N. Georgas. 2015. Hudson River Physical Forces Analysis: Data Sources and Methods. Davidson Laboratory, Stevens Institute of Technology Prepared for the Hudson River Sustainable Shorelines Project. NYSDEC Hudson River National Estuarine Research Reserve.

Miller, J., A. Rella, A. Williams, and E. Sproule. 2015. Living Shorelines Engineering Guidelines, Hoboken, NJ: Stevens Institute of Technology, Davidson Laboratory SIT-DL-14-9-2942.

New Jersey Department of Environmental Protection. 2015. N.J.A.C. 7:7 Coastal Zone Management Rules. [Online] Available at: http://www.nj.gov/dep/rules/rules/njac7_7.pdf [Accessed August 14, 2015].

New York State Sea Level Rise Task Force Report. 2010. http://www.dec.ny.gov/docs/administration_pdf/slrtf finalrep.pdf.

Northeast Regional Planning Commission. 2003. Lake Champlain Sea Grant Publication LCSG-04-03. *The Shoreline Stabilization Handbook for Lake Champlain and Other Inland Lakes.* 54 pp.

ny.gis.gov. Hudson River Estuary Data and Maps. New York State GIS Clearinghouse. GIS Data Set Details, http://gis.ny.gov/gisdata/inventories/details.cfm?DSID=1136.

NY/NJ Baykeeper. 2005. *Oyster Gardening Program Manual. Keyport*, New Jersey: Raritan Baykeeper, Inc.

NY/NJ Baykeeper. 2006. Baykeepr Volunteer Oyster Restoration Program Achieves Historic Milestone. Estuarian.

Partnership for the Delaware Estuary. 2012. Living Shorelines: Healthy Shores, Healthy Communities, Wilmington, DE: Partnership for the Delaware Estuary, Inc.

Partnership for the Delaware Estuary; Rutgers University Haskin Shellfish Research Lab. 2012. Delaware Living Shorelines Possibilities, s.l.: s.n.

Rella, A., and J. Miller.2012a. Engineered Approaches for Limiting Erosion along Sheltered Shorelines. In association with and published by Stevens Institute the Hudson River Sustainable Shorelines Project, Staatsburg, NY 12580, https://www.hrnerr.org/doc/?doc=240189605.

Rella, A., and J. Miller. 2012b. A Comparative Cost Analysis of Ten Shore Protection Approaches at Three Sites Under Two Sea Level Rise Scenarios. In association with and published by the Hudson River Sustainable Shorelines Project, Staatsburg, NY 12580, https://www.hrnerr.org/doc/?doc=240186100.

Rella, A., and J.K. Miller. 2015. Forensic Analysis Methodology Report. In association with and published by the Hudson River Sustainable Shorelines Project, Staatsburg, NY 12580, https://www.hrnerr.org/doc /?doc=240203442.

Strayer, D.L. 2011. Managing Shore Zones for Ecological Benefits. Published by the Hudson River Sustainable Shorelines Project, Staatsburg, NY 12580, https://www.hrnerr.org/doc/?doc=273743856.

Urban Waterfront Adaptive Strategies. 2013 by NYC Department of City Planning, Urban Waterfront Adaptive Strategies, http://www.nyc.gov/html/dcp/html/sustainable_communities/sustain_com7.shtml.

Van Luven, D. 2011. Economic Tradeoffs between Shoreline Treatments: Phase I—Assessing Approaches, https://www.hrnerr.org/doc/?doc=240189637.

WEDG. 2014. Shape Your Waterfront, How to Promote Access, Resiliency, and Ecology at the Water's Edge, an Introduction of the Waterfront Edge Design Guidelines, Produced by the Waterfront Alliance, http://waterfrontalliance.org/wp-content/uploads/delightful downloads/2015/06/WEDGprogrambrochure.pdf.

Whalen, L. et al. 2011. Practitioner's Guide: Shellfish-Based Living Shorelines for Salt Marsh Erosion Control and Environmental Enhancement in the Mid-Atlantic, Wilmington, DE: Partnership for the Delaware Estuary, Report # 11-04.

Zhao, H., H. Roberts, J. Ludy, A. Rella, J. Miller, P. Orton, G. Schuler, L. Alleman, A. Peck, R. Shirer, J. Ong, M. Larson, K. Mathews, K., Orff, G. Wirth, and L. Elachi. 2014. Coastal Green Infrastructure Research Plan for New York City. Prepared for the Hudson River Estuary Program, New York State Department of Environmental Conservation.

CHAPTER 6

Overcoming Barriers to Living Shoreline Use and Success

Lessons from Southeastern Virginia's Coastal Plain

Kevin R. Du Bois

CONTENTS

6.1 INTRODUCTION

The use of "living shoreline" erosion control techniques is not necessarily a new idea. Circa 1991, the Virginia Department of Conservation and Recreation's (DCR) Shoreline Erosion Advisory Service published an educational booklet entitled "Shoreline Erosion Problems? Think Green..." (Figure 6.1). The booklet described ways to stabilize shorelines using natural vegetation, stating: "If you have low to moderate shoreline recession problems, establishing marsh vegetation can provide long-term shoreline stabilization at a fraction of the cost of conventional structures such as bulkheads and rock revetments." "A significant benefit to this 'green' approach is the enhancement of Chesapeake Bay water quality and habitat availability." In coastal Virginia, the present-day concept of the use and applicability of living shorelines is not new, but, rather, reinforces these earlier advisories.

However, for some reason, vegetative stabilization techniques fell out of favor or were considered by the public to be inferior to shoreline hardening techniques. It is unclear how many Virginia wetland regulators were familiar with the DCR guidance, but whether unfamiliarity with the guidance, lack of personal experience designing and constructing vegetative stabilization projects, or lack of a champion promoting the technique, these factors conspired to make the publication's effect on vegetative shoreline stabilization minimal.

The documentation of net losses in wetlands from permitted activity in the early 2000s and the need for minimum 1:1 compensation to meet no net loss goals led Virginia to refocus on strategies that would disincentivize unnecessary shoreline hardening and incentivize the protection of on-site wetlands.

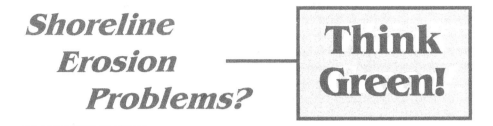

Shoreline Erosion Problems? — **Think Green!**

Control shoreline erosion help restore the Chesapeake Bay and save money all at the same time.

If you have low to moderate shoreline recession problems, establishing marsh vegetation can provide long-term shoreline stabilization at a fraction of the cost of conventional structures such as bulkheads and rock revetments. Additionally, no permits are required in many cases. A significant benefit to this "green" approach is the enhancement of Chesapeake Bay water quality and habitat availability.

Figure 6.1 1991 publication promoting vegetative shoreline stabilization in Virginia.

6.1.1 First Living Shoreline Summit (NC-MD), December 6–7, 2006

In 2005, a grant from the Keith Campbell Foundation for the Environment funded the creation of the Living Shoreline Stewardship Initiative with their ultimate goal to have: "Maryland and Virginia shorefront property owners routinely consider and frequently choose living shoreline alternatives as their preferred shoreline management treatment." Interest in the topic and science of living shorelines grew, and that culminated in the first regional living shoreline summit held in Williamsburg, Virginia, in 2006.

For those still relatively new to the concept of living shorelines, some of the interesting conceptual highlights from the summit included the following:

- Living shorelines were mentioned as part of a sea level rise (SLR) strategy.
- The need for contractor training and certification.
- A discussion of the ecosystem services provided by beaches, bluffs, and intertidal wetlands.
- Donna Bilkovic of the Virginia Institute of Marine Science (VIMS) spoke about the ecological response to percent nearshore development.
- Living shoreline options were defined to include beach nourishment and dune restoration.
- A discussion of the report "Why Landowners Restore Wetlands," which highlighted the role of social marketing in ecosystem restoration and the attempt to fund a study on the landowner perceptions of living shoreline barriers to implementation.
- The concept that living along the shoreline conveyed both rights and responsibilities for stewardship.
- The discussion of design considerations: slope, substrate, shade, salinity, plant material, fertilizer, plant timing, spacing, invasives control, maintenance, and site suitability (fetch, bathymetry, marsh presence, beach presence, bank stability, and presence of tree canopy).
- VIMS reconsideration of their shoreline management guidelines—especially the subaqueous guidelines regarding the placement of fill.
- Gene Slear of Environmental Concerns Inc. related wave attenuation to marsh width.
- A discussion of the role of the bank slope in shoreline stability.
- General Permits (GPs) were mentioned as a potential living shoreline incentive in 2006.

6.1.2 The Commonwealth of Virginia Legislates a Preference for Living Shorelines to Address Eroding Shorelines

Beginning in 2010, the Virginia General Assembly passed a series of resolutions and changes to existing Acts that ultimately led to the adoption of a State preference for living shoreline erosion control strategies and the development of a Living Shoreline General Permit (LSGP). However, because living shorelines are not required by regulation, the lack of a mandate has become a significant barrier to living shoreline implementation and has contributed to their variable and inconsistent use throughout the state.

In April 2011, the Virginia General Assembly amended and reenacted Chapters 15 and 28 of the Code of Virginia to require that VIMS Coastal Resource Management Guidance be incorporated into a locality's next scheduled Comprehensive Plan update. As outlined in Senate Bill 964, the coastal resource management guidance for local governments was to provide preferred options for shoreline management taking into account, among other things, forecasting of the condition of the Commonwealth's shoreline with respect to SLR.

6.1.3 Living Shorelines and Sea Level Rise (SLR)

The Hampton Roads region of southeastern Virginia has received national attention regarding SLR and faces one of the highest rates of SLR on the East Coast of the United States. SLR will affect communities with overland flooding from high tide events and "nuisance flooding" from the combined effect of higher high tides on the capacity of stormwater systems and more frequent and intense rainfall events.

While municipalities may consider some land to be too valuable to abandon, other lands may be considered too dangerous or economically unsustainable to maintain. Every government entity has limited resources, so each will have to balance their needs and investment in areas that meet their social and economic needs, and position themselves to be resilient and be able to "bounce forward" following stressors.

As an example of the potential benefit of strategic retreat, decisions to remove unsustainable development opens up opportunities to implement living shorelines and restore the functions and values of wetlands and floodplains. In this context, living shorelines can help municipalities take advantage of multiple benefits, increasing the following:

- Stormwater management potential (retention, runoff reduction, controlled discharge, etc.)
- Pollution abatement and water quality improvement
- Wildlife habitat enhancement and other ecosystem services
- Erosion control
- Green infrastructure connectivity
- The potential for flood insurance credits
- Waterfront access and passive or active waterfront recreation opportunities for entire communities that can generate economic development, promote social connections, and improve quality of life and overall societal resilience.

In a study of the Fort Norfolk area, in Norfolk, Virginia, the Urban Land Institute said: "Open space paired with temporary and low-intensity uses at the water's edge can fill the community's need for interaction with the waterfront while avoiding the cost and risk of development" (http://uli .org/wp-content/uploads/ULI Documents/Norfolk_PanelReport_Fweb.pdf).

In recent collaborations between the City of Norfolk and its international colleagues, dubbed the "Dutch dialogues" (http://www.centerforsealevelrise.org/recent-news/story-roundup-the-dutch -dialogues-virginia/), planners from the Netherlands suggested retreating from filled river beds and creating living shorelines to restore floodplain services, adapt to SLR, and enhance resilience.

In this way, careful planning can ensure that neighborhoods are less vulnerable, more desirable, and better connected for all residents (Figure 6.2).

Figure 6.2   Intertidal wetland restoration protected by a rock sill at the Hermitage Museum and Gardens in Norfolk, Virginia. (Photo courtesy of the City of Norfolk, Virginia.)

6.2 CHALLENGES TO IMPLEMENTATION

There are a number of challenges to living shoreline implementation. VIMS and Restore America's Estuaries have both reported on these extensively. Two of the most recognized challenges include the development of an effective education and outreach strategy and a focus on comparative cost and are described below.

6.2.1 Education and Outreach

In order to breach the barriers to living shoreline implementation, a broad constituency of advocates and practitioners must be built. Experience in Virginia has shown that, in order to break the institutional inertia for shoreline hardening, all levels of the shoreline management regulatory process must be engaged and work in concert to provide the public with consistent and predictable permit decision outcomes.

In Virginia, there have been many efforts at the state and local level to prioritize and incentivize the use of living shorelines. Based on guidance and authority in Senate Bill 964, Fairfax County, Virginia's Wetlands Board adopted a policy to require the use of living shorelines in all cases unless it was not technically feasible. While lacking a similar policy, Norfolk, Virginia's Wetland Board approved living shorelines for 71% of the projects where they were technically feasible in 2013. The Commonwealth of Virginia established a regulatory preference for the use of living shorelines and developed a General Permit process to streamline and expedite the regulatory process. The Chesapeake Bay Program has established a pollution abatement credit for the use of living shorelines in managing shoreline erosion as part of the federally mandated Total Maximum Daily Load (TMDL) nutrient and sediment reduction requirements.

While some efforts have been made to incentivize living shorelines in the federal regulatory arena (e.g., living shoreline projects meet US Army Corps of Engineers (USACE) RP-17 permit exemptions), there has been no federal mechanism to require living shorelines when appropriate. Moreover, some of the provisions for federal regional permits actually incentivize, streamline, and allow for unmitigated wetland losses for shoreline hardening structures without mandatory adoption of lesser-impact alternatives (e.g., hardened erosion control structures with wetland losses less than 1 ft^2 per linear foot).

In Virginia, there appears to be disagreement regarding if and how federal regulations (Clean Water Act 404(b)(1)), TMDL, and Chesapeake Bay, Climate Change, and Floodplain Presidential Executive Orders, should favor and require living shorelines and if and how federal agencies should be disincentivizing hardened shorelines. In the Norfolk District (Virginia), the federal Army Corps of Engineers frequently approve shoreline hardening projects as local governments seek to implement the state preference for living shorelines. This leads to landowner confusion about overarching government wetland policy and thwarts local and state efforts to require living shorelines to protect and restore coastal ecosystems. Living shoreline education and training are needed within the federal regulatory system so that the benefits and desirability of living shoreline erosion control techniques are better understood, and needed changes in policy and practice can be made in Virginia.

Likewise, an understanding of the design and construction criteria necessary for successful LS project implementation can lead to incentivized LS permitting and disincentivized shoreline hardening. However, even if local, state, and federal wetland regulatory personnel were united in their resolve to maximize the permitting of living shoreline erosion control practices, implementation would still be hampered by the dearth of qualified project designers and contractors.

Uncertainty in living shoreline construction estimating leads some inexperienced contractors to build large contingency fees into bids and inflates perception of cost. More, better-trained, and experienced contractors should result in competition and reduced industry-wide cost. VIMS has provided living shoreline training for marine contractors. However, without licensing or certification requirements for contractors, or a legal requirement for the installation of living shorelines when feasible, contractors may perceive that there is no compelling incentive to pursue continuing education and learn about new stabilization methods that preserve or enhance ecosystem services. Since contractor advice appears to have a significant effect on the erosion control strategies homeowners choose, lack of continuing education on vegetative erosion control methods in the marine contracting trade is a barrier to living shoreline implementation in Virginia.

In addition to education and training aimed at traditional targets like project designers, regulatory decision makers, and marine contractors, we must build understanding and create advocates for living shorelines among diverse but interconnected groups that would benefit from living shoreline implementation, including the following:

- Those who could benefit from the business of living shorelines including landscape architects (LS design), landscape services companies (LS maintenance/species diversification/aesthetics), environmental consultants, stormwater professionals, living shoreline product suppliers (wetland plants, slow-release fertilizers, coir logs, erosion control products, etc.), real estate agents and coastal landowners benefitting from higher real estate assessments, and so on
- Those who benefit from the ecological services living shorelines provide including fishery managers, recreational anglers, commercial watermen, local governments (improved water quality), bird watchers, and so on
- State and federal wildlife agencies pursuing ecosystem-wide approaches to species protection
- Colleges, universities, and K-12 schools that benefit from research and outdoor educational opportunities

In their national report "Living Shorelines: From Barriers to Opportunities," Restore America's Estuaries (2015) reinforces the need to build a "culture of practice" to meet the pressing education and outreach needs and overcome implementation barriers.

In 2014, VIMS hosted a wetland workshop focused on living shorelines and asked participants to comment on challenges and solutions to living shoreline implementation. Many complained that shoreline hardening was preferred due to psychological inertia—neighbors had a bulkhead or riprap so they needed it too!

Throughout hundreds of documented responses, a common theme among homeowners was the overwhelming need for education. Respondents requested that landowners and other target audiences be provided with information on the following:

- Ways to reduce costs
- Hardened versus living shoreline cost comparisons (especially amortized over time)
- Cost incentives
- Information on living shoreline level of protection and durability, especially in the face of extreme coastal storm events (hurricanes, nor'easters, etc.)
- The difference between erosion and flooding
- Information on maintenance requirements (including invasive species control) and expectations of wetland change over time with wetland maturation
- The benefits of living shorelines: wildlife habitat (people respond positively to birds), water quality (nutrients, algal blooms), financial, flood protection, support of recreational and commercial fisheries, and others
- The environmental interdependency of wetlands and their adjacent riparian buffers (symbiosis), barriers to landward migration, and the role of proper riparian bank slope in storm resilience
- A list of qualified or certified contractors
- A list of demonstration sites available for review
- The effects of projected SLR on wetlands (especially those with structures that would limit or slow landward migration)

Interestingly, for many involved in wetland science, wetland regulation, ecological restoration, or more specifically in living shoreline erosion control techniques, much of this information has already been documented and is readily available. That would suggest that living shoreline information is not really the problem; the biggest barrier to building a constituency for living shoreline implementation is lack of an effective outreach strategy—delivering existing information in a persuasive manner that changes behavior and results in the implementation of living shoreline projects.

However before an effective outreach strategy can be enacted, the community of practice first needs to understand what motivates land-owning decision makers. Accordingly, a science-based social marketing study is needed as a precursor to an effective living shoreline outreach strategy.

6.2.2 Focus on Cost Instead of Value

Landowners and regulatory decision makers frequently cite project cost as a major consideration in the selection of any shoreline erosion control solution. Unfortunately, hardened versus living shoreline cost comparisons are virtually impossible. Site-specific conditions (slope, elevation, etc.) along with case-by-case evaluations of minimum marsh width to achieve project goals can cause costs to vary widely as volumes of construction materials (sand and rock), numbers of plants, and potential maintenance requirements increase. One-time construction costs, particularly for wooden bulkheads, do not take into account the durability of construction materials (wood shething and walers, metal tie rods, nuts, bolts, etc.) or the long-term costs of periodic structure replacement. Accordingly, some in the living shoreline community of practice are recommending that the conversation on preferred shoreline protection shift from cost to value. Valuation of shoreline erosion control strategies would take into consideration the maintenance of valuable ecosystem services that hardened shorelines often lack or even further degrade.

6.3 OVERCOMING BARRIERS TO LIVING SHORELINE IMPLEMENTATION (NO PRIORITY ORDER ASSUMED)

6.3.1 Partnerships

The most important factors in successful local implementation of living shoreline projects have been the development of partnerships and capacity building. Expertise in living shoreline science, design, and construction is isolated, and it is difficult to leverage that expertise on the scale needed to meet demand or opportunity. In fostering a community of practice, the ingredients discussed in Sections 6.3.1.1 through 6.3.1.5 are necessary for successful partnerships.

6.3.1.1 Champions

For most endeavors to succeed, people with drive, passion, persistence, and vision are needed—those who will doggedly pursue a goal to its fruition—champions. Champions can be found in government, in the private business sector, in academia, within nongovernmental organizations (NGOs), and in the citizenry. In southeastern Virginia, and in the Hampton Roads region in particular, the Lafayette Wetlands Partnership, the Elizabeth River Project, and Lynnhaven River Now have been successful in cultivating a volunteer workforce that is knowledgeable, experienced, and willing to participate in wetland restoration projects to restore and enhance their respective watersheds. Champions are rare, but their identification and encouragement are critical to project success. When champions from across various disciplines collaborate and share expertise, barriers to living shoreline implementation are minimized, and the potential for new projects is unleashed (Figure 6.3).

Figure 6.3 John Stewart, a local champion, leads the Lafayette Wetland Partnership and fosters a diversity of partnerships that leads to annual successful projects in wetland restoration and enhancement. (Photo courtesy of Norfolk, Virginia.)

6.3.1.2 Assets

To implement living shoreline projects, a variety of assets need to be available and ready for mobilization. Assets can include, but are not limited to the following:

- Knowledgeable people (regulatory, designers, contractors, volunteers, etc.)
- Specialized equipment (dibble bars, drills, bulb-planting augurs, goose fencing, smaller, nimble excavation equipment, geotextile fabrics, etc.)
- Suitable sites (public sites for demonstration, willing landowners, etc.)
- Funding sources like government grants, state and municipal funds, repurposed violation fees, NGO funds, donations, private funds, and so on

6.3.1.3 Group Cohesion

One of the things that supercharges the effectiveness of a partnership is group cohesion. Naturally, this comes from a common sense of purpose, but it is strengthened through the age-old concept of shared work. Working together, consistently, and time after time builds bonds and deepens commitment. An important part of group cohesion extends beyond the work. Interacting socially and celebrating project successes together reinforces commitment to the work and to each other.

6.3.1.4 Ownership

Recognizing contributions to living shoreline projects instills a sense of ownership that is important for long-term stewardship. Whether it is goose or muskrat damage, invasive plants, erosion, plant mortality, or some other challenge, all projects need some degree of monitoring and maintenance. Partners with a strong sense of ownership will be more likely to be engaged to make sure projects succeed over time. Local residents, who by their nature are close to and enjoy the benefits of the project, are often the best project stewards. Every attempt should be made to have adjoining neighbors be project partners.

6.3.1.5 Success Breeds Success

Completing and maintaining successful living shoreline projects is rewarding and increases the likelihood that partners will want to participate in additional projects. Successful projects also have a way of drawing skeptics, critics, or potential partners sitting on the fence to join the partnership.

An understanding of the value of these elements was gained by working over time with the Lafayette Wetlands Partnership (https://www.facebook.com/LafayetteWetlandsPartnership?_rdr=p). They have mastered the art of developing strong and cohesive partnerships resulting in successful fundraising and the construction of numerous living shoreline projects on both public and private land. Their annual gains in created, restored, and enhanced wetlands substantially exceed what could possibly be achieved with the knowledge and effort of individual or unconnected project champions.

6.3.2 Identifying Incentives

In 2014, VIMS hosted a wetland workshop focused on living shorelines and asked participants to comment on challenges and solutions to living shoreline implementation. Throughout hundreds of documented responses, a common theme was the overwhelming need for homeowner education, and many asked for incentives to reduce costs. Some possibilities include those discussed in Sections 6.3.2.1 through 6.3.2.6.

6.3.2.1 TMDL Nutrient Reduction Cost Sharing Strategies

In a 2015 report entitled "Recommendations of the Expert Panel to Define Removal Rates for Shoreline Management Projects," a multistate panel of experts presented examples, the rationale, and methods for calculating the nutrient reduction ability of tidal wetlands used to stabilize shorelines. Accordingly, implementation of new living shoreline projects can help those in the Chesapeake Bay watershed meet their federally mandated TMDL nutrient reduction goals.

However, it has been suggested that there is not enough public land to implement all the nutrient reduction strategies needed to meet municipal TMDL requirements. If living shoreline implementation on private land can help meet publicly mandated nutrient reduction goals, then there is the opportunity for public/private cost sharing agreements.

6.3.2.2 Community Rating System

Communities in flood-prone areas that participate in the National Flood Insurance Program can help their citizens take advantage of lower flood insurance premiums by participating in the Community Rating System (CRS). Under the CRS, local governments that take action to lower their flood risk and make their communities more resistant to flood damage are able to get insurance premium discounts for their citizens.

If living shorelines can be incorporated into a green infrastructure framework of protected land, then perhaps they could qualify for credits. As reported by Wetlands Watch (2015) in their publication "Flood Protection Pay-Offs: A Local Government Guide to the Community Rating System," the receipt of CRS credit for living shorelines is essentially untested. Living shorelines do not fit well into any activity, but under certain conditions, they might be eligible for credit in two categories: Activity 422g, Natural Shoreline Protection, and Activity 532, Flood Protection. For Activity 422g, if a local ordinance or adopted policy requires the use of living shorelines in place of hard structures for shoreline management, credit may be available. For Activity 532, while the primary purpose of a living shoreline is erosion protection, a project on an individual property may be eligible for credit if it also offers flood damage protection to at least the 25-year-flood level.

6.3.2.3 Virginia Living Shoreline General Permit

With an effective starting date of September 1, 2015, Virginia LSGP Type I Projects will lower costs associated with the regulatory permitting of living shoreline projects. Plans can be prepared by homeowners or other competent individuals without the need for engineer seals or certifications, and permit applications fees are no longer required. The LSGP process is also designed to be expedited, saving permit applicants and permit reviewers valuable time.

6.3.2.4 Tax-Based Incentives

The Elizabeth River Project (www.elizabethriver.org) proposed creating a "wetland front" versus waterfront property tax system to encourage softened shorelines. In their proposal, lowering the tax burden on portions of a property defined as tidal wetlands (or alternatively as "open space" in a use-based tax system) could encourage landowners to implement or maintain a "soft" shoreline and manage their wetlands for sustainability.

Perhaps in response, during the 2016 legislative session, the Virginia General Assembly passed House Bill HB 526, which "Provides that any living shoreline project approved by the Virginia Marine Resources Commission (VMRC) or the applicable local wetlands board and not prohibited by local ordinance shall qualify for full exemption from local property taxes."

6.3.2.5 Regulatory Incentives

In 2012, Lynnhaven River Now (http://www.lynnhavenrivernow.org/) recommended that the VMRC be prepared to overturn local Wetlands Boards' decisions when shoreline hardening was approved despite the feasibility of installing a living shoreline to reduce or eliminate wetland impacts. State oversight authority over local Wetlands Board actions could be used to reduce the barrier of the State's preference (as opposed to requirement) for living shoreline use.

In some Virginia localities, two sets of regulations have the potential to regulate living shoreline projects—the Virginia Tidal Wetlands Act and the Chesapeake Bay Preservation Act. With a growing recognition of the value and function of holistic coastal ecosystems, some have suggested that the laws that govern these systems be combined into a single comprehensive protection program or at least not be redundant. This would enhance regulatory predictability and customer service.

6.3.2.6 Financing Incentives

Since 1987, the Virginia Clean Water Revolving Loan Fund (VCWRLF) has provided low interest loans for water quality improvement projects. In December of 2015, the VCWRLF published proposed enabling legislation for its Draft Living Shorelines Loan Program Guidelines. "The purpose of the Virginia Living Shorelines Loan Program is to provide a long-term source of low interest financing for the purpose of establishing living shorelines to protect or improve water quality and prevent the pollution of state waters."

As stated in the guidelines, funds would be taken from the VCWRLF and distributed to qualifying governments or individual landowners to cover "reasonable" and "necessary" planning, design, and construction costs associated with building a living shoreline project. Loan interest rates could be as low as zero where financial hardship exists.

Public comments on the guidelines were sought and the State Water Control Board was expected to finalize the guidelines by the summer of 2016.

6.4 THREATS TO SUCCESSFUL LIVING SHORELINE IMPLEMENTATION (NO PRIORITY ORDER ASSUMED)

There are a number of threats to living shoreline implementation and adoption as a credible solution to shoreline erosion challenges. Threats to living shorelines are similar to those for wetland restoration and preservation in general and include those discussed in Sections 6.4.1 through 6.4.5.

6.4.1 Invasive/Nuisance Species Control

In coastal Virginia, the invasive plant *Phragmites australis* can invade or persist within newly created wetlands without careful and ongoing monitoring. Phragmites can invade from seed sources or vegetative propagation from pieces of live roots. Nutria (*Myocastor coypus*) are large (12–20 lb) rodents that are not native to Virginia but, according to the Virginia Department of Game and Inland Fisheries, are spreading throughout the state. Nutria will burrow in wetlands and eat wetland plants and can cause significant damage to wetland resources. The muskrat (*Ondatra zibethicus*) is a native mammal and, although smaller (2–4 lb), is sometimes confused with nutria. While it also eats aquatic vegetation, wetland damage is normally associated with burrowing activity, which can destabilize wetland soils and subject them to accelerated erosion. Anecdotally, the Lafayette Wetlands Partnership found that an unintended consequence of wetland restoration efforts could be a resurgence in urban populations of muskrats and the potential for damage to living shoreline projects designed specifically for shoreline stabilization (and perhaps to meet other multiple benefits as described previously). Decisions on how to manage native muskrat populations causing a nuisance in living shorelines has presented a conundrum for environmental NGOs. Although considered migratory, Canada geese (*Branta canadensis*) live in the Chesapeake Bay region year-round. They will pull wetland plants out of the ground to get at the nutritious roots, especially over the winter when other sources of food are scarce (Figure 6.4). This process leaves holes in the wetland substrate, removes stabilizing plant roots and subjects the wetlands to accelerated erosion.

Figure 6.4   Goose fence protects a newly planted intertidal wetland from Canada geese herbivory. (Photo courtesy of Norfolk, Virginia.)

6.4.2 The Effect of SLR

While the VIMS definition of living shoreline "does not include structures that sever the natural processes and connections between uplands and aquatic areas," the State of Virginia definition does not include this distinction. As a result, wetlands used to stabilize eroding shorelines seaward of a bulkhead, revetment, upland retaining wall or similar structure can be considered a "living shoreline" in Virginia. Since many factors affect landward wetland migration (slope, rate of SLR, sediment supply, structure heights [low vs. high], landowner tolerance for landward migration of wetlands, etc.), it remains to be seen if living shorelines without backstops will be able to resist drowning over time. However, "living shorelines" installed seaward of hardened structures elevated to keep pace with SLR are not likely to last as long as a result of structure-induced erosion combined with the drowning effects of SLR and have a guaranteed finite life span within the landscape.

6.4.3 Conflicts with Regulatory Program Implementation

In Virginia, State tidal wetlands regulation falls under the control of the VMRC. For willing localities, VMRC can delegate some authority to regulate projects in the intertidal and supratidal regimes, but they retain oversight authority over local decisions to maintain consistency of intent and application of State tidal wetland laws.

One of the challenges in implementing living shorelines through the local regulatory process is that, in Virginia, they are the preferred method of shoreline stabilization if feasible, but not the required method. This approach works well if an applicant voluntarily requests a living shoreline.

However, if living shoreline erosion control is a feasible method, but a landowner proposes a shoreline hardening project instead, whether that hardening project will be rejected in favor of a living shoreline depends on the knowledge, experience, and understanding of the local Wetland Board and their role in implementing State law.

The VIMS (2012) Regulatory Fidelity Study looked at public permit records and decisions made by local Wetlands Boards through their regulatory process. It showed that local Wetlands Boards in Virginia rarely denied permit applications for shoreline hardening projects even when the evidence suggested a living shoreline alternative was feasible or more appropriate.

Serving as a local role model, Fairfax County, Virginia, developed a policy based on the guidance and authority granted in Senate Bill 964 to adopt a local permitting preference for living shorelines. Importantly, their program requires that an applicant demonstrate that a living shoreline cannot accomplish the proposed erosion control goals before they can propose some other strategy involving shoreline hardening. Absent a compelling argument to the contrary, the living shoreline is assumed to be the appropriate choice.

At the State level, Senate Bill 964 called for updating the integrated shoreline management guidance that Wetlands Boards are obligated to follow in their decision making. Updating the State guidance and educating local Wetlands Boards about the benefits and intended State preference for living shorelines could address some of the education and outreach barriers that exist. Updated guidance and a robust education and outreach campaign could also reduce the State's reluctance to exercise its oversight authority and overturn local Wetlands Board decisions that allow shoreline hardening proposals when living shoreline projects are feasible. The integrated guidance was also supposed to recommend procedures to achieve efficiency and effectiveness by the various regulatory entities exercising authority over a shoreline management project.

The USACE, Norfolk District, allows landowners willing to propose living shoreline projects to take advantage of an expedited regional permit process #19 (http://www.nao.usace.army.mil /Portals/31/docs/regulatory/RPSPdocs/13-RP-19.pdf). Similar to VMRC, the USACE has been reluctant to dissuade shoreline hardening projects and does not deny shoreline hardening projects when living shoreline alternatives are feasible. This is especially confounding when USACE approvals for shoreline hardening are in direct conflict with state preferences and local or state permit approvals for living shorelines. It is unclear whether the USACE regulatory approach for shoreline stabilization in the Chesapeake Bay region is in conflict with Presidential Executive Orders 13508 (Chesapeake Bay Protection and Restoration) and 11988 (Flood Management).

Without a coordinated local, state, and federal approach to living shoreline stabilization that both incentivizes living shorelines and disincentivizes (denies) unnecessary shoreline hardening projects, wetland protection and restoration goals may not be able to exceed meaningful thresholds for successful program implementation. If living shoreline implementation is to be the norm, a coordinated, consistent, and predictable regulatory approach would be helpful.

6.4.4 Unintended Consequences and Ecological Trade-Offs

The implementation of living shoreline erosion control strategies is not without trade-offs. Projects may result in the loss of nonvegetated wetlands to increase the width of vegetated wetlands. Gains in wave energy abatement, marsh longevity, SLR resilience, and nutrient treatment may be offset by losses in habitat for nonvegetated wetland infauna, food resources for wading shorebirds and fishes, and other benefits. If sufficient planted or natural wetlands are incorporated into a project, the VMRC can consider a living shoreline erosion control project to be "self-mitigating" even though conversions of one type of wetland to another may result in changes to wetland type or function.

Erosion is a natural and, in some cases, a necessarily vital process to preserve marine habitats. As debated during the Chesapeake Bay Partnership's Expert Panel to Define Removal Rates for Shoreline Management Projects, some in the scientific community are concerned about incentivizing living shorelines projects (e.g., TMDL credits) with the potential to stop or diminish bank erosion and the delivery of sand needed for the establishment and persistence of submerged aquatic vegetation substrates.

However, while scientists and policy makers debate the value and wisdom of these trade-offs, living shoreline projects continue to be approved as a better alternative to shoreline hardening.

In their report, "Rethinking Living Shorelines" (http://www.wcu.edu/WebFiles/PDFs/PSDS _Living_Shorelines_White_Paper.pdf) (Pilkey et al. 2012), the authors warn of the inevitable tendency for the original goals of habitat enhancement to be abused by overaggressive permit applicants, engineers, and property owners. They call for "a renewed scientific effort to evaluate the cumulative impacts of all existing structures (bulkheads and Living Shorelines) on natural and physical processes and ecosystems."

6.4.5 Lack of Competent Living Shoreline Designers, Regulators, and Contractors

Each step, from site selection to project design, permitting, construction, and monitoring and maintenance, is fraught with potential peril. Mistakes made anywhere from cradle to grave can result in a failed living shoreline project. Failed projects add to a negative feedback loop of distrust and lack of confidence fulfilling suspicions that the science of living shorelines is shaky, misleading, and unreliable. As a result, knowledgeable designers, regulators, and contractors with direct hands-on experience in the construction and monitoring of living shoreline projects are absolutely critical. Grant funding for living shoreline design and construction is starting to include criteria and deliverables for a variety of capacity-building project components, and this is a positive step. Regardless, all projects should look to build the knowledge and experience and reach of living shoreline practitioners.

6.5 DESIGN PRACTICES FOR LIVING SHORELINE IMPLEMENTATION

Over the years, a number of good design practices have been gleaned from the growing number of living shoreline installations in southeast Virginia. These ideas have been shared at community of practice meetings, forums, and summits and will hopefully result in more resilient living shoreline construction. Two design practices are highlighted below.

6.5.1 SLR Modifications

After the 2006 Living Shoreline Summit in Virginia, projects began to be designed for added resilience to the effects of SLR. Design guidance at the time was to plant vegetated intertidal wetlands from mid-tide to mean high water. In Norfolk, Virginia, the average tide range is approximately 0.78 m (~2.5 ft). To add SLR resilience, instead of planting between mid-tide and mean high water, projects were designed and built within the top 25% of the tide range—in the upper 0.39 m. This was done under the assumption that, without any enhancements or modifications, vegetated wetlands built this way would take longer to drown or be forced to migrate upslope. Another advantage of this design technique is that, with the wetland toe at a higher elevation, a gentler slope could be achieved over the width of the available footprint, making the resulting landform more stable (see also Section 6.4.2).

6.5.2 Marsh Diversity and the Use of Wetland Flowers

One simple improvement in the design and installation of living shorelines is to increase plant diversity. Because living shorelines are, first and foremost, a method for shoreline erosion control, designers often use a familiar and narrow pallet for plant selection. In the Mid-Atlantic, *Spartina alterniflora* is the plant of choice for vegetated intertidal wetlands and *Spartina patens* is commonly specified for use in high marsh situations. If *Distichlis spicata* is present in the preexisting plant community, it too will sometimes be specified for use in the constructed high marsh. Once a wetland is planted, natural colonization will automatically begin to increase plant diversity. For this reason, earlier fixations on planting wetland shrubs has given way to stabilizing high marshes with grasses with sparse shrub cover and letting nature supply the rapidly colonizing wetland shrubs for free.

There is at least one drawback to this practice. Natural colonization will only supply seeds from plants that already exist within the distribution area. If there are only five plant species that can deliver seeds to a site, then, generally, one only has the potential for five plant species at the site without human intervention.

Perhaps especially in urban areas, like Norfolk, Virginia, where there has been significant historical loss of wetlands through filling and dredging activity, much of the plant diversity has been lost. As a result, there has been a focus in Norfolk on restoring plant diversity in public demonstration sites highlighting living shoreline designs. In particular, flowering wetland plants have been specified for a number of recent projects. While not common in the Norfolk urban landscape, plants such as *Solidago sempervirens* (seaside goldenrod), *Hibiscus moscheutos* (marsh hibiscus), *Kosteletskya virginica* (rose mallow), *Borrichia frutescens* (sea oxeye daisy), and *Saururus cernuus* (lizard's tail) have all been planted. Others, like *Pluchea purpurascens* (saltmarsh fleabane), *Rosa palustris* (swamp rose), *Asclepias incarnata* (swamp milkweed), *Eupatorium maculatum* (spotted joe-pye weed), *Impatiens capensis* (jewelweed), and *Lobelia cardinalis* (cardinal flower), provide additional opportunities for experimentation in southeast Virginia.

One factor that should aid in the acceptance and adoption in the use of flowering wetland plants is the fact that many of the wetland flowers can be repeated in the upland landscape, creating a visual continuity and bond between the aquatic and upland landscapes.

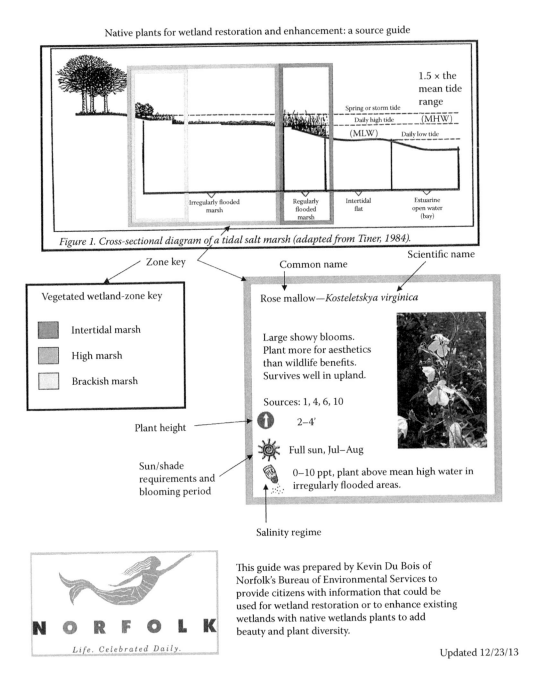

Native plants for wetland restoration and enhancement: a source guide

Figure 1. Cross-sectional diagram of a tidal salt marsh (adapted from Tiner, 1984).

Zone key

Common name

Scientific name

Vegetated wetland-zone key

Intertidal marsh

High marsh

Brackish marsh

Rose mallow—*Kosteletskya virginica*

Large showy blooms.
Plant more for aesthetics
than wildlife benefits.
Survives well in upland.

Sources: 1, 4, 6, 10

Plant height

2–4'

Sun/shade
requirements and
blooming period

Full sun, Jul–Aug

0–10 ppt, plant above mean high water in
irregularly flooded areas.

Salinity regime

This guide was prepared by Kevin Du Bois of
Norfolk's Bureau of Environmental Services to
provide citizens with information that could be
used for wetland restoration or to enhance existing
wetlands with native wetlands plants to add
beauty and plant diversity.

Updated 12/23/13

Figure 6.5 A City of Norfolk Publication distributed throughout the Commonwealth of Virginia to foster the integration of flowering wetland plants and enhance species diversity in wetland restoration projects. (Photo courtesy of Norfolk, Virginia.)

Norfolk's "Native Plants for Wetland Restoration and Enhancement—A Source Guide" can be found at http://www.norfolk.gov/DocumentCenter/View/3827 (Figure 6.5).

6.6 CONTINUING NEEDS/CRITICAL NEXT STEPS (NO PRIORITY ORDER ASSUMED)

There are a number of critical steps still needed for living shorelines to become the norm for shoreline protection along sheltered coasts in southeast Virginia. Such steps are discussed in Sections 6.6.1 through 6.6.16.

6.6.1 Update and Provide for Compatible Regulatory Guidance

Seek or develop regulatory guidance to explain why federal requirements (401 and (404(b)(1)) should favor or require living shorelines and should disincentivize hardened shorelines. Update the Virginia integrated shoreline management guidance. Guidance should include and compare the long-term costs of shoreline hardening options as well as the potential for secondary damage from hardened shorelines (e.g., loss of wetlands from wave reflection, etc.). Guidance from the perspective of value rather than cost would be desirable. Ensure that the guidance contains consistent messaging and train local, state, and federal regulatory decision makers on the new guidance so that the stated Virginia preference for living shorelines is realized.

6.6.2 Improve Regulatory Consistency and Predictability

During their workshop on barriers to living shoreline implementation, one VIMS participant wrote: "Streamline [the] permit application and response—[It] needs to be consistent from local, state, and federal agencies." As described previously, local, state, and federal regulatory agencies should be working in concert to promote living shorelines and improve permit decision consistency, predictability, and customer service.

6.6.3 Develop Implementation Capacity

Develop living shoreline knowledge and implementation capacity (design and construction) across all disciplines including engineers, planners, contractors, regulators, landscape architects/ landscape service professionals, erosion control plan preparers, environmental NGOs, master gardeners, master naturalists, citizen volunteers, and commercial interests (real estate, commercial fisheries, etc.). To meet the intended need for living shoreline project implementation, many more knowledgeable and conscientious contractors trained to work with specialized plants and in ecosystem restoration are needed. Catalog existing design manuals, scientific literature, and other information. Develop training that includes active participation in the design, construction, and monitoring of public demonstration projects.

6.6.4 Design a Living Shoreline Professional Certification

Given that so much effort and attention are being given to LS implementation, professionals with documented knowledge and experience are needed in all aspects of design, construction, regulation, and maintenance. The Society of Wetland Scientists uses a certification program to credential Professional Wetlands Scientists (PWS) and Professional Wetland Scientists in Training (PWSIT). Perhaps, in a similar way, some certifying body could credential living shoreline professionals. Restore America's Estuaries has proposed the development of a living shoreline academy to address this need. Grant funding sources could incentivize the development of certified living shoreline professionals by requiring their involvement in grant-funded projects.

Certification should include hands-on training. Regulators, project designers, contractors, community activists, and NGOs, anyone that could benefit from living shoreline projects, need cradle-to-grave, hands-on involvement in the implementation of successful living shoreline projects to build experience and expand the pool from which innovation will be derived.

6.6.5 Build Demonstration Projects in Every Community

The VIMS publication, *Regulatory Fidelity to Guidance in Virginia's Tidal Wetlands Program*, pointed out weaknesses in Virginia's wetland permitting process showing that shoreline hardening projects in low–wave energy areas are often approved by local Wetlands Boards. Analysis showed that the majority of projects, regardless of whether they were consistent with VIMS technical guidance or not, were approved as submitted.

One of the things that stymie implementation of new ideas is the fear of the unknown. For living shorelines, online photo galleries, with before and after photographs of successful projects, can be useful (Figure 6.6a and b). Dated project completion photographs and periodic updates, along with lists of major localized storm events, can speak to project durability. Another way to address living shoreline uncertainty is to provide good examples of living shoreline projects on accessible public land. In Norfolk, Virginia, the Wetlands Board frequently refers to the City's successful public demonstration sites to justify permit decisions that favor living shorelines on private land. For landowners, seeing successful projects firsthand, in one's own community, is a powerful and compelling way to convince them of living shoreline efficacy, durability, and beauty. Throughout Virginia, if more landowners proposed living shorelines voluntarily, more would be installed. For governments, leading by example, installing living shorelines builds the case that these erosion control techniques are expected and should be the norm.

Demonstration projects are also needed to study assumptions about the benefits of living shorelines including the effects of "hard" components over time (e.g., sills as navigational hazards with SLR).

Demonstration projects should provide the permitting flexibility necessary to experiment and innovate. For example, some living shoreline designers are beginning to consider adding woody debris, adding organics to soil to jumpstart chemistry, seeding marshes with mussels, and are using natural oyster reefs or fabricated reefballs for sills. With experimentation, some degree of failure should be

(a)   (b)

Figure 6.6   Before (a) and after (b) photographs of a living shoreline erosion control and wetland restoration project at Colley Bay in Norfolk, Virginia. (Photo courtesy of Norfolk, Virginia.)

acceptable as long as there is mandatory monitoring, reporting, and dissemination of experimental results back to the community of practice. Grant-funded projects should always include capacity-building elements and community involvement.

6.6.6 Conduct Social Marketing

Local government staff, schools, VIMS, local and bay-wide NGOs, and many others conduct education and outreach efforts designed to persuade landowners to voluntarily preserve, protect, and restore tidal wetlands. Local, state, and federal regulatory programs encourage the use of living shoreline techniques through expedited general and regional permits. However, even though these combined education and outreach programs have been going on for a decade or more, and living shoreline techniques are beginning to be adopted, many landowners still reject living shoreline implementation on their own properties.

Wetlands continue to be filled, dredged, or otherwise destroyed in southeast Virginia. Routine mowing of wetlands above and below mean high water is commonplace. Clearly, lack of information and outreach is not the cause of such widespread indifference to wetland function and values and the overwhelming public benefits they provide. An honest assessment of our collective educational efforts would suggest they are not meeting desired goals. Simply put, we do not understand our audience well enough to craft outreach campaigns that result in meaningful behavior change that leads to significant improvements in wetland protection and restoration.

Accordingly, there is a need to fund and conduct regional, scientific, social marketing campaigns that lead to an understanding of the beliefs, attitudes, and mores that guide decisions to adopt or reject living shoreline erosion control solutions on private property.

Since waterfront landowners are not found within one demographic group, studies should investigate factors influencing the diverse spectrum of decision makers to determine if there are unique barriers to living shoreline implementation within distinct groups (ethnic, socioeconomic, etc.).

Study results should be used to develop specific education and outreach campaigns to target the variety of user groups. Governments, landowners, landscape architects, real estate agents, commercial fishermen, and other living shoreline benefactors can tailor messaging to their clients. Existing agencies and personnel already delivering wetland outreach should substitute new messaging based on the social marketing research and new and effective modes of outreach should also be investigated (e.g., social media including Instagram, blogging, websites, etc.). Social marketing research should be used to develop effective incentives for living shoreline implementation and outreach strategies should be designed for measurable results and be managed adaptively.

6.6.7 Value Engineer All Aspects of Living Shoreline Construction

Landowners and LS advocates agree that more needs to be done to incentivize LS by lowering costs. Research on value-engineering the construction of living shorelines is needed along with dissemination of cost-saving measures to the entire community of practice.

6.6.8 Development of Materials for "Do-It-Yourselfers"

Living shoreline information needs to be created and formatted specifically for those who want to build their own. Instruction manuals could cover subjects including construction methods, simple monitoring, how to address common problems, helpful tools to buy or make, wetland plant sources and selection criteria, predation controls, and so on.

6.6.9 Develop Information on the Economics of Living Shoreline Construction

On the basis of regulatory trends preferring or requiring living shorelines, develop information to assist the emerging need for trained living shoreline designers and contractors. Develop contractor training on how to make money on living shoreline projects identifying necessary materials, suppliers, bidding strategies, value-engineering, contingency planning, and so on. Engage suppliers of living shoreline products (coir logs, silt fence, and other specialized erosion and sediment control materials) since their businesses will grow with expansion of living shoreline use. Identify other related economies that could benefit including environmental construction jobs and economies derived from healthy and improved ecosystem services. Identify the economic detriments to end users if living shoreline projects are not implemented.

6.6.10 Provide Grant Funding for Demonstration Projects

Private property projects and some of the smaller neighborhood scale projects can be done with private funds or augmented with fees collected from wetland violations. However, paying for larger-scale public demonstration projects, which cost hundreds of thousands of dollars, without grant funds, is hard to conceive. In Virginia, significant funding has been available from the Chesapeake Bay Trust Living Shoreline and Hurricane Sandy Coastal Resiliency Competitive Grant Programs, National Fish and Wildlife Fund grants, and the Chesapeake Bay Restoration Fund Advisory Committee. The VMRC has a Marine Habitat and Waterways Improvement Fund that appears to have gone untapped (see Section 28.2-1204.2 of the State Code). New and creative sources of funding will be needed to build public demonstration projects, conduct social marketing studies, train living shoreline professionals, help private landowners build living shorelines that serve the common good, and so on. Existing sources of funding are insufficient for the task at hand.

6.6.11 Encourage Partnerships

Regulators, designers, and contractors need to work together to develop a mutual understanding of the challenges involved in implementing successful living shoreline projects. Cradle-to-grave involvement will ensure the greatest level of understanding and empathy, build positive feedback loops, and result in innovation, a sense of shared accomplishment, and improved projects. Working with wetland-building NGOs and expanding partnerships to nontraditional advocates (landscape professionals, recreational and commercial fishing interests, fishery managers, etc.) will help build the "community of practice" needed to overcome barriers to implementation of living shorelines (Figure 6.7).

6.6.12 Activate Government Leadership

Governments need to act as role models for their citizens by installing living shorelines on public property. They have an important role in building institutional capacity and awareness and in making the behavior to preserve natural resources acceptable and normative. Governments need to build partnerships throughout the range of living shoreline beneficiaries, which could provide secondary benefits in the pursuit of other environmental initiatives. Governments need to hire professional wetland staff and support their regulatory agencies to incentivize living shorelines. In Virginia, the VMRC needs to overturn local Wetlands Board decisions that are inconsistent with the state preference for living shoreline implementation. Likewise, federal agencies should act in concert with state and local governments to achieve the maximum number of feasible and justifiable living shoreline projects.

Figure 6.7 Members of the Old Dominion University Marine Biology Club—one of the many partnerships that have been a source of living shoreline implementation success in Norfolk, Virginia. (Photo courtesy of Norfolk, Virginia.)

6.6.13 Disincentivize or Deny Shoreline Hardening Projects When Living Shoreline Projects Are a Viable Alternative

If living shoreline erosion control is feasible and consistent with land use, then hardening the shoreline should be made difficult if not impossible. Incentivizing living shoreline techniques alone may not be sufficient to meet goals for wetland restoration and the revitalization of ecosystem services.

6.6.14 Capitalize on Synergistic Effects

There needs to be a better understanding of ways to synergize nutrient, stormwater, and storm damage reduction programs (TMDL, MS4, and CRS) to maximize "triple bottom line" benefits for local governments.

6.6.15 Improve Living Shoreline Value and Function

There needs to be research and better dissemination of information on ways to improve the value that living shoreline projects can provide as part of a larger ecosystem restoration or green infrastructure strategy. The value and function of living shoreline projects might improve if more was known about the incorporation of mussels, woody debris, plant variety, soil amendments for infaunal development, and so on.

6.6.16 Treat Living Shorelines Like Other Managed Landscapes

Wetlands are part of the continuum that makes up an aesthetic coastal landscape. We must convince landowners that, while functional for erosion control, living shorelines are worthy of maintenance and beautification like their upland landscape. Landowners must be convinced that, when properly maintained, living shorelines provide both aesthetic beauty and real economic property value. Managing and enhancing the "wetlandscape" is an untapped potential business opportunity.

6.7 CONCLUSION

The coastal environment is not static. Climate change and SLR pose significant threats to the stability of coastal shorelines and the wetlands that provide so many valuable ecosystem services. Some in the scientific community scoff at living shorelines as a "solution" to either long-term erosion control or the restoration or maintenance of ecosystem function.

However, one thing is certain. Wise stewardship of coastal ecosystems and living marine resources will need to be adaptive. While communities debate elevating structures and land to defend development in place, hardening shorelines with bulkheads and revetments will remain ecologically worse than implementing living shoreline techniques. As coastal communities learn to adapt and decide how to live with encroaching water levels, living shorelines can act as a bridge to future shoreline management strategies maintaining better ecosystem services along the way (Figure 6.8).

Figure 6.8   Along a sheltered coast, biodegradable coir logs, aka biologs, protect the leading edge of a living shoreline erosion control project until the wetland plants become established and their roots can hold and stabilize the substrate. (Photo courtesy of Norfolk, Virginia.)

APPENDICES

Appendix of All VIMS Workshop Proposed Solutions (suggestions for grant-funded initiatives)

- MOU between federal agencies—DOD, EPA, DOI, DOC, and so on—to create a federal council to focus on moving LS concept across federal agencies.
- Require LS grants to explicitly focus on building knowledge and institutional capacity across the board (identify LSPITs that will get credit for design, plan review, or construction experience).
- Develop general guidelines for when LS is not appropriate (for use by regulators).
- Promote hiring of certified wetland or LS professionals (e.g., SWS, PWS) in all jobs that involve wetland permitting or plan review.
- Provide nationwide training
    - LSPIT Certification: 2- to 4-day workshop on design, construction, and monitoring for functional analysis
    - LSP Certification: 2- to 4-day workshop + documented working experience requirement (design, plan review, construction, monitoring)
- Have LS included in undergraduate and graduate marine science, wetlands, landscape design/architecture curricula, including mandatory hands-on design and construction practicum requirements.
- Provide continuing education credits across applicable disciplines for living shoreline training.
- Create a repository/list/map of LS projects using a uniform system and standardized reporting criteria.
- Create a space for the conversation on adaptive management/lessons learned (use existing LS Linked-In group?).
- Produce videos showing LS projects being built, emphasizing before and after footage, benefits (wildlife use, aesthetics, recreation, etc.), and community involvement. Attach videos to maps or lists of LS projects.
- In the Chesapeake Bay watershed, publicize TMDL credits available for LS projects and provide training on the process of obtaining credits.
- Get media to report on LS projects, especially on community participation aspects.
- Have social media studies suggest national LS "branding" logos, catch phrases, and so on.
- Have a group of knowledgeable engineers/designers "value engineer" living shoreline designs to lower cost. Look at minimum stone sizes necessary for a given fetch, minimum top width of sills, sill geometry and minimum sill slopes, coir log diameters and maximum fill heights, sill opening design, and so on. Suggest/support volunteer or landowner plantings (develop planting guidance document).
- Develop financial incentives for landowners—State grants (especially for replacement of hardened shorelines), tax breaks/credits, low-interest loans (Studied by the VA Middle Peninsula Planning District Commission), FEMA flood insurance discounts (CRS credits), reduced stormwater fees, special taxing districts, cost share (funding?), SLR preparedness, and so on. Use incentives to "level the playing field" on issues of cost.
- Increase the life span of living shoreline permits (6 years?) to allow for phased construction, adequate monitoring, repairs, and TMDL recertification.
- Develop a registry of certified or qualified contractors. Before the development of training for certification, develop an "Angie's List" type of satisfied customer registry.
- Link LS contractor certification to regulatory requirements.
- Develop a nationwide "master list" of living shoreline educational resources.
- Research and develop PR on how healthy wetlands provide real estate value (hedonic pricing) similar to the value of mature trees and other beneficial or aesthetic landscaping. Increase land value means increased tax revenue.
- Develop strategies for SLR adaptation (creative, paradigm-shifting): landward migration, maintaining in place through periodic fill to maintain proper elevation.
- Build public demonstration sites in every community.

- Develop promotional materials extolling the value of LS for each stakeholder: landowner, contractor, real estate agent, business.
- Engage not-for-profits to grow and install wetland plants (e.g., Eggleston Services that works to employ those with disabilities).
- Development of State-based permit guides that help landowners understand the multijurisdictional process. Use this document as a basis to ensure consistency of decisions, improve predictability, eliminate any duplication of effort, and improve and streamline processes.
- Have NGOs with similar interests in LS develop a strategic plan for outreach (e.g., Chesapeake Bay Foundation, Elizabeth River Project, Lafayette Wetlands Partnership, Wetlands Watch, etc.).
- Explain the current state of the environment relative to pre-European settlement conditions and build an understanding that maintaining current conditions is a lowering of the original benchmark. Provide a clear vision of environmental restoration sought and LS's place in that vision.
- Clearly state and promote the economics of restoration in an effective PR campaign.
- Develop a paper that identifies ways LS support compliance with other necessary regulatory requirements. "Co-benefit" awareness, flood prevention, stormwater permits, pollution abatement, erosion control, and so on.
- Get FEMA to provide CRS credits for wetland restoration or enhancement.
- Dispel the myth of "burdensome" LS maintenance. Maintenance level depends on desired project goals and project expectations. With proper design, after two growing seasons, little to no maintenance should be required (with the exception of the potential for invasive species control).
- Develop public and NGO programs to engage in phragmites control.
- Develop a media campaign that explains the damage and loss of value caused by shoreline hardening (perhaps part of a larger social marketing strategy).
- In Virginia, provide guidance to local wetlands boards that they can deny shoreline hardening proposals in favor of living shoreline erosion control protection and consistent with the Commonwealth's preference for LS. Publicize that they can require landowners to justify why a living shoreline solution is not appropriate before approving shoreline hardening proposals. Publicize Fairfax County example (Task: VMRC).
- Require governments to track and report on the number of LS projects approved and the percentage of possible LS projects approved to monitor progress.
- Reevaluate and reorder the multijurisdictional wetland permit decision-making process. Without usurping all necessary approvals, local approval should be obtained first, followed by state and federal approvals.
- Provide information on replacing dilapidated hard structures with living shorelines and how to transition living shorelines into adjacent hardened shorelines (contractors' photos).
- Develop public information on the relationship between wetlands and their adjacent riparian buffers for the supply of ecosystem services and their interdependence.
- Develop LS design guidance for SLR adaptability with an understanding of ecological trade-offs and design planning horizons.
- Use money from environmental violations and fines to financially incentivize LS projects.
- Require notice covenants or promote notice practices so that real estate agents have to disclose regulations governing shoreline stabilization.
- Disincentivize shoreline hardening by requiring higher mitigation ratios (how to prevent relocation of hard structures just landward of wetland jurisdiction?).
- Living shoreline lottery (winner gets a restored shoreline).

REFERENCES

Chesapeake Bay Partnership. 2014. Recommendations of the Expert Panel to Define Removal Rates for Shoreline Management Projects. 192 pp.

Pilkey, O. et al. 2012. Rethinking Living Shorelines. 10 pp. http://www.wcu.edu/WebFiles/PDFs/PSDS_Living_Shorelines_White_Paper.pdf.

Restore America's Estuaries. 2015. Living Shorelines: From Barriers to Opportunities. Arlington, VA. 55 pp. https://www.estuaries.org/images/stories/RAEReports/RAE_LS_Barriers_report_final.pdf.

Virginia Institute of Marine Science. 2012. Regulatory Fidelity to Guidance in Virginia's Tidal Wetlands Program. 59 pp. http://ccrm.vims.edu/publications/pubs/Permit_Fidelity_2012.pdf.

Wetlands Watch. 2015. Flood Protection Pay-Offs: A Local Government Guide to the Community Rating System. 192 pp. http://www.wetlandswatch.org/Portals/3/WW%20documents/insurance/WW-crs-review -draft-3-10-15.pdf.

CHAPTER 7

Green Shores
Using Voluntary Ratings and Certification Programs to Guide Sustainable Shoreline Development

Brian Emmett, D.G. Blair, and Nicole Faghin

CONTENTS

7.1 INTRODUCTION

People love to live, work, and play in places where water and land meet. Shorelines provide work, recreation, living space, mild climates, and wonderful views. People are not the only ones drawn to biologically rich and productive places. Unfortunately, many of the natural features that make shorelines so attractive are often the casualties of human activities. Native trees, shrubs, and grasses are cleared to make way for houses, lawns, and views. Bulkheads, docks, and piers displace beaches and alter natural shoreline processes. Loss of shoreline vegetation allows contaminants to flow directly into the water. Prime wildlife habitats disappear, taking with them birds, mammals, fish, and beneficial insects. The good news is that people are finding new strategies for protecting waterfront properties while also protecting and restoring habitats. Instead of concrete and sheetpile, these practices use a combination of plantings, gravel and sand, logs, stones, setbacks, and slope modification to protect against shoreline erosion and provide access while respecting the ecological attributes of the shoreline (adapted from Green Shorelines, City of Seattle, 2011).

The Green Shores program described in this chapter is one of many initiatives in the Salish Sea region with the broad objective of increasing our capacity to address impacts of shoreline development and climate change on coastal ecology and human well-being.

7.2 REGIONAL CONTEXT—THE SALISH SEA

The Salish Sea is a large inland sea of more than 17,000 km, encompassing Puget Sound in the State of Washington and the Strait of Georgia in British Columbia and the Strait of Juan de Fuca between British Columbia and Canada (see Figure 7.1). Although the Puget Sound/Strait of Georgia portion of the Salish Sea is considerably smaller than Lake Ontario, this area has over 10 times more coastline (14,000 km) because of the complexity of coastal fjords and island groups throughout the inland sea, including the Gulf Islands in British Columbia and San Juan Islands in Washington State. The Salish Sea is home to more than 7 million people, with most residing in the Greater Seattle and Vancouver areas. Beamish and McFarlane (2014) and DeLella Benedict and Gaydos (2015) are two recent books on the Salish Sea region that provide excellent overviews of the area.

The inland nature of the Salish Sea provides protection from large, open ocean wave energy but the complex seabed topography and island groups generate areas of high tidal currents, particularly in the southern Strait of Georgia. Most of the Puget Sound shoreline is formed of unconsolidated sediments, including coastal bluffs, beaches, and sand/gravel spits.* These shore types often occur as drift cells; units of eroding bluffs, shore sections with net directional sediment transport and depositional beaches and spits (see Figure 7.2). The degree of rocky, noneroding shoreline in the Strait of Georgia is greater than that in Puget Sound; however, similar drift cell units occur in the Strait of Georgia, particularly on the more heavily populated east coast of Vancouver Island and the Vancouver area.

The Fraser River, one of the major river drainages on the west coast of North America, flows into the Salish Sea at Vancouver and supports major stocks of all five Pacific salmon species, including chinook salmon, an important prey species for the endangered southern resident killer whale population of the Salish Sea. Salish Sea shorelines provide important rearing and migratory pathways for juvenile salmon, spawning habitat for Pacific herring, sandlance, and surf smelt as well as numerous invertebrate species.

Human settlement–associated coastal development has altered the physical character of Salish Sea shorelines and affected important nearshore habitat functions. More than 27% of the Puget Sound shoreline has been altered by development, with up to 90% armoring in the highly urbanized areas of Seattle and Tacoma.† Over the past several decades, there has been a strong effort on the part of the State of Washington to slow and ultimately reverse the degree of shoreline armoring by supporting bulkhead removal initiatives and alternative (soft shore) protection methods.‡ The degree of shoreline armoring is less in the Strait of Georgia portion of the Salish Sea, in part owing to the greater amount of rocky shoreline and lower population density; however, in highly developed areas such as the southeast coast of Vancouver Island and the Vancouver area, the degree of shoreline armoring approaches levels seen in some areas of Puget Sound.

* Retrieved from http://www.eopugetsound.org/articles/shoreline-formation-puget-sound
† Retrieved from https://sites.google.com/a/uw.edu/shoreline-armoring/background/shoreline-armoring
‡ Retrieved from http://www.psp.wa.gov/vitalsigns/shoreline_armoring.php

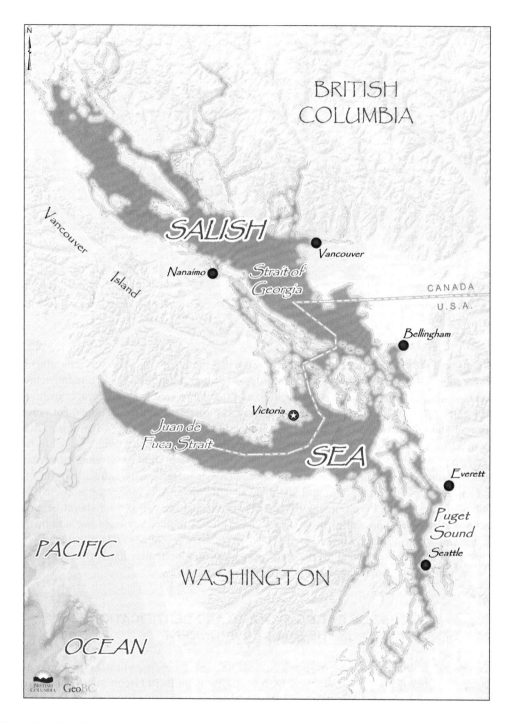

Figure 7.1 Salish Sea.

Figure 7.2   Schematic of a drift cell showing area of sediment sources (bluff), zone of transport (beach), and deposition (spit).

Throughout the United States and Canada, there is increasing recognition that shoreline development practices—particularly shoreline armoring—are a significant source of cumulative impacts to nearshore coastal processes and habitat values and that there are also considerable institutional barriers that hinder the development and adoption of sustainable shoreline development practices, including a lack of professional expertise and leadership (Restore America's Estuaries 2015). In addition, regional sea level rise and increased storm surge are seen as important issues in British Columbia and Washington State that could increase flooding and damage coastal infrastructure, resulting in property loss from erosion, habitat loss, decreasing biodiversity, saltwater intrusion into coastal aquifers, and loss of cultural and historical sites (Climate Action Secretariat [CAS] British Columbia Ministry of the Environment 2013). This is a compelling concern for shoreline property owners and shoreline communities throughout the Salish Sea region. Sea level rise and increased or more frequent storm surge will increase the need to stabilize shorelines and protect development from flooding; however, traditional engineered approaches (predominately hard armoring) to shoreline development may be maladaptive.

7.3 USE OF RATING SYSTEMS AND CERTIFICATION IN THE BUILT ENVIRONMENT

For several decades, development professionals have used credit and rating systems such as Leadership in Energy and Environmental Design (LEED) and Built Green to guide sustainable building practices from both an environmental and human health perspective. These rating and certification programs are tools to reduce the environmental impact of the built environment and transform the development market to a higher level of environmental performance and sustainable design. While providing clear guidance on improved environmental design, these programs also serve as an educational tool for builders and property owners. Reviewing and evaluating the methodology of these programs provide insight as to how rating and certification programs can be used in waterfront construction projects to harmonize the built and natural environment.

An ideal rating system tool includes several key elements (adapted from Cole 2005):

- Simple and practical in application
- Providing value for money
- Credible and transparent
- Challenging to reach higher score levels
- Addressing critical environmental and human well-being issues
- Regionally specific but globally applicable
- Encouraging innovation and education
- Providing useful design guidance
- Capable of evolving to reflect new knowledge

The US Green Building Council established the LEED for Building Design and Construction (LEED) program in 2000 to encourage the development of environmentally responsible buildings that are healthy places to live and work. This program is well established, with more than 16,000 certified projects in the United States and 1800 in Canada in 2015.* LEED also operates related rating and certification programs for single-family residences (LEED for Homes) and neighborhood design (LEED for Neighborhood Development). By 2015, more than 60,000 projects were participating in LEED programs worldwide.†

The Home Builders Association of Metro Denver with a focus on energy efficiency established Built Green in 1995. The program expanded to several US states in 2000 with a broadened goal of promoting and enabling environmentally responsible home construction practices. Built Green programs are operated by local home builders associations, and building credits can vary in different regions depending on regional priorities, such as water conservation, as well as regional programs that incorporate energy efficiencies. Built Green Canada has programs in five provinces and has certified more than 24,000 homes.‡

The American Society of Landscape Architects established the Sustainable Sites Initiative (SSI) in 2006 to "foster a transformation of land development and management practices that will bring the essential importance of ecosystem services to the forefront" (SSI 2009). In part, SSI was initiated to address the environmental aspects of development siting that were not part of LEED green building certification program. Recently, several aspects of the SSI program, including specific credits, have been incorporated into LEED programs.§

Typically, these credit and rating programs share some common elements:

- A series of mandatory requirements (prerequisites) as well as a number of optional credits usually grouped into specific categories such as energy, indoor environmental quality, and so on.
- A credit point system that addresses the relative environmental or human health value of each credit.
- One or more certification thresholds, based on credit points, that establish a rating or certification level; for example, LEED offers four certification levels (Certified, Silver, Gold, and Platinum).
- A structured format for submitting a project for rating or certification.
- A third-party certification or verification process.
- Associated training and professional accreditation programs; for example, more than 20,000 building professionals in Canada have participated in LEED training programs.¶

Although the main objective of these programs is reduced impact on ecosystem services, most also include a number of credits addressing human health and well-being, design innovation, and

---

* Retrieved from http://www.usgbc.org/, and https://www.cagbc.org/
† Retrieved from http://www.usgbc.org/
‡ Retrieved from http://www.builtgreencanada.ca/
§ Retrieved from http://www.sustainablesites.org/
¶ Retrieved from https://www.cagbc.org/

professional and public education. Some programs provide regionally specific credits and many, including LEED, rely on a continuous evaluation cycle to maintain and improve the program credits and certification standards. Innovation, education, and program audits are key elements of this evaluation process.

Most ratings and credit programs are voluntary but rely on support from the building sector and government for their adoption. Typically, they are also nonprofit and membership based, drawing support from the environmental and building sectors. The voluntary nature of these programs enables participants to distinguish themselves as sustainability leaders in their sectors. Increasingly, these volunteer programs take on mandatory aspects when adopted by local jurisdictions on a regional basis for specific types of buildings (e.g., schools) or as a condition of rezoning or a master development agreement; for example, the University of Victoria in British Columbia now requires all new buildings to meet the LEED Gold standard for building design and construction.* In this way, jurisdictions are promoting and enabling new ways to build and develop within their community.

7.4 THE GREEN SHORES PROGRAM

In 2003, the Stewardship Centre of British Columbia (SCBC) released the guide *Coastal Shore Stewardship: A Guide for Planners, Builders and Developers* as part of its ongoing Stewardship Series of technical guides.† Workshops were conducted throughout coastal British Columbia, and feedback from these workshops indicated that planners, builders, and developers appreciated the information provided by the document but needed more specific planning and development guidance to address coastal management and development issues. The Green Shores program resulted from conversations about how best to address this need.

Green Shores was initiated in 2005, with the support of multiple funding partners, to address coastal shore stewardship. The program provides tools for industry professionals in the planning, design, and construction fields, as well as shoreline property owners interested in minimizing the environmental impacts of their projects in a cost-effective manner. The program is underpinned by four guiding principles:

1. Preserve the integrity and connectivity of shoreline processes
2. Maintain and enhance shoreline habitat diversity and function
3. Minimize and reduce pollutants to the shoreline environment
4. Reduce and reverse cumulative impacts to shoreline systems

The first three principles are easily understood by planners, builders, and the public, while the fourth conveys the concept that impacts to shoreline habitats usually do not result from one project alone but rather from many projects acting in concert, emphasizing that developments need to reduce or eliminate their contribution to cumulative effects.

Recognizing the opportunity for rating and certification programs to influence environmentally sensitive design of shorelines, the Green Shores program and its partners have developed two certification programs for commercial, residential, and park properties, using the Green Shores guiding principles as the backbone of the program. Green Shores for Coastal Development (GSCD) applies to a broad range of types of coastal shoreline properties, while Green Shores for Homes (GSH) focuses on residential properties. Both programs are under development in British Columbia and Washington State.

* Retrieved from http://www.uvic.ca/campusplanning/about/green-buildings/index.php
† Retrieved from http://stewardshipcentrebc.ca/PDF_docs/StewardshipSeries/Coastal.pdf

7.4.1 Green Shores for Coastal Development

Using the LEED credit framework, work began to develop GSCD in 2005. With funding from the Real Estate Foundation of British Columbia and others, an interdisciplinary technical team of planners, marine ecologists, coastal geologists, and engineers developed the credit and rating framework. GSCD targets mixed residential and commercial waterfront development projects as well as public spaces (shoreline parks and recreational areas). A pilot version of the GSCD was subject to peer review and tested on a number of shore development projects ranging from a highly urbanized shoreline (South False Creek Olympic Village in Vancouver, BC) to a shoreline park (Tyee Spit on Vancouver Island). Version 1 of the GSCD was released by the SCBC in 2010.* GSCD is voluntary and relies on support from industry, government, nongovernmental organizations, building owners, and the building sector for its adoption. It is intended to be applicable to all coastal systems and ultimately be national or international in scope.

GSCD consists of 5 Green Shores prerequisites and 11 optional Green Shores credits (Figure 7.3). The prerequisites address the most critical issues of sustainable shoreline development, including siting of building structures, conservation of critical and sensitive habitats, coastal riparian values, and shoreline physical processes (sediment supply, transport and deposition). GSCD-certified projects must meet all five prerequisite requirements. Although no points are awarded for these prerequisites, they ensure that the four principles of the Green Shores program outlined above are met; providing a basic "greening goal" for a waterfront development project. Project designs that do not meet all five prerequisites can be quickly screened out of the certification process. By contrast, the 11 optional credits provide a range of opportunities for projects that meet the prerequisites to go beyond basic "green" goals to further reduce the cumulative impacts of waterfront development. The three certification levels—Certified, Silver, and Gold—are intended to incentivize proponents to higher performance levels than initial design might indicate.

GSCD credits address a number of critical aspects of coastal development design, including the following:

1. Climate change adaptation (Credit 4)
2. Habitat conservation and enhancement (Credits 1, 2, 5, and 7)
3. Conservation or renewal of coastal processes (Credits 1 and 6)
4. Pollutant input (Credits 8 and 9)

An innovation credit is included to encourage exceptional performance and sharing of innovative approaches to coastal design. As with LEED, this credit is only available to projects that are willing to make the information publicly available. The Outreach and Public Education Credit is intended to enable a broader uptake of sustainable coastal design and the Green Shores program by coastal property owners, developers, contractors, and local governments.

The prerequisites and credits are structured using a standard format:

1. Intent: a clear statement of the purpose or objective for the credit
2. Context: background information on the rationale for the credit from an ecological and, for certain credits, a human well-being perspective
3. Requirements: conditions that must be met to achieve the credit (and the number of points available for meeting specific requirement levels)
4. Submittals: information that must be provided to the certifier in order to apply for the credit
5. Strategies and technologies: guidance on how to achieve the credit
6. Resources: a list of relevant resources

---

* The GSCD Rating System Guide is available from SCBC at http://stewardshipcentrebc.ca/PDF_docs/greenshores /GreenShoresCDRS.pdf

Prerequisites	
Prerequisite 1	Siting of permanent structures
Prerequisite 2	Conservation of critical or sensitive habitats
Prerequisite 3	Riparian zone protection
Prerequisite 4	Conservation of coastal sediment processes
Prerequisite 5	On-site environmental management plan

Credits		
Credit 1	Site design with conservation of shore zone	1 to 3 points
Credit 2	Shore friendly public access	1 point
Credit 3	Re-development of contaminated sites	1 point
Credit 4	Climate change adaptation plan	1 to 5 points
Credit 5	Rehabilitation of coastal habitats	0.5 to 4 points
Credit 6	Rehabilitation of coastal sediment processes	2 to 3 points
Credit 7	Enhanced riparian zone protection	0.5 to 4 points
Credit 8	Light pollution reduction	1 point
Credit 9	Integrated stormwater planning and design	1 to 4 points
Credit 10	Innovation	1 to 2 points
Credit 11	Outreach and public education	1 point

Ratings levels	
GSCD bronze	All prerequisites plus 5 points
GSCD silver	All prerequisites plus 10 points
GSCD gold	All prerequisites plus 15 points

Figure 7.3 GSCD: list of prerequisites, credits, and certification levels.

Many of the prerequisites and credits require the project to achieve a certain performance level; for example, Prerequisite 4 requires a project to be designed such that the need to install shore protection works is unlikely over the lifetime of the project. Others have more prescriptive requirements, such as stating a required width and coverage for coastal riparian area (Prerequisite 3 and Credit 7). Most credits (including all credits based on performance criteria) require signoff by a qualified coastal or environmental professional as defined by the GSCD credit guide. As with LEED, application for certification is made using a standard submittal template provided by the certifying organization.

7.4.1.1 GSCD Pilot Project

To encourage use of the GSCD, a pilot project was initiated in 2013, with the following objectives:

- Demonstrate use of GSCD on three shoreline developments
- Provide outreach and professional development
- Promote Green Shores as a preferred standard for shoreline design and certification in British Columbia and beyond

The three properties selected for the pilot project represented different types of shoreline development:

- Commercial/Institutional Development: Deep Bay Marine Research Station, Vancouver Island University
- Public Park: Jericho Beach Park, Vancouver Parks Board
- Multifamily Residential Development: Squamish Oceanfront Phase One, Squamish Oceanfront Development Corporation

Working closely with the proponent, a Green Shores verifier assessed each property. The proponent collected documentation, using the GSCD Guide and Submittal Template, and submitted to the verifier team for assessment. One property achieved a Green Shores Gold rating, one achieved a Silver rating, and one did not have sufficient documentation to complete a rating during the pilot project timeline.

Overall, the verifier teams found that the prerequisites and credits detailed in the GSCD guide offered clear, quantifiable assessment criteria for projects. All proponents and verifier teams found the information in the guide useful and easy to understand; however, there were comments by all three proponents that they found the submittal template difficult to use. Suggestions for improving the submittal template included the following:

- Use of simple summary checklists to assist proponent in compiling the required information
- Revise the template and provide for an online submittal system
- List typical documents normally required by a shoreline development and link them to each prerequisite or credit

7.4.2 GSH Program Development

In 2009, the US Environmental Protection Agency (EPA) proposed grant funding for projects to support protection and restoration of highly valued Puget Sound aquatic resources in areas threatened by growth. In response to this call for proposals, the City of Seattle, in partnership with San Juan County and Washington Sea Grant, submitted a application for a project titled "Incentivizing Low Impact Shoreline Development: Developing and Piloting Green Shores for Homes on the City of Seattle's Lake Washington Shorelines and in San Juan County." Proposal partners included the SCBC in Canada and a Technical Team consisting of 12 members from Washington State and British Columbia with a broad range of expertise (landscape architects, planners, marine biologists, coastal geomorphologists and engineers, as well as homebuilders). Several members of this group were also the developers of the Green Shores Coastal Development rating system.

The City of Seattle proposed development and testing a program designed to incentivize protection and improvement of ecosystem function and processes along shorelines of single-family waterfront homes. The assessment framework, GSH, was based on the existing GSCD certification system developed in British Columbia and the Green Shorelines guidelines developed by the City of Seattle.* Many of these lessons learned from the development and piloting GSCD were brought forward to the development of GSH.

The City of Seattle proposed piloting GSH credits and locally developed incentives on Lake Washington, a highly developed freshwater urban area, building on the previous Green Shoreline efforts to facilitate alternatives to shoreline hardening such as armor and bulkheads. San Juan County participated to pilot test GSH in rural, marine waterfront locations.

A unique aspect of this program was its transboundary component with the Canadian partners, SCBC as well as the Islands Trust, a federation of local governments serving the Canadian islands of the Salish Sea. These organizations participated in order to expand GSH from Puget Sound to the

* Retrieved from http://www.govlink.org/watersheds/8/action/greenshorelines/

Strait of Georgia, in recognition of the interconnected nature of the Salish Sea marine environment. This program called for a single credit and rating system for the Canadian and American portions of the Salish Sea as well as collaborative efforts for sharing of technical expertise across the border. In response to funding opportunities, the American partners focused on the technical aspects of the program as well as identification of appropriate incentives while Canadian partners focused on development of educational materials, outreach, and community engagement.

In 2010, the US EPA awarded a grant to the City of Seattle, funded over 4 years, to develop the GSH program with an overall goal of developing a transboundary rating and credit system for both lake and marine environments. The scope of work for the project included the following components:

- Develop the GSH credit system
- Identify incentives for property owners to participate
- Engage homeowners and builders
- Train GSH Verifiers
- Pilot the GSH program
- Develop a delivery model that can be expanded to other regions

The GSH project team would produce a standard, science-based credit and rating system to be implemented in partnership with local jurisdictions responsible for applying locally appropriate incentives for participating waterfront property owners.

Concurrent with the EPA-funded development in Washington State of the GSH Credit and Rating system, SCBC sought funding to implement the GSH program in British Columbia. In 2014, SCBC began work with four local governments to design and test an implementation model for the GSH in British Columbia with the following project goals:

- Increase the ability of local governments, homeowners, and shoreline professionals to utilize GSH to meet upcoming climate change adaptation challenges
- Reduce the impacts of shore developments on water quality and habitat values
- Create a GSH initiative in BC reaching both freshwater and marine shoreline communities

7.4.2.1 Developing the GSH Credit and Rating System

A Steering Committee team guided the process of creating the GSH Credit and Rating System. Team members included representatives from City of Seattle, San Juan County, SCBC, Islands Trust, and Washington Sea Grant and the technical team coordinator. This team solicited applications and created the technical team that oversaw the technical aspects of credit development. In order to develop the GSH Credit and Rating System, the technical team evaluated other rating and incentive programs to determine their applicability to shoreline homeowners including GSCD along with other programs including LEED for Homes, Built Green, and SSI (Rueggeberg et al. 2012). The review also evaluated homeowner incentive programs and checklists to identify elements appropriate for a GSH program. Each program provided insights about specific credit, rating and certification structures, program delivery, and elements of interest to a GSH Program.

Three common components emerged: (1) each program included a third-party administrative and operational entity, typically a nongovernmental organization; (2) the programs included a checklist or submittal form used by applicants to apply for a rating or certification; and (3) an independent reviewer assessed the merits of the project to determine whether or not certification should be awarded.

Key findings from the review provided the framework for the development of the GSH program. Important elements included flexibility in order to address a wide range of shoreline development activities and a relatively simple application and evaluation process so that stakeholders can understand the program. The review also noted that certification usually required some form of applicant incentive to be effective, such as grant programs, low-interest loans, or permit fee reductions (Rueggeberg et al. 2012, p. 38). In addition, programs with prescriptive checklists were easier to implement than programs based on performance metrics. Training and education of contractors was recommended, although challenges were acknowledged owing to costs and potentially insufficient demand. While GSH is intended to create a voluntary, nonregulatory program, the review noted the value of municipal, regional, and state or provincial governments adapting aspects of the rating program into regulatory frameworks to advance the capacity of the program.

The GSH technical team met over the course of 2 years to develop and pilot test the GSH credit and rating structure. The team created four credit categories based on the Green Shores four guiding principles encompassing all aspects of shoreline development:

- Shoreline physical processes: Protect or restore natural physical processes that are vital to the health of shoreline environments
- Shoreline habitat: Protect, restore, and enhance aquatic and riparian habitats
- Water quality: Eliminate the amount of sediment, chemical, and organic pollutants discharged to lakes and marine waters in rainwater runoff
- Shoreline stewardship: General best management practices that help support public values of shorelines

During the development of the credit and rating system, a number of common issues emerged for all credit categories.

1. How to address whole site development as compared to renovation or modifications to existing shoreline such as bulkhead removal?
2. What is an appropriate building setback criterion?
3. Should credits differ for freshwater and marine shorelines?
4. How to determine the relative weight (points) of each credit?

To address these issues, for a number of specific credits, the program makes a distinction between the following:

- *Whole site development*, where the proposed project involves the entire waterfront lot, and *riparian or shoreline development*, where the proposed project occurs only in the area from the intertidal zone (in a marine environment) or from the littoral zone (in a freshwater environment) to the upper edge of the riparian zone.
- *Lake* (freshwater) and *marine* (saltwater) shorelines.
- *Greenfield* (not previously developed) and *redevelopment* (previously developed) sites.
- *Urban* and *rural* sites, which are distinguished primarily on the basis of lot size.
- *Rock-* and *sediment*-based shorelines. Although there are several detailed systems for classifying shores according to their physical type in both Washington and British Columbia, for the purpose of the GSH rating system, a simple distinction was made between sediment and rock shores.

The technical team developed specific credits within each of the four categories through an iterative process. In contrast to the GSCD and several of the LEED rating systems, GSH does not include prerequisite actions. Instead, there are four distinct general application requirements:

1. An existing conditions plan that maps site characteristics
2. A proposed site design showing the proposed development in the context of existing site conditions
3. An environmental management plan
4. A critical or sensitive habitat report showing avoidance of defined critical or sensitive habitats

Table 7.1   GSH Credits and Credit Points

Credit Category		Credit	Maximum Points Available		Total Points Available
			Base	Bonus	
Shoreline processes	1.1	No shoreline protection structures	15	–	15
	1.2	Setback/impact avoidance	10	4	14
	1.3	Bulkhead removal	15	8	23
	1.4	Groin removal	5	2	7
	1.5	Soft Shore protection or enhancement	12	5	17
	1.6	Managed retreat	10	3	13
Shoreline habitats	2.1	Riparian vegetation	10	5	15
	2.2	Trees and snags	5	1	6
	2.3	Invasive species	4	–	4
	2.4	Woody material	3	–	3
	2.5	Overwater structures	10	–	10
	2.6	Access design	3	–	3
Water quality	3.1	Site disturbance	5	–	5
	3.2	Reduce and treat runoff	6	2	8
	3.3	Environmental friendly building products	4	–	4
	3.4	Creosote material removal	4	0	4
	3.5	Herbicides, pesticides, and fertilizers	2	–	2
	3.6	Onsite sewage treatment	2	1	3
Shore Stewardship	4.1	Shoreline collaboration	8	–	8
	4.2	Public information and education	1	1	2
	4.3	Conservation easement or covenant	6	–	6
	4.4	Shoreline stewardship participation	2	–	2

These application requirements tend to coincide with typical permit application requirements for local jurisdictions, minimizing additional document requirements by the program participants.

These credits and associated rating points are fully documented in the *Green Shores for Homes Credit and Ratings Guide.** For each credit, the guide describes the situation or conditions to which the credit is applied, the benefits to the homeowner and environment, the points available for the credit, guidance on how to address the credit at the site level, and provides useful references.

A total of 22 specific credits were developed for the four credit categories described in Table 7.1. The Shoreline Process credits address sites with no shoreline protection, building setbacks and actions to reduce or reverse cumulative impacts from existing developments (bulkhead and groin removal). The Soft Shore Protection credit encourages alternative approaches to shoreline hardening but requires the applicant to demonstrate that shoreline protection is needed. A number of these credits (Setback, Managed Retreat) include consideration of projected sea level rise.

The Shoreline Habitat credits provide a variety of ways to incorporate ecological design features and functions into shoreline properties (riparian vegetation, tree retention or planting, invasive species removal and management). The Overwater Structure credit not only delivers the greatest number of credit points for properties with no docks and piers but also provides guidance and credit for best practices with respect to dock design and materials.

The Water Quality credits address potential contamination from site runoff, including sedimentation, removal of existing sources of contaminants (creosote), and minimizing inputs from building materials, landscape maintenance, and septic treatment.

* Available from the Green Shores for Homes website. See http:/www.greenshoresforhomes.org

Table 7.2 GSH Rating Levels

GSH 1 "ORCA" The project exhibits exceptional design regarding improvement/conservation of the natural features and processes of the shoreline.	Minimum 40 points of which a minimum of 20 points (collectively) is acquired from Shoreline Process and Shoreline Habitat credit categories.
GSH 2 "CHINOOK" The project results in recognizable improvement or conservation of the natural features and processes of the shoreline.	20–39 points of which a minimum of 10 points (collectively) is acquired from Shoreline Processes and Shoreline Habitats credit categories.

The Shore Stewardship credits include recognition of the value of collaboration among waterfront property owners on collective approaches to shoreline protection and enhancement and also include a number of credits that encourage public education and participation in Green Shores approaches to shoreline management. Considerable point recognition is given to projects that establish a conservation easement or covenant to protect the natural features of the property's shoreline.

Points are provided to meet specific credit conditions. The weighing of credit points is based on relative environmental benefit as determined by the technical team. Many of the credits have multiple point levels depending on the extent of the action; for example, between 2 and 10 points are awarded for bulkhead removal depending on how much of the property shoreline is subject to bulkhead removal. This approach is intentional and meant to encourage incremental improvement in situations where full bulkhead removal may not be possible or supported by the homeowner. Many of the credits include bonus points, awarded in addition to meeting specific credit requirements. Examples include implementing monitoring plans, applying the credit under specific situations, such as bulkhead removal in a documented forage fish spawning area, or providing a plan for managed retreat that considers and accommodates future sea level rise.

Although there are a total of 22 available credits, most shoreline projects will only qualify for 5 to 10 as many of the credits, particularly the higher point Coastal Process credits, are mutually exclusive.* This format allows the credit and rating system to apply to a broad range of project types and situations. Using the findings of pilot testing, two rating levels were defined based on the number of points achieved by an applicant. These are shown in Table 7.2.

In addition to the Credit Guide, an applicant checklist was developed for use by the homeowner or their contractor. The checklist provides a summary of the submittal requirements for each credit and identifies the points available including any bonus points. The applicant uses this form to apply for the credit points applicable to their shoreline project. In addition, there is a section of the form for use by the GSH Verifier to conduct an independent review of the project (see below). An example of the checklist for the Soft Shore Protection Credit is provided in Figure 7.4.

Resource materials found in the appendices of the *Green Shores for Homes Guide* provide additional information to assist in applying the various credits. One of the primary references is the *Marine Shoreline Design Guidelines* developed by the Washington Department of Fish and Wildlife (Johannessen et al. 2014). This document provides a comprehensive framework for determining the need for shore protection and the techniques that best suit the conditions at sites in the Salish Sea region.

7.4.2.2 GSH Pilot Properties

After completing an initial draft of the *Green Shores for Homes Credit and Rating Guide*, the GSH Steering Committee identified pilot properties to test the effectiveness of the credits. The

---

* For example, a property can only qualify for one of the Bulkhead Removal, Soft Shore Protection, or No Shoreline Protection Structures credits.

Credit 1.5: Soft shore protection				
To qualify for this credit, construct soft shore protection rather than hard shore protection structures anywhere shoreline erosion control is needed. Note that points cannot be earned for both this Credit and Credit 1.3 "bulkhead removal" for the same length of shoreline, except if a bulkhead is removed from a portion of the shoreline (Credit 1.3) and another portion of the shoreline that was previously unprotected is treated with soft shore methods (this credit).				
€	Submission requirements (see Credit 1.5 in guide for details)	Points available		Points applied for	Points verified
Basic:	a. On the **existing conditions plan**, show the extent of eroding shoreline and the major structures or improvements that are at risk from shoreline erosion. b. On the **site design plan**, show: • The extent and type of soft shore measures taken (added beach material, LWD, hard elements if needed, etc.), • A cross section of the soft shore protection design showing all soft and hard elements. c. Provide verification by a **qualified professional** that the soft shore treatment as shown is necessary to address the erosion. d. Provide **before and after photos** showing the eroding shoreline before and after soft shore treatment. Before and after photos should be taken from same vantage points and sight lines.	Soft shore measures used over 95–100% of shoreline; or	12		
		Soft shore measures used over 75–94% of shoreline; or	8		
		Soft shore measures used over 50–74% of shoreline; or	6		
		Soft shore measures used over 25–49% of shoreline; or	4		
		Soft shore measures used over 10–24% of shoreline.	2		
Bonus:	In areas where the beach and nearshore habitat have been degraded, provide **documentation** from a reliable source (e.g., scientific report, local fisheries authority, marine biologist, reputable stewardship organization) confirming that the soft shore measures **recreate, restore or enhance spawning habitat** for marine or freshwater fish and/or invertebrate species.		3		
Bonus:	Provide a **plan for monitoring** the soft shore project that may include (but not limited to) • Schedule for taking photos and measurement of key features site plan showing location of photo and measurement points • The features to be measured or observed; e.g., changes in vegetation line, beach substrate at specific spots (sand, gravel, cobble), log movement • The mid- to long-term maintenance measures required		2		
	Maximum points available/total points applied for:		**17**	**0**	**0**
Comments:				

Figure 7.4 GSH checklist for the soft shore protection credit.

pilots included two properties in a freshwater environment on Lake Washington and two marine waterfront properties in the San Juan Islands. In each case, completion of the projects occurred at least 1 year before the testing date. See Lake Washington example in Figure 7.5.

The purpose of the pilot was to test (a) whether the credits could be interpreted and applied at a site level, (b) whether the credits could lead to improved waterfront development design, and

(a)

(b)

Figure 7.5   Lake Washington pilot property (a) before and (b) after.

(c) whether the relative weighing of the credit points was allocated accurately. In addition, the pilot tests would help determine whether the credit checklist provided easily understood information for both applicants and reviewers, and whether it requested the necessary information to meet the credit requirement.

Technical team members conducted the pilot testing, first by receiving background information on the pilot properties, and then conducting site visits to evaluate each project. During the site visit, team members used the GSH checklist to determine an overall point "score" for each property. Afterward, they provided feedback on how the credits had applied to the properties, and reviewed whether the documentation provided had allowed for an adequate review.

The pilot testing program revealed a number of areas for improvement with both the credits and credit checklist, including a need to provide more freshwater-oriented credit points, better distinction between small urban and larger rural properties for certain credits, and a need to provide credit recognition for taking no shore protection action on a site.

Each of the four pilot properties achieved a score high enough to qualify as a GSH property, with three of the four achieving a Level 1 (Chinook) rating and one property achieving a Level 2 (Orca) rating. The Level 2 property was built on a high rocky bank where every effort had been made to minimize site disturbance during construction. The house on the Level 2 property had also achieved Platinum status under LEED for Homes; this provided extra points for the GSH assessment.

7.4.2.3 Role of Verifiers

As with other certification programs, independent, third-party review of an application is an important aspect of the GSH program. The GSH verifier conducts this review and may also act as a resource for the applicant during the application process. The homeowner (or contractor or engineer working with the homeowner) is responsible for the actual design of a project. It is not the role of a verifier to provide design or permitting advice to the homeowner; rather, the verifier is trained to help the homeowner understand the GSH program, including the credits and submittal process.

Verifiers should be locally based with a basic knowledge of the property development process, shore processes, and ecology, as well as an in-depth knowledge of the GSH credit and rating system. Verifiers are not required to be experienced coastal professionals. The prescriptive nature of the credits, with clearly defined, measurable outcomes, limits the amount of professional interpretation required of a verifier.

The technical team designed a 2-day program to train verifiers for the GSH program. Before the training course, each participant completed a pre-course assignment to familiarize themselves with the GSH Credit Guide and Checklist. Using the pilot properties, participants applied the GSH credits in both classroom and field exercises. Learning outcomes from the training session included the following:

- A working knowledge of the credit system
- The ability to conduct a screening of GHS credits applicable to a specific waterfront plan or design
- An understanding of the resources available to advise homeowners and their contractors to prepare necessary documentation for a GSH project
- The ability to verify credit submittals through review of application materials and site inspections

Participants were required to pass an examination based on an application submittal in order to demonstrate the above learning outcomes. Ten participants completed the 2-day training program and subsequent examination to become the first GSH verifiers.

7.4.2.4 Use of Incentives

As identified in the review of certification programs, offering a certification is typically insufficient to motivate a homeowner to take action (Rueggeberg et al. 2012, p. 38). Incentives can be provided in a number of different ways by government or nongovernmental entities. The GSH credit and rating program concept has always included some form of incentive for homeowner participation and achievement of specific rating levels, most likely delivered by the local jurisdiction participating in GSH.

Barriers to taking action and incentives to encourage property owners to change were the topic of a series of community meetings in Washington State (San Juan County and the City of Seattle) and British Columbia. These meetings included governmental regulators, professionals, and home-owners interested in the GSH program. Barriers identified at these meetings include permitting processes, lack of understanding of the ability to employ alternative shoreline protection designs, the perceived cost of these designs over traditional bulkheads or riprap, impacts on property value, and perceptions of what might be effective alternative techniques to address erosion on the shoreline.*

Incentives recommended for the GSH fell into four categories:

1. Technical Assistance: assistance with alternative protection design and riparian planting plans
2. Financial Incentives: grants and tax incentives
3. Education and Outreach: property owner demonstrations, technical training for consultants, architects, and landscape designers
4. Permitting Facilitation: permit exemptions, mitigation banks, streamlined, and consolidated permitting processes

Incentives explored in Washington during the pilot phase of the GSH program focused on education, outreach, and permit exemptions; however, in order to offer any permit-related incentives, each individual jurisdiction would need to agree to participate in the program. In BC, the four pilot communities focused their initial efforts on voluntary adoption through education and promotion of the positive benefits of GSH.

At the time of writing, no participating jurisdictions in Canada or the United States have implemented incentives programs. One BC local government included GSH as a policy in their Official Community Plan and others are considering its adoption at the planning level. Other jurisdictions have started processes to improve permitting efficiency for soft shoreline projects, and efforts are underway to coordinate these efforts with the GSH program.

7.4.2.5 Relationship to Regulatory Programs

GSH was conceived to be voluntary and incentive driven, rather than regulatory based or operated as a governmental program; however, many of the actions addressed by the GSH credits, such as bulkhead removal or soft shore protection, are highly regulated activities that require a permit in both the United States and Canada. Agencies at local, state/provincial, and federal levels review shoreline projects to ensure that they are safe, protect aquatic habitats and species, maintain water quality, and preserve public lands and interests.

In developing the GSH credits, attention was given to consistency and compliance with many of these key regulatory programs, particularly those in Washington State. Review of the credits by regulatory authorities provided a level of confidence that actions taken to meet credit requirements would not run contrary to permit authority. Since the two primary recipients of the EPA grant were local jurisdictions (Seattle and San Juan County), the GSH program was developed to coordinate with their local permit process.

7.4.2.6 GSH Program Delivery

The GSH Credit and Rating Guide and Checklist form the technical backbone for the GSH program. The long-term goal is to deliver the program through a third-party administrator similar to the LEED, Built Green, and Sustainable Sites models. Key aspects of program delivery include registration in the program, preparation of the GSH submittal, and postconstruction verification of the submittal.

---

* These findings are consistent with information from the B.C. meeting documented in a report by Modus (2015). Green Shores for Homes Pilot Project Summary Report, prepared for Stewardship Centre for B.C. Retrieved on September 22, 2015 from http://www.stewardshipcentrebc.ca/green-shores-for-homes-pilot-project-summary-report/

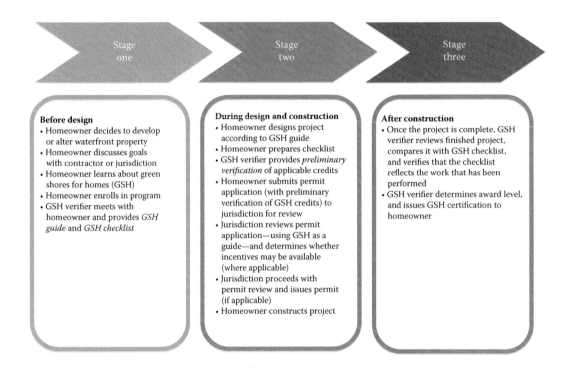

Figure 7.6   GSH design and rating process.

Homeowners will be able to register in the GSH program either through an online process available on the GSH website or by filling out the application form and e-mailing it to the GSH administrators.* Once an application has been received, a verifier is assigned to work with the homeowner to support their GSH application. The homeowner is responsible for compiling all documentation required for any credit for which they will apply. The verifier may meet with the homeowner to screen available credits; however, the verifier does not provide design advice or permitting assistance. This allows the verifier to retain an independent relationship with the design and regulatory aspects of the project as well as their subsequent verification of the project submittal. Figure 7.6 outlines the three stages of program delivery and describes how the GSH program intersects with a typical local permitting process.

7.5 GREEN SHORES AS A TOOL FOR CLIMATE CHANGE ADAPTATION

While the initial impetus for the Green Shores program focused on reducing impacts on near-shore habitat and coastal processes, adaptation to climate change is another reason for reducing hardscape such as bulkheads and armoring as hardscape solutions may not be as effective at preventing catastrophic storm damage as the softer approaches promoted by the Green Shores program.† In addition, Green Shores provides a number of credit points for climate change adaptation actions under the GSH building setback and managed retreat credits. The GSCD rating system has a dedicated credit for the production of a climate change adaptation plan that addresses one or more of the four approaches to adaptation (avoid, protect, accommodate, and retreat), with higher point allocations given to avoidance or retreat strategies.

* Retrieved from http:/ www.greenshoresforhomes.org
† These are the general conclusions found in the report Living Shorelines: From Barriers to Opportunities, Restore, America's Estuaries 2015.

In order to better understand the implications for the use of the Green Shores program for climate change adaption, SCBC—with funding from Natural Resources Canada—commissioned an engineering study to test several case examples and evaluate the effectiveness of soft shore armoring alternatives compared to an equally appropriate hard alternative, based on the following criteria (Lamont et al. 2014):

- Adaptability to climate change and related sea level rise
- Effectiveness in protecting the shoreline against flooding
- Effectiveness in providing ecological resilience
- Relative cost, considering initial capital cost, maintenance cost, and long-term replacement cost

The study evaluated three soft shore alternatives described as follows:

- Case Study #1: Use of beach nourishment, consisting of supply and placement of typical beach intertidal materials, ranging from sand to gravel/cobble mixture
- Case Study #2: Use of nearshore rock features, including boulder clusters and intertidal and subtidal rock habitat reefs
- Case Study #3: Use of a typical headland-beach system to maintain a conventional beach intertidal substrate in an area exposed to waves from more than one primary direction

The evaluations outlined in the report demonstrate that soft shore approaches can provide effective flood protection against climate change–related sea level rise and related issues. The three case examples demonstrated that soft alternatives for flood protection provide significant cost advantages over hardscape alternatives. The margin of cost saving varies, ranging from 30% to 70% of the cost of a comparable hardscape alternative. Moreover, an ecological services evaluation framework developed for this study demonstrated that soft alternatives provided similar or improved assessment scores for shoreline ecological resilience.

This study provided compelling reasons to consider design alternatives as put forward by the Green Shores program in the context of climate change and sea level rise adaptation. As a result of this study, there is ongoing work with multiple levels of governments, conservation organizations, and professional associations* to raise awareness of the use of Green Shores as a tool when addressing climate change adaptation.

7.5.1 The Need for Multidisciplinary Approaches to Shore Development

Development design and construction in the shoreline environment requires a multidisciplinary approach to address a range of complex issues. In order to develop a meaningful Green Shores rating system, credit development required the expertise of marine biologists, coastal geomorphologists, landscape architects, coastal engineers, shoreline planners, and ecologists.

This approach ensured that the broad range of issues associated with shoreline development could be addressed. Marine and freshwater biologists are critical for understanding the impacts of shoreline development on marine and riparian habitat. Coastal engineers and geologists provide structural design and physical process considerations unique to the shoreline environment. Landscape architects understand the design principles inherent in the integration of land and water and also bring knowledge of the role vegetation has on a healthy shoreline environment.

---

* For example, Green Shores is referenced in the British Columbia Sea Level Rise Primer and the BC Ministry of Environment's Climate Action Secretariat regularly participates on Green Shores Advisory Committees. SCBC is also working with the Professional Association Climate Change Adaptation Working Group—a group of professionals from the engineering, planning, biology, nongovernmental, and forestry sectors—to link these professionals to Green Shores training opportunities.

Land use planners understand the interplay between regulatory requirements and development standards.

It is not always feasible to assemble a full multidisciplinary team for a shoreline development project, particularly for single resident waterfront development. For this reason, the GSH guide and credit descriptions focus on design outputs (e.g., percentage of riparian zone conserved or restored) and provide design guidance to help achieve the required outputs. The credit guide identifies key areas or situations where professional advisement is recommended.

For larger shoreline development projects, such as those addressed by the GSCD, there are requirements for professional signoff on specific coastal process and habitat credits. This approach recognizes the sustainability benefit of forming multidisciplinary team in the early project design phase. The Green Shore program encourages the use of an integrated design process, which involves inclusive, collaborative, and holistic elements at all aspects of the design process.* The application of this approach and the guidance offered by the GSCD and GSH credit systems at the project level ultimately leads to positive human and ecological outcomes.

7.5.2 Professional Development Training and Outreach

Outreach and education efforts based on GSCD and later GSH have been ongoing in British Columbia over the program development period. Four workshops held in 2013 involved more than 150 participants including government agencies, developers, shoreline professionals, conservation organizations, and shoreline property owners. Workshop participants had the opportunity to learn about the GSCD credit and rating system and its application, receive updates on GSH, and share shoreline experiences. Some workshops included field trips to projects that had used Green Shores approaches. Participants recognized the multidisciplinary nature of the program and discussed the challenges to implementation. Workshop attendees noted the need for education and outreach to increase understanding of the issues in the community and secure greater buy-in on the value of the program. In 2014, SCBC initiated a multijurisdictional study to develop strategies for implementation of GSH programs in British Columbia. The study, *Green Shores for Homes; BC Pilot (2015)*, identified two key implementation strategies focused on education and outreach:[†]

1. A Green Shores educational and promotional initiative
2. A Green Shores professional training and certification program

After completion of this study, SCBC began work to develop training courses for Green Shores in British Columbia. The target audience will include landowners, local government staff, elected officials, conservation organizations, and coastal professionals. SCBC received funding for this new initiative to work with the University of Victoria and British Columbia Institute of Technology to develop and deliver Green Shores training in three regions of British Columbia. In addition, accessible, creative, and highly visual training materials will be developed, including videos, posters, FAQs, info graphics, and other community-based social marketing materials. All materials will be linked to the Green Shores[‡] and GSH[§] websites. In this way, the educational materials and learning outcomes will be directly transferrable to the Washington State Green Shores partners.

Two levels of training will be developed and delivered: Introduction to Green Shores (Level I) and Advanced Green Shores (Level II). The courses will introduce participants to the purpose

* For a full description of the IDP, see the *Integrated Design Process Guide*, Canadian Mortgage and Housing Corp. https://www.cmhc-schl.gc.ca/en/inpr/bude/himu/coedar/upload/Integrated_Design_GuideENG.pdf
† Other key strategies include working with local government and others to streamline approvals and developing targeted financial incentives.
‡ Retrieved from http:/www.greenshores.ca
§ Retrieved from http:/www.greenshoresforhomes.org

and application of Green Shores practices, how these practices can aid in climate change adaptation, and the roles and responsibilities of government, landowners, and professionals in shoreline development. Professionals completing the Level II training will form the initial registry of trained professionals available on the Green Shores websites.

Outreach and education in Washington on the use of soft shoreline alternatives began before the initiation of the GSH program. These efforts, spearheaded by local and state agencies, have begun to incorporate GSH into training and outreach efforts. During the initial phase of program development, Washington participants in GSH focused on development of the credit document and the website. Later phases for the program in both Washington and BC will include outreach and education to jurisdictions to encourage wider support and willingness to offer incentives such as reduced permit fees or streamlined permitting processes. In addition, later stages of program development will include outreach to homeowners to encourage their participation. These efforts will be coordinated with other outreach efforts to reduce the amount of armoring in the region.*

7.6 CONCLUSION

Green Shores provides a flexible program that encourages homeowners to protect or restore their shorelines, and address coastal impacts associated with climate change through the use of natural or "soft" solutions. Although voluntary, the Green Shores program provides options for regulatory authorities to adopt the credit criteria within their jurisdictional regulations or requirements. As a voluntary program, the development and coastal professional community rather than regulators can drive it. Unlike LEED, Green Shores has yet to establish widespread brand recognition in order to deliver a marketing advantage. As a technique to promote alternative shoreline stabilization, property owners, developers, and coastal professionals need a demonstration of the value-added component of Green Shores to justify the extra effort required to obtain certification; however, the benefit of a rigorous and transparent certification process provides credibility and will increase market demand and brand recognition over time.

The GSCD and GSH program development and pilot phase testing is largely completed and has been well received by land use professionals, governments, academia, and the public in both Washington State and British Columbia† and there is strong interest in continued growth of the program by local governments. The key challenge for the next phase of Green Shores is moving from the development and piloting phase to a fully operational program providing all the elements of the rating and verification process outlined in Figure 7.6. Key issues to be addressed to achieve a fully operational program include the following:

- Greater recognition of Green Shores approaches to ecological and human well-being values. This can be achieved by continued outreach to the existing Green Shores network as well to real estate markets, the green building community, and other professional networks.
- Specific incentives offered to property owners for achieving specific GSH rating levels (Chinook or Orca). To date, partner jurisdictions have considered use of incentives but none have implemented a dedicated GSH incentive program.
- Development of organizational leadership and capacity for coordinating and delivering Green Shores programs in British Columbia and Washington State. The Stewardship Centre for British Columbia and Washington Sea Grant are leading program delivery on each side of the border;

---

* Two significant programs in place that have partnered with the GSH program include the "Shore Friendly" effort initiated by the Washington Department of Fish and Wildlife and the Green Shorelines effort promoted by the Lake Washington, Cedar River, and Sammamish Watershed.
† Green Shores was recognized as a finalist in the Land Awards by the Real Estate Foundation of British Columbia in 2010 and also won Best Environmental Idea at Simon Fraser University's RISE competition in 2014 for ways to address sea level rise in coastal British Columbia.

however, development of a sustained operation and business model along with funding mechanisms remains to be completed and implemented.

- Need for local government champions. Successful implementation of the GSH programs requires a close partnership with a coastal community or region. The GSH program needs to be perceived within coastal communities as being supported by their municipal or regional (county) government in terms of both incentive delivery and harmonization, to the degree possible, with local regulations.

A recent report by Restore America's Estuaries (2015, p. 6) lists three major barriers to the implementation and broader use of Living Shoreline approaches to shoreline development and management:

1. Institutional Inertia: both regulators and coastal professionals being more locked into traditional approaches
2. Lack of a Holistic Context: regulators, planners, and designers do not consider system-wide or cumulative impacts when considering shoreline development on a project-by-project basis
3. Lack of an Advocate: there is no single agency or jurisdiction committed to promoting and facilitating Living Shoreline approaches to shoreline management

Recommended strategies to address these barriers include education and outreach, regulatory reform, improved institutional capacity, and public agencies leading by example.

The Green Shores program offers opportunities to address these barriers, including the following:

1. A sound technical framework focused on holistic (integrated) design approaches with a clear emphasis on cumulative impacts
2. Guidance and a process for Living Shoreline approaches that agencies and local jurisdictions can adopt to overcome "institutional inertia," which is caused in great part by lack of a clearly articulated alternative
3. A focus on a diverse outreach and education program at several levels to meet the needs of homeowners, the real estate industry, planners, and regulators, as well as coastal professionals

There is a growing awareness in the Salish Sea region by local, state, and provincial governments about the impacts of hardscape shore protection on the natural environment and the need to address adaptation to climate change. Washington State has implemented regulations at several agency levels to restrict bulkhead installation and encourage bulkhead removal.* In British Columbia, the recognition of the need to respond to climate change impacts has motivated local governments to seriously consider alternative approaches to traditional shoreline hardening. The Green Shores program offers these jurisdictions with guidance, a process, and potentially incentives to achieve these goals.

The Green Shores program chose to adopt the ratings and certification model as a method of facilitating change in shore development and management practices—an approach modeled on successful Green Building initiatives such as LEED. However, this approach presents implementation challenges including demonstration of the market advantage of certification, ability for local governments to provide viable incentives for homeowners to participate in GSH certification, and the need for a financially viable and sustainable organization to operate the program.

---

* The State of Washington Shoreline Management Act directs local governments to adopt shoreline master programs (SMPs) that "protect and restore the ecological functions of shoreline natural resources." SMPs should allow structural shoreline modifications only where they are demonstrated to be necessary to protect a primary structure [WAC 173-26-176 (3)(c)]. In 2014, the Washington Department of Fish and Wildlife revised their Hydraulic Project Approval to require more stringent review of proposed bulkhead proposals. The new guidelines require single-family homeowners to "use the least impacting technically feasible alternative" to address bank protection [WAC 220-660-370 (3)(b)].

Soft shore alternatives are an important and critical way for property owners to address erosion in an ecological manner while also mitigating the impacts of increased coastal inundation attributed to climate change. The Green Shores program provides an important pathway for property owners, with the support of local jurisdictions, to protect and restore their shorelines, conserving and enhancing coastal ecology and human well-being for future generations.

REFERENCES

Beamish, R. and G. McFarlane. 2014. *The Sea among Us: The Amazing Strait of Georgia*, Harbour Publishing, Madeira Park, B.C.

Cole, R.J. 2005. Building environmental assessment methods: Redefining intentions and roles. *Building Research & Information* 33: 455–467.

Climate Action Secretariat (CAS) British Columbia Ministry of the Environment. 2013. *Sea Level Rise in BC: Mobilizing Science into Action*. Victoria, B.C.

DeLella Benedict, A., and J.K. Gaydos. 2015. *The Salish Sea: Jewel of the Pacific Northwest*, Sasquatch Books, Seattle, Washington.

Johannessen, J., A. MacLennan, A. Blue, J. Waggoner, S. Williams, W. Gerstel, R. Barnard, R. Carman, and H. Shipman. 2014. *Marine Shoreline Design Guidelines*. Washington Department of Fish and Wildlife, Olympia, Washington. Retrieved from http://wdfw.wa.gov/publications/01583/

Lamont, G., J. Readshaw, C. Robinson, and P. St-Germain. 2014. *Greening Shorelines to Enhance Resilience, an Evaluation of Approaches for Adaptation to Sea Level Rise*, prepared by SNC-Lavalin Inc. for the Stewardship Centre for B.C. and submitted to Climate Change Impacts and Adaptation Division, Natural Resources Canada (AP040).

Modus. 2015. Green Shores for Homes Pilot Project Summary Report, prepared for Stewardship Centre for B.C.

Restore America's Estuaries. 2015. Living Shorelines: From Barriers to Opportunities. Retrieved from https://www.estuaries.org/images/stories/RAEReports/RAE_LS_Barriers_report_final.pdf

Rueggeberg, H., B. Emmett, and K. Litle. 2012. Review of Rating System and Incentive Programs for Residential Development, prepared for Green Shores for Homes Steering Committee.

Sustainable Sites Initiative. 2009. Guidelines and performance benchmarks. American Society of Landscape Architects, Lady Bird Johnson Wildflower Center and U.S. Botanic Garden.

CHAPTER 8

Building with Nature as Coastal Protection Strategy in the Netherlands

**Bas W. Borsje, Sierd de Vries, Stephanie K.H. Janssen,
Arjen P. Luijendijk, and Vincent Vuik**

CONTENTS

8.1 INTRODUCTION

Building with Nature solutions can contribute to the reduction of flood risks and at the same time increases ecological and socioeconomical values in coastal areas (De Vriend et al. 2014). Building with Nature is a Dutch term and many more similar concepts are being introduced worldwide, all focusing on the interplay between physics, ecology, and governance to guarantee safety against flooding: Ecological Engineering (Costanza et al. 2006), Ecosystem-based Management (Barbier et al. 2008), Working with Nature (PIANC 2011), Managed Realignment (French 2006), Nature-based

Flood Protection (Vuik et al. 2016), and Socio-ecological Resilience (Adger et al. 2005). Despite the varying terminology, all approaches are struggling with the implementation of nature-based solutions as innovative flood risk reduction measures, because of the fundamental uncertainties in the stability, efficiency, and especially long-term sustainability of these solutions and the multifunctional nature involving different types of stakeholder coalitions (Borsje et al. 2011).

In this chapter, we aim to draw lessons for managers and scientists involved in the implementation of nature-based solutions for coastal protection by focusing on three Building with Nature case studies within the Netherlands: shoreline nourishments, a mega-nourishment, and vegetated foreshores (salt marshes). Within these case studies, we (1) present a background and motivation, (2) discuss field observations, (3) focus on the lessons learned during and after the implementation phase, and (4) discuss paths forward in which we focus on research needs and strategies for working across different disciplines and overcoming implementation barriers. Next, we discuss the general transition in coastal protection policy in the Netherlands and the main challenges in implementing Building with Nature as a coastal protection strategy in the Netherlands. Finally, some general conclusions are presented.

8.2 DYNAMIC PRESERVATION OF THE COASTLINE: SHORELINE NOURISHMENTS

8.2.1 Background and Motivation

In the Netherlands, the coastal zone largely consists of vegetated dunes, beaches, and a mildly sloping foreshore toward the relatively shallow North Sea (see Figure 8.1 for geographic reference). The almost 120-km-long Holland coastal zone mainly consists of noncohesive sediment. The Holland

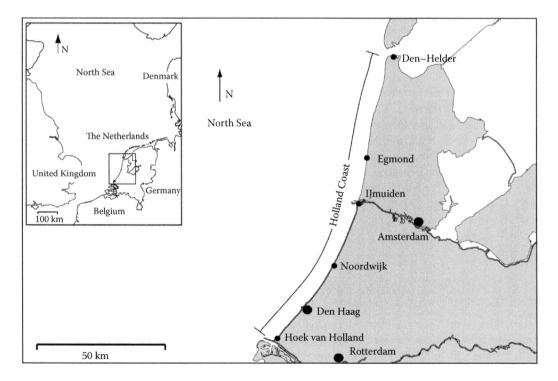

Figure 8.1 Overview of the almost 120-km-long Holland Coast. Nearly the entire Holland Coast consists of sandy beaches and vegetated dunes.

coast forms an important barrier that protects large parts of the Netherlands from flooding. It is in the interest of national safety that this barrier is maintained.

The coastal system of the Holland coast is characterized by an erosive trend. This natural erosion has led to loss of land and infrastructure in the past. Consequently, it has also led to the initiation of significant mitigation measures and maintenance strategies. During the last few decades, a maintenance strategy using sand nourishment has been adopted to compensate for the natural coastal erosion (Davison et al. 1992). This strategy is called *dynamic preservation* and was implemented in national law in 1990. This dynamic preservation strategy is typically a Building with Nature solution and aims to combine safety against floods with sustainable preservation of other services and values (ecological and recreational) of the dunes and beaches (de Ruig 1998). Changes after the implementation of the dynamic conservation policy manifested themselves in a stable or seaward migrating coastline location while allowing for dynamic coastline development. Dynamic preservation requires new methods for evaluating the evolution of the coastal system.

Dutch national government relies on engineering tools such as numerical models, field data analysis, and expert judgment for assessing coastal safety and designing mitigation measures for coastal erosion. These assessments and mitigation measures are adapted to the latest scientific knowledge. In parallel, there has been significant scientific development for predicting coastal evolution along the Holland coast on different time scales.

8.2.2 Field Observations

Scientific evidence considering different time scales helps explain and quantify the natural erosion along the Holland coast. Regarding geological time scales, Beets and Van der Spek (2000) explained the ongoing erosion by studying changes in sea level rise and sediment budgets in the Holocene period. They showed that in the current interglacial period, sediment supply to the coastal system has decreased significantly, resulting in a shift from long-term sediment accumulation in the coastal system to long-term erosion. On shorter time scales, the erosion of the coastal system is visible where a yearly sediment deficit at many locations along the Holland coast is measured (see, for instance, van Rijn 1997).

The awareness of the erosion problem of the Holland coast led to an increasing interest in measuring and predicting the evolution of the coastal system under natural and anthropogenic forcing. Such knowledge development is partly facilitated by a government-funded large measurement program called JARKUS (JAaRlijkse KUStmeting; yearly coastal measurement). JARKUS provides yearly measurements of the coastal morphology covering the subaqueous and aeolian coastal zones since 1965. Before JARKUS and since the 1860s, the coastline location with respect to a reference was measured at 1-km alongshore intervals. These long-term data sets are unique in the world and allow for studies on sediment budgets as well as geomorphology of the Dutch coastal system. By using these data, the development and behavior of the Holland coast are described in detail and sediment budgets are estimated and related to environmental conditions (see, for instance, van Rijn 1997).

When describing the Holland coastal system, the development of the coast owing to hydrodynamic forcing is often separated from development owing to aeolian forcing. The problem of coastal erosion is mainly driven by hydrodynamic forcing. Moreover, sediment budgets attributable to hydrodynamic forcing are relatively large with respect to aeolian forcing. Therefore, the subaqueous development of the coastal system has received much more attention compared to the aeolian development of the coastal system. For instance, significant variability in the behavior of the subaqueous system of sandbars is described comprehensively using the JARKUS data set (e.g., Wijnberg and Terwindt 1995). However, the specific effects of bar behavior on sediment budgets of the coastal systems and the link between foreshore, beach, and dune development remain unclear (de Vries et al. 2012; Ruessink and Jeuken 2002).

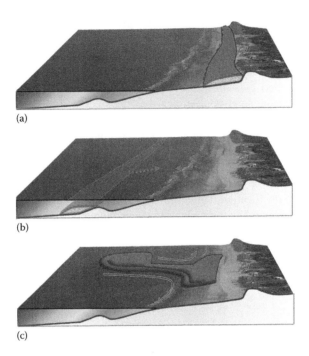

Figure 8.2 Change in nourishment strategies from (a) traditional beach and dune nourishments to (b) shore-face nourishments, and (c) localized mega-nourishments. (After Stive, M. J. F., De Schipper, M. A., Luijendijk, A. P., Aarninkhof, S. G. J., Van Gelder-Maas, C., Van Thiel de Vries, J. S. M., De Vries, S., Henriquez, M., Marx, S., and Ranasinghe, R. (2013). A new alternative to saving our beaches from sea-level rise: The sand engine. *Journal of Coastal Research*, 29.)

While gaining knowledge, experience, insights, and associated tools with respect to coastal systems and nourishments in combination with increasing technological possibilities in the dredging industry, nourishment strategies at the Dutch coast have changed over time. An overview is given in Figure 8.2. Traditionally, relatively small amounts of sand were placed on the beach and the dunes after an erosive event such as a storm. With the introduction of dynamic preservation in the 1990s, the policy has changed from a reactive to a proactive strategy to mitigate erosion. Consequently, nourishment volumes increased and foreshore nourishments were introduced as a measure to protect the coast from erosion attributed to future sea level rise (Stive et al. 1991). Foreshore nourishments are applied outside the breaker zone with the expectation that natural processes will cause the sediment to spread over the full cross-shore profile. Typically, a shoreface nourishment consists of approximately $1–2 \times 10^6$ m^3 of sand and has a lifetime of approximately 3–5 years (Hamm et al. 2002). It is expected that the nourishment volumes required will increase in the near future because of accelerated relative sea level rise and increasing storminess. To accommodate larger nourishment volumes and to experiment with a more sustainable approach to nourishing the coastline, the sand motor pilot project was implemented in 2011 (Stive et al. 2013). The sand motor pilot project is discussed in detail in Section 8.3.

8.2.3 Lessons Learned

The dynamic preservation policy is considered successful from different points of view. Coastal erosion is halted while dynamics and functions of the coastal system are maintained. It has also proven to be affordable, and the use of coastal nourishments to facilitate a seaward migrating coastline has been suggested and sporadically explored.

As a result of the dynamic preservation strategy, the behavior of the Dutch coastal system is inseparably connected to anthropogenic forcing. The change from erosion control and stabilizing the coastline toward the policy of dynamic preservation in 1990 has influenced the development of the coast significantly. In the subaqueous domain, the behavior of the subtidal bar systems can be locally disturbed by the nourished sediment (Grunnet and Ruessink 2005; van Duin et al. 2004). In the aeolian domain, it has been reported that the transition from an eroding coast to a stable and accreting coast had significant effects on habitat development of the dune vegetation (Arens et al. 2013).

Nevertheless, the multifunctional nature of shoreline nourishments require broadening of the traditional actor coalitions, since functional integration in Dutch coastal management is limited at the moment (Hermans et al. 2013).

8.2.4 Path Forward

Despite important steps being taken in the prediction of the behavior of the Dutch coastal system and coastal systems in general, there are many unknown processes governing changes in the coastal system. From a reductionist point of view, hydrodynamic modeling has reached a certain maturity. The accurate description of sediment transport processes is however still largely empirical and site specific. Many formulations for sediment transport rely on the extrapolation of empirical knowledge based on laboratory studies or occasional field data. The effects of ecology and the difficulties imposed by the land–water interface that characterizes the coastal system are especially poorly understood and complicate predictions of coastal evolution even more. A general interdisciplinary data set and model approach on coastline evolution as a function of environmental processes is needed to gain knowledge on these topics.

The Netherlands is an example of a society that depends on a safe and healthy coastal system. This societal relevance has resulted in coastal maintenance to be implemented in national law. Moreover, strong knowledge alliances have been established between universities of different disciplines, research institutes, industry, and government. An important organizational platform for this alliance is the Dutch network of coastal science (NCK) that organizes regular thematic days and yearly conferences to discuss collaboration and progress at the different organizations. This alliance plays an important role in facilitating the longevity of the dynamic preservation approach to Dutch coastal maintenance and associated research. Moreover, this alliance is instrumental in establishing innovative and applied research programs regarding coastal development in the Netherlands and beyond, such as the Dutch Building with Nature program (De Vriend et al. 2014).

Pilot projects and field studies are an ideal means to facilitate interdisciplinary coastal research to increase predictability of coastal systems. Shared and open strategies for data collection during pilots and field studies allow for a generic approach to coastal research. This should ideally not be limited to an institutional or even a national level. Modern techniques allow for an international and integrated approach to measure, predict, and manage coastal systems. Therefore, a global approach to knowledge development with respect to coast-related problems and solutions is both desired and possible.

8.3 MEGA-NOURISHMENTS AS AN INNOVATIVE NOURISHMENT STRATEGY: THE PILOT PROJECT SAND ENGINE

8.3.1 Background and Motivation

In 2008, a national committee, the Delta Committee, provided critical advice for protecting the Dutch coast and the low-lying hinterland from the consequences of climate change in the 21st century (Deltacommissie 2008). In line with a key recommendation of the Delta Committee, an innovative pilot project was developed to achieve a more efficient and sustainable nourishment approach: the Sand Engine (Stive et al. 2013). This mega-nourishment, built in 2011 along the Delfland coast (see Figure 8.3a), consists of a total sediment volume of 21 million m³. The pilot project Sand Engine is a combined beach/shoreface nourishment and consists of a man-made peninsula of approximately 128 ha (see Figure 8.3b). This new coastal maintenance strategy was designed to use the power of winds, waves, tides, and currents to help protect part of the Dutch coast from storms (van Slobbe et al. 2013), while encouraging the development of new dunes and the valuable flora and fauna associated with them and stimulate recreation. It is expected that over the next 20 years, these natural coastal processes will redistribute the sand in the peninsula along the 16-km-long coastal stretch between Hoek van Holland and Scheveningen, leading to an increase of the footprint of the dunes of 33 ha (Mulder and Tonnon 2010).

The Sand Engine pilot consists of a large peninsula of approximately 2 km alongshore while the most seaward position protrudes approximately 1 km into the sea (see Figure 8.4a). The main peninsula part is hook shaped with the outer tip curved toward the north. The crest of the peninsula rises up to 5 m above mean sea level. This design and location best fulfilled the multidisciplinary and multi-stakeholders requirements of safety in combination with recreation, nature development, and scientific innovation (Stive et al. 2013). The cross-shore slope of the peninsula is 1:50, such that the toe of the nourishment is positioned at MSL −8 m and ~1500 m from the original coastline. The northern tip of the peninsula creates a sheltered nurture area for different biotic species. A small lake of approximately 8 ha is intended to prevent the freshwater lens in the dunes from migrating seaward and endangering the groundwater extraction from the existing dune area. Sediment for the nourishment was mined at two borrow sites just beyond the 20-m-depth contour at a distance of approximately 9 km offshore. The sand was mined by Trailing Hopper Suction Dredgers and placed at the location of the Sand Engine by a combination of dumping through the doors in the hull, rain bowing, and pumping onto the beach. The Sand Engine was constructed in a period of approximately 3 months in the summer of 2011.

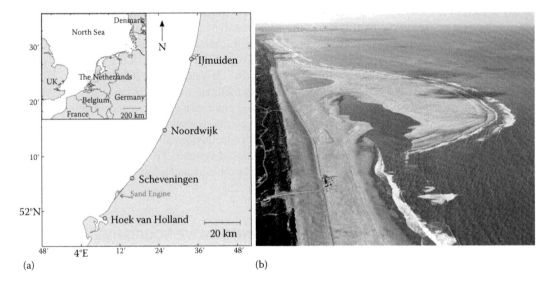

Figure 8.3 (a) Overview of the Dutch coast and the location of the Sand Engine and (b) aerial photograph of the Sand Engine after completion (July 2011).

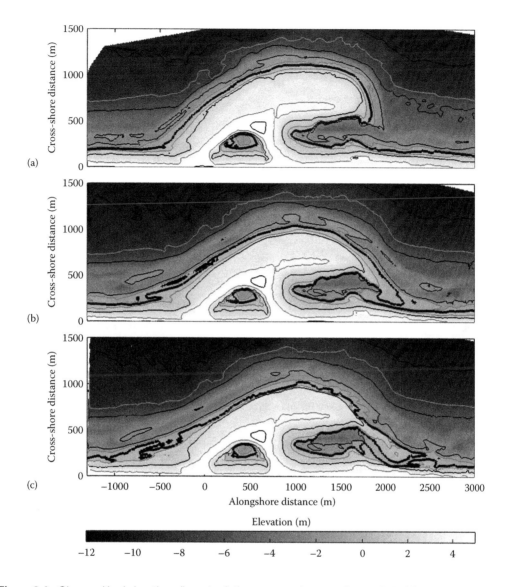

Figure 8.4 Observed bed elevations (in meters); three surveys between August 2011 (a) and August 2012 (c).

8.3.2 Field Observations

Monthly bathymetric surveys show a rapid, predominantly alongshore redistribution of sediment (Figure 8.4). The head of the peninsula eroded rapidly, leading to accretion in both northerly and southerly direction. In the first half-year after implementation, a spit developed from the northerly tip of the peninsula pinching the lagoon entrance. The maximum elevation of the spit and shoal were slightly below the high water level, such that they flooded during high tide (and storms). The channel landward of the shoal discharged the flow into and from the lagoon and strong flow velocities of over 1 m/s were observed here during rising and falling tides in the spring of 2012.

Over time, the transition between the mega-nourishment and the adjacent coast became more gradual (see Figure 8.5). After 2.5 years, approximately 2.5 million m^3 of sand was moved by

Figure 8.5   Aerial photograph of the Sand Engine approximately 4 years after construction (taken in May 2015).

natural forces. Most of the sand (1.14 million m³) was transported to the north, while approximately 680,000 m³ accreted south of the original peninsula. The remaining 740,000 m³ is transported outside the survey area: either cross-shore (to deeper water or into the dunes) or alongshore (past the limits of the survey area). Further analyses and additional monitoring will focus on understanding these sediment fluxes and balance.

The Sand Engine is a dynamic geomorphological feature: more than 10% of the sand has been moved in the first 2.5 years. Observations revealed different sediment pathways depending on sediment grain sizes. The most seaward part of the Sand Engine has eroded significantly. In this area, relatively coarse sand has been found, while northeast and southwest of the head of the Sand Engine, fine sand patches are observed owing to deposition of the material eroded from the head. The spatial variation in sediment sizes could, for example, affect the decolonization of benthic species over time. On a longer term, the coarsening of the sediments at the head of the Sand Engine could also affect the overall morphological evolution.

Ecologically speaking, the Sand Engine exhibits interesting developments (Linnartz 2013). Soon after construction of the Sand Engine, pilot vegetation started to grow at the crest combined with the development of embryo dunes (see Figure 8.6). Pioneer plant species have been found such as Sea Rocket (*Cakile maritima*) and Glasswort (*Salicornia maritima*). Even a very rare species, the Frosted Orache (*Atriplex laciniata*), unique for this part of the Dutch coast, has been found. As zonation has not been implemented at the Sand Engine, plants and embryo dunes are regularly affected by human behavior related to recreation and regular safety patrols with 4 × 4 cars. Dune growth has been observed behind the Sand Engine and just north and south of the pilot area, while the crest has become very stable as a result of armoring effects; shells stabilize the bed and prevent it from eroding owing to aeolian transport. The Sand Engine turns out to be a favorite resting area for birds and seals, and the lagoon is full of juvenile fish.

Figure 8.6 Photographs of the Frosted Orache (*Atriplex laciniata*) and embryo dune development at the crest of the Sand Engine.

8.3.3 Lessons Learned

In order to learn from this pilot project, research projects have been implemented to analyze the observations in detail. In the first years, the Sand Engine has changed drastically. The sand has been gradually redistributed by natural processes over the shoreface, beach, and dunes. Bathymetric measurements combined with numerical modeling illustrate the redistribution mechanism of the Sand Engine. After the first 2.5 years, the sand has been redistributed alongshore over a coastal stretch up to 8 km (De Schipper et al. 2016). The erosion of the peninsula is mostly dominated by waves—during large storms and otherwise. Analysis has shown that the 12 largest storm events of the first year were responsible for approximately 60% of the total erosion observed in that year (Luijendijk et al. 2015). The milder wave conditions, with a large probability of occurrence, are therefore almost as important to the erosional behavior of the peninsula as the storm conditions.

The strategy of introducing concentrated nourishments is seen as a climate-robust and eco-friendly means of countering coastal erosion, while the (temporary) presence of surplus sand also creates new areas for nature and recreation. The Sand Engine with its specific dimension forms a unique solution for the Delfland coast and is therefore not easily transferable to other locations in the world. However, the concepts underpinning this multifunctional coastal protection are applicable to other locations and environments, but are highly dependent on the local governance setting and ecological and physical system. An inventory on the relevant ecosystem services and a stakeholder analysis are crucial starting points when exploring sandy strategies at other locations.

8.3.4 Path Forward

After implementation of the Sand Engine, a collaborative effort between public authorities, private companies, and research institutes has resulted in the fact that the Sand Engine has become a focal point for coastal research and an innovative coastal management solution. In the coming years, the Sand Engine will be closely monitored and extensive research programs that will include detailed studies of the evolution of the Sand Engine and the driving mechanisms behind it—physical, ecological, and social–economical—will be defined. An example is the science project NatureCoast (www.naturecoast.nl), in which 15 PhD students and 6 postdocs are studying the temporal behavior of this mega-intervention along six different disciplines: coastal safety, marine and terrestrial ecology, hydrology, geochemistry, and governance. The new knowledge will be beneficial for future nourishment strategies and design. This research will enable an integral evaluation of the effectiveness of local mega-nourishments as a countermeasure for the anticipated enhanced coastal recession caused by accelerated sea level rise in the 21st century.

8.4 ADOPTING SALT MARSHES AS COASTAL PROTECTION MEASURE

8.4.1 Background and Motivation

Salt marshes are coastal ecosystems in the upper coastal intertidal zone between land and water that are regularly flooded by tides and surges. They are covered with dense stands of salt-tolerant plants, such as herbs and grasses. Salt marshes can exist in areas that roughly comply with two criteria. First, the hydrodynamic forcing by waves and currents should be limited to facilitate the settlement and accumulation of fine sediments and the establishment of pioneer vegetation. Second, there should be a sufficient supply of sediment, to allow the surface of the salt marsh to rise during high tides and storm surges (Van Loon-Steensma 2015). Vegetation plays a key role in the development of salt marshes. The presence of the canopy accelerates the sediment settlement by reducing the wave forces on the bed material. Additionally, the root systems of the plants stabilize the accumulated sediments and amplify the process of subsoil drainage, consolidation, and compaction (Deegan et al. 2012). Salt marshes and the neighboring intertidal flats form a coherent system with many mutual dependencies.

Salt marshes occur worldwide, particularly in middle to high latitudes (Figure 8.7). They are a common habitat in estuaries and barrier coasts. For example, extensive salt marsh systems can be found in the Dutch, German, and Danish Wadden Sea and along the Gulf Coast of the United States.

Vegetation requires a certain time span for seedling establishment, during which disturbance is low or absent. Such periods, characterized with mild wave conditions and limited flow velocities, are often referred to as windows of opportunity (Balke et al. 2011; Romme et al. 1998). A distinction can be made between salt marshes that have developed in areas where windows of opportunity occur by nature and salt marshes that can only persist because of artificial sheltering from waves and currents. An example of the latter can be found along the Dutch Wadden Sea coast, where an extensive system of earthen dams, brushwood dams (Figure 8.8), and drainage channels has led to the presence of 3000 ha of salt marsh habitat (Figure 8.9). Starting from the 18th century, this technique was initially aimed at land reclamation, but nowadays, it is applied for nature conservation

Figure 8.7   Map of global salt marsh abundance. (Taken from http://www.prweb.com/releases/The_Atlas_of /Global_Conservation/prweb3839504.htm.)

Figure 8.8 Brushwood dams to create artificial sheltering of salt marsh vegetation against waves and currents. (Photo courtesy of Vincent Vuik.)

Figure 8.9 Extensive salt marshes with brushwood dams along a dike bordering the Dutch Wadden Sea. (https://beeldbank.rws.nl, Rijkswaterstaat.)

purposes (Dijkema et al. 2011). Such salt marshes are classified as engineered ecosystems according to Van der Nat et al. (2016).

The benefits that ecosystems offer to humankind are known as ecosystem services. The most frequently quoted ecosystem services of salt marshes are storm protection for coastal communities, shoreline stabilization, nutrient removal, carbon sequestration, fisheries, and recreation (Deegan et al. 2012; Temmerman et al. 2013). In the context of this book, we focus on ecosystem services related to nature-based coastal protection. Salt marshes can act as a buffer zone against natural hazards such as floods, cyclones, tidal surges, and storms, by reducing storm waves and storm surges (Gedan et al. 2010; King and Lester 1995). Additionally, salt marshes may protect shorelines from erosion by buffering wave action and trapping sediments (Shepard et al. 2011).

At the moment, in the Netherlands, salt marshes are not explicitly considered as an official part of the flood defense itself. At most, the influence of the salt marshes is implicitly taken into account, when their presence influences the wave conditions at the dike, as computed with numerical wave models with the salt marshes included in the model bathymetry (e.g., the increased bed levels near the dike of Figure 8.9). For this reason, no case study has been elaborated in this section. The focus is on describing the potential of vegetated foreshores for flood protection purposes, and the challenges to overcome.

8.4.2 Field Observations

This section gives an overview of the current knowledge base regarding the ecosystem services storm wave attenuation, surge attenuation, and shoreline stabilization by salt marshes based on field observations.

Energy dissipation of storm waves on salt marshes is predominantly caused by depth-induced wave breaking and wave attenuation by vegetation. Waves, propagating toward shallower water, decelerate and increase in amplitude. Depth-induced wave breaking occurs when the wave height (actually the wave steepness) reaches a critical value. The fraction of breaking waves increases with decreasing depth. Naturally, because of the deposition of sediments by the tide, salt marshes have an elevation around the high tide level. That means that the water depth on top of the salt marsh surface is very low during normal high tides. A significant water depth is only found under storm surge conditions. The corresponding storm waves, approaching the shoreline, will decrease in height because of the process of depth-induced wave breaking.

The second process that plays an important role in dissipating wave energy on salt marshes is wave attenuation by vegetation. Waves propagating through vegetation fields lose energy as a result of the forces exerted by vegetation stems, branches, and leaves (Dalrymple et al. 1984). The decay in wave height increases with vegetation stem density (stems per surface area), stem diameter, vegetation height, and vegetation stiffness (Bouma et al. 2005). Waves with large heights and short periods generate the highest forces on the vegetation and, consequently, experience relatively high energy dissipation (Anderson and Smith 2014). This mechanism has appeared to be very efficient in dissipating wave energy for (nearly) emergent conditions (e.g., Knutson et al. 1982). However, measurements in the field (Möller et al. 1999; Vuik et al. 2016; Yang et al. 2012) and in a large wave flume (Möller et al. 2014) have proven that the vegetation also significantly contributes to wave energy dissipation for water depth of several meters, as can occur during storms.

Extensive salt marsh systems have a significant influence on surge propagation and surge levels. The rough salt marsh surface retards the fluid motion. For a rapidly varying wind and a relatively short surge wave, this effect can lead to lower surge levels at the coast behind such large salt marshes. However, the retardation of the fluid motion might lead to higher surge levels at the coastline at the windward side of the marshes. That is why Wamsley et al. (2010) concluded that the magnitude of surge attenuation by salt marsh wetlands is dependent on the surrounding coastal landscape and the strength and duration of the storm forcing. Therefore, the application of a constant attenuation rate (in terms of surge reduction per kilometer salt marsh) is misleading and not appropriate (Resio and Westerink 2008).

The third main ecosystem service of salt marshes is the stabilization of coastlines. First, the presence of vegetation leads to a reduction of wave heights, current velocities, and turbulence and, as a consequence, to an increase of sediment deposition and a decrease of sediment entrainment. Belowground, plant roots increase the shear strength of the substrate, since plant roots tend to enhance the cohesion and tensile strength of their substrate (Gedan et al. 2010). The bottom surface of salt marshes is mostly very stable, according to poststorm observation (Spencer et al. 2015) as well as large-scale wave flume experiments (Möller et al. 2014). All in all, surface erosion of salt marshes is not very likely to occur, even during severe storm surges. However, salt marshes might erode laterally, starting with the formation of a cliff at the salt marsh edge. These then retreat because of the generation of tension cracks in response to the erosive attack of waves and tidal currents, which subsequently leads to toppling failures and rotational slips (Allen 1989; Francalanci et al. 2013). Such cliff erosion is the main mechanism for the retreat of many salt marshes.

8.4.3 Lessons Learned

For coastal protection purposes, stable and static salt marsh characteristics are preferred during the lifetime of the flood defense. However, natural salt marshes inherently have a dynamic character. Conceptual models describe the long-term development of salt marshes as a cyclic process with a time scale of several decades, with alternately a period with expansion and a period with retreat. In the expanding phase, the dynamics on the mudflat in front of the salt marsh are below a certain threshold, and there are windows of opportunity for new vegetation seedlings to settle. Tides and surges supply sediment to the vegetated surface, and the elevated salt marsh platform extends in seaward direction. In systems with sufficient sediment supply, salt marshes are able to adapt to sea level rise by trapping the required volume of sediment to raise the marsh platform with the same rate. As a consequence of the salt marsh expansion, the slope of the mud flat steepens, and the wave and flow dynamics on the mud flat intensify. At a certain moment, a cliff develops: the tipping point from expansion to retreat. Breaking waves and tidal currents exert relatively high stresses on the marsh edge cliff, and usually the cliff height exceeds the rooting depth of the vegetation. The cliff retreats, until the steepness of the mud flat declines to such an extent that seedlings can settle beyond the cliff. In a favorable environment, this cyclic behavior leads to regular rejuvenation of the ecosystem and to relatively high biodiversity (van de Koppel et al. 2005). For that reason, many ecologists detest the construction of stone protections and groins, which aim to fix the position of the salt marsh edge.

Natural developments and human interventions can disrupt the cyclic salt marsh dynamics and can lead to irreversible salt marsh retreat. For example, dredging of shipping channels as well as natural migration of tidal channels can lead to a steepening of the intertidal mud flats, with decreasing windows of opportunity for vegetation seedlings as a result. Another example is the (partial) damming of estuaries, which results in a decline in tidal current velocities. As tidal currents are the main driver of the buildup of intertidal flats, the intertidal landscape might flatten out. During storms, the wind waves can propagate easier over the drowned tidal flats, and higher waves hit the coastline, including the salt marshes bordering the estuary. Such developments can lead to irreversible salt marsh retreat, and nature conservation agencies might consider the protection of these ecosystems with artificial measures, like brushwood dams (Figure 8.8) or stone protections.

The persistence and health of characteristic salt marsh plant species depend on abiotic drivers such as water salinity and water quality, which makes these quantities of crucial importance for the longevity of salt marshes. The dominance of certain plant species is determined on the basis of competition. In salty environments, salt-tolerant species like cordgrass (*Spartina anglica*) dominate, while in brackish environments, they are outcompeted by other species such as grassweed (*Scirpus maritimus*) or common reed (*Phragmites australis*) in freshwater. Vegetation properties such as shoot density, stem flexibility, and rooting depth determine the physical properties of the salt marsh to a large extent. Changes in salinity will lead to changes in ecosystem characteristics and, consequently, changes in the ecosystem services they offer. For example, Howes et al. (2010) suggest that the difference in rooting structure between low-salinity wetlands and high-salinity wetlands has led to the preferential erosion of the low-salinity wetlands during hurricanes Katrina and Rita. Another important aspect is coastal eutrophication as a driver of salt marsh loss (Deegan et al. 2012). They found that high nutrient levels lead to rapid growth of vegetation, relatively high aboveground leaf biomass, and limited belowground biomass of stabilizing root systems. Consequently, geomorphic stability is reduced, resulting in collapse of the creek-banks with significant areas of salt marsh converted to areas of bare mudflat.

8.4.4 Path Forward

For the flood protection services of salt marshes to be actually incorporated in flood risk assessments, a thorough reliability assessment should be carried out. There should be empirical or process-based knowledge of the processes that determine the most important characteristics of the salt marsh in view of surge and wave attenuation and shoreline stabilization, such as cliff erosion, seasonal biomass variations, uprooting, and stem breakage attributed to wave impact (Bouma et al. 2014). The inherent dynamic character of ecosystems and the required extrapolation to extreme storm conditions make exact predictions of the functioning of the salt marshes impossible. Quantification of uncertainties is required. Before nature-based flood defenses can be considered as a reliable supplement to conventional flood defenses in densely populated areas, they need to be tested according to engineering standards for probability of failure (Van Wesenbeeck et al. 2014).

Salt marshes, bordering a dike, lead to a reduction of wave forces on the dike revetments during periodic storm events, elongating the expected lifetime of the revetments. This leads to a reduction of maintenance efforts and costs. On the other hand, dike managers should spend more time removing organic debris from the dike after storm events. In the end, governmental issues are at least of the same importance as knowledge of the physical functioning of the system, to ensure the official incorporation of salt marshes in flood protection strategies. Interdisciplinary research should facilitate the decision making.

8.5 DISCUSSION: TRANSITION IN COASTAL
PROTECTION POLICY IN THE NETHERLANDS

The first coastal protection measures inspired by the Building with Nature philosophy now become visible in the Dutch landscape. A range of pilots has been initiated to test the potential of the Building with Nature solutions in practice as discussed in this chapter. Pilots form an important tool in exploring or evaluating innovations, and the lessons learned from these pilots may form the first step toward actual implementation on the landscape scale. In order to place the lessons learned within the current coastal protection policy, we review the general transition in coastal protection policy in the Netherlands for the last decades.

The Netherlands is worldwide known for its flood protection management. A culture of protecting the land and its inhabitants against floods already emerged in the 13th century (Van Koningsveld et al. 2008). In the 20th century, two iconic projects illustrate this flood protection practice: the damming of the Zuiderzee Sea in the north and the Delta project in the southwestern part of the Netherlands, which closed a number of tidal inlets. These major interventions allow today for safe living in a country where two-thirds is vulnerable for river or sea flooding and approximately 25% lies below sea level. The second Delta committee successfully raised awareness in Dutch society and politics of the potential impacts of climate change (Deltacommissie 2008). Today, protection against flooding is high on the political agenda as ever: a new "delta program" has been initiated to protect the Netherlands against flooding in the future and includes flood protection works for more than 700 km of dikes and dams. The management approach, however, and anticipated solutions have undergone fundamental change over the last decades.

Traditionally, Dutch flood protection management is dominated by a technocratic–scientific regime, which promotes "hard" engineering structures to control and keep out flooding (Lintsen 2002). Since the 1970s, there has been a growing appreciation and understanding of the negative environmental effects of such hard structures (Airoldi et al. 2005; Van Wesenbeeck et al. 2014). In the Netherlands, the construction of the Oosterschelde Dam, which is part of the Delta project, indicates the start of an ecological transformation (Disco 2002). While initially planned to close off the tidal inlet, the dam was designed as a semiopen barrier to allow for tidal dynamics. Over the past 40 years, a fundamental shift has taken place from a technocratic engineering strategy to integral and participatory water management (Van der Brugge et al. 2005). Water management is not just assessed from a flood protection point of view, but other social, ecological, and physical components are valued as well. The integrated approach, increased environmental awareness, and changing climate have enabled a new pathway to emerge in which Building with Nature is the central concept (De Vriend et al. 2014).

Implementation in Dutch mainstream flood protection practice is one of the main challenges of the Building with Nature approach. Even projects with an explicit Building with Nature ambition have failed to be realized in practice (Janssen et al. 2014). The multifunctional nature and inclusion of dynamics structurally differs from conventional hard solutions, which are designed in a static and monofunctional way. Building with Nature solutions introduce uncertainties arising from the inherent unpredictability of natural dynamics and this is at odds with the prevailing idea that flood should be controlled and strict criteria should be employed to assess flood protection solutions. Adaptive management solutions are needed to foster implementation (Borsje et al. 2011; Korbee et al. 2014; Van Wesenbeeck et al. 2014). Building with Nature requires integration among nature and flood protection policy domains, which is a challenge given the dominant nature of the flood protection domain in terms of institutions, resources, and actors (Janssen et al. 2014). Moreover, the multifunctional nature of Building with Nature requires broadening of the traditional participatory actors (Korbee et al. 2014; van Slobbe et al. 2013). Another challenge relates to the assessment of Building with Nature solutions, which should support its multifunctionality. Traditional assessment tools, however, such as those used in cost–benefit analyses or in environmental impact assessments, do not automatically support this integrated nature (De Jonge et al. 2012).

For the three case studies discussed in this chapter, we draw the following general lessons learned for implementation: (1) in order to design Building with Nature solutions, we should not only integrate the different disciplines (physics, ecology, and governance) but also focus on the integrated coastal system (both subaqueous and aeolian); (2) in order to extrapolate the short-term fluctuations in stability of a certain Building with Nature solution into long-term behavior, we should quantify the uncertainties and set up an adaptive management strategy; and (3) in order to successfully implement Building with Nature solutions worldwide, we should set up new actor coalitions within a certain governance setting.

8.6 CONCLUSIONS

The three promising Building with Nature solutions for coastal protection discussed in this chapter inherently have a dynamic nature. Therefore, there is a relatively large degree of uncertainty with respect to their adaptive behavior. Consequently, harnessing the complex and nonlinear dynamics of natural systems is an essential challenge in Building with Nature solutions. Moreover, designing Building with Nature solutions in an interdisciplinary research approach allows for transferring Building with Nature concepts to other locations in the world. Finally, developing adaptive management strategies in close collaboration with local stakeholders, decision makers and scientists are of utmost importance. The three general lessons learned on the Building with Nature pilots discussed in this chapter may form a first step toward actual implementation of these solutions on the landscape scale both in the Netherlands and worldwide.

ACKNOWLEDGMENTS

This work is part of the research program BE SAFE and FORESHORE, which is financed by the Netherlands Organisation for Scientific Research (NWO), Deltares, Boskalis, Van Oord, Rijkswaterstaat, World Wildlife Fund, and HZ University of Applied Science. We would like to acknowledge Leon Hermans for fruitful discussions on the governance aspect. Furthermore, the European Research Council of the European Union is acknowledged for the funding provided for this research through the ERC Advanced Grant 291206-NEMO. Also the Dutch Technology Foundation STW is acknowledged, as part of the Netherlands Organisation for Scientific Research (NWO), which is partly funded by the Ministry of Economic Affairs (project no. 12686; NatureCoast).

REFERENCES

Adger, W.N., T.P. Hughes, C. Folke, S.R. Carpenter, and J. Rockstrom. 2005. Social–ecological resilience to coastal disasters. *Science* 309(5737): 1036–1039.

Airoldi, L., M. Abbiati, M.W. Beck, S.J. Hawkins, P.R. Jonsson, D. Martin, P.S. Moschella, A. Sundelof, R.C. Thompson, and P. Aberg. 2005. An ecological perspective on the deployment and design of lowcrested and other hard coastal protection structures. *Coastal Engineering* 52: 1073–1087.

Allen, J.R.L. 1989. Evolution of salt-marsh cliffs in muddy and sandy systems: A qualitative comparison of British West-coast estuaries. *Earth Surface Processes and Landforms* 14: 85–92.

Anderson, M.E. and J.M. Smith. 2014. Wave attenuation by flexible, idealized salt marsh vegetation. *Coastal Engineering* 83: 82–92.

Arens, S.M., J.P. Mulder, Q.L. Slings, L.H. Geelen, and P. Damsma. 2013. Dynamic dune management, integrating objectives of nature development and coastal safety: Examples from the Netherlands. *Geomorphology* 199: 205–213.

Balke, T., T. Bouma, E. Horstman, E. Webb, P. Erftemeijer, and P. Herman. 2011. Windows of opportunity: Thresholds to mangrove seedling establishment on tidal flats. *Marine Ecology Progress Series* 440: 1–9.

Barbier, E.B., E.W. Koch, B.R. Silliman, S.D. Hacker, E. Wolanski, J. Primavera, E.F. Granek, S. Polasky, S. Aswani, L.A Cramer, D.M. Stoms, C.J. Kennedy, D. Bael, C.V. Kappel, G.M.E. Perillo, and D.J Reed. 2008. Coastal ecosystem-based management with nonlinear ecological functions and values. *Science* 319(5861): 321–323.

Beets, D. and A. Van der Spek. 2000. The Holocene evolution of the barrier and the bank-barrier basins of Belgium and the Netherlands as a function of Late Weichselian morphology, relative sea-level rise and sediment supply. *Geologie en Mijnbouw—Netherlands Journal of Geosciences* 79(1): 3–16.

Borsje, B.W., B. van Wesenbeeck, F. Dekker, P. Paalvast, T.J. Bouma, and M.B. de Vries. 2011. How ecological engineering can serve in coastal protection—A review. *Ecological Engineering* 37: 113–122.

Bouma, T.J., M.B. De Vries, E. Low, G. Peralta, I.C. Tánczos, J. van de Koppel, and P.M.J. Herman. 2005. Trade-offs related to ecosystem engineering: A case study on stiffness of emerging macrophytes. *Ecology* 86(8): 2187–2199.

Bouma, T.J., J. van Belzen, T. Balke, Z. Zhu, L. Airoldi, A.J. Blight, ... P.M.J. Herman. 2014. Identifying knowledge gaps hampering application of intertidal habitats in coastal protection: Opportunities and steps to take. *Coastal Engineering* 87: 147–157.

Costanza, R., W.J. Mitsch, and J.W. Day. 2006. A new vision for new Orleans and the Mississippi delta: Applying ecological economics and ecological engineering. *Front Ecol Environ* 4(9): 465–472.

Dalrymple, R.A., J.T. Kirby, and P.A. Hwang. 1984. Wave diffraction due to areas of energy dissipation. *Journal of Waterway, Port, Coastal, and Ocean Engineering* 110(1): 67–79.

Davison, A.T., R.J. Nicholls, and S.P. Leatherman. 1992. Beach nourishment as a coastal management tool: An annotated bibliography on developments associated with the artificial nourishment of beaches. *Journal of Coastal Research* 984–1022.

De Jonge, V.N., R. Pinto, and R.K. Turner. 2012. Integrating ecological, economic and social aspects to generate useful management information under the EU Directives' 'ecosystem approach.' *Ocean and Coastal Management* 68: 169–188.

De Ruig, J.H. 1998. Coastline management in the Netherlands: Human use versus natural dynamics. *Journal of Coastal Conservation* 4(2): 127–134.

De Schipper, M.A., S. de Vries, B.G. Ruessink, R.C. de Zeeuw, J. Rutten, C. van Gelder-Maas, and M.J.F. Stive. 2016. Initial spreading of a mega feeder nourishment: Observations of the sand engine pilot project. Submitted to *Coastal Engineering* 111: 23–38.

De Vriend, H.J., M. Van Koningsveld, S.G. Aarninkhof, M.B. De Vries, and M.J. Baptist. 2014. Sustainable hydraulic engineering through building with nature. *Journal of Hydro-environment Research*.

De Vries, S., H. Southgate, W. Kanning, and R. Ranasinghe. 2012. Dune behavior and Aeolian transport on decadal timescales. *Coastal Engineering* 67: 41–53.

Deegan, L.A., D.S. Johnson, R.S. Warren, B.J. Peterson, J.W. Fleeger, S. Fagherazzi, and W.M. Wollheim. 2012. Coastal eutrophication as a driver of salt marsh loss. *Nature* 490: 388–392.

Deltacommissie. 2008. Samen werken met water. Een land dat leeft, bouwt aan zijn toekomst. Bevindingen van de Deltacommissie 2008.

Disco, C. 2002. Remaking "nature": The ecological turn in Dutch water management. *Science Technology and Human Values* 27(2): 206–235.

Dijkema, K.S., W.E. Van Duin, E.M. Dijkman, E. Nicolai, H. Jongerius, H. Keegstra, … J.J. Jongsma. 2011. Vijftig jaar monitoring en beheer van de Friese en Groninger kwelderwerken: 1960–2009. Wageningen.

Francalanci, S., M. Bendoni, M. Rinaldi, and L. Solari. 2013. Ecomorphodynamic evolution of salt marshes: Experimental observations of bank retreat processes. *Geomorphology* 195: 53–65.

French, P.W. 2006. Managed realignment: The developing story of a comparatively new approach to soft engineering. *Estuar Coast Shelf Sci* 67: 409–423.

Gedan, K.B., M.L. Kirwan, E. Wolanski, E.B. Barbier, and B.R. Silliman. 2010. The present and future role of coastal wetland vegetation in protecting shorelines: Answering recent challenges to the paradigm. *Climatic Change* 106(1): 7–29.

Grunnet, N.M., and B. Ruessink. 2005. Morphodynamic response of nearshore bars to a shoreface nourishment. *Coastal Engineering* 52(2): 119–137.

Hamm, L., M. Capobianco, H.H. Dette, A. Lechuga, R. Spanhoff, and M.J.F. Stive. 2002. A summary of European experience with shore nourishment. *Coastal Engineering* 47(2): 237–264.

Hermans, L.M., J.H. Slinger, and S.W. Cunningham. 2013. The use of monitoring information in policy-oriented learning: Insights from two cases in coastal management. *Environmental Science and Policy* 29(1): 24–36.

Howes, N.C., D.M. FitzGerald, Z.J. Hughes, I.Y. Georgiou, M.A. Kulp, M.D. Miner, … J.A. Barras. 2010. Hurricane-induced failure of low salinity wetlands. *Proceedings of the National Academy of Sciences of the United States of America* 107: 14014–14019.

Janssen, S.K.H., J.P.M. van Tatenhove, H.S. Otter, and A.P.J. Mol. 2014. Greening flood protection—An interactive knowledge arrangement perspective. *Journal of Environmental Policy & Planning* 1–23. DOI: 10.1080/1523908X.2014.947921.

King, S.E. and J.N. Lester. 1995. The value of salt marsh as a sea defence. *Marine Pollution Bulletin* 30(3): 180–189.

Knutson, P., R. Brochu, W. Seelig, and M. Inskeep. 1982. Wave damping in *Spartina alterniflora* marshes. *Wetlands* (1978), 87–104.

Korbee, D., A.P.J. Mol, and J.P.M. Van Tatenhove. 2014. Building with nature in marine infrastructure: Toward an innovative project arrangement in the Melbourne Channel Deepening Project. *Coastal Management* 42(1): 1–16.

Linnartz, L. 2013. The Second Year Sand Engine: Nature Development on a Dynamic Piece of the Netherlands. Technical Report. ARK Natuurontwikkeling, p. 33.

Lintsen, H. 2002. Two centuries of central water management in the Netherlands. *Technology and Culture* 43(3): 549–568.

Luijendijk, A.P., B. Huisman, and M. De Schipper. 2015. Impact of a storm on the first year evolution of the sand engine. In Proceedings of the Coastal Sediments 2015 conference.

Möller, I., M. Kudella, F. Rupprecht, T. Spencer, M. Paul, B.K. van Wesenbeeck, … S. Schimmels. 2014. Wave attenuation over coastal salt marshes under storm surge conditions. *Nature Geoscience* 7(10): 727–731.

Möller, I., T. Spencer, J.R. French, D.J. Leggett, and M. Dixon. 1999. Wave transformation over salt marshes: A field and numerical modelling study from North Norfolk, England. *Estuarine, Coastal and Shelf Science* 49(3): 411–426.

Mulder, J.P.M. and P.K. Tonnon. 2010. Sand engine: Background and design of a mega nourishment pilot in the Netherlands. *Proceedings of the International Conference on Coastal Engineering* 32.

PIANC. 2011. Working with Nature. Position paper, World Association for Waterborne Transport Infrastructure, Brussels, Belgium. http://www.pianc.org/wwnpositionpaper.php.

Resio, D.T. and J.J. Westerink. 2008. Modeling the physics of storm surges. *Physics Today* 61(9): 33–38.

Romme, W.H., E.H. Everham, L.E. Frelich, M.A. Moritz, and R.E. Sparks. 1998. Are large, infrequent disturbances qualitatively different from small, frequent disturbances? *Ecosystems* 1(6): 524–534.

Ruessink, B.G. and M.C.J.L. Jeuken. 2002. Dunefoot dynamics along the Dutch coast. *Earth Surface Processes and Landforms* 27(10): 1043–1056.

Shepard, C.C., C.M. Crain, and M.W. Beck. 2011. The protective role of coastal marshes: A systematic review and meta-analysis. *PLoS One* 6(11), e27374.

Spencer, T., S.M. Brooks, B.R. Evans, J.A. Tempest, and I. Möller. 2015. Southern North Sea storm surge event of 5 December 2013: Water levels, waves and coastal impacts. *Earth-Science Reviews* 146: 120–145.

Stive, M.J.F., M.A. De Schipper, A.P. Luijendijk, S.G.J. Aarninkhof, C. Van Gelder-Maas, J.S.M. Van Thiel de Vries, S. De Vries, M. Henriquez, S. Marx, and R. Ranasinghe. 2013. A new alternative to saving our beaches from sea-level rise: The sand engine. *Journal of Coastal Research* 29.

Stive, M.J., R.J. Nicholls, and H.J. de Vriend. 1991. Sea-level rise and shore nourishment: A discussion. *Coastal Engineering* 16(1): 147–163.

Temmerman, S., P. Meire, T.J. Bouma, P.M.J. Herman, T. Ysebaert, and H.J. De Vriend. 2013. Ecosystembased coastal defence in the face of global change. *Nature* 504(7478): 79–83.

van de Koppel, J., D. van der Wal, J.P. Bakker, and P.M.J. Herman. 2005. Self-organization and vegetation collapse in salt marsh ecosystems. *Am. Nat.* 165(1): E1–E12. DOI:10.1086/426602.

Van der Brugge, R., J. Rotmans, and D. Loorbach. 2005. The transition in Dutch water management. *Regional Environmental Change* 5(4): 164–176.

van der Nat, A., P. Vellinga, R. Leemans, and E. van Slobbe. 2016. Ranking coastal flood protection designs from engineered to nature-based. *Ecological Engineering* 87: 80–90.

van Duin, M., N. Wiersma, D. Walstra, L. van Rijn, and M. Stive. 2004. Nourishing the shoreface: Observations and hindcasting of the Egmond case, the Netherlands. *Coastal Engineering* 51(8–9): 813–837.

Van Koningsveld, M., J.P.M. Mulder, M.J.F. Stive, L. Van der Valk, and A.W. Van der Weck. 2008. Living with sea-level rise and climate change: A case study of the Netherlands. *Journal of Coastal Research* 24(2): 367–379.

Van Loon-Steensma, J.M. 2015. Salt marshes to adapt the flood defences along the Dutch Wadden Sea coast. *Mitigation and Adaptation Strategies for Global Change* 20(3): 929–948.

Van Rijn, L.C. 1997. Sediment transport and budget of the central coastal zone of Holland. *Coastal Engineering* 32: 61–90.

Van Slobbe, E., H.J. De Vriend, S. Aarninkhof, K. Lulofs, M. De Vries, and P. Dircke. 2013. Building with nature: In search of resilient storm surge protection strategies. *Natural Hazards* 66(3): 1461–1480.

Van Wesenbeeck, B.K., J.P.M. Mulder, M. Marchand, D.J. Reed, M.B. de Vries, H.J. de Vriend, and P.M.J. Herman. 2014. Damming deltas: A practice of the past? Towards nature-based flood defenses. *Estuarine, Coastal and Shelf Science* 140: 1–6.

Vuik, V., S.N. Jonkman, B.W. Borsje, and T. Suzuki. 2016. Nature-based flood protection: The efficiency of vegetated foreshores for reducing wave loads on coastal dikes. *Coastal Engineering* 116: 42–56.

Wamsley, T.V., M.A. Cialone, J.M. Smith, J.H. Atkinson, and J.D. Rosati. 2010. The potential of wetlands in reducing storm surge. *Ocean Engineering* 37(1): 59–68.

Wijnberg, K.M. and J.H.J. Terwindt. 1995. Extracting decadal morphological behaviour from high-resolution, long-term bathymetric surveys along the Holland coast using eigenfunction analysis. *Marine Geology* 126(1–4): 301–330.

Yang, S.L., B.W. Shi, T.J. Bouma, T. Ysebaert, and X.X. Luo. 2012. Wave attenuation at a salt marsh margin: A case study of an exposed coast on the Yangtze Estuary. *Estuaries and Coasts* 35(1): 169–182.

CHAPTER 9

Managed Realignment in Europe
A Synthesis of Methods, Achievements, and Challenges

Luciana S. Esteves and Jon J. Williams

CONTENTS

9.1 INTRODUCTION

In Europe, there is a growing trend of "building with nature" initiatives to improve the long-term, environmental, and socioeconomic sustainability of coastal management strategies (e.g., Luisetti et al. 2011; Rijkswaterstaat and Deltares 2013; Spalding et al. 2014; Temmerman et al. 2013). In particular, managed realignment approaches are increasingly considered as a no-regret option (Elliott et al. 2014; van Loon-Steensma and Vellinga 2013) bringing social (improved flood risk management), economic (lowering costs of flood protection maintenance), and environmental benefits through habitat restoration (e.g., Committee on Climate Change 2013; Defra 2002; Spalding et al. 2014; Spencer and Harvey 2012). These are common drivers underpinning the implementation of managed realignment in Europe. However, the form in which managed realignment has been implemented varies between countries as it will be illustrated by the examples presented in this chapter.

In the literature, the terminology used to describe managed realignment varies regionally, through time, and between authors. Many terms have been used as synonyms of managed realignment, including setback, managed retreat, de-embankment, and depoldering. A review of the terminology and definitions can be found in Esteves (2014). Here, managed realignment means taking planned actions to create space for natural dynamic processes, usually involving relocation of river embankments, estuary, or open coast shorelines to shorten the overall length of protected shores (therefore reducing maintenance costs) and to provide opportunity for habitat restoration (Defra 2002; Esteves 2014). Key objectives of this habitat restoration are the provision of (i) a range of ecosystem services, including natural storm-buffering capacity, and (ii) compensation for habitat loss resulting from land reclamation or coastal squeeze, as required by environmental legislation (e.g., Defra 2002; Esteves 2014).

The first managed realignment projects in Europe (implemented in France in 1981 and in Germany in 1982*) were isolated initiatives meeting local needs. At the end of 2015, 140 projects have been completed or were under construction, most of them driven by national and European environmental legislation. Technical capacity to improve projects design and the scientific understanding of physical and ecological changes at realignment sites have greatly advanced (see Section 9.5). However, the wider implementation of managed realignment is still hindered by a number of challenges, such as public acceptance, funding constraints, availability of suitable land, and uncertainties related to natural coastal evolution.

Recent national and regional strategies (e.g., in the United Kingdom, Belgium, the Netherlands, and France) give an important role to managed realignment. Therefore, it is timely and relevant to identify lessons learned and assess how to overcome current challenges. This chapter provides an overview of the current state of play related to managed realignment in Europe. The examples provided demonstrates that a combination of project types can be implemented in both rural and urban areas as a long-term and sustainable strategy to reduce flood risk and promote the provision of other ecosystem services. The first sections introduce and discuss basic concepts, including the main underlying drivers (Section 9.2), the five most common methods of implementation (Section 9.3), and examples of existing strategies (Section 9.4). Then, a summary of lessons learned (Section 9.5) is presented, followed by a summary of key findings (Section 9.6).

9.2 UNDERLYING DRIVERS

There are two main drivers underpinning the need for managed realignment in Europe: (1) environmental legislation aiming to prevent the loss and degradation of coastal habitats and associated biota (and the consequent impacts on society) and (2) the need to reduce flood and erosion risk to people and property and manage the increasing maintenance costs, especially due to climate change impacts. The implementation of managed realignment in Europe is greatly influenced by European Union (EU) Directives, in particular: (a) the Birds Directive,[†] (b) the Habitats Directive,[‡] (c) the Water Framework Directive,[§] and (d) the Floods Directive.[¶] Each EU country is obliged to adopt the EU Directives into national legislation.

---

* A list of managed realignment projects, providing date of implementation, location, and references is provided in Esteves (2014).

[†] Directive 79/409/EEC (April 1979) of the European Parliament and of the Council on the conservation of wild birds is the oldest EU environmental legislation, amended in 2009, it became the Directive 2009/147/EC, http://ec.europa.eu /environment/nature/legislation/birdsdirective/.

[‡] Directive 92/43/EEC of the European Council on the conservation of natural habitats and of wild fauna and flora, http:// ec.europa.eu/environment/nature/legislation/habitatsdirective/.

[§] Directive 2000/60/EC of the European Parliament and of the Council establishing a framework for the Community action in the field of water policy, http://ec.europa.eu/environment/water/water-framework/.

[¶] Directive 2007/60/EC on the assessment and management of flood risks, http://ec.europa.eu/environment/water /flood_risk/.

The Birds Directive and the Habitats Directive have been fundamental instruments for nature conservation in the EU, including the restoration of coastal habitats (Pontee 2014). Under these Directives, each EU member state is responsible for taking all necessary measures to protect designated habitats and species of European importance, mainly through the establishment of an EU-wide network of designated conservation sites (called Natura 2000). Most intertidal flats and saltmarshes in Europe are within Natura 2000 sites. Therefore, any human-induced loss or damage must be prevented. Exceptions exist due to imperative reasons of overriding public interest (i.e., when certain development is required for the greater benefit of society). In these cases, any damage or loss to designated habitats must be compensated by restoration or creation of habitats with equivalent ecological function.

Compensation is also required to offset long-term habitat loss attributed to coastal squeeze, which is the loss of intertidal habitat caused by rising sea levels in front of an artificially fixed shoreline (Esteves 2016; Pontee 2013). In practical terms, this means that any new or improved flood-risk management measures unavoidably causing habitat loss or degradation within Natura 2000 sites require creation of habitat as a compensatory measure.

The extent of compensatory habitat that needs to be created takes into consideration direct and indirect historical and future losses expected to occur during the lifetime of the development (Thomas 2014). For example, the Defra Flood Management Division (2005) estimates that an average of approximately 100 ha of intertidal habitat needs to be created per year to compensate loss due to coastal squeeze and development projects within Natura 2000 sites, especially in south and east England. Managed realignment is often implemented to create the habitats required as compensatory measures.

Under the Floods Directive, EU countries must map flood risk to people and assets from coasts and inland waters and establish flood risk management plans focused on prevention, protection, and preparedness. The Floods Directive must be implemented in coordination with the Water Framework Directive, taking due consideration of the potential impacts of flood protection measures on water quality and the ecological status of coast and estuaries. In the United Kingdom, for example, the River Basin Management Plans, in combination with Shoreline Management Plans (SMPs), are key instruments supporting sustainable water management, with managed realignment being a preferred option to restore the natural functions of estuary and coastal systems (Thomas 2014).

9.3 TYPES OF MANAGED REALIGNMENT PROJECTS

A review of published literature enables the identification of different methods of managed realignment being implemented in Europe, which generally will fit into one of five types: (1) removal of coastal protection structures, (2) breach of seawalls, (3) realignment of the coastal protection line, (4) controlled tidal restoration, and (5) managed retreat (Esteves 2014). The primary characteristics of each category are summarized in Table 9.1. Categorizing the projects into these five types helped quantify the preferred methods of implementation, how they relate to different objectives (Table 9.1), and how they vary geographically (Table 9.2). Details and examples of the relationships between the type of implementation, the primary project objective, and the physical characteristics of the site are provided in Sections 9.3.1 through 9.3.5.

The categories are also a means to standardize the terminology used to describe the projects. For example, in the literature, the terms "breach" and "realignment" have been used indiscriminately to describe projects implemented in the same way, independently whether the design involved the construction of new coastal protection structures or not. Using the five categories, a clear distinction is made as all projects involving the construction or upgrading of a new coastal protection line are categorized as "realignment" (Table 9.1). On the other hand, the categories "breach" or "removal"

Table 9.1 Primary and Secondary Characteristics of the Five Managed Realignment Methods of Implementation

	Removal	Breach	Realignment	Controlled Tidal Restoration RTE	Controlled Tidal Restoration CRT	Managed Retreat
Large sections of coastal protection are removed	■		▨			▨
Seawalls/embankments are artificially breached		■	▨			▨
Coastal protection is allowed to breach naturally			▨			▨
Project design includes new or upgraded structures			■	▨		
Tidal flow is restored through sluices/culverts			▨	▨	■	
Project involves flood control areas				▨	▨	▨
Planned removal of people and assets at risk	▨	▨	▨	▨	▨	■
Primary and Secondary Objectives						
Creation of habitat	■	■	■	■	■	▨
Improved flood risk management	▨ *	▨ *	■	▨ *	■	■
Other ecosystem services	▨	▨	▨	▨	▨	▨
Climate change adaptation	■	■	▨	▨	■	■

Note: Shading: black, primary; gray, secondary; white, not applicable.
*Improved flood risk management depends on the habitat that will be created and therefore it should be considered either a secondary outcome or a long-term primary objective.

Table 9.2 Number, Type, and Size of Managed Realignment Projects Implemented or under Construction in Europe (in December 2015)

Country	Number of Projects	Types of Managed Realignment Projects	Area (ha)[a]
Belgium	18	Controlled tidal restoration: 7; realignment: 10; removal: 1	3530
Denmark	2	Breach: 2	206
France	5	Breach: 1; controlled tidal restoration: 2; managed retreat: 1; removal: 1	511
Germany	30	Breach: 13; controlled tidal restoration: 3; realignment: 3; removal: 11	5036
The Netherlands	13	Breach: 9; controlled tidal restoration: 2; realignment: 2	1090
Spain	3	Controlled tidal restoration: 1; removal: 2	3272
UK	69	Breach: 16; controlled tidal restoration: 23; realignment: 27; removal: 3	2162
Total	140	Breach: 41; controlled tidal restoration: 38; managed retreat: 1; realignment: 42; removal: 18	15,807

will only be used to describe projects not involving the upgrade or construction of new structures. Some projects use a combination of methods, especially to better address multipurpose objectives (e.g., creation of specific habitats and reduction of flood risk). A combination of managed retreat and other methods of managed realignment is probably the best alternative when considering the long-term sustainability of multipurpose projects. However, such combination has not yet been widely implemented because of complex socioeconomic issues associated with long-term planning and private property rights, as discussed in Section 9.5.

9.3.1 Removal of Coastal Protection Structures

At some locations, entire sections of coastal protection structures are removed to restore the space required for the coast to respond more dynamically to environmental change (waves, tides, and sediment supply). Often, the structures removed are failing, poorly maintained, or not offering the expected level of protection. It is expected that the removal of coastal protection will result in the landward or seaward realignment of the shoreline position, depending on site-specific conditions. For example, seaward displacement of the shoreline may occur where the removal of a seawall may reactivate cliff erosion restoring sediment supply to the adjacent beach (if the cliff is formed by beach-quality sediment).

This type of managed realignment may increase exposure to waves, tides, and storm surges to inland areas, and it is important to take into consideration how erosion and flood risk may change in the future (e.g., due to sea level rise). Additionally, removal of coastal protection might not be the most suitable method if the objective is to create habitats requiring sheltered environments (e.g., saltmarshes), as higher-energy conditions might prevent the development of such habitats (Nottage and Robertson 2005). Figure 9.1 illustrates some of the conditions more suitable for the removal of coastal protection structures: (a) risks are controlled by natural topography (e.g., higher grounds) or (b) by the presence of existing coastal protection further inland, or (c) the potential increase in erosion and flooding hazard can be tolerated (e.g., where critical infrastructure and people are not affected or they are resilient to the expected impact). On the other hand, depending on site-specific characteristics, benefits created may include a wide range of ecosystem services, including (a) provision of sediment to replenish adjacent areas, (b) creation of flood-water storage space to reduce risk of flooding elsewhere, (c) habitat creation, (d) enhanced biodiversity, (e) and improvement of recreation opportunities.

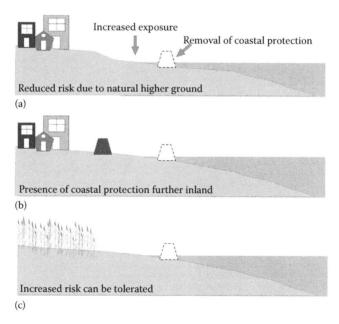

Figure 9.1 Managed realignment through the removal of coastal protection is more suitable for locations where flood and erosion risks are controlled either by (a) natural topography (e.g., higher grounds) or (b) the presence of existing coastal protection further inland and where (c) the potential increase in erosion or flooding hazard can be tolerated (e.g., where critical infrastructure and people are not affected or they are resilient to the expected impact).

Figure 9.2 (a) Failing coastal protection structures in Brownsea Island (b) were removed in 2011, and (c) by 2013, the shore profile was similar to pre-protection times. (Photos courtesy of Tony Flux.)

Public and political acceptance of removing coastal protection is still challenging in Europe and elsewhere. Nordstrom and Jackson (2013) suggest that a number of demonstration projects may be required to evidence the benefits of such approaches before they become more widely acceptable. The involvement of willing landowners is an important facilitator. The National Trust* is a UK-based charity devoted to protect historic places and open spaces for the enjoyment of the public. Currently, the National Trust owns more than 1240 km of coastline across England, Wales, and Northern Ireland where they adopt a "making space for nature" approach to coastal management (see Dyke and Flux 2014). An example of this management approach is the removal of coastal protection structures along the south shoreline of Brownsea Island (Dorset, UK), illustrated in Figure 9.2.

9.3.2 Breach of Coastal Protection Structures

This method involves removal of one or more small sections of the existing coastal protection structures to restore tidal flow into previously protected land. Therefore, the considerations about the suitability of sites (illustrated in Figure 9.1) are also applicable to managed realignment through breaching. The remaining sections of the coastal protection line offer a certain degree of shelter within parts of the realignment site (Figure 9.3a). The extent of the level of exposure within the realigned site will depend on the local topography, the width of the breach, the characteristics of tidal flow, wave conditions, and so on. Ideally, the sheltering effect can reduce flooding and erosion risk to inland areas, promote sedimentation, and favor the development of habitats, such as saltmarshes.

The long-term management of the remaining coastal protection will change with time and must be considered with care. A lack of maintenance will lead to structural degradation, which, in turn, will increase exposure within the realigned site, jeopardizing the habitat created under sheltered conditions. To avoid such undesired impacts, it may be necessary to provide regular maintenance of the remaining sections of the breached coastal protection (which will add additional costs to the project). However, to the knowledge of the authors, this type of measure has not yet been attempted.

9.3.3 Realignment of the Coastal Protection Line

Unlike the removal or breaching methods, realignment of the coastal protection line (seaward or landward) involves the construction of new structures as part of the project design (Figure 9.3b through f). The construction of a new line of coastal protection allows implementation of managed realignment in areas where the control of erosion and flood risk to inland areas requires active intervention. Tidal restoration can be implemented through breach (Figure 9.3b and c) or removal (Figure 9.3d and e) of the existing line of coastal protection. Realignment seaward (Figure 9.3f) can be achieved through mega-size sediment nourishment projects, which provide protection against storms and create space for the development of coastal habitats. An example is the "Sand Motor" project (also known as Sand Engine; see Chapter 8, Section 8.3), built south of The Hague, in the Netherlands (De Schipper et al. 2014; Stronkhorst and Mulder 2014).

Inland realignment of the coastal protection line through breaching (Figure 9.3b) is the most common type of managed realignment in the United Kingdom (Table 9.2). Projects may involve one or multiple breaches; their dimensions and location control the tidal exchange and the level of exposure within the realignment site. The characteristics of the flow across the breaches determine the erosion and sedimentation patterns within the site and in the adjacent areas. The use of numerical modeling is essential to assess how breach design will alter the hydrodynamics and sediment dynamics at the site.

---

* More about the National Trust is found at: http://www.nationaltrust.org.uk/what-we-do/.

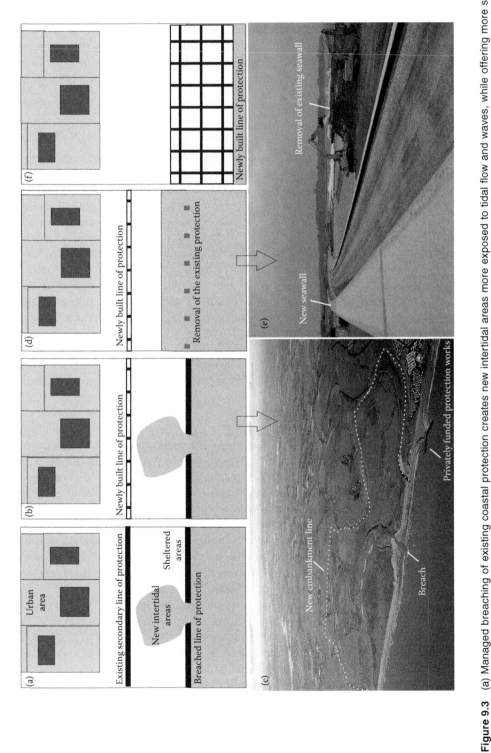

Figure 9.3 (a) Managed breaching of existing coastal protection creates new intertidal areas more exposed to tidal flow and waves, while offering more sheltered conditions in areas further away from the breach. Realignment of the coastal protection line involves the construction of a new line of coastal protection either inland, through (b and c) breach or (d and e) breach or (d and e) removal of the existing line of coastal protection, or (f) seaward. (c) Medmerry, West Sussex, England, illustrates inland realignment through breach (photo by John Akerman) and (e) Littlehaven Beach, South Tyneside, England, illustrates inland realignment through removal of existing coastal protection (photo by Steve Burdett, courtesy of Royal HaskoningDHV).

Not all realignment projects are designed to create habitats. Continued coastal erosion or inadequate design/positioning of seawalls can result in shorelines too exposed to waves. Realignment of coastal protection structures may be implemented to reduce exposure and create opportunities for recreation. In Littlehaven Beach, South Tyneside, Northeast England, seawall realignment (Figure 9.3e) changed the shoreline from a protruding to a concave planform, reducing maintenance costs and increasing amenity value (Cooper et al. 2013). The Kreetsand project* along the River Elbe (Germany) combines flood risk reduction to the port of Hamburg with the creation of recreational areas.

9.3.4 Controlled Tidal Restoration

Controlled tidal restoration methods involve the maintenance of the existing line of coastal protection and the restoration of tidal flow into the protected area through the installation of culverts and sluices. The size and elevation of sluices and culverts determine the characteristics of the tidal inundation and sedimentation patterns within the site. In controlled tidal restoration, the high water line moves landwards within an embanked area. Therefore, in this case, realignment refers to the position of the high water line. This method offers more control over erosion and flood risk than other types of managed realignment and is therefore a suitable alternative at locations with high coastal development pressure and where land availability is scarce (Cox et al. 2006).

Two types of controlled tidal restoration methods are described in the literature: (a) regulated tidal exchange (RTE) and (b) controlled reduced tide (CRT). CRT schemes differ from RTE for having greater control on the tidal exchange to maximize the use of flood control areas to create intertidal habitat (Meire et al. 2005). Therefore, CRT allows managed realignment of coastlines heavily reliant on flood protection and where flood risk mitigation is a serious concern.

RTE schemes are widely implemented in France and the United Kingdom, often with the primary objective of habitat restoration and a secondary function of floodwater storage (Table 9.1). For example, the Polder de Sébastopol (Figure 9.4a), Vendée, France, was reclaimed from the sea in 1856, the existing dike breached during a storm in 1978, and the installation of culverts in 1999 restored tidal flows into the diked area. In 2008, the Polder de Sébastopol Regional Natural Reserve was created to protect 133 ha of wetlands that support protected migratory birds and other fauna and flora species of interest. The sheltered conditions and controlled tidal flow at RTE sites favor sediment deposition and the chances for saltmarsh development. Culverts can be designed to reduce the inundation depth at low-lying sites, but variations of water levels tend to be similar at neap and spring tides, limiting the restoration of the full spectrum of intertidal gradient (Beauchard et al. 2011).

CRT schemes are being implemented in Belgium as part of the *Sigma Plan* (see Section 9.4). Using a combination of high inlet culverts and low outlet gravitational valves (Figure 9.4b), the CRT creates a wide neap-spring range of inundation levels required for the establishment of the full spectrum of intertidal habitats within flood control areas (Cox et al. 2006; Maris et al. 2007). Flood control areas are enclosed by dikes, which are higher inland and lowered along the estuary or coast (Figure 9.4b). During high water level events, a large volume of water can overtop the lowered dike, creating a temporary (one tide) floodwater storage area, alleviating flood risk to nearby areas (Cox et al. 2006).

Jacobs et al. (2009) summarize the tidal exchange in CRT sites as follows. During normal high tides, the volume of water entering the site is limited by the height and size of the inlet culvert. The water retained within the CRT site will start to drain only when the ebb tide lowers to the internal inundation level, allowing the low outlet culvert to open. The control of water levels allows saltmarshes to develop on elevations considerably lower than it would be possible under natural tidal conditions (Vandenbruwaene et al. 2011).

---

* http://www.iba-hamburg.de/en/projects/elbe-islands-dyke-park/pilot-project-kreetsand/projekt/pilot-project-kreetsand.html.

Figure 9.4 Two examples of controlled tidal restoration projects: (a) The Polder de Sébastopol (Vendée, France) illustrates an RTE scheme, in which tidal flows were restored into an embanked reclaimed land through culverts implemented in 1999 (photo by Jacques Oudin, courtesy of Communauté de Communes de île de Noirmoutier); (b) Lippenbroek polder (Belgium) is a CRT pilot project where tidal flows enter a Flood Control Area through a high culvert during high water levels, and site drainage is controlled by a gravitational valve installed in a low culvert (photo from Olivier Beauchard).

The Lippenbroek CRT (Figure 9.4b) construction started in 2004 and finished in March 2006 (Teuchies et al. 2012). The tidal amplitude is reduced from 5.2 m in the estuary to 0.9 m within the CRT site (Beauchard et al. 2011). Sediment accumulation was rapid, especially at lower elevations, where agricultural soils were covered by ~30 cm estuarine sediments in 3 years (Vandenbruwaene et al. 2011). Therefore, tidal inflow enhances sediment accumulation within the CRT, which may lead to (potentially undesirable) reduction of storage capacity through time. Conversely, emptying of floodwater may result in localized high erosion rates affecting habitat restoration (Cox et al. 2006). Numerical modeling simulations indicate that, because of differences in sedimentation patterns, CRT marshes may be less able to cope with rising sea levels than natural marshes, thus affecting their long-term sustainability (Vandenbruwaene et al. 2011).

9.3.5 Managed Retreat

Managed retreat involves the relocation of people, property, and infrastructure from hazard-prone areas. If the objective is to reduce the number of people and assets at risk, managed retreat is perhaps the only option available for developed coasts threatened by rising sea levels (Alexander et al. 2012; Reisinger et al. 2015). However, its implementation requires long-term and strategic planning, which are difficult to achieve because of the complex nature of private property rights, social acceptance, and uncertainties concerning future climate conditions. As a result, managed retreat initiatives tend to be slow and of limited scale, often not facilitating land-use changes within time scales that preclude the potential increase in risk (e.g., because of climate change or population growth). A typical example is the relocation of single structures at threat. For example, in 1999, the Belle Tout lighthouse (East Sussex, UK), then converted to a private home, was moved 17 m away from the edge of the eroding cliff, an effort privately funded.

Recently, a more strategic implementation of managed retreat has been promoted in national policies, such as the Protocol on Integrated Coastal Zone Management in the Mediterranean and the French national strategy for shoreline management (which is described in Section 9.4). Legislation supporting managed retreat usually involves alterations to private property rights based on defined thresholds of risk. It might include, for example, restrictions to restoration or reconstruction of properties affected by flooding or erosion if they are within high-risk zones. Although setback lines are not direct instruments for managed retreat, they are often associated with some measures that facilitate its implementation.

In Spain, the Coastal Law of 1988 established an area of public domain (up to the most landward reach of waves during storms) and a "zone of protection," in fact a setback, extending a further 100 m inland, or 20 m in areas developed before 1988 (Sanò et al. 2011). According to the Coastal Law, properties located within public domain would be considered illegal (and therefore requiring demolition). Properties within the zone of protection were granted a concession of use until 2018. Under the Coastal Law, these properties cannot be sold, transferred, or upgraded. In the long term, this legislation has the potential to decrease the number of property and people in risk areas, reflecting a mechanism of managed retreat.

The Spanish Coastal Law was amended by Law 2/2013 on protection and sustainable use of the coast (ratified in 2014). This amendment expanded the definition of public coastal domain, causing many cases of appeal. However, it also increased the conditions in which the zone of protection is reduced from 100 to 20 m and extended the concessions for a further 30 years, which allows a period of grace to request a permit to transfer the property deed. These amendments must comply with the Protocol on Integrated Coastal Zone Management in the Mediterranean,* which is part of EU law and is legally binding. The Protocol requires the implementation of a setback zone, at least 100 m wide, "from the highest winter waterline…where construction is not allowed." As the Protocol indicates certain conditions in which the setback zone may be "adapted," national legislators have some flexibility to adjust their setback zones, and not necessarily to impose stricter measures.

The Spanish example illustrates the complexity involved in the implementation of managed retreat measures owing to the influence of social and political pressures. Public support for managed retreat depends on a number of factors related to the underpinning cultural values of communities and individuals (Alexander et al. 2012) and social justice (e.g., Reisinger et al. 2015), including (a) who will pay the costs, (b) how changes in existing rights of use are managed, and (c) who will benefit or lose with the changes.

9.4 EXISTING STRATEGIES

Emerging national and regional strategies are driving the present-day increase in managed realignment projects being implemented in Europe. The four strategies described here have the common objective of providing a more sustainable flooding risk management along coasts, estuaries, and rivers, through innovative "building with nature" approaches, including management

* http://ec.europa.eu/environment/iczm/barcelona.htm.

realignment, as a sustainable climate change adaptation measure. Although other relevant initiatives exist (e.g., Cities and Climate Change program promoted by IBA Hamburg, Germany), examples from the United Kingdom, Belgium, the Netherlands, and France are selected to be detailed here as they illustrate the scale in which managed realignment is playing a role to achieve the strategies' objectives and the range of project types that are being considered.

9.4.1 Making Space for Water and Making Space for Nature, UK

In the United Kingdom, the public bodies most directly involved in managed realignment are (a) the Department for Environment Food and Rural Affairs (Defra), responsible for policy-making concerning nature conservation and flood risk; (b) the Environment Agency (EA), responsible for implementing policy related to coastal erosion and flood risk management; and (c) local authorities, which are the designated Lead Local Flood Authority under the Flood and Water Management Act (2010), having the duty for managing local flood risk.

Defra (2005) published the Making Space for Water strategy with the aim to "reduce the threat to people and their property" and "deliver the greatest environmental, social and economic benefits." Managed realignment was promoted as the preferred approach for managing flood risk in rural areas and to create habitat to offset or compensate loss. Opportunities for managed realignment are identified by coastal management instruments, such as SMPs and Catchment Flood Management Plans (Thomas 2014).

SMPs are developed by coastal partnerships formed by local authorities and the EA and must consider four management policies: (a) no active intervention (no planned investment in coastal protection), (b) hold the line (investments on coastal protection will be made to maintain the shoreline position), (c) managed realignment, and (d) advance the line (new coastal protection will advance the shoreline seawards). Although not legally binding, SMPs recommend the most suitable policy to be implemented at each coastal segment looking at three time frames into the future: 0–20 years, 20–50 years, and 50–100 years. The most recent SMPs were published in 2012.

Looking at the recommendations of the last SMPs, there is an "ambition" to realign, in England and Wales, a total of 550 km by 2030, resulting in the creation of 6200 ha of intertidal habitat at a cost of £10–15 million per year (Committee on Climate Change 2013). However, between 1991 and 2013, only around 66 km of coastline has been realigned. To achieve the target, government plans require an eightfold increase in realignment in the next 15 years.

The document "Making Space for Nature" (Lawton et al. 2010) presents results from an independent assessment on the sustainability of natural environments in England. The report identifies that natural habitat sites are too small and fragmented and unable to provide all ecosystem functions, especially when climate change impacts on existing sites are considered. Additionally, the report makes 24 general recommendations that intend to steer future habitat restoration strategies without being prescriptive on how they should be implemented. Relevant to the context of this chapter, two key points are made by Lawton et al. (2010): (a) natural coastal protection will be critical to manage climate change impacts related to coastal erosion and flooding (e.g., sea level rise and increased storminess), and (b) biodiversity offsetting and payment for ecosystem services should be considered as a way of enhancing nature and creating wider benefits through planning. It is possible to deduct from the report that habitat creation and a more effective ecological network is required if society is to benefit from ecosystem services related to water quality, flood and erosion control, and carbon storage.

Managed realignment can deliver multiple functions that satisfy the need to adapt to climate change, compensate for habitat loss, and provide sustainable coastal protection. The land-use changes resulting from managed realignment projects are an important element of the United Kingdom's National Adaptation Programme (Defra 2013) to provide improved climate and flood regulation. Implementation of managed realignment at the scale and rate planned by the UK government requires (a) securing land at locations showing conditions favorable to the development of the habitats to be created, (b) better understanding of how realigned sites evolve in the long term, (c) increasing public acceptance, and (d) attracting external funding. Either collectively or individually, these challenges are known to delay or hinder the wider uptake of managed realignment projects.

9.4.2 Sigma Plan (Belgium)

Flood risk mitigation is a serious concern in Belgium, where estuarine and open coastlines are heavily engineered by flood protection structures. Restoration of intertidal habitats is a legal requirement as in other EU countries. The Sigma Plan* was devised in 1977 by the Waterways and Sea Canal (which is responsible for flood protection and navigation) in partnership with the Agency for Nature and Forests. It is a major regional strategy aiming to deliver improved flood risk management in the Flanders region through enhancement of environmental conditions along the river Scheldt and its tributaries.

The Scheldt Estuary Development Plan, a Netherlands-Flanders agreement, establishes an integrated long-term vision for the estuary's accessibility, environmental conservation, and flood protection. Human interference in the Scheldt estuary has reduced intertidal habitat by 50% over the past century (Meire et al. 2005). The economic and ecological importance of estuary management interventions has caused historic cross-border conflicts between Belgium and the Netherlands owing to enhanced flood risk and environmental degradation, resulting particularly from land reclamation and channel dredging undertaken by the Netherlands (Esteves 2014). More recently, such conflicts involved the managed realignment of the Hedwige polder, which the Netherlands agreed to implement as a compensation measure for cross-border habitat loss (Stronkhorst and Mulder 2014).

A revision of the Sigma Plan in 2005 aimed to integrate the "Room for the River" concept (see Section 9.4.3) and its multifunctionality (i.e., recreation, nature restoration, climate change adaptation, and sustainability of economic activities). The Sigma Plan involves the construction or upgrading of 512 km of dikes to the agreed "sigma" height (i.e., to provide protection against water levels with a return period of 1000 years), including the creation of flood control areas and 15 CRT projects to be implemented by 2030 resulting in the creation of 4000 ha of intertidal habitat. A good overview of the Sigma Plan is provided in De Beukelaer-Dossche and Decleyre (2013). Lippenbroek (Figure 9.4b) was the first CRT project piloted in Belgium (Beauchard et al. 2011; Jacobs et al. 2009; Teuchies et al. 2012).

Land availability and cost are significant issues in Belgium. To reduce the impact of the revised Sigma Plan on urban and agricultural areas, most of the CRT projects are planned to take place in predefined preferred areas. Nevertheless, the impact on land value in the preferred areas is considerable and the government has implemented a policy of expropriation and freehold purchasing that is currently being tested in the Kalkense Meersen (Van Rompaey and Decleyre 2013). Other measures to facilitate land purchase include (a) creation of a land bank (so land is available to offer as exchange to owners affected by the Sigma Plan), (b) €2000/ha in financial incentives (above market value) for willingness to sell, (c) relocation, (d) compensation for loss of production, and (e) low interest loans.

9.4.3 Room for the River and the Delta Programme (The Netherlands)

Coastal management strategies in the Netherlands vary along the coast, with complex dike systems protecting the low-lying land around the Rhine–Meuse–Scheldt delta and a range of soft engineering and managed realignment methods implemented along the sandy coast of the North Sea and

* http://www.sbe.be/en/reference/sigma-plan-river-scheldt.

the silty shorelines of estuaries and the Wadden Sea (Stronkhorst and Mulder 2014). Flood protection is of paramount importance to the Netherlands, where approximately two-thirds of the land area is below sea level (Brouwer and van Ek 2004). Dutch policies have evolved through time and recognize the importance of naturally evolving coasts by incorporating concepts of eco-engineering (Rijkswaterstaat and Deltares 2013). Many of the projects described as "building with nature" in the Netherlands could be classified as managed realignment. However, the perception of "retreat" associated with the term "managed realignment" restricts its use in the Netherlands (Eertman et al. 2002). The Dutch experience demonstrates that a combination of managed realignment and other coastal protection approaches can be strategically implemented to provide the desired level of protection and coastal uses, taking into consideration physical characteristics and socioeconomic objectives of the sites.

The Rijkswaterstaat (Ministry of Infrastructure and Environment) is responsible for the main waterways and water systems in the Netherlands, including flood protection. In the aftermath of the catastrophic 1953 floods, the Delta Plan was devised to enhance flood protection through the construction of a robust dyke–dam system. As a result of hard engineering conducted under the "deltaworks," the Dutch shoreline was shortened by 700 km. From 1990, under the "Dynamic Preservation of the Coastline" policy, the shoreline is not allowed to retreat inland of its 1990 position (Hillen and Roelse 1995; Van Koningsveld and Mulder 2004), with management efforts focusing on beach nourishment along the sandy coasts of the North Sea. Annually, 12 million m^3 of sand are used in foreshore, beach, and dune nourishments to keep pace with the present sea level rise of 2 mm/year (Stronkhorst and Mulder 2014). This approach has resulted in a slight seaward shift of the Dutch coastline (Giardino et al. 2014).

Climate change adaptation and environmental concerns have led to new policy developments, such as the *Room for the River** (2007–2016) and the *Delta Programme*.[†] The Room for the River comprises 30 projects (at a cost of €2.3 billion), aiming to restore floodplains and their marshlands to improve flood safety and environmental quality. Realignment of river dikes is one of the measures used to tackle the combined effects of high river discharge and rising sea level that exacerbate flood risk.

The Delta Programme aims to improve flood risk management, securing freshwater supply, and promoting climate-proof spatial planning for the delta area. Underpinning the Programme is the decision to reduce "the probability of individual mortality due to floods anywhere in the Netherlands to a maximum of 1:100,000 per annum" by 2050. The implementation of the Programme will be based on predefined preferential strategies, and these include a combination of measures, for example, beach nourishment along the sandy coast, dyke realignment along the rivers (and other measures identified in the Room for the River), and reinforcement of dykes in densely occupied or strategic areas. Up to 2028, the Delta Programme has a secured average annual budget of €1.2 billion.

9.4.4 National Integrated Coastline Management Strategy (France)

The French National Integrated Coastline Management Strategy[‡] published in 2011 explicitly promotes managed retreat from areas at risk. The Strategy is based on eight principles, which include (a) acknowledging that the coast is dynamic and cannot be fixed everywhere, (b) the need to stop occupation in coastal areas where risk of flooding and erosion are high, (c) planning for long-term relocation of activities and property exposed to coastal risks taking into consideration how risks will change because of climate change, and (d) wide dissemination of knowledge on coastal ecosystems and hazards to all stakeholders. The strategy was developed by the Ministry of Ecology,

---

* https://www.ruimtevoorderivier.nl/english/.
[†] https://www.government.nl/topics/delta-programme.
[‡] An English version of the strategy is available at: http://www.developpement-durable.gouv.fr/IMG/pdf/12004_Strate
   _ugie_gestion_trait_de_co_ete_GB_140326_BD.pdf.

Sustainable Development and Energy, and the responsibility for its implementation is shared with local authorities.

The strategy recommends that planning considers predictions of coastal evolution at time frames of 10, 40, and 90 years and anticipates relocation of property and infrastructure as a medium- and long-term alternative to reduce coastal risks, justified through cost–benefit and multicriteria analyses. Additionally, hard engineering protection should be considered only for densely populated areas or sectors of national strategic importance. Generally, flexible management options, allied with opportunities for ecological engineering to enhance natural protection, are preferred. Other managed realignment measures, such as the extension of the 100-m setback and removal of coastal protection structures reaching the end of their concession time, are included in the strategy's plan of actions.

A call for innovative ecological engineering was launched in July 2011 to select demonstration projects where recommendations of the national strategy could be tested. In 2013, five projects were selected, each representing a different coastal typology and receiving €600,000 from the central government, and with the task to identify and evaluate coastal relocation measures within 2 years. The selected demonstration projects are (1) Ault (affected by a rapidly eroding cliff at the margins of the English Channel), (2) Hyères les Palmiers (low-lying coastal plain along the Mediterranean), (3) three locations along the Aquitaine coast (Lacanau, La Teste-de-Buch, and Labenne have sandy beaches affected by erosion and dune migration), (4) Petit-Bourg (in the Caribbean Guadeloupe affected by hurricanes), and (5) Vias (a rapidly growing coastal population at risk of flooding in the Mediterranean).

In May 2014, a seminar was organized in Paris to discuss mid-project progress* and results indicate that the main issues so far include (a) public acceptance (associated with poor understanding of coastal risks and how they may change in the future), (b) the definition of temporal and spatial scales in which the projects' costs and benefits should be evaluated, (c) how to incorporate the uncertainty of predictions, and (d) the lack of governance instruments that support the types of actions required to relocate properties and infrastructure. Lack of awareness, poor acceptance of coastal communities, and mistrust of government actions are often cited as constraints to the implementation of managed realignment in France (Bawedin 2004; Goeldner-Gianella 2007; SOGREAH 2011) and in other countries (e.g., Roca and Villares 2012).

At present, the *Conservatoire du Littoral*† is a main player in nature conservation and habitat recreation in France, usually through implementation of RTE projects. It is a government agency with an action plan based on the philosophy of the British charity National Trust. In its 40th anniversary, the Conservatoire du Littoral has an annual budget of €50 million and currently owns 1450 km of coastline and 160,000 ha, a good part managed as areas of conservation.

9.5 LESSONS LEARNED

"Building with nature" approaches are now underpinning an increasing number of national and regional strategies as described above. Considering the increase in flooding and erosion risks associated with climate change, the benefits from natural coastal protection and other ecosystem services are becoming increasingly important to the sustainability and resilience of coastal communities (Temmerman et al. 2013). Managed realignment offers great opportunities for the creation of multifunctional areas that are able to benefit the wider society through a range of ecosystem services (e.g., Luisetti et al. 2011). The provision of these services depends on the size and type

---

* Documents (in French) describing the five projects and results of the seminar are found at: http://www.developpement-durable.gouv.fr/Strategie-nationale-de-gestion.html. At the time of writing, no further information was found about these demonstration projects in the scientific literature available in English.
† http://www.conservatoire-du-littoral.fr/.

of managed realignment project, its adjacent environments, connectivity with water (Schleupner and Schneider 2013), and the previous land use in the realigned site (Spencer and Harvey 2012). The range of conditions in which managed realignment has been implemented makes each project almost unique and site specific. Therefore, generalizations of lessons learned need to be considered carefully and adjusted to specific needs, as they may not necessarily be a good recipe for success elsewhere.

Considering the important role of managed realignment in existing coastal management policies, it is timely to identify lessons that may be more widely applicable to improve current practices and maximize social, economic, and environmental benefits. These lessons are listed in Table 9.3 in three broad groups, factors important at the (a) high strategic level and (b) project level, and key aspects concerning (c) public perception and stakeholder engagement. There are overlaps in the factors identified in Table 9.3, as some aspects are important both at the strategic and local level. In particular, aspects of public and stakeholder engagement are fundamental to the wider uptake of managed realignment at the strategic and local level, and therefore these are emphasized as a separate group in Table 9.3. This section describes the key aspects identified at the strategic and project level, including the importance of public perception and stakeholder engagement in each.

9.5.1 Governance and High-Level Strategy

Adaptive management is often considered a good practice, particularly when outcomes of decisions may be affected by inherent uncertainties (Williams 2011), including time and magnitude of climate change impacts, sea level rise, and so on. A key element of adaptive management is a systematic assessment of performance, so gained knowledge can be used to inform adjustments or changes where and when required. To ensure that performance can be properly evaluated, it is essential that clear targets and their time frames are defined both for the overall strategy and for individual projects (Garbutt et al. 2006; Williams 2011).

Strategies should be guided by a well-justified strategic vision aiming to achieve clearly defined targets and time frames, which are widely disseminated and understood (e.g., the Sigma Plan and the Room for the River). Although strategic targets are often set out, it is less common that strategies identify the time frames in which specific targets must be achieved and the criteria that should be used to measure performance. For example, the French National Integrated Coastline Management Strategy indicates that coastal local authorities should have plans developed by 2020 but does not identify exact targets (and time frames) that should be achieved by the plans (e.g., the level of protection to be provided or the number/percentage of properties that should be removed from high risk areas). It is understandable that targets are less prescriptive because of the difficulty in predicting the course of nature and the variability of socioeconomic, physical, and environmental conditions along the coast. However, if objectives are unclear and not linked to time frames, performance cannot be adequately measured; thus, adaptive management becomes impractical.

The success of national strategies depends on governance capacity; existing legislative mechanisms facilitating land acquisition, the licensing process, and controls related to individual property rights; and public acceptance. Clearly, the success of strategies largely depends on how well individual projects are delivered and managed at the local level, and these aspects are discussed in Section 9.5.2.

Land availability is a common issue in Europe, in both urban and rural areas (e.g., high-grade agricultural land), owing to the increasing demand for locally sourced produce and the strong cultural values related to farming found in many countries, such as the United Kingdom, the Netherlands, and Spain. Not only legal and financial mechanisms must exist so that land can be acquired, the land must have suitable conditions for managed realignment to deliver the desired strategic objectives. The locations where managed realignment may be the most sustainable option are often identified at the regional/national level. However, the delivery of the strategy depends on the willingness of

Table 9.3 Requisites for Facilitating the Implementation and Wider Uptake of Managed Realignment

Governance and high-level strategy	Adaptive management (i.e., based on regular assessments)
	Clear and well-justified strategic vision (targets and time frames are widely disseminated and understood)
	Availability of suitable land to deliver regional and local targets
	Funding and institutional mechanisms that facilitate the implementation of the strategy across the national to local levels (e.g., land purchase; to fund educational campaigns)
	Ensure implementation mechanisms do not conflict with private property rights
	Strong knowledge basis about associated uncertainties and potential benefits accruing from the strategy
	Robust public dissemination and stakeholder engagement strategy
Delivery at the project level	Clear targets and well-defined time frames for each project
	Capacity building for practitioners concerning project uncertainties, socioeconomic implications, and how to transfer this knowledge to the public
	Tailor project design to maximize benefits relevant to local communities
	Project design based on modeling outputs considering worst-case scenarios
	Better understanding of long-term evolution of realigned sites
	Systematic monitoring of relevant parameters until rates of change/conditions stabilize
	Independent and science-based data analysis to provide evidence of performance
Public perception and stakeholders engagement	Good understanding of national, regional, and local targets
	Transparent decision making and a legitimate participatory process to increase trust in government and nongovernment players
	Education efforts to reduce negativity associated with "give in to the sea" perception
	Increased awareness about ecosystem services, climate change adaptation needs, the concept of managed realignment
	Long-term dissemination and engagement plan to reduce the "novelty effect"
	Bottom-up approach to determine local targets
	Focus on multiple functions and benefits (to reduce not-in-my-backyard attitude)
	Dissemination of evidence about the wider benefits gained from existing projects
	Working with the media to disseminate consistent messages and reduce influence of misinformation or unfounded perception

landowners to sell the land or to form partnerships with relevant government or nongovernmental organizations at the local level.

Compulsory purchase mechanisms have been used in France in the aftermath of the Xynthia storm to force relocation from high-risk areas. In Belgium, some incentives are identified in the Sigma Plan to stimulate interest of landowners (see Section 9.4.2). In England, land purchase is negotiated case by case and subjected to high price variability, sometimes resulting in acquisition of less suitable sites if willing landowners are identified in areas of high demand for habitat compensation (Esteves and Thomas 2014). In support of managed retreat strategies, it is paramount that existing legal mechanisms deal effectively with private property rights (e.g., the Coastal Law in Spain) and the time frames of execution are faster than the increased risk posed by climate change or population growth. Otherwise, the strategy may not effectively reduce the number of people at risk.

In the United Kingdom, the Environment Agency has formed partnerships with the Royal Society for the Protection of Birds and The Wildlife Trusts to deliver projects focusing on nature conservation and able to provide wider environmental and socioeconomic outcomes (e.g., Abbotts Hall, Freiston Shore, and Wallasea Island). Working in partnership with landowners and relevant organizations can reduce the overall costs (e.g., eliminating the need for land purchase) and expedite implementation (e.g., less local opposition). Partnerships are facilitated when landowners can see the potential benefits on offer, such as diversification of land use, lower costs to maintain flood protection, or grants/subsidies offered by the government (e.g., payment for ecosystem services schemes).

Most projects in the United Kingdom have been implemented in areas where flood protection structures are in poor state of repair and land prices are lower. However, if site conditions are not ideal, managed realignment may not offer the best return for the investment of public money (Thomas 2014). Additional issues arise where managed realignment may result in flooding of freshwater habitats within designated areas of conservation, as these will need to be compensated through habitat recreation to comply with the EU Habitats Directive.

Farlington Marshes (Portsmouth, southern England) were reclaimed in the 1770s and the protection of a seawall allowed the development of marshes and other freshwater habitats that are now within Natura 2000 sites. A plan to realign the seawall at Farlington Marshes is favored by the local authority (the landowner) but has faced strong public opposition due to impacts on locally important recreational space, habitats supporting internationally important bird populations, and flood risk to more than 500 homes, a major access road, and the rail line. The creation of intertidal habitats cannot be considered as offering "equivalent value" to the freshwater habitats that will be lost.

Managed realignment in such areas creates conflict between the uncertain gains (which will depend on the type and quality of intertidal habitats that might develop) and the certain loss of the services provided by the established freshwater habitats (which cannot be recreated locally or in the short term). Such a conundrum is well described by Maltby (2006, p. 93): "We are then confronted by the contradictory situation of ecosystem destruction and re-establishment both featuring prominently in society's agenda. The challenge is to manage the processes of change so that we do not irretrievably lose assets difficult or impossible to replace."

As in Farlington Marshes, public opposition has delayed or prevented the implementation of managed realignment at other locations, such as Donna Nook on the Humber estuary (UK); Bas-Champs de Cayeux in Picardy, France (SOGREAH 2011); and the Ebro Delta in Spain (Roca and Villares 2012). In the UK, public acceptance may delay or prevent the implementation of managed realignment projects, especially when projects involve loss of access rights to existing footpaths or there is a perception that increased flood risk will occur elsewhere in their area. The recent demonstration projects of managed retreat in France have exposed the lack of awareness of certain community groups about coastal risks and how existing coastal protection works may exacerbate risks under certain conditions.

Public perception is clearly a knowledge and communication issue (e.g., Goeldner-Gianella 2007; Esteves and Thomas 2014) that needs to be addressed as part of strategies (e.g., through robust educational campaigns) and delivered at the local level (i.e., where projects will be implemented). At the strategic level, it is important to convey a consistent message through educational campaigns and clearly explain why managed realignment is needed here and now, the expected gains and losses, supported by quantitative evidence of benefits realized from existing projects. This evidence-based and consistent message is particularly necessary to convince and engage stakeholders in situations involving a change in practice, such as where the government moves from a hard engineering hold-the-line policy toward managed realignment.

Gathering quantitative evidence of benefits requires projects to be systematically monitored through time and results measured against the intended objectives. Typically, monitoring of vegetation and macro invertebrates colonization, bird counts, and sedimentation rates is undertaken in most projects. Often, surveys and data analyses are conducted by the parties involved in project design, resulting in restricted data availability and very few published independent studies. In fact, published studies and data monitoring reports are available for only a small number of management realignment projects; thus, results disseminated in peer-reviewed publications tend to be limited in scope, space, and time (Esteves 2013). Fortunately, the CRT pilot project Lippenbroek (Belgium) has been extensively studied and published by researchers of the University of Antwerp and collaborators (e.g., hydrology, sedimentation, nutrient cycling, vegetation, macroinvertebrates, fish habitat, metal concentration, etc.), providing a good background for future CRT projects.

9.5.2 Project Design and Site Evolution

The lack of predetermined targets and performance indicators greatly compromises the sign-off of projects designed as compensatory measures for habitat loss or damage. The complications to achieve the sign-off of compensatory managed realignment are a concern of consultancies involved in project design and monitoring and stakeholders who want to ensure that compensation is actually being achieved.* It is in the public interest that measures are in place to ensure that individual projects are fulfilling their objectives (e.g., actually compensating habitat loss) and contributing to the overall impact of national strategies.

Although 140 managed realignment projects exist in Europe, they vary greatly in their method of implementation and local specificity. The design of the first managed realignment projects was relatively simple, involving mainly the planning of how coastal protection structures were going to be removed or breached. More recently, projects also involve the design of drainage channels and landscaping of the realigned area to produce a range of elevations and topography that would facilitate the creation of the desired types of habitats.

Technical expertise and modeling capability are key to project design and considerable advances have been achieved in the last decade. However, important challenges remain, such as modeling hydrodynamic–sediment–vegetation interactions in mixed-grain size environments and dealing with uncertainties related to the changing physical environment (e.g., due to climate change and climate variability) and how it will affect restoration of ecosystems and the services they provide. These challenges associated with a deficient field-based knowledge still limit our understanding of and ability to predict the long-term evolution of realignment sites (e.g., Esteves and Thomas 2014; Ni et al. 2014; Rotman et al. 2008).

Depending on project location, type, and size, managed realignment can cause important changes to local hydrodynamic conditions, altering the tidal prism and the erosion and deposition patterns in the intertidal zone. After breaching, the drainage system evolves rapidly; the main channel usually deepens and enlarges until a dynamic equilibrium with the new tidal prism is reached. Predictions of channel evolution depend on empirically defined coefficient and exponent (e.g., Hughes 2002), which

---

* This topic was discussed in the 2013 ABPmer Conference "Coastal Habitat Creation—Are We Delivering?"; conference presentations can be downloaded from: http://www.omreg.net/conference-papers/.

vary with the scale of the system, tidal range, salinity, vegetation, and sediment characteristics (e.g., Williams et al. 2002). The required field data rarely exist (Vandenbruwaene et al. 2011), and values defined for other areas are often applied for largely different systems (e.g., cohesive vs. noncohesive). Such practice may lead to large errors in the prediction of optimal breach width and channel cross-sectional area in managed realignment projects, as reported by Friess et al. (2014) for Freiston Shore.

The design of the managed realignment scheme at Freiston Shore was informed by modeling, which included three breaches (each 50 m wide), upgrading of an existing embankment further inland, and the creation of an artificial tidal creek system. Two years from breaching, tidal creeks were still growing landward at rates of 400 m/year or approximately 20 times greater than observed at natural conditions (Symonds and Collins 2007). The rapid erosion and deposition associated with the evolution of the tidal creeks at Freiston Shore were not anticipated by model results. Perhaps model runs did not include the extreme water levels experienced a few days after breaching. Considering the availability of suitable data (e.g., Friess et al. 2014; Ni et al. 2014; Symonds and Collins 2007), it would be relevant to test whether models are able to reproduce the observed hydrological sediment response to the extreme conditions at Freiston Shore.

An increasing number of publications discuss the functional equivalency of recreated sites compared with natural ecosystems concerning vegetation (e.g., Mossman et al. 2012) and macroinvertebrates (e.g., Beauchard et al. 2013; Pétillon et al. 2014), the potential for nutrient cycling and carbon storage, and metal mobility (e.g., Teuchies et al. 2012, 2013). Although there are large variations in site conditions and results, it is possible to briefly summarize the current state of the art as follows.

Colonization by macroinvertebrates occurs fast in areas of new sedimentation (i.e., the new layer deposited on top of more consolidated old soils), where in a few years after tidal restoration assemblages may be similar or richer than control sites (e.g., Beauchard et al. 2013). However, the size of individuals may be smaller and biomass may be lower (e.g., Mazik et al. 2010), which is likely to affect the diversity of bird assemblages using the realigned sites (e.g., Atkinson et al. 2004) or requiring greater feeding effort from individual birds (e.g., Mander et al. 2013). Additionally, compaction and geochemistry of older agricultural soils might slow or prevent colonization by invertebrates (Garbutt et al. 2006). Looking at arthropods, results reported by Pétillon et al. (2014) indicate that complete functional equivalency (including structure of trophic guilds and the potential for fish nursery) was not achieved at managed realignments.

Colonization of saltmarsh species within realigned sites can be fast (within 1–2 years). The types of assemblages are controlled by site elevation in relation to the tidal level, with pioneer and low marsh species dominating in lower areas (Garbutt et al. 2006) and even in higher ground (Mossman et al. 2012). Even after a long time, species diversity tends to be lower than adjacent natural saltmarshes (Wolters et al. 2005); influencing factors include poor drainage and seed availability (Spencer et al. 2008), the small extent of sites, and poor range of elevations between mean high water of neap and spring tides (Wolters et al. 2005). The functioning of recreated saltmarshes was found to be "significantly impaired" when compared with natural systems, affecting their ability to deliver ecosystem services (Spencer and Harvey 2012). Therefore, they may not satisfy the requirements of the EU Habitats Directive (Mossman et al. 2012).

Concerning nutrient cycling and metal mobility, results are geographically variable because of site-specific conditions and change through time as realigned sites evolve. The nitrogen and carbon storage capacity depends on vegetation density and sedimentation rates (Adams et al. 2012) and the chemistry of the floodwaters and soils (Blackwell et al. 2010). As an example of the variability across sites, realignment sites on the Blackwater Estuary were found to be net sources of methane and nitrous oxide (Adams et al. 2012), while on the River Torridge (southwest England), they are a net source of nitrous oxide and a sink for methane. The chemistry of the soils and the floodwater also determines whether the realigned sites may act as a sink (e.g., Teuchies et al. 2012) or source (e.g., Emmerson et al. 2001) of metal contaminants; however, metal availability and release are expected to occur within the first months after tidal restoration (e.g., Teuchies et al. 2013) and expected to change through time.

Only a few publications refer to long-term morphological evolution of managed realignment sites (Spearman 2011; Vandenbruwaene et al. 2011; Ni et al. 2014). Very little attention has been given in the literature to how site evolution may affect flood risk (Esteves and Williams 2015). In the United Kingdom, both the media and the published literature place a greater focus on ecological aspects of managed realignment, while there is a lack of evidence on other potential benefits. This imbalance has created a public perception, often detrimental, that managed realignment is an expensive nature conservation measure aiming to create habitats for birds while not enough effort is made to reduce flood risk to people (e.g., Esteves 2014). Morris (2012) argues that managed realignment will only attract wider public support if there is a shift in focus toward its wider societal benefits, especially those related to coastal erosion and flood risk management.

Indeed, public perception is influenced by the lack of understanding about the benefits local communities might accrue from managed realignment. Demonstrating the multiple functions and ecosystem services that can be provided through managed realignment is likely to be more appealing to the public than emphasizing single objectives or achievements. By engaging with local communities to identify how they are likely to benefit from future managed realignment, a greater sense of ownership might be created, leading to increased uptake of managed realignment in general and a better acceptance of projects near homes and businesses.

9.6 CONCLUDING REMARKS

In Europe, managed realignment is increasingly promoted as a more sustainable coastal management approach able to deliver improved flood risk management and creation of habitats. In practice, the term "managed realignment" reflects a number of initiatives aiming to create space to restore the natural adaptive capacity of coastal environments and their ability to provide a range of ecosystem services. A total of 140 managed realignment projects have been implemented or are underway in Europe, which generally involve at least one of the following: removal, breach, or realignment of existing coastal protection; controlled tidal restoration; and managed retreat (i.e., relocation from risk areas).

European Directives (e.g., Habitats, Birds, Floods, and Water Framework) and climate change adaptation needs are the key drivers leading to the wider promotion of managed realignment approaches in recent national strategies. The French National Integrated Coastline Management Strategy, for example, supports relocation of economic activities and assets from high-risk areas. In the United Kingdom, SMPs suggest that removal, breach, or realignment of existing coastal protection may be the best management option for approximately 550 km of the coast by 2030. In Belgium, the Sigma Plan includes 15 CRT projects to be implemented along the Scheldt estuary, resulting in the creation of 4000 ha of intertidal habitat by 2030.

It is of paramount importance that managed realignment projects are carefully planned, taking into account local characteristics (social and environmental), and that site evolution is systematically monitored. The impact of managed realignment must be objectively measured against objectives set for each individual project and at the high strategic level. Evidence of benefits gained will help attract public support and lessons learned can be used to improve future practice.

The effective implementation of managed realignment requires the integration of (a) improved scientific knowledge, (b) efficient mechanisms of governance, and (c) robust public engagement. The knowledge about how the many social, economic, and technical aspects interact is evolving fast as new policies are formulated, more projects are implemented, and new monitoring data become available.

REFERENCES

Adams, C.A., J.E. Andrews, and T. Jickells. 2012. Nitrous oxide and methane fluxes vs. carbon, nitrogen and phosphorous burial in new intertidal and saltmarsh sediments. *Science of the Total Environment* 434(15): 240–251.

Alexander, K.S., A. Ryan, and T.G. Measham. 2012. Managed retreat of coastal communities: Understanding responses to projected sea level rise. *Journal of Environmental Planning and Management* 55(4): 409–433.

Atkinson, P.W., S. Crooks, A. Drewitt, A. Grant, M.M. Rehfisch, J. Sharpe, and C.J. Tyas. 2004. Managedrealignment in the UK—The first 5 years of colonization by birds. *IBIS* 146:101–110.

Bawedin, V., 2004. La dépoldérisation, composante d'une gestion intégrée des espaces littoraux? Prospectivesur le littoral picard et analyse à la lumière de quelques expériences: Baie des Veys (Normandie), Aber de Crozon (Bretagne), Tollesbury (Essex) et Freiston shore (Lincolnshire). *Cahiers Nantais* 6:11–20.

Beauchard, O., S. Jacobs, T.J.S. Cox, T. Maris, D. Vrebos, A. Van Braeckel, and P. Meire. 2011. A new technique for tidal habitat restoration: Evaluation of its hydrological potentials. *Ecological Engineering* 37: 1849–1858.

Beauchard, O., S. Jacobs, T. Ysebaert, and P. Meire. 2013. Sediment macroinvertebrate community functioning in impacted and newly-created tidal freshwater habitats. *Estuarine, Coastal and Shelf Science* 120: 21–37.

Blackwell, M.S.A., S. Yamulki, and R. Bol. 2010. Nitrous oxide production and denitrification rates in estuarine intertidal saltmarsh and managed realignment zones. *Estuarine, Coastal and Shelf Science* 87(4): 591–600.

Brouwer, R. and R. van Ek. 2004. Integrated ecological, economic and social impact assessment of alternative flood control policies in the Netherlands. *Ecological Economics* 50: 1–21.

Committee on Climate Change. 2013. Managing the land in a changing climate. Chapter 5: Regulating services—Coastal habitats. Available from: http://www.theccc.org.uk/publication /managing-the-land-in-a-changing-climate/, 92–107.

Cooper, N.J., S. Wilson, and T. Hanson. 2013. Realignment of Littlehaven Sea Wall, South Tyneside, UK. *Proceedings Institution of Civil Engineers Coastlines, Structures and Breakwaters Conference*, 10 pp.

Cox, T., T. Maris, P. De Vleeschauwer, T. De Mulder, K. Soetaert, and P. Meire. 2006. Flood control areas as an opportunity to restore estuarine habitat. *Ecological Engineering* 28: 55–63.

De Beukelaer-Dossche, M. and D. Decleyre (Eds.), 2013. Bergenmeersen, Construction of a Flood Control Area with Controlled Reduced Tide as Part of the Sigma Plan. Available from: http:// www.vliz.be/imis/docs/publications/250657.pdf, last accessed on September 18, 2015.

Defra (Department for Environment, Food and Rural Affairs). 2002. Managed Realignment Review. Policy Research Project FD 2008, Flood and Coastal Defence R&D Programme. Available from: http://randd.defra.gov.uk/Document.aspx?Document=FD2008_537_TRP.pdf, last accessed on January 4, 2015.

Defra. 2005. Making space for water—Taking forward a new Government strategy for flood and coastal erosion risk management in England. First Government response to the autumn 2004 consultation exercise. London.

Defra. 2013. The National Adaptation. Programme. Making the country resilient to a changing climate. Norwich: The Stationery Office, 182 pp. Available from: https://www.gov.uk/government /uploads/system/uploads/attachment_data/file/209866/pb13942-nap-20130701.pdf/.

Defra Flood Management Division. 2005. Coastal Squeeze Implications for Flood Management, the requirements of the European Birds and Habitats Directives. Defra Policy Guidance.

De Schipper, M.A., S. De Vries, M.J.F. Stive, R.C. De Zeeuw, J. Rutten, B.G. Ruessink, S.G.J. Aarninkhof, and C. Van Gelder-Maas. 2014. Morphological Development of a Mega-Nourishment: First Observations at the Sand Engine. Proceedings of 34th International Conference on Coastal Engineering (ICCE 2014), doi.org/10.9753/icce.v34.sediment.73.

Dyke, P. and T. Flux. 2014. The National Trust Approach to Coastal Change and Adaptive Management., In: Esteves, L.S. (Ed.), *Managed Realignment: Is It a Viable Long-Term Coastal Management Strategy?* New York: Springer.

Eertman, R.H.M., B.A. Kornman, E. Stikvoort, and H. Verbeek. 2002. Restoration of the Sieperda Tidal Marsh in the Scheldt Estuary, The Netherlands. *Restoration Ecology* 10: 438–449.

Elliott, M., N.D. Cutts, and A. Trono. 2014. A typology of marine and estuarine hazards and risks as vectors of change: A review for vulnerable coasts and their management. *Ocean & Coastal Management* 93: 88–99.

Emmerson, R.H.C., M.D. Scrimshaw, J.W. Birkett, and J.N. Lester. 2001. Solid phase partitioning of metals in managed realignment soils: Laboratory studies in timed soil sea-water batch mixtures. *Applied Geochemistry* 16(14): 1621–1630.

Esteves, L.S. 2013. Is managed realignment a sustainable long-term coastal management approach? *Journal of Coastal Research* SI 65: 933–938.

Esteves, L.S. 2014. *Managed Realignment: Is It a Viable Long-Term Coastal Management Strategy?* New York: Springer.

Esteves, L.S. 2016. Coastal squeeze, In: Kennish, M.J. (Ed.), *Encyclopedia of Estuaries*, Dordrecht: Springer.

Esteves, L.S. and K. Thomas. 2014. Managed realignment in practice in the UK: Results from two independent surveys. *Journal of Coastal Research* SI 70: 407–413.

Esteves, L.S. and J.J. Williams. 2015. Changes in coastal sediment dynamics due to managed realignment. Coastal Sediments 2015 (11–15 May 2015, San Diego, USA), DOI: 10.1142/9789814689977_0165.

Friess, D.A., I. Möller, T. Spencer, G.M. Smith, A.G. Thomson, and R.A. Hill. 2014. Coastal salt-marsh managed realignment drives rapid breach inlet and external creek evolution, Freiston Shore (UK). *Geomorphology* 208: 22–33.

Garbutt, R.A., C.J. Reading, M. Wolters, A.J. Gray, and P. Rothery. 2006. Monitoring the development of intertidal habitats on former agricultural land after the managed realignment of coastal defences at Tollesbury, Essex, UK. *Marine Pollution Bulletin* 53(1–4): 155–164.

Giardino, A., G. Santinelli, and V. Vuik. 2014. Coastal state indicators to assess the morphological development of the Holland coast due to natural and anthropogenic pressure factors. *Ocean & Coastal Management* 87: 93–101.

Goeldner-Gianella, L. 2007. Perceptions and attitudes towards de-polderisation in Europe: A comparison of five opinion surveys in France and in the UK. *Journal of Coastal Research* 23(5): 1218–1230.

Hillen, R. and P. Roelse. 1995. Dynamic preservation of the coastline in the Netherlands. *Journal of Coastal Conservation* 1: 17–28.

Hughes, S.A. 2002. Equilibrium cross sectional area at tidal inlets. *Journal of Coastal Research* 18: 160–174.

Jacobs, S., O. Beauchard, E. Struyf, T. Cox, T. Maris, and P. Meire. 2009. Restoration of tidal freshwater vegetation using controlled reduced tide (CRT) along the Schelde Estuary (Belgium). *Estuarine, Coastal and Shelf Science* 85(3): 368–376.

Lawton, J.H., P.N.M. Brotherton, V.K. Brown, C. Elphick, A.H. Fitter, J. Forshaw, R.W. Haddow, S. Hilborne, R.N. Leafe, G.M. Mace, M.P. Southgate, W.A. Sutherland, T.E. Tew, J. Varley, and G.R. Wynne. 2010. Making Space for Nature: A review of England's Wildlife Sites and

Ecological Network. Report submitted to Defra. Available from: http://archive.defra.gov.uk/environment/biodiversity/documents/201009space-for-nature.pdf, last accessed on January 15, 2014.

Luisetti, T., R.K. Turner, I.J. Bateman, S. Morse-Jones, C. Adams, and L. Fonseca. 2011. Coastal and marine ecosystem services valuation for policy and management: Managed realignment case studies in England. *Ocean & Coastal Management* 54(3): 212–224.

Maltby, E., 2006. Wetland conservation and management: Questions for science and society in applying the ecosystem approach. In: Bobbink, R., B. Beltman, J.T.A. Verhoeven, and D.F. Whigham. (Eds.), Wetlands: Functioning, Biodiversity Conservation, and Restoration. *Ecological Studies* 191(2): 93–116.

Mander, L., L. Marie-Orleach, and M. Elliott. 2013. The value of wader foraging behaviour study to assess the success of restored intertidal areas. *Estuarine, Coastal and Shelf Science* 131: 1–5.

Maris, T., T. Cox, S. Temmerman, P. De Vleeschauwer, S. Van Damme, T. De Mulder, E. Van den Bergh, and P. Meire. 2007. Tuning the tide: Creating ecological conditions for tidal marsh development in a controlled inundation area. *Hydrobiologia* 588: 31–43.

Mazik, K., W. Musk, O. Dawes, K. Solyanko, S. Brown, L. Mander, and M. Elliott. 2010. Managed realignment as compensation for the loss of intertidal mudflat: A short term solution to a long term problem? *Estuarine, Coastal and Shelf Science* 90(1): 11–20.

Meire, P., T. Ysebaert, S. Van Damme, E. Van den Bergh, T. Maris, and E. Struyf. 2005. The Scheldt estuary: A description of a changing ecosystem. Hydrobiologia 540: 1–11.

Morris, R.K.A. 2012. Managed realignment: A sediment management perspective. *Ocean & Coastal Management* 65:59–66.

Mossman, H.L., A.J. Davy, and A. Grant. 2012. Does managed coastal realignment create salt-marshes with 'equivalent biological characteristics' to natural reference sites? *Journal of Applied Ecology* 49: 1446–1456.

Ni, W., Y.P. Wang, A.M. Symonds, and M.B. Collins. 2014. Intertidal flat development in response to controlled embankment retreat: Freiston Shore, The Wash, UK. *Marine Geology* 355: 260–273.

Nordstrom, K.F. and N.L. Jackson. 2013. Removing shore protection structures to facilitate migration of landforms and habitats on the bayside of a barrier spit. *Geomorphology* 199: 179–191.

Nottage, A. and P. Robinson. 2005. The saltmarsh creation handbook: A project manager's guide to the creation of saltmarsh and intertidal mudflat. London: RSPB and CIWEM.

Pétillon, J., S. Potier, A. Carpentier, and A. Garbutt. 2014. Evaluating the success of managed realignment for the restoration of salt marshes: Lessons from invertebrate communities. *Ecological Engineering* 69: 70–75.

Pontee, N. 2013. Defining coastal squeeze: A discussion. *Ocean & Coastal Management* 84: 204–207.

Pontee, N. 2014. Factors influencing the long-term sustainability of managed realignment. In: Esteves, L.S. (Ed.), *Managed Realignment: Is It a Viable Long-Term Coastal Management Strategy?* New York: Springer.

Reisinger, A., J. Lawrence, G. Hart, and R. Chapman. 2015. From coping to resilience: The role of managed retreat in highly developed coastal regions of New Zealand. In: Glavovic, B., R. Kaye, M. Kelly and A. Travers (Eds.), *Climate Change and the Coast: Building Resilient Communities.* London: CRC Press.

Rijkswaterstaat and Deltares. 2013. *Eco-engineering in the Netherlands: Soft interventions with a solid impact*, 44 pp. Available from: http://dtvirt35.deltares.nl/products/30490.

Roca, E. and M. Villares. 2012. Public perceptions of managed realignment strategies: The case study of the Ebro Delta in the Mediterranean basin. *Ocean & Coastal Management* 60: 38–47.

Rotman, R., L. Naylor, R. McDonnell, and C. MacNiocaill. 2008. Sediment transport on the Freiston Shore managed realignment site: An investigation using environmental magnetism. *Geomorphology* 100: 241–255.

Sanò, M., J.A. Jimenez, R. Medina, A. Stanica, A. Sanchez-Arcilla, and I. Trumbic. 2011. The role of coastal setbacks in the context of coastal erosion and climate change. *Ocean & Coastal Management* 54: 943–950.

Schleupner, C. and U.A. Schneider. 2013. Allocation of European wetland restoration options for systematic conservation planning. *Land Use Policy* 30: 604–614.

SOGREAH. 2011. Etude de Faisabilite—Depolderisation partielle et eventuelle des Bas-Champs du Vimeu—La recherché d'un avenir sur un territoire perenne. Phase 1: Etat des lieux—Diagnostic du territoire. Chapitre 8: Approche géographie sociale—Perceptions.

Spalding, M.D., A.L. McIvor, M.W. Beck, E.W. Kock, I. Möller, D.J. Reed, P. Rubinoff, T. Spencer, T.J. Tolhurst, T.V. Wamsley, B.K. van Wesenbeeck, E. Wolanski, and C.D. Woodroffee. 2014. Coastal ecosystems: A critical element of risk reduction. *Conservation Letters* 7(3): 293–301.

Spearman, J. 2011. The development of a tool for examining the morphological evolution of managed realignment sites. *Continental Shelf Research* 31(10): S199–S210.

Spencer, K.L., A.B. Cundy, S. Davies-Hearn, R. Hughes, S. Turner, and C.L. MacLeod. 2008. Physicochemical changes in sediments at Orplands Farm, Essex, UK following 8 years of managed realignment. *Estuarine, Coastal and Shelf Science* 76(3): 608–619.

Spencer, K.L. and G.L. Harvey. 2012. Understanding system disturbance and ecosystem services in restored saltmarshes: Integrating physical and biogeochemical processes. *Estuarine, Coastal and Shelf Science* 106: 23–32.

Stronkhorst, J. and J. Mulder. 2014. Considerations on managed realignment in the Netherlands. In: Esteves, L.S. (Ed.), *Managed Realignment: Is It a Viable Long-Term Coastal Management Strategy?* New York: Springer.

Symonds, A.M. and M.B. Collins. 2007. The establishment and degeneration of a temporary creek system in response to managed coastal realignment: The Wash, UK. *Earth Surface Processes and Landforms* 32: 1783–1796.

Temmerman, S., P. Meire, T.J. Bouma, P.M. Herman, T. Ysebaert, and H.J. De Vriend. 2013. Ecosystembased coastal defence in the face of global change. *Nature* 504: 79–83.

Teuchies, J., O. Beauchard, S. Jacobs, and P. Meire. 2012. Evolution of sediment metal concentrations in a tidal marsh restoration project. *Science of the Total Environment* 419: 187–195.

Teuchies, J., G. Singh, L. Bervoets, and P. Meire. 2013. Land use changes and metal mobility: Multi-approach study on tidal marsh restoration in a contaminated estuary. *Science of the Total Environment* 449: 174–183.

Thomas, K. 2014. Managed realignment in the UK: The role of the Environment Agency. In: Esteves, L.S. (Ed.), *Managed Realignment: Is It a Viable Long-Term Coastal Management Strategy?* New York: Springer.

Vandenbruwaene, W., T. Maris, T.J.S. Cox, D.R. Cahoon, P. Meire, and S. Temmerman. 2011. Sedimentation and response to sea-level rise of a restored marsh with reduced tidal exchange: Comparison with a natural tidal marsh. *Geomorphology* 130: 115–126.

Van Koningsveld, M. and J.P.M. Mulder. 2004. Sustainable coastal policy developments in the Netherlands. A systematic approach revealed. *Journal of Coastal Research* 20(2): 375–385.

van Loon-Steensma, J.M. and P. Vellinga. 2013. Trade-offs between biodiversity and flood protection services of coastal salt marshes. *Current Opinion in Environmental Sustainability* 5: 320–326.

Van Rompaey, M. and D. Decleyre. 2013. Public support. In: De Beukelaer-Dossche, M. and D. Decleyre (Eds.), *Bergenmeersen, Construction of a Flood Control Area with Controlled Reduced Tide as Part of the Sigma Plan.* Available from: www.vliz.be/imisdocs/publications/250657.pdf, last accessed on September 1, 2015.

Williams, B.K. 2011. Passive and active adaptive management: Approaches and an example. *Journal of Environmental Management* 92(5): 1371–1378.

Williams, P.B., M.K. Orr, and N.J. Garrity. 2002. Geomorphic design tool for tidal marsh channel evolution in wetland restoration projects. *Restoration Ecology* 10(3): 577–590.

Wolters, M., A. Garbutt, and J.P. Bakker. 2005. Salt-marsh restoration: Evaluating the success of de-embankments in north-west Europe. *Biological Conservation* 123(2): 249–268.

PART III

Synthesis of Living Shoreline Science
Physical Aspects

CHAPTER **10**

Practical Living Shorelines
Tailored to Fit in Chesapeake Bay

Walter I. Priest III

CONTENTS

10.1 INTRODUCTION

Living Shorelines are a mix of both art and science. They are always a balancing act that strives to provide the dynamic equilibrium between natural forces and structures that can deliver both shoreline erosion protection and increased habitat values. This dynamic equilibrium allows a Living Shoreline to be able to shape and reshape itself in response to external processes yet remain stable enough to provide long-term erosion control and habitat enhancement benefits to a reach of shoreline.

For something that seems so simple, why is it so difficult to achieve? The reason is that everything must be balanced and in harmony for all of the processes to come together to provide the physical and ecological harmony that makes everything look so simple.

People have strived for decades to develop a uniform set of criteria for designing Living Shorelines that would be applicable under every set of circumstances. This has proven to be an elusive goal, primarily because every site is different and requires a measure of experience, the art, to tailor the science of the design criteria to fit a particular site and stabilize the complex interactions that provide a true Living Shoreline.

The use of Living Shorelines is not a panacea for all shoreline problems. Living Shorelines are not always the best solution to an erosion problem. Also, there is no "one size fits all." The basic techniques must be adapted to a particular site through a balancing process that minimizes any adverse impacts and maximizes erosion protection and habitat enhancement.

Let us begin with a definition for Living Shorelines. Definitions of Living Shorelines are many and varied. For the purposes of this paper, let me propose the following definition: Living Shorelines are a suite of best management practices that are applied to combine the use of natural materials and stone to provide erosion protection and enhanced ecological value for a site. The design must provide the dynamic equilibrium among natural forces necessary to achieve long-term stability and viability for the project site and not sever the contiguity between estuarine waters and the adjacent upland.

In reality, Living Shorelines are not really a new concept. They trace their history back to the efforts of W.W. Woodhouse, Ed Seneca, and Steve Broome (Woodhouse et al. 1974) and Ed Garbisch (Garbisch et al. 1975) to stabilize dredged material for the Corps of Engineers in the early 1970s. They pioneered the techniques for propagating and planting *Spartina alterniflora* and *Spartina patens* to stabilize shorelines that are the foundation of Living Shoreline techniques today.

Similarly, marsh toe revetments, commonly used to protect existing eroding fringe marshes, have been in use in the Chesapeake Bay for many years. These are a frequent component of Living Shoreline projects but also have evolved into more offshore structures that have been converted into low-profile sills in Living Shorelines that are then backfilled with sand and planted with marsh grass to enlarge and enhance existing fringe marshes or to restore eroded marshes. Innovative materials such as coir logs, oyster shell bags, Reef Balls, Oyster Castles, and other proprietary concrete structures have also recently found utility in the Living Shoreline process in low-energy areas as both marsh toe protection and offshore sill construction to contain sandy backfill for planting with wetlands vegetation.

Another technique that evolved independently in high-energy areas, breakwaters with beach construction, is based on observations of naturally stable beaches between headland features (Hardaway and Gunn 2010). These breakwater systems are much larger, more robust structures designed to modulate high-energy waves in a manner consistent with the development of sandy beach and dune structures on their landward side. These structures require sophisticated and detailed engineering designs to be successful. Consequently, they are beyond the scope of this paper.

Living Shoreline implementation involves a three-step process, including screening criteria, site suitability, and design and construction. Each step must be followed in order to achieve an effective, affordable, and acceptable project. These steps lead to a project that is effective at reducing erosion and enhancing habitat values at the site, can be affordably constructed by the landowner, and has an acceptable level of environmental impact that will make the project permissible. For example, there may be a case where a project has an effective design and is affordable to build but may have such an unacceptable level of adverse impacts on existing natural resources or adjacent property owners that a permit for the project cannot be obtained. Again, here is where the balancing process comes into play where the design must be adjusted to minimize adverse impacts, perhaps making the project more expensive, but results in a permissible project.

Similarly, each project should also undergo an analysis of alternatives that includes the gamut of options from doing nothing to Living Shoreline to breakwaters to riprap to bulkhead to find the least environmentally damaging alternative.

10.2 SCREENING PROCESS

The screening process evaluates the potential impacts of a project on critical resources or conditions that can preempt or significantly restrict the opportunity for the construction of a Living Shoreline. First of all, the site must be screened for potential impacts to adjacent natural resources including, but not limited to, wetlands, shellfish resources including aquaculture, submerged aquatic vegetation (SAV), threatened and endangered species, critical essential fish habitat including anadromous fish spawning areas, and cultural resources as well as the presence of toxic materials. The site must pass muster on all of these areas with no significant adverse impacts anticipated on these resources.

This process must also identify whether there is active detrimental erosion occurring at the site that necessitates measures to protect the eroding shoreline. This is critically important for Living Shorelines where structural components such as rock sills are proposed. There can be instances where, even though there is minimal erosion, nonstructural Living Shoreline techniques like trimming trees for additional sunlight and planting vegetation or limited filling with coir log or oyster shell bag containment might be appropriate to enhance existing or restore lost resources. These techniques have a very limited potential for adverse impacts and are routinely permitted. In fact, these are the types of projects that the first phase of Virginia's General Permit for Living Shorelines addresses.

It must also have limited impacts on navigation, for example, in a narrow waterway or adjacent to an established navigation channel where encroachment would hinder or impede navigation or maintenance of the channel.

Downdrift impacts of the Living Shoreline on littoral processes that would adversely affect erosion rates on adjacent properties must also be taken into account.

Impacts to adjacent riparian buffer areas must also be considered. Major issues include clearing of riparian trees and vegetation for construction access and bank grading, if required, in the design of the Living Shoreline.

Something else that needs to be addressed in the screening phase is habitat conversion. This involves converting one habitat type to another, for example, converting nonvegetated intertidal habitat to marsh or shallow subtidal to rocky intertidal and marsh. While these habitats provide habitat for a wide range of species, some species may be excluded or habitat functions may be compromised for other species. This can be a major issue in instances where critical habitats such as SAV, shellfish resources, or fish spawning areas are being affected. Thus, it is important to avoid these areas or limit these conversions to the minimum needed to provide an effective project without affecting its acceptability.

Much of the screening process can be accomplished with a desktop review of available resource maps, shoreline erosion studies, topographic maps, navigational charts, soil surveys, Google Earth, and aerial photography. Additionally, state and federal natural and cultural resource agencies may have to be consulted for specifics on a particular site. Also, a site visit must be made to evaluate active detrimental erosion and specific site conditions that would affect the design of the project.

The entire screening process is designed to ensure that a proposed project has avoided potential adverse impacts to the maximum extent possible and make the project acceptable for the issuance of the appropriate permits. Also, care must be taken to avoid constraining or limiting the scope of the project to the extent that it compromises the effectiveness of the Living Shoreline. This is another example of the balancing process inherent in the design of Living Shorelines.

10.3 SITE SUITABILITY REVIEW

The Site Suitability Review gauges the applicability of Living Shoreline techniques at a particular site. It takes into account all of the physical, geological, and biological parameters that must be considered to effectively establish a Living Shoreline. It also provides guidance on what type of Living Shoreline would be most appropriate.

Site Suitability Review includes the following topics that will be addressed in detail below:

1. Storm surge
2. Fetch
3. Bank height
4. Bank condition
5. Nearshore depths
6. Sediment type
7. Width of waterway
8. Tide range
9. Infrastructure proximity
10. Buffer condition
11. Shoreline orientation
12. Erosion rate
13. Shoreline configuration

Some of this information can be determined from map and aerial photography utilized in the screening process. Other data must be obtained from additional research and a site visit.

10.3.1 Storm Surge

Storm surge is one of the most critical factors in designing a Living Shoreline because it dictates the level of protection desired for a site. Storm events are usually predicted in terms of 10-year, 50-year, and 100-year recurrence intervals with a 10%, 2%, and 1% chance of happening in any given year. Each storm event is associated with an increasing level of inundation (Table 10.1). The storm surge is typically given as a still water elevation. The anticipated wave height must be added to the storm surge elevation to determine the actual potential for inundation. This increased water level can allow for larger waves to reach the shoreline, often requiring more robust structures. This means that the storm surge plus anticipated wave heights, which can typically range from 1.5' to 3' in areas typically suited to Living Shorelines, must be added to the tide range in order to determine the elevation of the desired level of protection. Significant storm surges in the Chesapeake Bay region are given in Table 10.2.

Table 10.1   Height Frequency Levels of Total Tide at Selected Chesapeake Bay Stations

Western shore stations					Eastern shore stations				
Station	Annual frequency				Station	Annual frequency			
	10 years	50 years	100 years	500 years		10 years	50 years	100 years	500 years
Havre De Grace, Maryland	5.3	9.6	11.5	14.6	Betterton, Maryland	5.1	8.7	10.5	13.4
Baltimore, Maryland	4.1	6.8	8.1	10.7	Tolchester, Maryland	4.3	7.2	8.7	11.5
Annapolis, Maryland	4.0	6.2	7.2	9.4	Love Point, Maryland	4.3	6.4	7.4	9.7
Chesapeake Beach, Maryland	3.5	5.2	6.1	7.9	Matapeake, Maryland	4.2	6.2	7.2	9.2
Cove Point, Maryland	3.4	4.5	5.2	6.6	Cambridge, Maryland	3.9	5.1	5.9	7.5
Solomon's Island, Maryland	3.4	4.8	5.5	7.0	Hoopers Island, Maryland	3.5	4.7	5.3	6.6
Cornfield Harbor, Maryland	3.2	4.2	4.6	5.8	Chance, Maryland	4.2	5.4	5.8	6.8
Windmill Point, Virginia	3.2	3.7	3.9	4.4	Crisfield, Maryland	3.9	4.8	5.1	6.1
Gloucester Point, Virginia	3.2	3.7	3.9	4.4	Guard Shores, Virginia	4.2	5.6	6.3	7.8
					Gaskins Point, Virginia	3.5	4.0	4.2	4.6
					Eastville, Virginia	3.6	4.1	4.4	4.9
					Kiptopeke Beach, Virginia	4.1	4.8	5.2	6.2

Heights in feet above NGVD.*

*National Geodetic Vertical Datum

Source: Hardaway, Jr., C.S. and R.J. Byrne. 1999. Shoreline Management in Chesapeake Bay. Special Report in Applied Marine Science and Ocean Engineering Number 356. Virginia Institute of Marine Science, College of William and Mary, Gloucester Point, VA. 54 pp.

Table 10.2   Storm Surges at Sewells Point, Virginia

Date	Storm	Storm surge (ft)
23-Aug-33	Hurricane (unnamed)	8.02
18-Sep-03	Hurricane Isabel	7.89
12-Nov-09	Northeaster	7.75
7-Mar-62	Ash Wednesday Storm	7.22
18-Sep-36	Hurricane (unnamed)	6.72
22-Nov-06	Thanksgiving Northeaster	6.63
5-Feb-98	Twin Northeaster (#2)	6.58
7-Oct-06	Columbus Day Northeaster	6.52
27-Apr-78	Northeaster	6.41
11-Apr-56	Northeaster	6.32
16-Sep-33	Hurricane (unnamed)	6.12
28-Jan-98	Twin Northeaster (#1)	6.04
16-Sep-99	Hurricane Floyd	5.97
27-Sep-56	Hurricane Flossy	5.92
12-Sep-60	Hurricane Donna	5.92

10.3.2 Fetch

Fetch is the average distance the wind blows across water toward the project site. Fetch, water depth, dominant wind direction, and duration are the primary factors that dictate the wave energy at a site. Wave energy is the key element in the design that dictates material sizing and structure dimensions. Fetch is calculated by measuring the distance to the opposite shore perpendicular to the shoreline and at angles of 22.5° and 45° on both sides of the shore normal vector (Figure 10.1). The average of these distances is the average fetch at the site. Average fetch must also be evaluated in terms of the longest fetch, which usually produces the largest waves and the direction of the prevailing winds. Typically, the longest fetch is the criterion most used to determine whether and what type of Living Shoreline might be applicable to the site. The range of fetches has been categorized by Hardaway and Byrne (1999). They described three ranges of fetch: low, 0 to 1 mile; medium, 1 to 5 miles; and high, 5 to 10 miles. These can be further divided into four categories for Living Shorelines: very low, <0.5 miles; low, 0.5 to 1 mile; medium, 1 to 5 miles; and high, 5 to 15 miles. The first three categories are the most amenable to Living Shoreline–type projects with the upper end of the medium and the high categories typically requiring breakwater projects. A typical summary of wind conditions at the Norfolk, Virginia, Airport are given in Table 10.3.

10.3.3 Bank Height

Bank height describes the height of the upland bank landward of the shoreline. Protection of this bank from erosion is often the reason for constructing a Living Shoreline; the higher the bank, the higher the level of protection that is needed. Low banks can be protected with lower elevation sills and backfill while higher banks need higher sills to hold the sand backfill higher on the face of the bank to give the desired level of storm protection. A vital component of bank evaluation is whether it can be graded back to a stable slope. This is usually dictated by the presence of infrastructure,

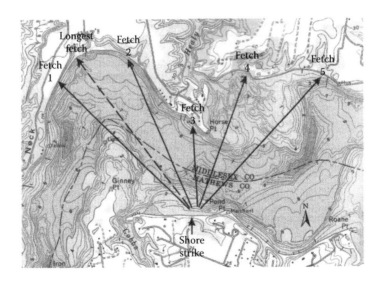

Figure 10.1   Fetch determination diagram. (From Hardaway, Jr., C.S. and R.J. Byrne. 1999. Shoreline Management in Chesapeake Bay. Special Report in Applied Marine Science and Ocean Engineering Number 356. Virginia Institute of Marine Science, College of William and Mary, Gloucester Point, VA. 54 pp.)

Table 10.3 Summary of Wind Conditions at Norfolk, Virginia, Airport between 1960 and 1990

Wind speed (mph)	Mid range (mph)	Wind direction								Total
		South	South west	West	North west	North	North east	East	South east	
<5	3	5497* 2.12+	3316 1.28	2156 0.83	1221 0.47	35,748 13.78	2050 0.79	3611 1.39	2995 1.15	56,594 21.81
5–11	8	21,083 8.13	15,229 5.87	9260 3.57	6432 2.48	11,019 4.25	13,139 5.06	9957 3.84	9195 3.54	95,314 36.74
11–21	16	14,790 5.70	17,834 6.87	10,966 4.23	8404 3.24	21,816 8.41	16,736 6.45	5720 2.20	4306 1.66	100,572 38.77
21–31	26	594 0.23	994 0.38	896 0.35	751 0.29	1941 0.75	1103 0.43	148 0.06	60 0.02	6487 2.5
31–41	36	25 0.01	73 0.03	46 0.02	25 0.01	162 0.06	101 0.04	10 0.00	8 0.00	450 0.17
41–51	46	0 0.00	0 0.00	0 0.00	1 0.00	4 0.00	4 0.00	1 0.00	0 0.00	10 0.00
Total		41,989 16.19	37,446 14.43	23,324 8.99	16,834 6.49	70,690 27.25	33,133 12.77	19,447 7.50	16,564 6.38	259,427 100.00

*Number of occurences   +Percent

Source: Hardaway, Jr., C.S., D.A. Milligan and K. Duhring. 2010. Living Shoreline Design Guidelines for Shore Protection in Virginia's Estuarine Environments, Version 1.2. Special Report in Applied Marine Science and Ocean Engineering Number 421. Virginia Institute of Marine Science, College of William and Mary, Gloucester Point, VA. 43 pp. with Appendix.

dwellings, wells, and septic systems that cannot be affected or a well-developed forested buffer that cannot be disturbed. The ability to grade the bank can have demonstrable effect on the design, often allowing a less robust Living Shoreline than would otherwise be needed.

10.3.4 Bank Condition

Bank condition generally falls into three categories: stable, transitional, and eroding. Stable banks are those with relatively gentle slopes that are uniformly vegetated and show no signs of erosion from either wave attack or upland runoff. An eroding bank is one with a steep to vertical face with little to no vegetation and signs of undercutting at the base from wave action. Intermediate or transitional banks are partially stable along portions of the bank but evidence undercutting at the base and some slumping of the bank face (Hardaway et al. 2010).

10.3.5 Nearshore Depth

Nearshore depths have a tremendous impact on incident wave energy at a site. The worst condition is when deep water, over −6′ mean low water (MLW), is immediately offshore with little intervening shallow water. This allows large breaking waves to impinge directly on shore. Extensive shallows or offshore bars substantially reduce wave energy by forcing waves to break further offshore. The intermediate condition is a site with a narrow shallow nearshore with deeper waters offshore. Nearshore bathymetry can affect design in another aspect as well. When the depth offshore exceeds minus one foot MLW at the location of the proposed sill, the dimensions of the sill dramatically increase in order to reach the elevations required for marsh establishment landward of the sill. In some cases, the dimensions of the sill will become so large that the affordability of a Living Shoreline is severely compromised. In these instances, where some type of bank protection is indicated, the situation might necessitate the consideration of a revetment, either oyster shell bags in low-energy areas or rock in higher-energy areas.

10.3.6 Substrate/Sediment Type

The type of bottom substrate at the site is important from a construction viewpoint. Soft muddy substrates can limit Living Shoreline options. If the mud is soft enough, even sand fill for wetland plantings can consolidate the underlying mud, lowering the fill below optimal planting elevations. An intermediate mix of mud and sand can often support sand fill, coir logs, and oyster shell bags. In these areas, the elevation of the Living Shoreline should be designed high enough to accommodate some consolidation and settling of the underlying sediments. The best sediment type for Living Shoreline construction is a sandy substrate or stiff clay that will allow the use of a rock sill, if needed, with little anticipated settling. These conditions can be approximated by using the 200 lb/ft^2 rule of thumb. This can be determined, roughly, by a 200-lb person putting their feet together, approximately 1 ft^2, and gently hopping in place to see if they sink in the sediment. If they do not sink, the bearing capacity is very roughly 200 lb/ft^2. This is an indication that the substrate is firm enough to support a small sill structure.

10.3.7 Tide Range

The tide range at the site is a critical parameter in designing a Living Shoreline. There are three tidal elevations that are important: mean high water (MHW), mean low water (MLW), and mean tide level (MTL). MHW is the average of all of the high tides at the site and is the elevation that determines the transition from low marsh to high marsh plant species. MLW is the average of all of the low tides at the site and is often the appropriate location for a sill structure. MTL is the elevation halfway between MHW and MLW and typically delimits the lower limit for planting low marsh species. These elevations can be determined from the nearest National Oceanic and Atmospheric Administration (NOAA) tide station based on the MLW datum. MHW and MTL can often be approximated on site by using the elevations of the low marsh and high marsh plants in adjacent marshes using the concurrent water levels as a means of transferring these elevations to the project site. If water levels cannot be used to transfer planting elevations to the site, biological benchmark elevations may need to be obtained from a nearby marsh. This involves surveying a nearby marsh to determine the upper and lower limits of the plant communities to be planted. It is imperative that multiple elevations (20–30) be obtained for the upper and lower limits of each planting zone. These can then be averaged to give a good representation of the appropriate planting elevations. An example of a biological benchmark determination is given in Table 10.4.

Site plans prepared by a surveyor typically use the geodetic datum, NAVD 88, as the datum on their surveys. This is not a tidal datum and can only be used in design of Living Shorelines when the proper conversion factors have been applied. These conversions can be obtained either from the benchmark datum of the NOAA tidal reference station or from a Tidal Datum Conversion Diagram prepared by a knowledgeable surveyor. An example of a tidal datum conversion diagram is given in Figure 10.2. In any case, NAVD 88 elevations must be converted to the local MLW tidal datum for Living Shoreline design purposes. The tide range drives every aspect of the design from the determination of storm surge levels to sill heights to planting zones. It is critically important that an accurate tide range and tidal elevations be determined for the site.

Table 10.4   Example of a Biological Benchmark Determination

Biological benchmark form								
Site:		Libertyville Wetland Bank						
Date:		17-Aug-05						
Investigator:		W. I. Priest						
Datum:		NAVD 88						
	Lower limit *Spartina alterniflora*	Upper limit *Spartina alterniflora*	Lower limit *Phragmites australis*	Upper limit *Phragmites australis*	*Juncus roemarianus*	Saltbush	Misc. species	Note
	1.12	2.45	2.45		2.73	3.8	0.65	Creek bottom
	0.93	2.8	2.17		2.48	3.86	1.36	Ditch bottom
	0.8	2.77	2.1		2.83	3.61	2.58	Panne
	0.91	2.65	2.65		3.03	3.59	3.3	Existing grade
	0.82	2.81	2.82	3.43	2.93	3.46		
	1.24	3.04	2.83		2.9		2.81	S.r.
		2.93	2.97		2.7		2.86	S.cyno.
		2.89	3.03				3.04	D.s.
			2.9				3.22	D.s.
Mean	**0.97**	**2.79**	**2.66**		**2.80**	**3.66**		
Max	1.24	3.04	3.03		3.03	3.86		
Min	0.80	2.45	2.10		2.48	3.46		
N	6	8	9		7	5		
Ave dev	0.14	0.13	0.28		0.14	0.13		
Std dev	0.17	0.18	0.34		0.18	0.16		

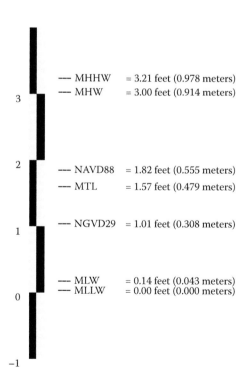

— MHHW	= 3.21 feet (0.978 meters)
— MHW	= 3.00 feet (0.914 meters)
— NAVD88	= 1.82 feet (0.555 meters)
— MTL	= 1.57 feet (0.479 meters)
— NGVD29	= 1.01 feet (0.308 meters)
— MLW	= 0.14 feet (0.043 meters)
— MLLW	= 0.00 feet (0.000 meters)

Figure 10.2   Tidal datum conversion diagram for Money Point, Virginia.

10.3.8 Erosion Rate

The existing erosion rate is a strong indicator of the need for erosion control and the "robust-ness" of the structures needed (Table 10.5). A site in a creek may be only experiencing minimal erosion and may be suited for very minimalist applications such as plantings and sand fill with plantings with coir log toe support. More exposed sites on larger creeks and rivers with longer fetches and higher erosion rates will need higher and more robust structures such as rock sills. Bank

Table 10.5   Average Shoreline Erosion Rates in Tidewater Virginia

York River

North side	Erosion rates	Average
Gloucester Co.	−0.5 ft/year	−0.4 ft/year
King and Queen Co.	−0.3 ft/year	−0.4 ft/year
South side	Erosion rates	Average
York Co.	−0.9 ft/year	
James City Co.	−1.8 ft/year	−1.2 ft/year
New Kent Co.	−0.9 ft/year	

James River

North side	Erosion rates	Average
Newport News	−0.8 ft/year	
James City	−0.1 ft/year	−0.45 ft/year
South side	Erosion rates	Average
Isle of Wight Co.	−1.8 ft/year	
Surry Co.	−1.2 ft/year	−1.5 ft/year

Rappahannock River

North side	Erosion rates	Average
Lancaster Co.	−0.6 ft/year	
Richmond Co.	−0.6 ft/year	−0.6 ft/year
South side	Erosion rates	Average
Middlesex Co.	−1.0 ft/year	
Essex Co.	−1.2 ft/year	−1.1 ft/year

Chesapeake Bay

Western shore	Erosion rates	Average
Gloucester Co.	−0.6 ft/year	
Hampton	−1.0 ft/year	
Lancaster Co.	−1.4 ft/year	
Mathews Co.	−0.8 ft/year	
Northumberland Co.	−1.0 ft/year	
York Co.	−1.5 ft/year	−0.9 ft/year
Eastern shore	Erosion rates	Average
Accomack Co.	−1.5 ft/year	
Northampton Co.	−0.7 ft/year	
Fisherman's Is.	+11 ft/year	−1.0 ft/year*
Southern shore	Erosion rates	Average
Virginia Beach	−1.7 ft/year	
Norfolk	−1.2 ft/year	
Nansemond	−1.2 ft/year	−1.4 ft/year

*Does not factor in Fisherman's Island.

Source: Hardaway, Jr., C.S. and G.L. Anderson. 1980. Shoreline Erosion in Virginia. Sea Grant Program, Marine Advisory Service, Virginia Institute of Marine Science, Gloucester Point, VA. 25 pp.

condition detailed above can be a good indicator of erosion potential. Also, cut vertical scarps on the face of existing stands of marsh can indicate modest erosion rates.

10.3.9 Shoreline Orientation

Shoreline orientation can be of particular concern, especially when facing north for two reasons. One, north-facing shorelines receive the least amount of direct sunlight during the year, which can make vegetation establishment problematic. This is particularly true when the adjacent upland bank supports tall and overhanging trees that shade the immediate shoreline most of the day. In urban areas, this situation can also arise when tall buildings are located near the shoreline. Second, the highest winds typically come from the north and east quadrants (Table 10.3). The increased wave activity from wind events can make it difficult to establish vegetation. South-facing shorelines are typically the best for vegetation establishment because they receive the most sunlight during the growing season and generally milder wind conditions. East- and west-facing shorelines are the intermediate orientations with a slight edge to the west-facing shoreline because of the afternoon sun and milder winds.

10.3.10 Shoreline Configuration

Shoreline configuration can also play a strong role in the selection of Living Shoreline applications. Concave shorelines such as coves and small creeks offer the best protection from winds because of typically shorter fetches and limited exposure to prevailing winds and shallower water. Straight or linear shorelines can present different issues depending on its orientation. Sometimes, these shorelines have long multidirectional fetches that can require specific adaptations to sill designs to accommodate different directional impacts such as angled segments to effectively modify waves from different directions. Convex shorelines like points and headlands can be problematic because of extreme exposures to fetches and waves from different directions. Also, points tend to be the shorelines most affected by boat wakes (Zabawa and Ostrom 1980).

10.3.11 Infrastructure

Infrastructure location, including buildings, wells, and septic systems, is also a concern when designing Living Shorelines. These features often limit bank grading activities, which can be an important component of Living Shoreline design. If the bank cannot be graded without affecting adjacent infrastructure, this can push the design offshore or increase the size and height of structures to achieve the necessary level of protection for the backshore or upland bank. Infrastructure locations can also affect construction access to a site in that they can allow only limited pathways for construction equipment and material.

10.3.12 Waterway Width

The width of the waterway at the site can be an issue when it is very narrow or constrained by the presence of a navigation channel. In some cases, it can severely limit the option of extending the Living Shoreline channelward, especially where bank grading for proper slopes is not possible because of infrastructure location or bank height. When navigation channels are present, hardened structures like rock sills must be able to be situated so as not to cause a hazard to navigation. This may require a buffer distance from the channel that may complicate design of the project. The structures must also be designed to accommodate increased wave action from boat wakes in areas of high boating activity.

Table 10.6 Living Shoreline Site Evaluation Criteria

Parameter	Score	Criteria Values		
		1	2	3
Storm surge		2′	2′–4′	>4′
Fetch		<0.5 mi	0.5–1 mi	1–5 mi
Bank height		<3′	3′–6′	>6′
Bank condition		Stable	Transitional	Eroding
Nearshore depths		<1′	1′–2′	>3′
Sediment type		Mud	Mud/sand	Sand
Tide range		1′–2′	2′–4′	>4′
Erosion rate		1′	2′	>3′
Shoreline orientation		South	East or west	North
Shoreline configuration		Cove	Linear	Point
Infrastructure proximity		>100′	50′–100′	<50′
Width of waterway		>300′	300′–100′	<100′
Buffer condition		Lawn	Natural grasses	Forest
Total score	13–18	Low energy, trim trees and plant marsh		
	19–32	Medium energy, sill system		
	33–39	High energy, breakwater system		

10.3.13 Riparian Buffer

Riparian buffer condition can also affect Living Shoreline designs, especially when there is a well-developed forested buffer immediately adjacent to the shoreline. Forested riparian buffers have well-documented benefits to water quality by intercepting nutrient-laden ground water with uptake from their root systems. Consequently, disturbance of these systems for construction access and necessary bank grading can become a regulatory issue. When a Living Shoreline is to be sited immediately adjacent to a forested riparian area, it can present problems with vegetation establishment, particularly on north-facing shorelines. In some cases, this problem can be ameliorated by judicious pruning and limbing up of tree branches.

10.3.14 Site Suitability Matrix

I have attempted to evaluate all of these parameters associated with the suitability of a site for construction of a Living Shoreline in the matrix table (Table 10.6). Some of the parameters address constructability while others address ancillary issues such as potential environmental constraints. The scoring is a very simplistic 1, 2, or 3, but the scoring should help lead one to the more acceptable, affordable, and effective Living Shoreline options for a particular site. It is not intended to be definitive but to be used as a guide in the analysis of the alternatives available to the property owner. In order to evaluate the matrix score, it is important to look and see whether constructability or environmental factors are driving the score.

10.4 DESIGN CONSIDERATIONS AND CONSTRUCTION

So far, we have looked at a site in terms of its environmental acceptability and potential suitability to support a Living Shoreline that will provide functional erosion protection with the added value of viable wetland and fisheries habitat. Having made these decisions, we move into the design and construction phase of our project. This involves evaluating the site to determine which of the Living Shoreline techniques is most applicable to our site.

There is a specific suite of design alternatives that typically constitute a Living Shoreline project. We will begin with the least intrusive and work our way up to the more intensive designs needed in higher-energy areas. A number of these considerations will be applicable to more than one Living Shoreline application. The techniques we will consider include the following:

1. Shoreline plantings only
2. Sand fill and plantings
3. Shoreline plantings with bank grading
4. Marsh toe protection with low impact materials, for example, oyster shell bags
5. Sand fill with low-impact toe protection, for example, coir logs or oyster shell bags
6. Rock sills with sand fill and plantings

10.4.1 Shoreline Plantings

Shoreline plantings with wetlands vegetation alone are applicable only in very low energy environments on existing nonvegetated intertidal areas within the planting zone for a particular species. Typically, the only modification needed for this option is, perhaps, limbing up adjacent trees to provide sufficient sunlight for plant growth, a minimum of 6 h of direct sunlight per day during the growing season. Appropriate elevations can normally be determined by using adjacent marshes at the same elevation as a guide. Plant the same species at the same elevations as the adjacent marshes. Substrate can be almost anything from mud to sand. If the appropriate elevations do not exist, sand will need to be added to the intertidal area to raise the area into the intertidal zone. This involves the placement of clean sand fill along the shoreline with a slope flat enough and wide enough (15′–30′) within the intertidal area to provide a stable platform that will support enough wetlands vegetation to control erosion.

10.4.2 Sand Fill

Good quality sand is one of the most important components of any Living Shoreline design. The sand needs to be coarse-grained and free from a high percentage of fines, either very fine sand or silts and clays. There are several ways of specifying sand. One is to specify a median grain size. The preferred median grain size is 0.6 mm (±0.25 mm). Another method that can be used is screen mesh size. The typical specification is sand with less than 10% fines passing a #100 sieve. The problem with using the screen size method is that you really do not know how coarse the sand is. Potentially, the majority could be fine sand retained by the #100 sieve. Fine sand is easily moved by wave action and can result in loss or relocation of the sand sufficient to alter the planting elevations or result in the plants being dislodged and washed away. In general, use the coarsest grained sand you can find.

10.4.3 Bank Grading

Bank grading can also be used to achieve the appropriate elevations for planting marsh where they do not exist, and filling out into the waterway with sand is not an option. It can also be used in situations where the landward expansion of an existing marsh is desired. It is generally only an option in very low energy areas with low banks, minimal tree cover, and no infrastructure constraints. However, it can also be used in combination with other techniques to limit encroachment into shallow water habitat or increase the functionality of a Living Shoreline by reducing the slope of the upland bank. It involves excavating the existing upland area down to intertidal elevations suitable for planting. The slope of the area should be as flat as possible with a width between 15′ and

30′ in the intertidal area. The soil type at the planting elevations is very important. If it is a hard plastic clay or other material that is unsuitable for plant growth, the area must be overexcavated by at least 1 ft and backfilled with sand to the appropriate elevations for planting.

10.4.4 Marsh Toe Protection

Marsh toe protection can be an effective Living Shoreline technique in areas with an existing fringe marsh that is experiencing erosion at its seaward edge. This can often be achieved with oyster shell bags stacked against the eroding scarp to reduce the impact of wave action (Figure 10.3). In higher-energy areas, stone may be needed to afford effective protection. These structures are typically low profile, barely exceeding the elevation of the adjacent marsh. Coir logs are not an option in this situation because of their temporary nature.

10.4.5 Low Impact Sills

Low impact sills made of coir logs or oyster shell bags with sand fill may be an appropriate option in low-energy areas with short fetches or minimal boat wake issues. Coir logs, made of coconut fiber, are only a temporary measure designed to provide one or two growing seasons of protection for the plantings. The coir logs are used to provide support for the sand and plantings only long enough for the plants to become established and cannot be part of a long-term remedy (Figure 10.4). Oyster shell bags are another low-impact option in these areas. The bags of shells are typically stacked in a triangular configuration to an elevation sufficient to support the desired elevation of the sand fill for planting. The use of oyster shell bags also allows for a component of oyster restoration in areas with good oyster recruitment (Figure 10.5).

Coir logs are constructed of coconut fiber stuffed in woven mesh of coconut fiber rope. These logs come in 12″, 16″, and 20′ diameters in 10′ lengths. They also come in two densities: 7 lb/ft^3 and 9 lb/ft^3. The 12″ diameter size is the most often used toe protection because they usually can contain enough sand to raise an area into the intertidal planting zone. The 12″ size is also more easily

Figure 10.3   Oyster shell bag marsh toe structure, York River, Virginia.

Figure 10.4 Coir log marsh toe support, Lafayette River, Virginia.

Figure 10.5 Oyster shell bag sill, Sarahs Creek, Virginia.

manhandled into place by property owners. The higher-density logs are recommended for most applications because of their increased durability. These logs need to be staked in place because they float and can become dislodged by high tides or storm events and float to unwanted areas such as channels and adjacent properties. The logs should be staked on both sides of the log every 3′–4′ with wooden stakes long enough to provide firm placement. These stakes are then connected above the logs with twine to hold the logs in place. The ends of the bags should be overlapped to help prevent loss of the sand fill at the junctures. They also have a very finite life span before they begin to deteriorate and must not be a functional part of the final design. My experience with several projects has shown that in low-energy areas when deployed in the spring, they will last approximately two growing seasons. In most instances, this should be sufficient time for the wetland plantings to become established and for the logs to no longer be needed.

The next most durable type of toe protection for sand fills and plantings or marsh toe protection is oyster shell bags. Oyster shell bags are a technique borrowed from the oyster aquaculture industry where they are used to produce spat-on-shell for oyster production and restoration. They are produced by sliding a plastic mesh sleeve closed with a hog ring at one end over an 8″ PVC pipe. The pipe is then filled with oyster shell. Once filled, the pipe is removed, leaving the shell in the mesh bag. The bag is then closed with another hog ring and the bag is complete. A rack can be constructed to hold a number of these pipes vertically to facilitate filling when producing a large number (Figure 10.6). For design purposes, these bags are typically 18″ long by 8″ wide by 6″ high and weigh approximately 25 lb. Shell bags can be arranged in almost any configuration needed and can be easily handled by one person. They should be stacked so that the bags overlap at their ends to produce a staggered structure to improve their stability and structural integrity. They have been effectively used as marsh toe protection by placing two side by side and one on top, yielding a 1-ft-high structure that can be placed immediately adjacent to the eroding marsh scarp. They can also be stacked in a triangular cross section of reasonable height to produce a semipermanent sill structure to contain sand fill in low-energy areas. See Table 10.7 for the dimensions of typical shell bag structures. A shell bag sill structure should be underlain and backed with filter cloth to add structural integrity and to help prevent the loss of sand through the structure. An oyster shell bag sill is typically placed at or close to MLW.

Figure 10.6 Oyster shell bag construction

Table 10.7   Oyster Shell Bag Structural Dimensions

Layer Configuration	Number of Bags	Nominal Height	Nominal Base Width	Bags per Linear Foot
5–4–3–2–1	15	2.5′	3.4′	10
4–3–2–1	10	2.0′	2.7′	7
3–2–1	6	1.5′	2.0′	4
2–1	3	1.0′	1.4′	2
1	1	0.5′	0.7′	0.7

10.4.6 Rock Sills

Rock sills are typically required in medium-energy areas or areas with high boat wake energy to provide sufficient protection for sand fill with plantings. They are also designed to "trip" incoming waves to help dissipate wave energy. This can involve a fairly substantial structure to achieve the desired level of protection (Figure 10.7).

Rock used in sill construction is, preferably, quarry stone of granite or similar rock. Broken concrete can also be used as a core to reduce the quantity of rock needed. Broken concrete should not be large slabs but broken into pieces where the longest dimension is no longer than three times the smallest dimension. The size of rocks needed to construct a sill is based on the fetch, water depth, and the size of breaking waves expected at the site during a storm. Regardless of the rock size used, it must be underlain and backed with filter cloth to provide structural integrity and prevent loss of sand backfill. The stone must also be placed and not simply dumped. Placement involves placing the stones together so that they interlocked in place and not easily dislodged by wave action. The crest of the sill must be at least two rocks wide to provide effective wave attenuation and crest stability. The channelward slope of the sill should typically be at least 2:1 (H:V) to provide a stable slope. The landward side of the sill can usually be placed on a steeper 1.5:1 (H:V) slope to help reduce the quantity of stone and minimize the foot print of sill.

The size of the rock needed for sill construction is a critical element in the design. It must be sized to withstand the predicted wave energy at the site, which is based on fetch, water depths including storm surges, and wind speed. For this discussion of rock sizes, the Virginia Department of Transportation (VDOT) standard rock riprap classification will be used (see Table 10.8). In very

Figure 10.7   Rock sill, Ware River, Virginia.

Table 10.8   VA Department of Transportation Riprap Size Classes

Riprap Class	Weight Range (lb)	Stone Mixture Requirements
Class AI	25–75	Max. 10% > 75 lb
Class I	50–150	60% > 100 lb
Class II	150–500	50% > 300 lb
Class III	500–1500	50% > 900 lb

low fetch environments, less than 0.5 miles, rock sills are not usually necessary. If needed, though, for areas of high boat wakes, the VDOT Class 1A stone is typically sufficient. In low-energy areas with fetches up to a mile, the VDOT Class 1 riprap is often large enough. In medium-energy areas with fetches not exceeding 5 miles, at least, VDOT Class 2 needs to be used. These recommendations are based on my experience and should be confirmed with a contractor or engineer before construction is commenced. In high-energy areas, the size of the stone for a sill needs to be calculated by a coastal engineer or geologist and is beyond the scope of this paper.

The height of the sill and its placement offshore are a function of storm surge, wave climate, and the height of the backshore necessary to protect the bank. The level of protection required is predicated on the storm surge elevation and wave height of the design storm. The higher the storm surge, the higher the backshore has to be to protect the upland bank. Higher backshore elevations require the sills to be higher and more robust as well as further offshore to provide a stable slope for the sand fill. For example, if the desired level of storm protection on a low bank is 4′ in an area with a 2-ft tide range, the slope of the backfill would begin at +4′ MLW and slope channelward to the MTL elevation of +1′ MLW behind a +1′ MHW sill. The change in grade from +4′ MLW to +1 MLW equals 3′. A 3′ vertical change in grade on a 10:1 (H:V) slope equals 30′ horizontally. Therefore, the sill needs to be located approximately 30′ from the top of the bank. This would then allow for a 30′-wide planting bench to be planted with the appropriate vegetation to effectively protect the shoreline from future erosion. Similarly, a shoreline with a high unstable bank that needs protection from 6′ storm surge will need a 6′-high backshore that slopes on at 10:1 to MTL and would require a 50′-wide backshore and a sill at +1.5′ above MHW. Also, as banks get higher, bank grading may be required to provide effective protection during extreme storm events (Figure 10.8).

Where long sills are employed, periodic gaps in the sill are typically required to allow unfettered faunal access to the wetlands landward of the sill. Gaps can also be incorporated into sill design to provide beach habitat and access to the waterway for swimming or small boats. Gaps in the sill structure may also be necessary to provide drainage through the Living Shoreline for stormwater discharge outfalls or natural drainage for creeks located landward of a proposed Living Shoreline. The width of these gaps is quite variable, depending on the use and the orientation of the sill. The smaller the gap the better, in the range of 5′–10′ wide. This provides sufficient access for fishes and crabs but does not expose the backshore to excessive wave energy coming through the gap. Larger gaps for water access or creek drainage must be designed to prevent excessive exposure of the backshore and upland bank to wave energy and continued erosion of the backshore and loss of sand backfill. An empirical formula has been derived to predict the depth of the embayment landward of a gap of a given width (Hardaway et al. 1991) (see Figure 10.9). This formula, developed for larger headland breakwater systems, can be applied to sill gaps and gives the depth of the embayment as a ratio of the depth of the embayment to the gap width of 1:1.65. For example, a gap of 10′ will produce a pocket beach with a new MLW shoreline 6′ landward of the sill. Gaps can be configured in a number of ways to limit their exposure to wave action. The ends of a sill can be overlapped at a gap or the gap can be fronted with a short sill slightly offshore. In some instances when the bank cannot be graded back, a short revetment at the toe of the bank opposite the gap may be necessary to protect it from wave action. Other means of reducing sand movement at gaps include a low sill of stone at MTL across the mouth of the gap or placing a cobble-sized stone over the beach landward of the gap (Hardaway et al. 2007).

10.4.7 Planting

Plants used in the construction of Living Shorelines need to be specific to the appropriate planting elevations on the sand backfill and the salinity regime at the site. General guidance for plant selection in Chesapeake Bay is given in Table 10.9. In the vast majority of projects in typical

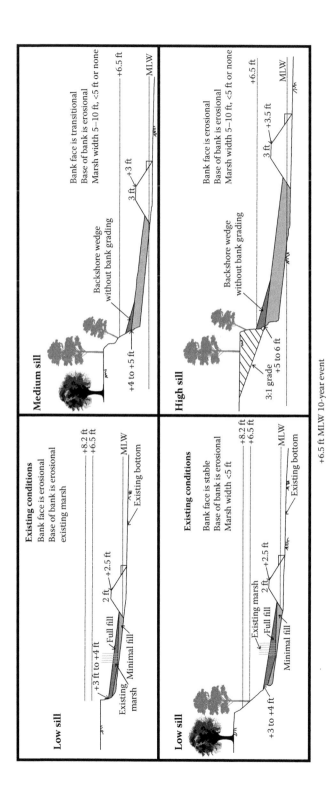

Figure 10.8 Schematic Living Shoreline cross sections with varying bank conditions. (From Hardaway, Jr., C.S., M. Berman, S. Killeen, D. Miligan, K. Nunez, K. O'Brien, T. Rudnicky and C. Wilcox. 2008. Occohannock Creek, Shoreline Erosion Assessment and Living Shoreline Options report. Virginia Institute of Marine Science, College of William and Mary, Gloucester Point, VA. 47 pp. with Appendix.)

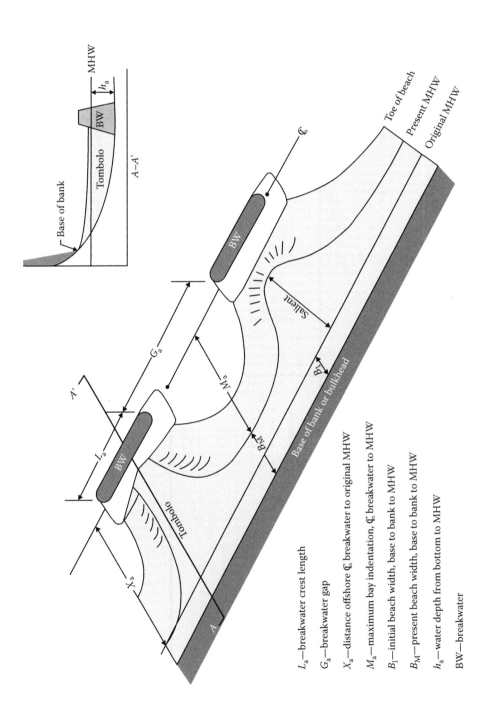

L_a—breakwater crest length

G_a—breakwater gap

X_a—distance offshore ₵ breakwater to original MHW

M_a—maximum bay indentation, ₵ breakwater to MHW

B_I—initial beach width, base to bank to MHW

B_M—present beach width, base to bank to MHW

h_a—water depth from bottom to MHW

BW—breakwater

Figure 10.9 Breakwater design parameters. (From Hardaway, Jr., C.S. and R.J. Byrne. 1999. Shoreline Management in Chesapeake Bay. Special Report in Applied Marine Science and Ocean Engineering Number 356. Virginia Institute of Marine Science, College of William and Mary, Gloucester Point, VA. 54 pp; After Suh, K.D. and R.A Dalrymple. 1987. *Journal of Waterway, Port, Coastal and Ocean Engineering* 113(2):105–121.)

Table 10.9   Zonation and Salinity Levels for Common Wetland Plants

Species	Inundation Zone	Salinity Range
Spartina alterniflora	MTL–MHW	5–30 ppt
Spartina patens	MHW–ULW	5–30 ppt
Spartina cynosuroides	MHW–ULW	0–5 ppt
Distichlis spicata	MHW–ULW	10–30 ppt
Scirpus pungens	MHW–ULW	0–15 ppt
Juncus roemarianus	Above MHW	10–25 ppt
Iva frutescens	Near ULW	5–30 ppt
Baccharis halimifolia	Near ULW	0–30 ppt
Panicum virgatum	Above ULW	0–25 ppt

estuarine systems in Virginia, smooth cordgrass, *S. alterniflora*, will be planted from MTL immediately landward of the sill or toe structure up to the elevation of MHW. Above MHW, planting should shift to saltmeadow hay, *S. patens*, up to the upper limit of tidal wetlands (ULW). The ULW is defined in Virginia as an elevation equal to 1.5× the mean tide range at the site above MLW. For example, if the tide range at the site is 2′, then the ULW is 2 × 1.5 = 3′. However, saltmeadow hay can be planted higher on the bank, if necessary. Above the ULW in the transitional area to the upland, switch grass, *Panicum virgatum*, is typically planted. The shrubs, marsh elder, *Iva frutescens*, and groundsel tree, *Baccharis halimifolia*, can also be planted in this transitional area. However, in many projects, these shrubs often volunteer from seeds. In freshwater areas, the plant species need to be adjusted to those adapted to the salinity conditions at the site, for example, using common threesquare, *Scirpus pungens*, in the intertidal zone and switchgrass, *P. virgatum*, above MHW. At all species transition areas, for example, MTL, MHW, and ULW, each species should be planted both above and below the transition elevation or alternated in the rows. This helps ensure that if the planting elevations are slightly off, the appropriate plants will be present at the right elevations for survival.

The best time for planting is in the spring because the plants have the entire growing season to get established, but as with any planting, there is always a measure of risk. Spring planting times are often dictated by plant availability. Unless the plants have been overwintered in the nursery, early spring, March, plantings often have to be done with very young plants that have not fully matured, which can have an impact on success. The best planting times are April, May, and June. This gives the young plants time to mature with the whole growing season to get successfully established. Summer plantings, July, August, and early September, can be hot and dry and difficult for high marsh plants to survive unless there is adequate rainfall, or they can be irrigated. Fall and early winter plantings can be successful if done in very protected areas. Planting in these seasons is typically not recommended because of the high probability of storm events, which can wipe out entire plantings in very short order. Late planting can also be very susceptible to damage and dislodgement from ice during the winter because they have not had time to establish strong root systems. The decision of when to plant has to be tempered with the amount of risk that is acceptable and whether replanting is an affordable option. Shrubs should not be planted until the late fall or late winter to avoid high transpiration stress during the warmer seasons.

Plants are typically obtained from a nursery that specializes in wetland plants. They are usually obtained in plastic flats or trays that contain 50 or 72 cells or plants per flat. The most commonly used size is the cell pack 72 flat. This provides a roughly 2″ × 2″ × 3″ plug for transplanting. The 50–cell pack flats are more expensive but give a larger plug. The plants should have uniformly green leaves with no extensive yellow areas with reddish splotches indicating a fungal rust infection. The plug should fill the cell with obviously white healthy-looking roots. Salt marsh plants are grown and maintained in freshwater in the nursery. If the plants are going to be transplanted to a high-salinity

Table 10.10   Plant Spacing and Specifications

Plant Spacing	Plants/SF	Plants/1000SF	70% Cover[a]
1′ on center	1	1000	1 year
1.5′ on center	2.25	444	1–2 years
2′ on center	4	250	2 years
3′ on center	9	111	5 years

[a] Based on April planting.

area, greater than 25 parts per thousand, you should tell your nurseryman so that he can salt harden the plants by slowly increasing the salinity before delivery. This will make them better able to adapt to their new environment. Sometimes, the plants received from the nursery have been clipped to a uniform height. This is done to facilitate transport and does not affect viability.

The plants can be planted on a number of grid spacings from 1′ on center to 3′ on center. The 1′ on center grid is only used in small areas where dense coverage is desired in a very short period. The most common grid sizes are 1.5′ and 2′ on center. The 1.5′ on center is used in higher-energy areas where good coverage is desired during the first growing season. The 2′ on center grid is also commonly used where good cover is expected after the second growing season. To achieve these optimal coverage rates, it is very important that the planting occurs early, by the end of April or first of May. The 3′ on center grid is only used on very large areas where 70% cover can only be expected after five growing seasons (Priest et al. 2010). The number of plants per square foot needed for each of these grid sizes is given in Table 10.10. Something worth noting, for design purposes, is that with each incremental increase in the spacing, the number of plants required is reduced by roughly one-half. This can be a factor in adjusting the cost of planting based on the desired level of performance.

Each plug should be planted in a hole at least 6″ deep with 3″ to 4″ of cover. It is important to plant the plants deep and firmly pack the cover over them. If not, because they float, they can be more easily dislodged if the sand shifts or by wave action during a storm event. You cannot plant them too deep or pack them in too tight. When planted, each plug should receive one-half ounce, one tablespoon, of high nitrogen slow release fertilizer in the hole. This type of fertilizer is contained within a special membrane that controls the release of the fertilizer based on temperature and moisture so it does not harm the plant and provides the nutrients slowly as the plant needs them. This is important for two reasons. One, the planting usually occurs in low nutrient soils, either sand or sterile subsoils in excavated areas. The other reason is that these plants are expected to grow fast so that their erosion control benefits can accrue in a timely fashion. If you want the plants to grow fast and work, you need to feed them. For additional information on planting guidelines, see Priest (2006).

10.4.8 Planting Substrate

The slope of the planting platform, that is, the sand backfill, is an important priority factor in the design of a Living Shoreline. It is important to keep these slopes at, at least, 10′ horizontally to each foot vertically (10:1, H:V) (Garbisch and Garbisch 1994). This helps provide for relatively broad planting zone widths for the different species of plants. It also allows for the main component of the wave energy dissipation component of a Living Shoreline by providing sufficient wave run-up to exhaust most of the wave's energy by the time it reaches the toe of the bank. This is in addition to the frictional loss of wave energy owing to its passage through the stems of vegetation. The minimum width for a planting zone should be at least 30′ wherever possible. A width of 30′ provides approximately 88% reduction in wave energy (Knutson et al. 1982). The wave attenuation of a sill, if present, is also a component of this equation. The only exception to this design criterion is when the width of the backfill is necessarily limited, and the 10:1 slope cannot be extended all the way to the top of the bank or up to the desired level of storm protection. In these cases, the 10:1 slope should be extended up to, at least, MHW. At MHW a "dogleg" in the slope can be created where the slope changes from 10:1 to no steeper than 3:1 that extends up to the desired elevation. In all types of Living Shorelines, it is important to allow the newly placed sand to adjust to tidal and wave action before planting. This will result in minor changes in grade as the sand shifts in response to the tides and waves. Optimally, 2 to 3 weeks should allow this process to complete itself. The planting plan should then be laid out in concert with the resulting elevations. If there is an excessive amount of sand movement or loss, it may be necessary to augment the sand fill before planting. In placing the sand, consolidation of the underlying sediments must also be taken into consideration. If this is an issue, the sand should be placed higher than needed to allow for some consolidation and still have the appropriate planting elevations.

10.4.9 Construction Access

Construction access for Living Shorelines can be from either the land or water and can range from relatively easy to difficult to impossible. *Easy* is defined as unencumbered access for machinery and materials that does not require extensive modifications to site by grading banks or removing trees and the ability to work around buildings, septic systems, and wells. Water access can be a viable option provided sufficient water depths are available offshore. *Difficult* means, extensive bank grading, tree removal, or infrastructure avoidance, are necessary to reach the shoreline with machines and materials. Very often, timber mats may also be necessary in difficult situations to accommodate soft soils and infrastructure. The size of a contractor's equipment may contribute to difficulty in construction access. Many types of Living Shorelines do not require large equipment to handle materials. A contractor using small tracked loaders, like Bobcats and smaller excavators, can make difficult access situations much more constructible. *Impossible* means there is no way to reasonably reach the shoreline with the necessary machines and materials either from the land or water. In some instances, this may necessitate the selection of another shoreline protection option.

10.4.10 Goose Fencing

Once a Living Shoreline is constructed and planted, another issue that may be encountered is grazing on the planted plants by Canada geese. When first planted, the plants provide an easily accessible, palatable source of food for a number of animals, particularly geese. If left unchecked, a small flock of geese can obliterate a sizable planting in a matter of days or weeks. To counter this threat in areas of high goose concentrations, it is vital to construct fencing to exclude geese from the newly planted area. This is accomplished by erecting plastic mesh fencing supported by hardwood stakes to completely enclose the planting site, even from the landward side. Plastic mesh fencing at least 3 ft high is strung between hardwood stakes placed 20' apart and attached to the stakes with plastic cable ties. If the site is large, the geese must be deterred from flying in and landing in the site. This is accomplished by stringing twine between the tops of the stakes. On broad sites, it will be necessary to place additional stakes within the interior of the enclosure to support the twine. Plastic flagging can be attached to this twine as an additional deterrent (Figure 10.10).

10.4.11 Site Restoration

The final component of construction is construction site restoration. This involves restoring the areas of the site that were disturbed during construction. It can include regrading and revegetating

Figure 10.10   Goose fencing at Trent Hall, MD.

both wetland and upland areas damaged by equipment during construction and repairing roads and driveways used by trucks to haul materials to the site.

10.4.12 Permitting in Virginia

Typically, permits from local, state, and federal regulatory agencies are required for the construction of Living Shorelines. Consulting with these agencies in the planning phase of a project is strongly advised, as it might help eliminate potential problems during the permitting. This is especially important if any environmental issues are discovered in the screening phase of the project. It also might reveal that some types of Living Shorelines have an abbreviated permit process. This is particularly true in Virginia where the recent passage of a State General Permit for Living Shorelines without rock sills dovetails with a Corps of Engineers Regional permit to significantly reduce permit requirements and streamline the process. These processes generally allow for impacts to existing wetlands at the site provided there is a net increase in wetlands resulting from the project. Permit conditions often require postconstruction monitoring. This can range from photographs taken at selected stations to reveal plant survival and cover along with a short narrative describing the success of the project or problem areas and proposed remedies to detailed percent survival and percent cover studies based on quadrat sampling data.

10.4.13 Postconstruction

Monitoring of a Living Shoreline is always necessary to ensure its successful and effective implementation. Monitoring is the key factor in detecting problem areas early when remedies are more readily available. This can be important if shifting sand alters the planting areas or there are areas of plant loss or mortality.

Replanting in Living Shorelines is not an uncommon occurrence and should not reflect adversely on the project. There are often conditions that do not reveal themselves until the project is completed. Sometimes, soil conditions in graded banks can be problematic for plant growth, changes in storm water runoff patterns may cause erosion of the sand backfill, or the planting elevations shift slightly, reducing plant survival. It is important to identify the cause of the problem, rectify the problem, and replant as necessary to promote rapid revegetation.

Another postconstruction problem can be the accumulation of debris; algal, marsh plant, or seagrass wrack; and flotsam and jetsam. The presence of these materials can smother newly planted marshes and can be significant causes of plant mortality. Debris accumulation needs to be monitored regularly and removed as expeditiously as possible. This becomes slightly less of a problem as the vegetation matures but can be a continuing issue that needs to be addressed as necessary.

10.4.14 Future Considerations

As with any wetlands restoration project, the impacts of sea level rise (SLR) must be factored into the design of a Living Shoreline. In many areas of the country, SLR is a major concern for those living along the coast, which can affect all types of development issues. It is an issue with Living Shorelines as it affects the long-term survival of wetlands plantings. As sea level rises, the planting zones for plant species gradually shift upward and landward where there is room for migration. Where there is no opportunity for landward migration, for example, in front of a high bank, riprap, or bulkhead, there will be a slow narrowing of the width of the planting zones as the seaward edge gets progressively deeper and can no longer support the planted vegetation. To ameliorate this problem within the context of the design of a Living Shoreline, it is best to design the planting zones of the planted areas to the upper range of the plant's ideal elevations. For example, smooth cordgrass, *S. alterniflora*, normally grows in the upper half of the tide range from MTL to MHW, +1.0′ to +2.0′ MLW in our representative marsh with a 2′ tide range. To better accommodate SLR and make the marsh viable for a longer period, it would be desirable to grade the smooth cordgrass planting zone from +1.3′ to +2.0′ MLW, that is, in the upper one-third of the tide range instead of the upper half. This will allow sea level to rise roughly 0.3′ before it starts flooding the smooth cordgrass too often for its continued survival. Concomitantly, the salt-meadow hay, *S. patens*, zone should be continued up the bank above the ULW where it is adapted to survive under most conditions. Planting as high as possible in the plant's normal range is the best way to ensure their continued survival in the face of SLR. It also must be remembered that Living Shorelines typically do not provide any significant benefits with regard to tidal flooding during storm events. This is because they are designed to work in concert with natural forces not to hinder or impede the effects of flooding.

Another future consideration of Living Shorelines is the role they may or may not play in the meeting of nutrient and suspended solids total maximum daily loading (TMDL) requirements. Marshes have well-documented abilities to sequester nutrients and suspended solids, which the TMDLs are aimed at reducing in waterways to help improve water quality. There are a number of modeling efforts underway to determine if credit can be given toward reaching TMDL goals by constructing Living Shorelines and other wetland restoration projects. Currently, there has been no definitive conclusion on what, if any, role Living Shorelines and other constructed wetlands may play in TMDL achievement. Hopefully, there will be a viable role that will provide an additional incentive for their construction (Drescher and Stack 2015).

ACKNOWLEDGMENTS

I would like to acknowledge the efforts of the reviewers of this paper, R. Neville Reynolds of Vanesse, Hangen and Brustlin, and C. Scott Hardaway of the Virginia Institute of Marine Science. Their comments led to a much improved paper. I would especially like to thank Mr. Hardaway for his years of counsel and advice. I could not have written this paper without his help.

REFERENCES

Drescher, S. and B. Stack. 2015. Recommendations of the Expert Panel to Define Removal Rates for Shoreline Management Projects. Chesapeake Bay Program. http://www.chesapeakebay.net/documents/Shoreline _Management_Protocols_Final_Approved_07132015-WQGIT-approved.pdf (accessed May 25, 2016).

Garbisch, E.W. and J.L. Garbisch. 1994. Control of upland bank erosion through tidal marsh construction on restored shores: Application in the Maryland portion of Chesapeake Bay. *Environmental Management* 18(5): 677–691.

Garbisch, E.W., Jr., P.B. Woller, and R.J. McCallum. 1975. Salt Marsh Establishment and Development. TM-52, U.S. Army Corps of Engineers, Coastal Engineering Research Center, Fort Belvoir, VA.

Hardaway, Jr., C.S. and G.L. Anderson. 1980. Shoreline Erosion in Virginia. Sea Grant Program, Marine Advisory Service, Virginia Institute of Marine Science, Gloucester Point, VA. 25 pp.

Hardaway, Jr., C.S., M. Berman, S. Killeen, D. Miligan, K. Nunez, K. O'Brien, T. Rudnicky, and C. Wilcox. 2008. Occohannock Creek, Shoreline Erosion Assessment and Living Shoreline Options report. Virginia Institute of Marine Science, College of William and Mary, Gloucester Point, VA. 47 pp. with Appendix.

Hardaway, Jr., C.S. and R.J. Byrne. 1999. Shoreline Management in Chesapeake Bay. Special Report in Applied Marine Science and Ocean Engineering Number 356. Virginia Institute of Marine Science, College of William and Mary, Gloucester Point, VA. 54 pp.

Hardaway, C.S. and J.R. Gunn. 2010. A brief history of headland breakwaters for shore protection in Chesapeake Bay, USA. *Shore and Beach* 78(4)/79(1): 26–34.

Hardaway, Jr., C.S., D.A. Milligan, and K. Duhring. 2010. Living Shoreline Design Guidelines for Shore Protection in Virginia's Estuarine Environments, Version 1.2. Special Report in Applied Marine Science and Ocean Engineering Number 421. Virginia Institute of Marine Science, College of William and Mary, Gloucester Point, VA. 43 pp. with Appendix.

Hardaway, Jr., C.S., J. Shen, D.A. Milligan, C.A. Wilcox, K.P. O'Brien, W.G. Reay, and S. Lerberg. 2007. Performance of Sills: St. Mary's City, St. Mary's River, Maryland. Virginia Institute of Marine Science, College of William and Mary, Gloucester Point, VA. 61 pp.

Hardaway, C.S., G.R. Thomas, and J.-H. Li. 1991. Chesapeake Bay Shoreline Study: Headland Breakwaters and Pocket Beaches for Shoreline Erosion Control. Special Report in Applied Marine Science and Ocean Engineering No. 313. Virginia Institute of Marine Science, Gloucester Point, VA. 153 pp., Appendices.

Knutson, P.L., R.A. Brochu, W.N. Seelig, and M. Innskeep. 1982. Wave damping in *Spartina alterniflora* marshes. *Wetlands* 2: 87–104.

Priest, W.I. 2006. Design Criteria for Tidal Wetlands. Management, Policy, Science, and Engineering of Nonstructural Erosion Control in the Chesapeake Bay, Proceedings of the 2006 Living Shoreline Summit.

Priest, W.I., J.W. Priest, J.P. Priest, and A.T. Priest. 2010. Libertyville Tidal Wetland Bank Fifth Annual Monitoring Report, Chesapeake, VA. Wetland Design and Restoration, Bena, VA. 21 pp.

Woodhouse, Jr., W.W., E.D. Seneca, and S.W. Broome. 1974. Propagation of *Spartina alterniflora* for Substrate Stabilization and Salt Marsh Development. TM-46, U.S. Army Corps of Engineers, Coastal Engineering Research Center, Fort Belvoir, VA. 155 pp.

Zabawa, C. and C. Ostrom. Editors. 1980. Final Report on the Role of Boat Wakes in Shoreline Erosion in Anne Arundel County, Maryland. Tidewater Administration, Maryland Department of Natural Resources, Tawes State Office Building, Annapolis, MD.

CHAPTER 11

Response of Salt Marshes to Wave Energy Provides Guidance for Successful Living Shoreline Implementation

Carolyn A. Currin, Jenny Davis, and Amit Malhotra

CONTENTS

11.1 INTRODUCTION

Estuarine shorelines are temporally dynamic, responding to sea level changes, storm events, and changes in sediment supply. The abundant natural resources provided by estuarine ecosystems support local economies and, as a result, homes and businesses are often located in proximity to estuarine shorelines. For these waterfront properties, and the communities they are part of, shoreline erosion is a constant challenge. Humans have historically combated erosion through a strategy of shoreline hardening. Anthropogenic modification of coastal areas is a primary cause of wetland loss in the United States over the last century (Dahl and Stedman 2013). In the United States today, 39% of the human population, or 123.3 million people, live in coastal counties (National Oceanic and Atmospheric Administration [NOAA] 2013). Nationally, human interest in maintaining the current shoreline position has resulted in 14% of the US estuarine shoreline being hardened (Gittman et al. 2015). In most cases, shoreline hardening involves the placement of vertical (concrete, wood, or steel) bulkheads or the deposition of loose rubble (rock or concrete) directly at the land–sea interface. While these structures can be effective at stopping the landward migration of the shoreline, they result in an abrupt disconnect between upland and intertidal habitats and thus a loss of the valuable ecosystem services associated with tidal wetlands including the provision of fishery habitat and water quality mediation (Currin et al. 2010; Gittman et al. 2016).

In recent years, there has been a growing awareness of the importance of wetlands to the long-term resilience of coastal communities (Temmerman et al. 2013). Vegetated shorelines are able to attenuate wave energy, making them natural agents of erosion control (Gedan et al. 2011). Vegetated shorelines also trap sediments that are suspended in the water column, leading to sediment accretion and an increase in surface elevation (Friedrichs and Perry 2001; Morris et al. 2002; Mudd et al. 2010). This ability can enable marshes to keep up with sea level rise (SLR), and therefore to function as self-sustaining erosion control devices. Recognition of this potential has led to an effort to incorporate salt marshes into estuarine shoreline protection efforts nationally and globally (Sutton-Grier et al. 2015; Temmerman et al. 2013).

There is a continuum of possible approaches to protecting shorelines from erosion. These approaches range from conservation of natural shorelines to creation or restoration of natural habitats (also known as natural infrastructure), including salt marshes, oysters, and mangroves; to hybrid shorelines that incorporate both natural habitats and built infrastructure; to built infrastructure alone, or hardened shorelines, which include seawalls, bulkheads, and riprap (Currin et al. 2010; NOAA 2015; Sutton-Grier et al. 2015). Both natural and hybrid approaches have also been termed Living Shorelines, and in this review, we limit our use of the term "living shoreline" to those projects where the footprint of natural vegetation exceeds that of built or hardened structures (NOAA 2015). Recent research indicates that living shoreline approaches like the use of salt marsh in conjunction with offshore stone sills can provide effective erosion control and valuable fishery habitat (Bilkovic and Mitchell 2013; Currin et al. 2008; Gittman et al. 2016). These results have helped encourage state and federal resource managers to embrace the use of natural infrastructure or living shoreline strategies (NOAA 2015; Sutton-Grier et al. 2015; SAGE http://sagecoast.org/), with nine states implementing at least partial bans on traditional shoreline hardening practices to date (Gittman, personal communication). One of the key issues limiting widespread utilization of natural infrastructure to stabilize estuarine shorelines is uncertainty about whether a nature-based shoreline will provide adequate erosion protection, and how resilient these shorelines will be to the predicted acceleration in SLR and to storm events. Although the ability of wetlands to accrete sediments, buffer wave energy, and reduce shoreline erosion is well documented, many marsh shorelines may experience significant erosion, indicating that marsh resilience varies with physical setting (Currin et al. 2015; Fagherazzi et al. 2013; Marani et al. 2011; Priestas et al. 2015). Guidance on the physical settings in which living shorelines offer a sustainable approach to shoreline stabilization is needed.

11.1.1 Objectives

The objectives of this chapter are to (1) review the existing literature on (1a) the energetic determinants of marsh habitat distribution, (1b) the relationship between shoreline wave energy and marsh erosion rates, and (1c) the ability of fringing marshes to attenuate waves and trap sediments; (2) describe results of a case study of natural and stabilized fringing salt marsh from central North

Carolina; and (3) combine results from the literature review and case study to provide guidance on the physical settings in which fringing marsh and hybrid living shorelines can be considered. The discussion presented here is specific to *Spartina alterniflora*–dominated estuarine marshes, which commonly occur along the US Gulf and Atlantic coasts.

11.2 BACKGROUND

11.2.1 Energetic Determinants of Marsh Habitat Distribution

Coastal salt marshes occur naturally along low-energy estuarine shorelines in temperate to subtropical zones. The current distribution of marshes is the result of dynamic equilibriums among sediment supply and surface accretion, wave erosion, and SLR (Fagherazzi et al. 2013). Wind waves are the primary factor driving erosion on salt marsh shorelines (Marani et al. 2011; Tonelli et al. 2010). For a given shoreline location, wave energy is controlled by wind speed, orientation relative to prevailing winds, bathymetry, and fetch. The present distribution of salt marsh on estuarine shorelines provides guidance on the conditions in which a living shoreline could be considered. An early examination of the energetic conditions that make a shoreline suitable for marsh restoration/creation was made by Knutson et al. (1981). This study characterized the stability of 86 created marshes ranging from Connecticut to California in relation to a suite of site physical characteristics including fetch, slope, nearshore water depth, sediment grain size, orientation to prevailing winds, and shoreline geometry. Their results indicated that success rates were greatest in cases where fetch was <1 km and marginal at 1–3 km and, further, that sediment grain size and shoreline geometry were also correlated with success. Knutson et al. (1981) concluded that "miles of shoreline along the coasts of the U.S. have been stabilized with rock and wooden structures under conditions suitable for vegetation."

Subsequent studies examined the relationship between shoreline vegetation distribution and "effective fetch," which sums the product of mean wind velocity, fetch, and percent frequency of wind from each of 16 compass directions and thus factors in the percent of time the wind blows from each direction and the strength of winds from that direction (Keddy 1982; Shafer and Streever 2000). Shafer and Streever (2000) further modified this into a Relative Exposure Index (REI) by dividing Keddy's exposure index by 100. Keddy (1982) found that direct fetch correlated more strongly with measures of vegetation and sediment type than did effective fetch. Shafer and Streever (2000) concluded that marshes created on dredged material in Texas were exposed to similar wave energy as natural marshes and that protective structures associated with created marshes may have been overbuilt.

The critical wave energy threshold for marsh stability was estimated by Shafer et al. (2003), using average water depth in addition to effective fetch to model the height of waves occurring along a given shoreline. Analysis of modeled wave heights along Gulf of Mexico shorelines representing a spectrum of energy levels indicated that the critical threshold for marsh existence is predicted by 20% exceedance wave heights between 0.15 and 0.3 m (viz., 80% of all waves are smaller than this; Shafer et al. 2003). Using this same approach, Roland and Douglas (2005) determined the critical wave height threshold for *S. alterniflora* in coastal Alabama to be 0.3–0.4 m. Shorelines that experienced waves of this height (even 5% of the time) did not support marsh. In the relatively shallow New River Estuary in North Carolina (mean depth is <2 m), Currin et al. (2015) examined the relationships among marsh distribution, shoreline change rate, and wave energy, as calculated by a Wave Energy Model (WEMo) that incorporates exposure to prevailing wind direction and bathymetry (Malhotra and Fonseca 2007). Wave energy is a function of wave height squared and can therefore be estimated from modeled wave heights. In the New River Estuary, significant wave heights > 0.3 m occur less than 1.25% of the time, with maximum wave heights of approximately 0.5 m. Marsh shorelines occurred across the wave energy spectrum in the New River Estuary, at fetches up to 6 km (Currin et al. 2015).

Early examinations of fetch and marsh distribution supported a few "rules of thumb" that are commonly used for living shoreline site selection by regulatory agencies. In many regions, state resource agencies and nonprofits that advocate living shoreline use have provided guidelines to help guide coastal landowners with decisions about how to protect their shorelines. In most cases, these criteria are based on some combination of fetch and water depth (Table 11.1). For example, in the Chesapeake Bay, it is recommended that only sites with a fetch of <1 km or nearshore water depth <1 ft (0.34 m) be considered for marsh planting without an offshore sill or other structural reinforcement (Hardaway and Byrnes 1999). These general rules are a helpful starting point, but site-specific assessments of wave energy and water depth will allow for nature-based infrastructure approaches to be used more widely, and with a greater degree of confidence. A large fetch, for example, is irrelevant if the wind rarely blows from that direction, and modeled wave regimes using average water depth may not accurately capture the importance of nearshore bathymetry to shoreline energy regimes. Further, while it is important to understand the critical threshold for marsh survival when selecting sites for marsh creation, there are many sites that fall in a gray area. That is, they are under the critical energy threshold for marsh existence but may still exhibit significant erosion rates; hence, marsh creation alone is an undesirable shoreline stabilization option. In some situations, these sites may be ideal for hybrid approaches that combine the use of native vegetation and some type of wave-reducing structure like an oyster reef or offshore sill.

In many estuaries, oysters are naturally abundant near the marsh edge, although their distribution is declining (Beck et al. 2011). Oyster reefs provide numerous ecosystem services (Coen et al. 2007), and the structure provided by fringing oyster reefs helps reduce wave energy and stabilize sediments (Fagherazzi et al. 2013; Grabowski and Peterson 2007). Created oyster reefs are effective at slowing marsh edge erosion (Meyer et al. 1997; Moody et al. 2013; Scyphers et al. 2011). Further, oysters have been shown to be capable of keeping pace with SLR, making them a natural partner for marshes when it comes to providing self-sustaining erosion control (Rodriguez et al. 2014). Similar to marshes, oysters do not occur in high wave energy settings and they are less effective at buffering against erosion in higher energy settings (Piazza et al. 2005). Thus, while oysters provide effective

Table 11.1 Current Recommended Guidelines for Living Shoreline Site Suitability

Region	LS Type	Fetch Criteria	Additional Comments
North Carolina[a]	Vegetation	<1 mile (1.6 km)	May be longer if sandbars/mudflat present
	Hybrid	1–3 miles (1.6–4.8 km)	
Virginia[b]	Vegetation	<1000 ft (<0.3 km)	Average and maximum fetch. Nearshore depth of <3 ft
	Hybrid	1000 ft to 5 miles (0.3–8.0 km)	
Gulf Coast[c]	Vegetation	<0.5 miles (<0.8 km)	Nearshore depth <1 ft
	Hybrid	1–2 miles (1.6–3.2 km)	Nearshore depth <2 ft
Delaware[d]	Vegetation	<0.5 miles (<0.8 km)	
	Hybrid	0.5–1.0 miles (0.8–1.6 km)	Vegetation with minimal structure like biologs
	Hybrid	>1 mile (<0.8 km)	Limited success without structural reinforcement
New Jersey[e]		None	Erosion history, tidal range, wave height, offshore depth, and other factors instead of fetch
Washington State[f]	Vegetation	1–5 miles (1.6–8.0 km)	With southerly fetch, multiply by 0.5 if north facing. May require log breakwater as well

[a] North Carolina Division of Coastal Management (2011).
[b] Hardaway et al. (2010).
[c] Gulf Alliance Training Program (2010).
[d] Partnership for the Delaware Estuary (2012).
[e] Miller et al. (2015).
[f] Johannessen et al. (2014).

erosion control in some settings (Cheong et al. 2013), there is currently little available information regarding the energetic settings under which marsh/oyster combinations are a viable erosion control option.

11.2.2 Wave Energy and Marsh Edge Erosion

The relationship between estuarine shoreline erosion rate and wave energy has received a good deal of attention in recent years. Several different parameters are used to characterize wave energy in shallow, nearshore waters, including wave height (m), representative wave energy (RWE; J m^{-1}) and wave power (J m^{-1} s^{-1} or W m^{-1}). The relationship between these parameters can be found in the US Army Corps of Engineers Shore Protection Manual (1977) and is described in the references provided in the discussion below. Briefly, wave energy is a function of wave height[2], RWE describes the energy in an average single wave per unit area, and wave power describes the energy in an average single wave per unit area per unit of time. A number of studies have described erosion rates associated with marsh shorelines and examined the empirical relationship between marsh edge erosion and measures of wave energy (Cowart et al. 2010, 2011; Currin et al. 2015; Leonardi et al. 2016; Marani et al. 2011; Priestas et al. 2015). Wave energy at eroding marsh shorelines is (or has recently been) low enough for marsh establishment but is high enough that the marsh edge is unstable under current conditions.

Attempts to define the relationship between shoreline change rate and wave energy (or a proxy for wave energy, like fetch) within an estuary have had variable success. Cowart et al. (2011) conducted a geographic information system (GIS) analysis of historical shoreline change and compared shoreline change rate to modeled REI values in the Neuse River Estuary, North Carolina. They found no discernable relationship when all points in the estuary were compared. However, when the data were grouped into eight shoreline regions (of increasing average fetch), significant correlations between mean shoreline change rate and mean REI were detected within groups. The lack of a trend in the larger data set highlights an important consideration: site-to-site variability in the way a shoreline responds to wave energy (as a result of underlying lithology, elevation, or differences in vegetative community) plays an important role in determining erosion rates. Indeed, these authors did document differences in land use, land cover, and elevation within the study region, and these differences were related to shoreline erosion rates.

Schwimmer (2001) compared wave power to shoreline erosion rate at nine salt marsh shoreline segments along Delaware Bay, Delaware, and found that wave power (computed from local wind speed, bathymetry, and fetch data) was significantly correlated with erosion rate. More recently, Currin et al. (2015) used NOAA's Wave Exposure Model (WEMo; Malhotra and Fonseca 2007) to investigate the relationship between shoreline erosion and wave energy in the New River Estuary, North Carolina. WEMo calculates RWE (J m^{-1}), which can be converted to wave power by dividing by average wave period. In this latter effort, there was no correlation between wave energy or power and erosion rate. However, the range of wave power calculated for New River Estuary shorelines is well below that of the sites investigated by Schwimmer (2001) in Delaware Bay (0.1 to 567 vs. 660 to 9200 W m^{-2}, respectively). A key finding in the New River Estuary was that narrow fringing marshes significantly reduced the erosion rate of sediment banks in the highest wave energy settings, lending support to the use of marsh vegetation for erosion control (Currin et al. 2015).

Ravens et al. (2009) found no evidence for a predictable relationship between shoreline erosion rate and wave energy in the Galveston Bay estuary. This system, like the New River Estuary, is characterized by relatively low wave energy. In all cases mentioned above, the assumed driver of shoreline erosion is wind-generated waves. In fetch-limited systems where the development of wind waves is limited, other sources of energy, such as boat wakes or tidal currents, may drive erosion.

A theoretical and empirical examination of the relationship between marsh edge erosion and wave power density (W m^{-1}) was provided by Marani et al. (2011). Their model, supported by data from

Venice Lagoon, demonstrated a linear relationship between the rate of edge erosion and wave energy for cliff, or scarped, marsh edges. In this analysis, time intervals in which the tide is above the marsh platform were excluded, highlighting the fact that wave erosion of the marsh edge primarily occurs during periods in which the unvegetated sediment or scarp seaward of the marsh edge is exposed to wave energy. The relationship between short- and long-term marsh erosion rates and wave energy and wave power was also examined in a shallow bay in Virginia (Priestas et al. 2015), at several sites with maximum fetches >5 km. A linear relationship between long-term salt marsh erosion rates and wave energy was found, and wave power was a good proxy for marsh erosion rates across several temporal scales (days to decades). Finally, a recent meta-analysis by Leonardi et al. (2016) provides additional support for a linear relationship between wave power and long-term marsh erosion rates. These authors note that the data demonstrate that marsh erosion continuously occurs, even at low wave energies. However, the linear relationship observed across a wide range of wave power suggests the lack of catastrophic marsh collapse at a particular threshold of wave energy, and the authors conclude that marsh erosion is not particularly associated with storm events (Leonardi et al. 2016).

One explanation for the failure of some studies to observe a significant relationship between erosion and wave energy is that, within a relatively small range of wave energy such as might exist within a region or water body, other factors may dominate variability in marsh erosion. Significant drivers of shoreline erosion include sediment type, shoreline geometry, and bioturbation. Sandy sediments have been shown to be more susceptible to erosion than fine-grained sediments (Feagin et al. 2009). This suggests that erosion rates may decrease as a restored or created marsh matures, because sediment grain size tends to decrease as marshes age (Craft et al. 2002; Wolf et al. 2011). In terms of shoreline geometry, shorelines located at a headland will receive wave energy from more directions than an embayed marsh, but the latter may receive more refracted wave energy (Priestas et al. 2015). Additionally, several studies have identified bioturbation as an important driver of marsh edge erosion (Paramor et al. 2004; Talley et al. 2001).

Other factors that contribute to the relative stability of the marsh edge include shoreline morphology or scarped versus sloped marsh edges and the availability of suspended sediments. Most living shoreline designs call for a gently sloped interface between the marsh and adjacent unvegetated flat (Currin et al. 2010), and there is some evidence that sloped or terraced marsh edges erode at slower rates than scarped or vertical edges (Theuerkauf et al. 2015; Tonelli et al. 2010). Models of marsh edge erosion often assume a scarped or vertical edge (Fagherazzi et al. 2012; Marani et al. 2011) and note that immersion of the scarp, before marsh flooding, is the period in which wave energy exerts the highest erosive force on the edge (Tonelli et al. 2010). Wave attack on vertical scarps can result in undercutting of the marsh platform and eventual slumping of the marsh surface. However, the ability of marshes to rebuild from an eroded or slumped edge has also been documented, and the transport of eroded sediment to the marsh platform can be an important process in maintaining marsh surface elevation relative to SLR (Allen 1989; Fagherazzi et al. 2012).

Fagherazzi et al. (2013) demonstrated that suspended sediment concentration is a key variable in estimating marsh erosion under variable SLR rates. Sediment concentrations of less than 30 mg L^{-1} result in lateral retreat of the marsh edge under most scenarios, while higher sediment concentrations can result in progradation. Other authors have identified suspended sediment concentrations of >20 mg L^{-1} as being critical to the vertical maintenance of salt marshes under predicted SLR rates (Kirwan et al. 2010; see below).

11.2.3 Ability of Marsh Vegetation to Attenuate Waves and Trap Sediments

The presence of vegetation on estuarine shorelines can increase resistance to shoreline erosion and increase resilience to SLR (Gedan et al. 2011; Temmerman et al. 2013). Although all sloping shorelines will dissipate wave energy, vegetation increases friction, dampens waves, and reduces turbulent mixing, resulting in increased wave energy attenuation as compared to unvegetated shorelines or mudflats (Möller 2006; Yang et al. 2012). Even fairly narrow bands of marsh (30 m wide), as might be incorporated into living shorelines, have been shown to be effective at attenuating up to 94% of incoming wave energy (Knutson et al. 1982). Yang et al. (2012) found that waves were completely eliminated in approximately 80 m of *S. alterniflora* marsh. Leonard and Croft (2006) analyzed changes in flow velocity and turbulence along shore-perpendicular marsh transects in the absence of waves and found that both flow velocity and total turbulent kinetic energy were reduced by approximately 50% within 5 m of the marsh edge in *S. alterniflora* marshes with stem densities of 150–300 stems m^{-2}. Several studies have also noted that wave energy reduction increases with increased marsh stem density or canopy height (Leonard and Luther 1995; Möller 2006; Yang et al. 2012). Möller (2006) further noted that seasonal changes in canopy height and biomass will result in decreased wave attenuation in winter months. The greatest wave energy attenuation occurs when water depth and canopy height are approximately equal; when the water level height exceeds canopy height, the ability of the marsh to attenuate wave energy is dramatically reduced (Möller 2006). This has major implications for the importance of fringing marshes during storm conditions. When storm surges exceed canopy height, the marsh's impact on waves is negligible (Gedan et al. 2011). Salt marshes are capable of reducing storm surge height and speed as the vegetation provides drag, but measurements are difficult, and modeling efforts demonstrate the complexity of landscape and storm forcing variables in determining the impact of vegetation on storm surge (Wamsley et al. 2010). Estimates of the width of salt marsh capable of a 1-m storm surge reduction range from 4 to 25 km (Shepard et al. 2011; Wamsley et al. 2010). In practice, the living shorelines we discuss here will have little impact on reducing the storm surge experienced by adjacent property.

As marsh plants attenuate incoming wave energy, and slow the movement of water, they cause an increase in the deposition of suspended sediments onto the marsh surface (Leonard and Croft 2006). Salt marsh vegetation influences sediment deposition both by the reduction of wave energy and by direct interception of particles by plant stems (Mudd et al. 2010). The trapped sediments contribute to both vertical and horizontal maintenance/growth of marshes. Several previous investigators have demonstrated a positive correlation between vegetative density and sediment deposition rates (Gleason et al. 1979; Morris et al. 2002), suggesting that greater stem density will confer greater resilience to both wave energy and SLR in fringing salt marshes. More recently, experimental results have demonstrated improved growth and survival of marsh plants in high-density clumps (Silliman et al. 2015), further supporting the use of denser plantings in living shoreline projects. A meta-analysis conducted by Shepard et al. (2011) revealed a significant positive effect of vegetation on shoreline sediment accretion, surface elevation increase, and erosion reduction, demonstrating the ability of fringing salt marshes to increase coastal resiliency to SLR via wave attenuation and sediment trapping. However, in recent decades, the damming of rivers has resulted in lower suspended sediment concentrations in coastal receiving waters (Weston 2013). This poses a problem for coastal marshes, which require sediments for maintenance and growth. Modeling efforts indicate that there is a minimum sediment load (20–50 mg L^{-1}) at which marshes can be expected to keep pace with SLR, and that those sites with greater suspended sediment loads are more resilient to changes in water levels (Kirwan et al. 2010; Mariotti and Carr 2014) and may demonstrate progradation rather than erosion.

While rivers are often important sources of suspended sediments, marsh growth can also be supported by erosion of nearby shorelines and resuspension of sediments from open water areas (Currin et al. 2015; Mariotti and Carr 2014). Further, although wave energy from storm events may be responsible for lateral erosion of the marsh edge, many investigators have documented increased marsh sediment deposition rates associated with storms (e.g., Baustian and Mendelssohn 2015; Reed 1989; Schuerch et al. 2012; Tweel and Turner 2012). During storm events, elevated wave energy resuspends sediments from the bottom of nearshore regions and incorporates that material into tidal waters inundating the marsh surface, where they may ultimately be deposited. Hurricanes have also been associated with elevation growth in fringing marshes and hybrid living shorelines (Currin et al. 2008; Gittman et al. 2014).

11.3 NORTH CAROLINA SENTINEL SITE CASE STUDY

11.3.1 Relationship between Wave Energy and Distribution of Natural Fringing Salt Marshes

To investigate the energetic constraints on the distribution of natural fringing marsh shorelines, we compared calculated RWE to marsh width for 197 randomly selected marsh shorelines within the North Carolina Sentinel Site (NCSS). We used WEMo to calculate RWE as the total wave energy in one wave length per unit crest in units of J m^{-1} (Malhotra and Fonseca 2007). The NCSS, one of five current members of the NOAA Sentinel Site Program, is located on the central coast of North Carolina and encompasses a National Estuarine Research Reserve, Cape Lookout National Seashore, Morehead City State Port, and a diverse coastal ecosystem including barrier islands, lagoons, riverine estuaries, salt marshes, and seagrass beds (Figure 11.1). RWE was calculated for the entire NCSS shoreline using WEMo, which produces a spatially registered GIS grid of wave energy (J m^{-1}) based on a shoreline vector data set, local bathymetry, and hourly wind data (Malhotra and Fonseca 2007). WEMo does not forecast far-field waves such as ocean-generated swells, focusing instead on wind events that generate waves within 10–100 km of a site. This characteristic makes the model ideal for estuaries, sounds, and other enclosed water bodies. For the current investigation, WEMo was run on a grid of points spaced at 50-m intervals along the shoreline and extending to 200 m offshore. Bathymetric inputs included data from various sources including LIDAR from the North Carolina Flood Plain Mapping Service, soundings data generated by the National Ocean Service, and bathymetric data collected by a small boat to fill in data gaps. For input into WEMo, these data were referenced to NAVD 88 and assimilated to create a 20-m-resolution bathymetry for the entire study region. The tidal range in the study region is approximately 1 m and *S. alterniflora* occupies the mid to upper intertidal elevations, approximately local mean sea level to local mean high water. To ensure that RWE measures represent flooded marsh conditions, we lowered the bathymetry data by 0.6 m for all WEMo runs. We used hourly wind data from 2012 to 2015 from the NOAA National Data Buoy Center station CLKN7 at Cape Lookout, North Carolina. Only the top 20% of all wind data were used to calculate RWE$_{20}$, as it is these higher-intensity, lower-frequency winds that are thought to be the primary drivers of shoreline erosion (Kelley et al. 2001, but see Leonardi et al. 2016).

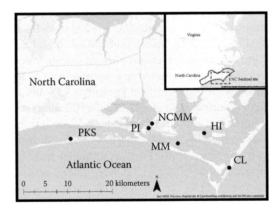

Figure 11.1   Study area. Representative wave energy was calculated for the entire shoreline of the NC Sentinel Site; the boundary is indicated on inset figure. Labeled points indicate sites where additional data were collected on vegetation dynamics and surface elevation changes in paired natural and sill, or natural and oyster fringing marshes.

Analysis of the entire NCSS at 50-m intervals resulted in calculated RWE_{20} values ranging from 0 to 1072 J m^{-1} ($n > 40{,}000$ shoreline points). This data set was trimmed to include only grid points within 50 m of the shoreline and overlaid with shoreline habitat maps generated by the North Carolina Division of Coastal Management. The habitat maps were used to further trim the data set to include only salt marsh shorelines. Grid points that intersected salt marsh shorelines were grouped into eight categories based on their RWE_{20} values using Jenks optimization, which maximizes variance between groups and minimizes variance within groups (Table 11.2). From each of these eight categories, up to 30 points were randomly selected for comparison of marsh width versus RWE. Random point selection was done in ArcGIS 10.2, and included only shoreline marshes that were not adjacent to hardened or developed structures. To avoid sampling the same marsh twice, points were selected that were at least 100 m apart. We were unable to locate 30 marshes in each category because of these constraints.

The results of this analysis indicate that width of shoreline marsh decreases with increasing wave energy (Figure 11.2), with average width decreasing to less than 100 m at RWE_{20} values of greater than 300 J m^{-1}. Fringing marsh was rare, and narrow, on shorelines experiencing RWE values of >583 J m^{-1}. Map-based examination of the 197 point data set highlighted another important point: many of the marshes in this region exist in areas of frequent boat traffic (e.g., along the Atlantic Intracoastal Waterway [AIWW]) and therefore are subject to exposure to boat-generated

Table 11.2 RWE_{20} Classes of Shoreline Marsh Habitat in Central North Carolina Calculated Using the Top 20% of Wind Data

Class	RWE_{20} Range (J m^{-1})	n
1	0–21	25
2	21.2–59.06	29
3	59.06–111	31
4	111–184	28
5	184–337	26
6	337–400	25
7	400–583	26
8	583–1072	7

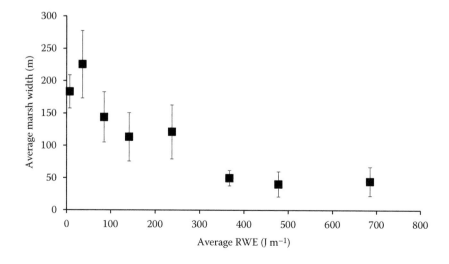

Figure 11.2 Average marsh width as a function of representative wave energy. Data represent a random subsample of fringing marshes within the NCSS region. Vertical error bars represent standard error.

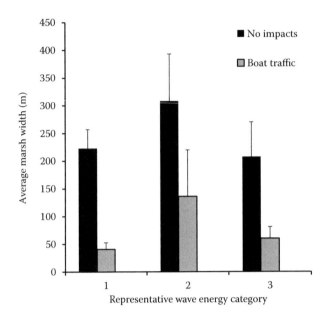

Figure 11.3 Impacts of boat traffic on shoreline marsh width exposed to low wind wave energy conditions. Marshes were classified as potentially affected by boat traffic if they occurred along a well-marked navigation channel or a shoreline with a large number of private docks.

waves in addition to wind waves. To investigate the importance of boat wake wave energy on marsh width, we classified each of the marshes in the three lowest RWE_{20} categories (1–3) based on their proximity to navigation channels. Classification was accomplished by visual inspection of aerial photography. A marsh was categorized as potentially affected by boat wakes if it existed along a frequently used navigation channel, or an area of shoreline that was populated by numerous private docks, indicating that a high volume of traffic is probable. Average marsh width was then separately calculated for marshes that were and were not likely affected by boat wakes. This analysis revealed that in low wind wave energy settings, marshes that are most likely to be affected by boat wakes are narrower than those that are not adjacent to areas of high boat traffic (Figure 11.3).

11.3.2 Wave Energy, Marsh Vegetation, and Surface Elevation Change in Natural and Stabilized Fringing Marshes

A study of natural and stabilized fringing salt marshes was initiated in 2005 within the NCSS, which is located at the southern end of the North Carolina Outer Banks (Figure 11.1). The goal of this study was to provide an assessment of the performance of living shorelines under a range of site conditions. The study sites included four locations with paired natural fringing marsh (Natural) and created or restored marshes stabilized with offshore granite sills (Sill). Two additional sites included paired natural fringing marshes with and without fringing oyster reef (Natural Oyster and Natural, respectively). At each site, long-term monitoring of surface elevation and vegetation change was established between 2004 and 2006, and wind wave energy experienced by the site during the study period was modeled using WEMo. The mean tide at the Beaufort tide station (8656483, which is centrally located within the study region) is 0.95 m, and salinity in the outer estuary, where our sites were located, typically ranges between 25 and 35 ppt.

11.3.2.1 Study Sites

The four locations with paired Natural and Sill marshes include Pivers Island (PI), Pine Knoll Shores (PKS), the North Carolina Maritime Museum (NCMM), and Harkers Island (HI) (Table 11.3). Further details of sill construction and site characteristics of the PI, PKS, and NCMM sites are provided in Currin et al. (2008). As at the other sites, the HI sill was constructed of granite, and followed the North Carolina Division of Coastal Management (NC DCM) permit conditions (Currin et al. 2010). No fill was added to the site, but it was planted with *S. alterniflora* and *Spartina patens* shortly after sill installation by the North Carolina Coastal Federation, a nonprofit environmental group. Natural reference marshes were located within 0.5 km of Sill marshes and were selected to match fetch, shoreline orientation, and distance to navigable waterways as much as possible. As reported in Currin et al. (2008), both Natural and Sill marshes at PI, PKS, and NCMM are characterized by sandy sediments (>74% sand) with low (<10%) organic matter content. In addition to the paired Natural and Sill marsh sites, we sampled two locations, Cape Lookout (CL) and Middle Marsh (MM), with paired natural fringing marshes, one of which was bordered by tidal flat and the other by a natural fringing oyster reef (Figure 11.1). The PI Natural reference marsh was also bordered by a live oyster reef. *S. alterniflora* was the dominant (essentially monotypic) marsh vegetation within 5 m of the shoreline edge at all study sites.

Table 11.3 Location and Descriptions for Sites Where Surface Elevation and Vegetation Data Were Collected in Carteret County, North Carolina

Site Marsh Types	Location (Latitude/Longitude)	Sill Installation Date	Built Sill Height, ft (m) NAVD88	Sill Length, ft (m)	Sample Period[a]
PI		2002	0.951 (0.290)[b]	300′ (91)	2005–2008
Natural Oyster	34° 43′ 14.28116″ N				
Marsh sill	76° 40′ 30.44430″ W				
NCMM		2001	1.1722 (0.525)[b]	70′ (21)	2005–2008
Natural	34° 43′ 47.58449″ N				
Sill	76° 40′ 5.62936″ W				
PKS		2002	1.097 (0.334)[c]	400′ (122)	2005–2008
Natural	34° 42′ 1.13429″ N				
Sill	76° 49′ 59.29379″ W				
HI		2004	1.469 (0.448)[c]	410′ (125)	2005–2008
Natural	34° 42′ 37.52731″ N				
Sill	76° 33′ 45.73896″ W				
CL		N/A	N/A	N/A	2006–2008
Natural	34° 38′ 29.70436″ N				
Natural Oyster	76° 30′ 44.82136″ W				
Middle Marsh		N/A	N/A	N/A	2005–2008
Natural	34° 41′ 25.47800″ N				
Natural Oyster	76° 36′ 57.13369″ W				

Note: PI, Pivers Island; NCMM, North Carolina Maritime Museum; PKS, Pine Knoll Shores; HI, Harkers Island; CL, Cape Lookout; MM, Middle Marsh; N/A, not applicable.

[a] Sample period describes the period in which hourly wind data were used to calculate site Representative Wave Energy, and during which SET readings were obtained in the Fall (October–November) and Spring (March).

[b] Data collected in June 2008.

[c] Data collected in November 2009.

11.3.2.2 Methods

Net change in marsh surface elevation was measured by deep-rod Surface Elevation Tables, or SETs (Cahoon et al. 2002). In each marsh, one SET was established within 1 m of the lower edge of *S. alterniflora* vegetation. The lower vegetation line was determined visually along a 30-m shoreline section of each marsh and a random location selected for placement of the lower SET along the shoreline (Figure 11.6a). Each of the 9 SET pins was read at 4 positions, for a total of 36 readings per SET, per sample interval. Readings that were affected by loose shells or crab holes were eliminated from the data set. An average elevation change was determined for each pin of each SET at each sampling interval, and each SET was treated as a single replicate in subsequent analyses to determine treatment (Sill, Natural) effects on surface elevation change. WEMo was run for each of the six site locations for the period encompassing SET measures (Table 11.3) as described previously, except that the bathymetry was raised by 1 m. RWE_{20} values were calculated to match the intervals between SET readings at each site.

Vegetation parameters were measured in August 2006, 2007, and 2011 at four sites (MM, NCMM, PI, and PKS). Vegetation at CL and HI was measured only in 2006 and 2007. Site and treatment comparisons of vegetation parameters are based on data collected in 2006–2007, when all sites were surveyed, while vegetation change data were calculated for four sites for the 2006–2011 interval (see Table 11.3). At each site, four or five transects were established along a 30- to 40-m section of shoreline using restricted random sampling (Elzinga et al. 1998). Each transect began at the lowest edge of *S. alterniflora* growth and continued perpendicular to the shoreline toward the upland border. In this study, data are reported from plots (1 m^2) established 1 m below the marsh edge (–1 m), at the marsh edge (0 m), and 5 m from the marsh edge (5 m, see Figure 11.6a). Measurements were made for *S. alterniflora* live stem density and stem height in each 1-m^2 plot. Live stems were counted in a 0.25-m^2 subplot, unless the plot contained very few stems (i.e., less than 25), in which case a count was done on the entire m^2. Mean stem height was obtained by measuring the height (from sediment surface to tallest green tissue) of the first 10 live plants that intersected a string bisecting the 1-m^2 plot.

11.3.2.3 Surface Elevation Change and Wave Energy in Natural and Sill Salt Marshes

The average RWE_{20} values calculated for the six shoreline sites ranged from 13 to 677 J m^{-1}. Marsh surface elevation change, as measured by SETs located within 1 m of the lower marsh edge, ranged between –10.1 and 5.4 mm year^{-1} in Natural fringing marshes. In sites with paired Natural and Sill marshes, surface elevation change was always greater in Sill marshes (0.2 to 7.1 mm year^{-1}) than in their paired Natural counterparts (–3.5 to 2.1 mm year^{-1}; Figure 11.4). Regression analysis of surface elevation change in Natural versus Sill marshes revealed a significant difference ($p < 0.0001$) in slope between the two marsh types. The two marsh sites that experience the lowest wind wave energy, MM and NCMM, exhibited greater loss in Natural marsh edge elevation than Natural marshes at higher wave energy sites, such as CL, HI, and PKS. The effect of fringing oyster reefs on surface elevation change was variable across the three sites for which we had SET measures. At only one of those sites, PI, did we observe a positive surface elevation change in Natural Oyster marshes (Figure 11.4). However, Sill marshes exhibited positive surface elevation change at all sites, with the highest rate (7.1 mm year^{-1}) at the highest wave energy site, PKS (Figure 11.4).

The positive effect of stone sills on marsh surface elevation change is consistent with previous reports of increased sediment accretion in stabilized North Carolina fringing marshes (Currin et al. 2008; Gittman et al. 2015). Here, we utilize SETs to report statistically significant higher elevation change in shoreline marshes fronted by stone sills, which had been established 2 to 4 years before the SET measurements. All SET measurements were conducted over an 18- to 32-month

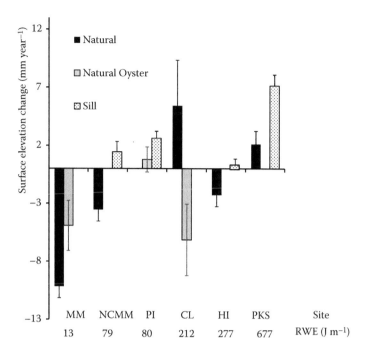

Figure 11.4 Average annual surface elevation change between 2005 and 2008 at six study sites. Each site included either paired Natural and Sill marshes, or paired Natural and Natural Oyster marshes. Sites are plotted in order of representative wave energy (RWE_{20}).

period. In addition to increasing sediment accretion at all four paired sites, the data suggest that sills may trap more sediment in high-energy areas than in low-energy areas. We hypothesize that tidal water in these high-energy areas carries greater sediment loads owing to resuspension of nearshore sediments (Mariotti and Carr 2014). When these waters flow onto the marsh, the sill and marsh vegetation combine to reduce velocity and turbulence to a greater degree than in fringing marshes without a sill, leading to greater deposition of suspended sediment onto the marsh surface. We note that each of these sills was constructed to include an unvegetated tidal flat between the sill and vegetation, but this bare area was not maintained, as salt marsh rapidly colonized the newly deposited sediment immediately behind the sill (see Vegetation Results, Section 11.3.2.4). We also note that we observed edge erosion at natural marshes adjacent to sills at two sites, NCMM and PKS (Mattheus et al. 2010). Construction of stone sills adjacent to natural marsh must be done with caution to minimize this result.

Only three of the natural fringing marshes exhibited a positive surface elevation change over the study period, and several SETs, including those adjacent to oyster reefs, showed significant surface elevation loss. Wind wave energy is not the only factor driving marsh sediment accretion and erosion processes, and in this study, boat wake energy and limited sediment supply likely contributed to surface elevation loss at some study sites. The lowest-energy sites sampled here, at MM, are shoreline sites adjacent to tidal creeks, in a protected relict flood-tide marsh, which have limited suspended sediment supply (Biber et al. 2008). Another low-energy site exhibiting elevation loss, NCMM, is located within 100 m of the AIWW, near the beginning of a "no-wake" zone, and thus experiences the large waves associated with sudden increases or decreases in boat speed. Therefore, despite their relatively low RWE_{20}, the lower edge of these fringing marshes lost up to 2 cm of elevation through the study period. At MM, the presence of a fringing oyster reef at one of the study locations reduced the elevation loss (Figure 11.4).

11.3.2.4 Marsh Vegetation in Natural and Sill Marshes

Marsh vegetation parameters collected from the shoreline (Plot 0; Figure 11.6a) at six sites in 2006–2007 exhibit variability by marsh type and by wave energy setting. Stem density at the lower marsh edge was nearly double behind sills (x = 172.8 stems m^{-2}; SD = 35.0) as compared to Natural (66.2; 24.4) and Natural Oyster (55.2; 18.7) fringing marshes (Figure 11.5a). Across the five Natural marsh sites sampled, there was a positive relationship between site RWE_{20} and stem density, while there was no pattern observed in the Natural Oyster marshes. There was also a positive relationship between site RWE_{20} and stem density for the three Sill sites established before 2003, while the newest Sill site (HI) had much lower stem density than other sites during the study period (Figure 11.5a). Average *S. alterniflora* stem height did not vary significantly by marsh type or site wave energy, and across all sites, mean stem height at the lower marsh edge was 46.2 cm (Figure 11.5b).

We used data from four sites sampled identically in 2006, 2007, and 2011 to examine changes in marsh vegetation over time. The plots that were originally established just below the marsh edge, Plot −1, showed little change over the 5-year period in the natural marshes, and overall, stem density remained below 25 stems m^{-2}. However, the −1 m plot in the Sill marshes showed a significant increase over time, from 58.7 stems m^{-2} in 2006 to 229 stems m^{-2} in 2011 ($p >$ |t| 0.0272, df = 37). Stem density did not change significantly from 2006 to 2011 in any of the marsh types in Plot 0 or Plot 5 (Figure 11.6b through d).

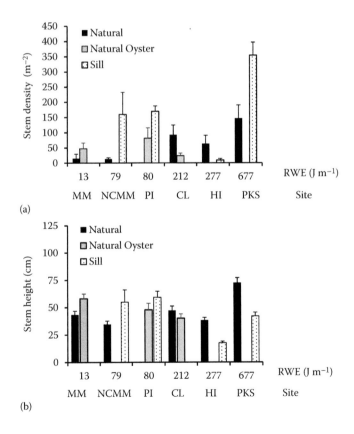

(a)

(b)

Figure 11.5 Average (a) stem density and (b) stem height of *S. alterniflora* measured in 0-m plots at the lower marsh edge for all paired study sites. Sites are plotted in order of representative wave energy (RWE).

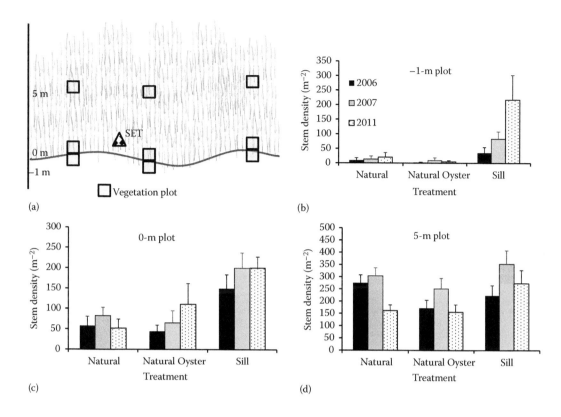

Figure 11.6 (a) Schematic showing relative locations of the −1-, 0-, and 5-m plots relative to shoreline and SET. (b through d) Average stem density as a function of treatment and year for the −1-, 0-, and 5-m plots, respectively.

The increase in stem density in the Sill marsh plots waterward of the 2006 marsh edge is coincident with an increase in sediment deposition in the area behind the sill, as captured by the SET data. In fact, in all the Sill sites sampled in this case study, *S. alterniflora* has colonized the bare substrate as constructed between the original marsh edge and the stone sill and, in many cases, is moving waterward into the Sill (Figure 11.7). Throughout the study period, stem density in the Sill marshes was also greater in both the 0- and 5-m plots (x = 219.8 and 321.5 stems m^{-2}, respectively) than in the Natural and Natural Oyster marshes (x = 75.5 and 191.4 stems m^{-2}, respectively) (Figure 11.6b through d). Relatively low measured stem densities at the lower edge of natural fringing marshes is, to some degree, a consequence of measuring an uneven marsh edge, as well as the result of their low placement in the tidal frame. The stem density, stem height, and aboveground biomass of shoreline marshes are important features in determining the ability of these marshes to trap sediments (Temmerman et al. 2005) and attenuate wave energy (Leonard and Croft 2006; Möller 2006; Möller et al. 2014). Overall, we observed greater stem density in Sill marshes, particularly at the lower marsh edge, regardless of wave energy setting, although this was not observed in the youngest (<4 years) Sill marsh. This is correlated with the observation of greater surface elevation increase at the lower edge of Sill marshes (Figure 11.4).

The decrease in marsh surface elevation observed in both Natural and Natural Oyster marshes between 2006 and 2008 (Figure 11.4) did not result in declines in marsh vegetation at the marsh edge through 2011. The stability of lower marsh vegetation at both natural and sill marshes suggests that there was little horizontal erosion in these marshes in the period 2005–2011. The preservation of the vegetated marsh edge in natural fringing marshes speaks to the resilience of these natural

(a)

(b)

Figure 11.7 Marsh sill type of Living Shoreline at Pivers Island (PI) in (a) 2002 (notice the large area of unplanted, bare substrate between sill and marsh edge) and (b) 2007. Marsh has completely filled in the bare area behind the sill.

shorelines and their short-term ability to withstand both waves and erosion of the marsh surface. However, if these marshes are not able to increase their surface elevation at a rate commensurate with local SLR, they will eventually drown and the lower marsh edge will move landward.

11.3.3 Living Shorelines and Hurricanes

A question that often concerns waterfront property owners is the ability of a living shoreline to withstand the wave energy and storm surge that accompany tropical storms or hurricanes. The response of marsh vegetation and surface elevation to the passage of Hurricane Isabel in 2003 at three of our case study sites (NCMM, PKS, and PI) was reported in Currin et al. (2008). At the time of the storm, the Sill marshes had been established less than 3 years, and *S. alterniflora* stem density was less than that in the Natural marsh sites. Nevertheless, post storm stem density in both Natural and Sill marshes was equal to or greater than that recorded before the storm (Currin et al. 2008). In addition, marsh surface elevation change measured at two sites (PKS and PI) increased in both Natural and Sill marshes.

Hurricane Irene passed through the case study area in August 2011, just outside our period of vegetation monitoring. Gittman et al. (2014) evaluated the response of North Carolina living shorelines, including several in our study area, to the passage of Hurricane Irene and found that marsh sills performed better than seawalls. These authors also examined differences in erosion, sediment accretion, and *S. alterniflora* stem density of several paired natural marsh and marsh sill sites before and after hurricane passage. These paired sites are located in the eastern end of Bogue Sound, approximately 5 km east of the depth versus RWE transect illustrated in Figure 11.9b, and include the PKS SET Natural and Sill sites analyzed in this paper. Calculated RWE_{20} values for the Bogue Sound sites described in Gittman et al. (2014) range from 310 to 677 J m^{-1}, using the same input conditions described for our RWE analysis. Average prestorm stem density in these paired sites was reported to be between 300 and 400 stems m^{-2} and did not vary between natural and sill marsh treatments. In these relatively high RWE sites, with fetches of >5 km (3.1 miles), marsh surface elevation was not altered by the hurricane passage, though stem density was decreased immediately in both marsh types, and as of 1 year post storm, it had rebounded in Sill marshes but not natural marshes (Gittman et al. 2014). These results suggest that living shorelines are resilient to storm events, consistent with the conclusions of Leonardi et al. (2016), and support our conclusion that living shorelines, with or without sills, may be appropriate at sites with fetches greater than those often found in current state guidance documents (Table 11.1).

11.4 USING WAVE ENERGY TO GUIDE LIVING SHORELINE SITE SELECTION

We utilize a desktop WEMo (http://products.coastalscience.noaa.gov/wemo/), which runs in ArcGIS and has been used by a number of researchers to investigate impacts of wave energy on estuarine shorelines (Cowart et al. 2010, 2011; Currin et al. 2015). We note that there are other models available for forecasting the resiliency of shoreline marshes to SLR and wind wave energy (Fagherazzi et al. 2012; Kirwan et al. 2010; Marani et al. 2011; Mariotti and Fagherazzi 2013) that incorporate nearshore bathymetry, suspended sediment concentration, tide range, wind wave energy, scarp height, and rate of SLR to predict marsh edge erosion. There are also advanced models for forecasting wind wave energy, which capture refracted wave energy, a variable that may be important in estuarine embayments (e.g., Simulating Waves Nearshore [SWAN]; Booij et al. 1999; Priestas et al. 2015). However, it will usually be beyond the scope of a waterfront property owner, or a regulatory agency, to run these models for small-scale (<300 m shoreline) proposed living shoreline projects.

WEMo offers several analysis options, including the percentage of wind data used to calculate wave energy (Malhotra and Fonseca 2007). The greater the percentage of wind events used for analysis, the lower the calculated RWE (Figure 11.8). Thus, reporting the exact percentage of wind data used is imperative if results are to be compared with other studies or used for decision making. We suggest that users of WEMo report RWE values with a subscript to indicate what percentage of

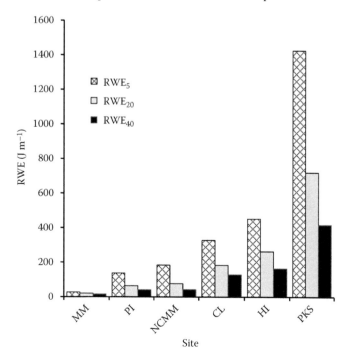

Figure 11.8 Representative wave energy at each study site as calculated using top 5% (RWE_5), 20% (RWE_{20}), and 40% (RWE_{40}) of the hourly wind data from 2005 to 2008.

wind events the data represent. We note that we reported RWE_5 results in our study of shoreline erosion in the New River Estuary (Currin et al. 2015) and report RWE_{20} results in this study. Although the top 5% of wind events may account for most of the wave energy a site experiences on an annual basis, smaller events, especially from prevailing wind directions, may also be important (Leonardi et al. 2016). Site distinctions become smaller with increasing percentage of wind data used, and the top 20% of wind events offers a compromise between incorporating more wind data and maintaining important site distinctions.

By taking prevailing wind direction and nearshore bathymetry into account, WEMo can provide a more accurate assessment of site wave energy conditions than fetch alone. Comparison of two potential living shoreline sites located on opposite sides of Bogue Sound, within the NCSS, illustrates this point (Figure 11.9a). Each site has a similar cross-sound fetch (approximately 3.6 km) and longest fetch (12 or 13 km). The site on the north, south-facing shore, is protected from the stronger prevailing winds from the northeast, which are typical of this area, and has a much lower calculated RWE_{20}. On this side of the sound, marsh vegetation may be sustainable without a stone sill, while a hybrid approach is likely needed on the opposite, north-facing shore. However, there are several important cautions to note in the application of WEMo to determine the suitability of a particular shoreline for a living shorelines project.

As with all models, the quality of WEMo output is heavily reliant on the quality of input data sets. For wind wave energy predictions, reliable wind data are generally available from buoys, airports, or other local weather stations. Shoreline spatial data are also generally available, although the accuracy should be checked for potential project sites. Accurate nearshore bathymetry is the input data set that is most difficult to locate, particularly in areas dominated by

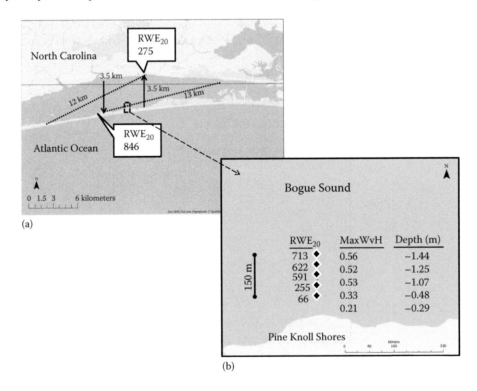

(a)

(b)

Figure 11.9 (a) Comparison of RWE_{20} values at two sites with similar fetch characteristics but different orientation to prevailing winds. (b) Analysis of the impact of water depth on calculated RWE_{20}. Maximum wave height (MaxWvH) and water depth (Depth) data correspond to 5 points (represented by diamonds) located along a 150-m transect perpendicular to the shoreline.

soft sediments, shoaling, or dredging. For larger-scale projects, local bathymetry may be worth obtaining with a small boat, to augment existing bathymetric data sets before WEMo analysis. To illustrate the sensitivity of wave energy to nearshore bathymetry, we analyzed wind wave energy across a depth transect in Bogue Sound, North Carolina (Figure 11.9b). Over a 150-m length, depth increased from −0.29 m to −1.44 m, and calculated RWE_{20} increased from 66 to 713 J m^{-1}. This difference in potential wind wave energy with a depth increase of a little over 1 m also illustrates the importance of calculating RWE at appropriate tide levels. In central North Carolina, tidal amplitude is approximately 1 m, and salt marshes occupy the mid to upper intertidal regions. Tidal bathymetry is provided relative to MLLW; thus, corrections must be made to the bathymetry to calculate wave energy during high tides when the marsh is submerged. We added either 0.6 or 1.0 m to the water depths to estimate wind wave energy during periods when the marsh edge would be inundated. In addition to RWE values, standard WEMo output includes average and maximum wave heights, wave period, and both horizontal and shear velocity at the sediment level. These outputs allow for easy conversion among units. Wave power, for example, a value that has often been reported in the literature, is calculated by dividing RWE by wave period. Thus, RWE values are readily comparable to previously published data and readily interpreted in terms of wave height. Further, velocity parameters at the sediment interface can be combined with knowledge of sediment characteristics to predict the likelihood of sediment resuspension for a given wave energy regime. It is also important to remember that, in narrow channels, including many parts of the AIWW, boat wake energy will far exceed wind wave energy. We found that proximity to navigation channels significantly reduced average fringing marsh width in areas with $RWE_{20} < 300$ J m^{-1}, where marsh vegetation might otherwise be an option for living shoreline design (Figure 11.3).

Previous research efforts, including the case study presented above, indicate that marsh sustainability is strongly predicted by wave conditions. To date, the challenge in using this information for predicting the sustainability of any given stretch of shoreline has been in accurately predicting wave conditions on a site-by-site basis. Many of the currently existing guidelines are based on fetch because fetch is simple to estimate; however, bathymetry and dominant wind direction both play pivotal roles in determining wave climate. WEMo offers one option to incorporate all of these variables into a GIS-based package that has the advantage of being easier to run than many of the more complex models. This approach can yield a product that is of value for planning decisions regarding appropriate citing of living shoreline projects. However, WEMo does not calculate wave refraction and currently can only be operated in ArcGIS 9.2.

11.5 SUMMARY AND RECOMMENDATIONS

The use of living shorelines, or fringing salt marshes with or without low oyster or rock sills, for erosion control is preferable to traditional hardening practices because they maintain the ecosystem services provided by vegetated shorelines. They can also be more cost-effective and resilient to storms and impacts of SLR. Recognition of the benefits of living shorelines has led many states to declare a preference for this approach to erosion control in low-energy settings. The challenge for shoreline property owners is to determine whether the wave energy characteristic of their shoreline will support a living shoreline approach, and if so, what design options can be considered. To date, these decisions are most commonly based on the fetch to which a shoreline is exposed. Our review of the existing literature and our results from the North Carolina case study suggest that while a complex suite of factors combine to control marsh shoreline erosion rate, measures of incident wave energy provide a useful guideline on which to base site suitability. However, we also find that existing recommendations based on fetch alone may be too conservative. For our analysis in the NCSS, we calculated shoreline wave energy values with WEMo, an ArcGIS-based wave energy model, using the top 20% wind events (RWE_{20}). We found that natural fringing marsh average widths exceeded 100 m in shorelines experiencing RWE_{20} values up to 250 J m^{-1}, which correlates to wave power values of 170 J m^{-1} s^{-1}. Our case study results demonstrated that Natural and Sill marshes maintained or increased *S. alterniflora* stem density at the lower marsh edge in wind wave settings experiencing RWE_{20} values up to 677 J^{-1}. These RWE_{20} values may be found on shorelines with maximum fetches of 5 km or more. Modeling efforts like the one presented here could be undertaken at the estuary-wide scale to provide homeowners with improved guidance on living shoreline implementation. We identified several conditions, including orientation of the site relative to prevailing winds, nearshore bathymetry, and proximity to navigation channels, which, in addition to fetch, should be factored into decisions concerning the placement of living shorelines.

It is important to note that, although we observed a stable marsh edge over a 6-year period in our case study of natural and sill marshes, most natural marsh shorelines exhibit erosion over decades (Currin et al. 2015; Leonardi et al. 2016). Long-term monitoring of living shoreline projects, including the quantitative assessment of plant density, surface elevation change, and horizontal erosion, is crucial to providing the best possible guidance for the utilization of living shorelines and other nature-based approaches to shoreline protection. This is particularly crucial given projections of accelerated RSLR, which may result in increased marsh erosion. Based on our literature review, case study results, and experience in implementing living shorelines, we present the following summary recommendations for assessment of the physical setting appropriate for living shorelines, and research results to guide site-specific design:

- Visual examination and elevation surveys of adjacent healthy, vegetated shorelines provide useful initial guidance for living shoreline design, as existing marsh shorelines integrate the numerous site-specific factors controlling marsh vegetation.
- Fetch from the prevailing wind direction(s) and nearshore bathymetry should be considered when using fetch alone as a guideline. Shallow nearshore depths (e.g., 0.25 m) will reduce the development of wind waves, and larger fetches may be considered in these conditions. In contrast, deep nearshore bathymetry will result in relatively greater wave energy affecting the shoreline.
- Shoreline stabilization projects utilizing fringing salt marshes should be considered for shorelines experiencing an average wave power of <200 J m^{-1} s^{-1} or a representative wave energy (RWE_{20}) of <300 J m^{-1}. Projects incorporating marshes with stone sills should be considered for shorelines experiencing an RWE_{20} of up to 700 J m^{-1} or an average wave power up to 450 J m^{-1} s^{-1}. Depending on bathymetry and exposure to prevailing winds, a living shoreline design may be appropriate for shorelines exposed to fetches of up to 12 km (~7 miles).
- Stone sills increase sediment accretion in fringing salt marshes. This increases resilience to SLR, but may result in the loss of low marsh habitat, and the introduction of hardened substrate to the marsh edge may increase erosion on adjacent properties. Sill height should be maintained below mean high water level in most settings. This increases the introduction of resuspended sediment from tidal waters to sill marshes, while reducing the potential for sill marshes to accrete sediment above the optimal elevation for marsh habitat. Additionally, taller sills require greater base areas for support; thus, reducing sill height also reduces the amount of shallow subtidal habitat affected by nonnative structural material. Low sills may also be colonized by native shellfish populations.
- Fringing or offshore oyster reefs add habitat complexity and some erosion protection to a living shoreline project. They should be considered as alternatives to rock sills in some settings. The presence of live oyster reefs along nearby shorelines is a useful indicator of the suitability of site conditions for oyster survival.
- In many settings, a 30-m-wide (98-ft-wide) fringing *S. alterniflora* marsh has been shown to reduce wave energy by 90% or more. Wave energy attenuation is positively correlated with stem density. Marsh restoration or creation associated with living shorelines should utilize denser plantings than might be recommended for restoration projects in non-shoreline settings.
- Boat wakes can be a greater source of wave energy than wind in many estuarine settings. Proximity to navigation channels is an important consideration when designing a living shoreline.

ACKNOWLEDGMENTS

We thank L. Cowart, P. Delano, M. Greene, and A. Hilting for excellent technical support over the course of this study. We appreciate the field support provided by staff from the North Carolina National Estuarine Research Reserve. We thank the NOAA Restoration Center and the NOAA National Centers for Coastal Ocean Science for funding support. The scientific results and conclusions, as well as any views or opinions expressed herein, are those of the authors and do not necessarily reflect the views of NOAA or the Department of Commerce. Additional support for this research was provided by the Defense Coastal/Estuarine Research Program (DCERP), funded by the Strategic Environmental Research and Development Program. Views, opinions, and findings contained in the report are those of the authors and should not be construed as an official US Department of Defense position.

REFERENCES

Allen, J.R.L. 1989. Evolution of salt marsh cliffs in muddy and sandy systems: A qualitative comparison of British west coast estuaries. *Earth Surface Processes and Landforms* 14: 85–92.

Baustian, J.J. and I.A. Mendelssohn. 2015. Hurricane-induced sedimentation improves marsh resilience and vegetation vigor under high rates of relative sea level rise. *Wetlands* 35: 795–802.

Beck, M.W., R.D. Brumbaugh, L. Airoldi et al. 2011. Oyster reefs at risk and recommendations for conservation, restoration and management. *BioScience* 61: 107–116.

Biber, P.D., C.L. Gallegos, and W.J. Kenworthy. 2008. Calibration of a bio-optical model in the North River, North Carolina (Albemarle-Pamlico Sound): A tool to evaluate water quality impacts on seagrasses. *Estuaries and Coasts* 31: 177–191.

Bilkovic, D.M. and M.M. Mitchell. 2013. Ecological tradeoffs of stabilized salt marshes as a shoreline protection strategy: Effects of artificial structures on macrobenthic assemblages. *Ecological Engineering* 61: 469–481.

Booij, N., R.C. Ris, and L.H. Holthuijsen. 1999. A third-generation wave model for coastal regions 1. Model description and validation. *Journal of Geophysical Research* 104: 7649–7666.

Cahoon, D.R., J.C. Lynch, P. Hensel et al. 2002. High precision measurements of wetland sediment elevation: I. Recent improvements to the Sedimentation-Erosion Table. *Journal of Sedimentary Research* 72: 730–733.

Cheong, S., B. Silliman, P. Wong, B. van Wesenbeeck, C. Kim, and G. Guannel. 2013. Coastal adaptation with ecological engineering. *Nature Climate Change* 3: 787–791.

Coen, L.D., R.D. Brumbaugh, D. Bushek et al. 2007. Ecosystem services related to oyster restoration. *Marine Ecology Progress Series* 341: 303–307.

Cowart, L., D.R. Corbett, and J.P. Walsh. 2011. Shoreline change along sheltered coastlines: Insights from the Neuse River Estuary, NC, USA. *Remote Sensing* 3: 1516–1534.

Cowart, L., J.P. Walsh, and D.R. Corbett. 2010. Analyzing estuarine shoreline change: A case study of Cedar Island, North Carolina. *Journal of Coastal Research* 26: 817–830.

Craft, C., S. Broome, and C. Campbell. 2002. Fifteen years of vegetation and soil development after brackish water marsh creation. *Restoration Ecology* 10: 248–258.

Currin, C. A., W.S. Chappell, and A. Deaton. 2010. Developing alternative shoreline armoring strategies: The living shoreline approach in North Carolina, in Shipman, H., Dethier, M.N., Gelfenbaum, G., Fresh, K.L., and Dinicola, R.S., eds., *Puget Sound Shorelines and the Impacts of Armoring—Proceeding of a States of the Science Workshop*, May 2009: U.S. Geological Survey Scientific Investigations Report 2010-5254, pp. 91–102.

Currin, C., J. Davis, L. Cowart Baron, A. Malhotra, and M. Fonseca. 2015. Shoreline Change the New River Estuary, North Carolina: Rates and Consequences. *Journal of Coastal Research* 31: 1069–1077.

Currin, C.A., P.C. Delano, and L.M. Valdes-Weaver. 2008. Utilization of a citizen monitoring protocol to assess the structure and function of natural and stabilized fringing salt marshes in North Carolina. *Wetlands Ecology and Management* 16: 97–118. DOI: 10.1007/s11273-007-9059.

Dahl, T.E. and S.M. Stedman. 2013. Status and trends of wetlands in the coastal watersheds of the Conterminous United States 2004 to 2009. U.S. Department of the Interior, Fish and Wildlife Service and National Oceanic and Atmospheric Administration, National Marine Fisheries Service. (46 pp.).

Elzinga, C.L., D.W. Salzer, and J.W. Willoughby. 1998. Measuring and monitoring plant populations. Technical Reference 1730-1. United States Department of the Interior, Bureau of Land Management, National Business Center, Denver.

Fagherazzi, S., M. Kirwan, S. Mudd, G.R. Guntenspergen, S. Temmerman, A. d'Alpaos, J. van de Koppel, J.M. Rybczyk, E. Reyes, C. Craft, and J. Clough. 2012. Numerical models of salt marsh evolution: Ecological, geomorphic and climatic factors. *Reviews of Geophysics* 50, RG1002, DOI: 10.1029/2011RG000359.

Fagherazzi, S., G. Mariotti, P.L. Wiber, and K.J. McGlathery. 2013. Marsh collapse does not require sea level rise. *Oceanography* 26: 70–77.

Feagin, R.A., S.M. Lozada-Bernard, T.M. Ravens, I. Möller, K.M. Yeager, and A.H. 2009. Does vegetation prevent wave erosion of salt marsh edges? *Proceedings of the National Academy of Sciences* 106: 10109–10113.

Friedrichs, C.T. and J.E. Perry. 2001. Tidal salt marsh morphodynamics: A synthesis. *Journal of Coastal Research* 27: 7–37.

Gedan, K.B., M.L. Kirwan, E. Wolanski, E.R. Barbier, and B.R. Silliman. 2011. The present and future role of coastal wetland vegetation in protecting shorelines: Answering recent challenges to the paradigm. *Climatic Change*. DOI: 10.1007/s10584-010-0003-7.

Gittman, R.K., F.J. Fodrie, A.M. Popowich et al. 2015. Engineering away our natural defenses: An analysis of shoreline hardening in the US. *Frontiers in Ecology and Environment* 13: 301–307.

Gittman, R.K., A.M. Popowich, J.F. Bruno, and C.H. Peterson. 2014. Marshes with and without sills protect estuarine shorelines from erosion better than bulkheads during a Category 1 hurricane. *Ocean & Coastal Management* 102: 94–102.

Gittman, R.K., C.H. Peterson, C.A. Currin, F.J. Fodrie, M.F. Piehler, and J.F. Bruno. 2016. Living shorelines can enhance the nursery role of threatened coastal habitats. *Ecological Applications*. DOI: 10.1890/14-0716.1.

Gleason, M.L., D.A. Elmer, N.C. Pien, and J.S. Fisher. 1979. Effects of stem density upon sediment retention by salt marsh cord grass, *Spartina alterniflora* Loisel. *Estuaries* 2: 271–273.

Grabowski, J.H. and C.H. Peterson. 2007. Restoring oyster reefs to recover ecosystem services, in Cuddington, K., Byers, J.E., Wilson, W.G., and Hastings, A., eds., *Ecosystem Engineers*. Elsevier Academic Press, New York, pp. 281–297.

Gulf Alliance Training Program. 2010. Living shoreline design methodologies. Available: http://www.gulfal liancetraining.com/dbfiles/Living%20Shoreline%20Design%20Methodology.pdf.

Hardaway, C.S. Jr. and R.J. Byrne. 1999. Shoreline Management in Chesapeake Bay. Report. College of William and Mary, Virginia Institute of Marine Science, Gloucester Point, VA, 12 pp.

Hardaway, C.S. Jr., D.A. Milligan, and K. Duhring. 2010. Living Shoreline Design Guidelines for Shore Protection in Virginia's Estuarine Environments. Special Report in Applied Marine Science and Ocean Engineering # 421, Virginia Institute of Marine Science, Gloucester Point, VA.

Johannessen, J., A. MacLennan, A. Blue et al. 2014. Marine shoreline design guidelines. Washington Department of Fish and Wildlife, Olympia, Washington.

Keddy, P.A. 1982. Quantifying within-lake gradients of wave energy: Interrelationships of wave energy, substrate particle size and shoreline plants in Axe Lake, Ontario. *Aquatic Botany* 14: 41–58.

Kelley, S.W., J.S. Ramsey, and M.R. Byrnes. 2001. Numerical Modeling Evaluation of the Cumulative Physical Effects of Offshore Sand Dredging for Beach Nourishment. U.S. Department of the Interior, Minerals Management Service, International Activities and Marine Minerals Division (INTERMAR), Herndon, VA. OCS Report MMS 2001-098, 95 pp. + 106 pp. appendices.

Kirwan, M.L., G.R. Guntenspergen, A. D'Alpaos, J.T. Morris, S.M. Mudd, and S. Temmerman. 2010. Limits on the adaptability of coastal marshes to rising sea level. *Geophysical Research Letters*. DOI: 10.1029/2010GL045489.

Knutson, P.L., J.C. Ford, and M.R. Inskeep. 1981. National survey of planted salt marshes. *Wetlands* 1: 129–157.

Knutson, P.L., R.A. Brochu, W.N. Seelig, and M. Inskeep. 1982. Wave damping in *Spartina alterniflora* marshes. *Wetlands* 2: 87–104.

Leonard, L.A. and A.C. Croft. 2006. The effect of standing biomass on flow velocity and turbulence in *Spartina alterniflora* canopies. *Estuarine Coastal and Shelf Science* 69: 325–336.

Leonard, L.A. and M.E. Luther. 1995. Flow hydrodynamics in tidal marsh canopies. *Limnology and Oceanography* 40: 1474–1484.

Leonardi, N., N.K. Ganju, and S. Fagherazzi. 2016. A linear relationship between wave power and erosion determines salt-marsh resilience to violent storms and hurricanes. *Proceedings of the National Academy of Sciences* 2016: 64–68. DOI: 10.1073/pnas.1510095112.

Malhotra, A. and M.S. Fonseca. 2007. WEMo (Wave Exposure Model): Formulation, Procedures and Validation. Beaufort, North Carolina: NOAA Technical Memorandum NOS NCCOS 65, 28 pp.

Marani, M., A. D' Alpaos, S. Lanzoni, and M. Santalucia. 2011. Understanding and predicting wave erosion of marsh edges. *Geophysical Research Letters*. DOI: 10.1029/2011GL048995.

Mariotti, G. and J. Carr. 2014. Dual role of salt marsh retreat: Long-term loss and short-term resilience. *Water Resources Research* 50: 2963–2974.

Marriotti, G. and S. Fagherazzi. 2013. Critical width of tidal flats triggers marsh collapse in the absence of sea level rise. *Proceedings of National Academy of Sciences* 110: 5353-5356. DOI:10.1073/pnas .1219600110/-/DCSupplemental.

Mattheus, C.R., A. Rodriguez, B. Mckee, and C.A. Currin. 2010. Impact of land-use change and hard structures on the evolution of fringing marsh shorelines. *Estuarine, Coastal and Shelf Science* 88: 365–376.

Meyer, D.L., E.C. Townsend, and G.W. Thayer. 1997. Stabilization and erosion control value of oyster cultch for intertidal marsh. *Restoration Ecology* 5: 93–99.

Miller, J.K., A. Rella, A. Williams, and E. Sproule. 2015. Living shorelines engineering guidelines. New Jersey Department of Environmental Protection SIT-DL-14-9-2942.

Möller, I. 2006. Quantifying saltmarsh vegetation and its effect on wave height dissipation: Results from a UK east coast saltmarsh. *Estuarine, Coastal and Shelf Science* 69: 337–351.

Möller, I., M. Kudella, F. Rupprecht et al. 2014. Wave attenuation over coastal salt marshes under storm surge conditions. *Nature Geoscience*. DOI: 10.1038/NGEO2251.

Moody, R.M., J. Cebrian, S.M. Kerner, K.L. Heck Jr., S.P. Powers, and C. Ferraro. 2013. Effects of shoreline erosion on salt-marsh floral zonation. *Marine Ecology Progress Series* 488: 145–155.

Morris, J.T., P.V. Sundareshwar, C.T. Nietch, B. Kjerfve, and D.R. Cahoon. 2002. Responses of coastal wetlands to rising sea level. *Ecology* 83: 2869–2877.

Mudd, S.M., A. D'Alpaos, and J.T. Morris. 2010. How does vegetation affect sedimentation on tidal marshes? Investigating particle capture and hydrodynamic controls on biologically mediated sedimentation. *Journal of Geophysical Research* 115: F03029. DOI: 10.1029/2009JF001566.

National Oceanic and Atmospheric Administration (NOAA). 2013. National Coastal Population Report: Population Trends from 1970–2020. http://stateofthecoast.noaa.gov/coastal-population-report.pdf.

NOAA Office of Habitat Conservation. 2015. Guidance for considering the use of living shorelines, 35 pp. Available at http://www.habitat.noaa.gov/pdf/noaa_guidance_for_considering_the_use_of_living_shore lines_2015.pdf.

North Carolina Division of Coastal Management. 2011. Weighing your options: How to protect your property from shoreline erosion. Available: http://portal.ncdenr.org/web/apnep/planningresources.

Paramor, O.A.L. and R.G. Hughes. 2004. The effect of bioturbation and herbivory by the polychaete *Nereis diversicolor* on loss of saltmarsh in south-east England. *Journal of Applied Ecology* 41: 449–463.

Partnership for the Delaware Estuary. 2012. PDE report no. 12-04. Delaware Living Shoreline Possibilities. Available: http://delawareestuary.org/delaware-estuary-living-shoreline-initiative-data-products-and-reports.

Piazza, B.P., P.D. Banks, and M.K. La Peyre. 2005. The potential for created oyster shell reefs as a sustainable shoreline protection strategy in Louisiana. *Restoration Ecology* 13: 499–506.

Priestas, A.M., G. Mariotti, N. Leonardi, and S. Fagherazzi. 2016. Coupled wave energy and erosion dyanamics along a salt marsh boundary, Hog Island Bay, Virginia, USA. *Journal of Marine Science and Engineering*, 3, 1041-1-65. DOI:10.3390/jmse3031041.

Ravens, T.M., R.C. Thomas, K.A. Roberts, and P.H. Santschi. 2009. Causes of salt marsh erosion in Galveston Bay, Texas. *Journal of Coastal Research* 25(2): 265–272.

Reed, D.J. 1989. Patterns of sediment deposition in subsiding coastal salt marshes, Terrebonne bay, Louisiana: The role of winter storms. *Estuaries* 12: 222–227.

Rodriguez, A.B., F.J. Fodrie, J.T. Ridge et al. 2014. Oyster reefs can outpace sea-level rise. *Nature Climate Change*. DOI: 10.1038/NCLIMATE2216.

Roland, R.M. and S.L. Douglas. 2005. Estimating wave tolerance of *Spartina alterniflora* in coastal Alabama. *Journal of Coastal Research* 21(3): 453–463.

Schuerch, M., J. Rapaglia, V. Liebetrau, A. Vafeidis, and K. Reise. 2012. Salt marsh accretion and storm tide variation: An example from a barrier island in the North Sea. *Estuaries and Coasts* 35: 486–500.

Schwimmer, R.A. 2001. Rates and processes of marsh shoreline erosion in Rehoboth Bay, Delaware, USA. *Journal of Coastal Research* 17(3): 672–683.

Scyphers, S.B., S.P. Powers, K.L. Heck, Jr., and D. Bryon. 2011. Oyster reefs and natural breakwaters mitigate shoreline loss and facilitate fisheries. *PLoS ONE*. DOI: 10.1371/journal.pone.0022396.

Shafer, D.J., R. Roland, and S.L. Douglass. 2003. Preliminary evaluation of critical wave energy thresholds at natural and created coastal wetlands. WRP Technical Notes Collection ERDC-TN-WRP-HS-CP-2.2, U.S. Army Engineering Research and Development Center, Vicksburg, MS.

Shafer, D.J. and W.J. Streever. 2000. A comparison of 28 natural and dredged material salt marshes in Texas with an emphasis on geomorphological variables. *Wetlands Ecology and Management* 8: 353–366.

Shepard, C.C, C.M. Crain, and M.W. Beck. 2011. The protective role of coastal marshes: A systematic review and meta-analysis. *PLoS ONE* 6: e27374. DOI: 10.1371/journal.pone.0027374.

Silliman, B.R., E. Schrack, Q. He, R. Cope, A. Santoni, T. van der Heide, R. Jacobi, M. Jacobi, and J. van de Koppel. 2015. Facilitation shifts paradigms and can amplify coastal restoration efforts. *Proceedings of the National Academy of Sciences* 112: 14295–14300.

Sutton-Grier, A.E., K. Wowk, and H. Bamford. 2015. Future of our coasts: The potential for natural and hybrid infrastructure to enhance the resilience of our coastal communities, economies and ecosystems. *Environmental Science and Policy* 51: 137–148.

Talley, T.S., J.A. Crooks, and L.A. Levin. 2001. Habitat utilization and alteration by the invasive burrowing isopod, *Sphaeroma quoyanum*, in California marshes. *Marine Biology* 138: 561–573.

Temmerman, S., T.J. Bouma, G. Govers, Z.B. Wang, M.B. De Vries, and P.M.J. Herman. 2005. Impact of vegetation on flow routing and sedimentation patterns: Three-dimensional modeling for a tidal marsh. *Journal of Geophysical Research*. DOI: 10.1029/2005JF000301.

Temmerman, S., P. Meire, T.J. Bouma, P.M.J. Herman, T. Ysebaert, and H.J. De Vriend. 2013. Ecosystem-based coastal defence in the face of global change. *Nature* 504: 79–83.

Theuerkauf, E.J., J.D. Stephens, J.T. Ridge, F.J. Fodrie, and A.B. Rodriguez. 2015. Carbon export from fringing saltmarsh erosion overwhelms carbon storage across a critical width threshold. *Estuarine Coastal Shelf Science*. DOI: 10.1016/j.ecss.2015.08.001.

Tonelli, M., S. Fagherazzi, and M. Petti. 2010. Modeling wave impact on salt marsh boundaries. *Journal of Geophysical Research*. DOI: 10.1029/2009JC006026.

Tweel, A.W. and R.E. Turner. 2012. Landscape-scale analysis of wetland sediment deposition from four tropical cyclone events. *PLoS ONE*. DOI: 10.1371/journal.pone.0050528.

U.S. Coastal Engineering Research Center (USCOE). 1977. Shore Protection Manual, Vol. 1. U.S. Army Coastal Engineering Research Center, Ft. Belvoir, Virginia. 514 p.

Wamsley, T.V., M.A. Cialone, J.M. Smith, J.H. Atkinson, and J.D. Rosati. 2010. The potential of wetlands in reducing storm surge. *Ocean Engineering* 37: 59–68.

Weston, N.B. 2013. Declining sediments and rising seas: An unfortunate convergence for tidal wetlands. *Estuaries and Coasts* 37: 1–23.

Wolf, K.L., C. Ahn, and G.B. Noe. 2011. Development of soil properties and nitrogen cycling in created wetlands. *Wetlands* 31: 699–712.

Yang, S.L., B.W. Shi, T.J. Bouma, T. Ysebaert, and X.X. Luo. 2012. Wave attenuation at a salt marsh margin: A case study of an exposed coast on the Yangtze Estuary. *Estuaries and Coasts* 35: 169–182.

CHAPTER 12

Lessons Learned from Living Shoreline Stabilization in Popular Tourist Areas
Boat Wakes, Volunteer Support, and Protecting Historic Structures

Linda Walters, Melinda Donnelly, Paul Sacks, and Donna Campbell

CONTENTS

12.1 INTRODUCTION

Estuaries support large coastal communities, and development has dramatically increased over the past century along these shorelines and waterbodies, threatening natural habitats. Coastal counties, in fact, occupy only 17% of the land area in the continental United States, yet contain 53% of the US population (US Census Bureau 2012). In response to this, many of these waterways are now protected to some extent through a range of regulations and designations, such as national parks, national wildlife refuges, estuaries of national significance, national estuarine research reserves, state parks, and state-protected waters. While such protection is economically and ecologically beneficial to communities, it can also greatly increase recreational use, especially the number of recreational boat users and their unintended impacts on the local ecosystems.

Anthropogenic damage to estuaries has resulted in losses of every major estuarine habitat type. Water quality reduction and loss of coastal wetlands to dredge-and-fill practices are well documented. Likewise, hard-armoring of shorelines via the construction of sea walls, breakwaters, or placement of riprap decreases connectivity and breaks biological and physical links between land–water boundaries by creating artificial transitions from marine to terrestrial habitats (e.g., Pilkey and Wright 1988). While hard-armored shorelines may reduce upland erosion, sediment losses to beach and intertidal areas seaward of such structures are enhanced (Bozek and Burdick 2005; Kraus and McDougal 1996; Pilkey and Wright 1988). Storm surge protection is also proposed as a reason for hard-armoring shorelines; unfortunately, ongoing sea level rise and increased storm intensity limit the strength of this argument.

Although less studied, recreational boating practices also cause direct and indirect negative impacts in estuarine ecosystems (e.g., Bejder et al. 2006; Fonseca and Malhotra 2012; Schroevers et al. 2011; Wasson et al. 2001; Williams et al. 2002; Zacharias and Gregr 2005). Direct physical impacts include changes in sedimentation and hydrology, boat strikes to aquatic flora and fauna, and chemical pollution associated with engine emissions, antifouling agents, and sewage dumping (Burgin and Hardiman 2011). Indirect biotic impacts include the spread of nonnative species (Carlton 2001) and interactions of boat wakes with organisms (Burgin and Hardiman 2011; Campbell 2015; Donnelly and Walters 2008; Garvis et al. 2015; Grizzle et al. 2002; Wall et al. 2005).

Wakes generated from boats have been shown to be detrimental to many diverse organisms (Bickel et al. 2011; Bishop 2005, 2008; Gabel et al. 2012; Lorenz et al. 2013). For example, Donnelly and Walters (2008) documented that boat wakes can disperse seeds of the nonnative Brazilian pepper tree (*Schinus terebinthifolius*) high enough in the intertidal zone to facilitate recruitment. Once present, the plants were able to outcompete native flora, especially native mangroves. Campbell (2015) found that boat wakes eroded intertidal oyster clusters around their bases, thereby facilitating cluster dislodgment with subsequent wakes. Water volumes filtered by mussels were decreased with increasing shear stress from boat wakes (Lorenz et al. 2013). There is also evidence that planktonic copepods had higher mortality rates in turbulent waters as a result of boat wakes, which influenced bottom-up trophic interactions in high boating activity areas (Bickel et al. 2011). Wakes also dislodge and displace epifauna and macroinvertebrates from shorelines, resulting in potential changes in faunal assemblages with increased boating pressure (Bishop 2008; Demes et al. 2012; Gabel et al. 2012). This could ultimately influence the ability of a shoreline community to act as a nursery ground for fisheries (Bishop 2008).

Florida has the most registered boats of any state (2012: 870,031), encompassing 7.1% of all registered boats in the United States (US Coast Guard 2013). Additionally, boating in Florida is becoming increasingly popular, which is exemplified by a 73% rise in recreational boat registrations between 1985 and 2005 on the east coast of central Florida in counties bordering the Indian River Lagoon (IRL) (Sidman et al. 2007). Campbell (2015) conducted a yearlong survey of boating demographics in the northern IRL, while Bowerman and DeLorme (2014) conducted quantitative and qualitative surveys to better understand boater values and needs in these same waters with a goal to understand the strategies needed to make people more ecologically responsible boaters. Campbell determined that flats and v-hull boats were the primary vessels used (more than 63% of boats), signifying the importance of recreational fishing in the area. The average boat speed was 24.9 kph, and for those that motored past predetermined shoreline sites, 75.2% were passing within a few meters of shorelines that already had obvious shoreline erosion. The remainder of the boaters went past control shoreline sites with no or very limited erosion. Bowerman and DeLorme (2014) conducted six focus groups (total participants = 60) and 404 phone surveys of boaters from within the central Florida population who recreate in the IRL. They found that local residents overwhelmingly rank protecting the IRL for future generations and future high-quality angling as high priorities, but were only aware of boaters being responsible for estuarine damage in the form of propeller scars in seagrass beds and by boaters dumping trash, monofilament, or bait in the water. Many individuals in these focus groups stated they felt inconvenienced by speed limits or wake regulations while boating on the lagoon. In areas of Florida where boat speed restrictions are in place to protect the West Indian manatee, many boaters admitted that they regularly do not observe the speed limits and would like to see these regulations repealed (L. Walters, personal observation).

Many estuarine intertidal ecosystems, including oyster reefs and coastal wetlands, form along low-energy shorelines. Increases in boating activity can result in higher wave energies, thus changing the basic characteristics of intertidal habitats and limiting the survival and recruitment of oysters and shoreline plants adjacent to significant boating channels. Restoration practices in high boating areas need to address this issue in order for the stabilization to be successful.

12.2 THE IRL AND CANAVERAL NATIONAL SEASHORE

The IRL, along the east coast of Florida, is one of 28 estuaries in the US Environmental Protection Agency's National Estuary Program (NEP) and was designated as an "Estuary of National Significance" in 1990. This designation indicates that the IRL's waters, natural ecosystems, and economic activities are critical to the environmental health and economic well-being of the United States (US Environmental Protection Agency 2014). The IRL has been valued at approximately $3.7 billion annually and supports 15,000 jobs (St. Johns River Water Management District [SJRWMD] 2014a). The IRL has also been recognized as one of the most biologically diverse estuaries in the United States, primarily as a result of its overlap between temperate and subtropical climatic zones (IRLNEP 2008).

Mosquito Lagoon, the northernmost region of the IRL, is one of the world's most popular fishing and boating locations ("Redfish Capitol of the World") and thus is essential for the local economy (Figure 12.1). This results in intensive boating traffic in this shallow-water, microtidal estuary (Scheidt and Gareau 2007). More than 40 recreational boats per hour regularly pass a few meters from shorelines and intertidal patch oyster reefs (Walters et al. 2007). Mosquito Lagoon has an average depth of less than 1.5 m and salinity between 25 and 45 ppt (Walters et al. 2001). Its waters are primarily dominated by wind-driven currents (Smith 1987, 1993). Mosquito Lagoon is composed of three important habitats: (1) seagrasses (primarily *Halodule wrightii*), (2) salt marshes, composed of both mangroves and marsh cordgrass (*Rhizophora mangle*, *Avicennia germinans*, *Laguncularia racemosa*, and *Spartina alterniflora*), and (3) intertidal oyster reefs (*Crassostrea virginica*). Each of these habitat types has been experiencing declines worldwide as a result of various stressors (Beck et al. 2011; Fletcher and Fletcher 1995; Garvis et al. 2015; Grizzle et al. 2002; Valiela et al. 2001; Waycott et al. 2009). Seagrass beds currently experience an average global decline of 1.5% annually, with a global coverage loss of 29% since 1879 (Waycott et al. 2009). Global declines of mangroves are at 35% (Valiela et al. 2001), and those of shellfish reefs are at 85% (Beck et al. 2011). In the IRL, the declines are even worse. Approximately 75% of saltmarsh habitat, including mangroves, was lost between the 1950s and the 1970s, primarily attributed to mosquito control impoundments (SJRWMD 2014b). Although seagrass abundances are highly variable, 11% of IRL seagrasses were lost from the 1970s to 1992 (Fletcher and Fletcher 1995) and approximately 60% of IRL seagrasses were then lost from 2009 to 2012 due primarily to abiotic factors, algal blooms, and decreases in water quality (SJRWMD 2014a). There has been a 24% loss (15 ha) of intertidal oyster habitat in Mosquito Lagoon since 1943, where the natural reefs were replaced by dead oyster reefs or dead seaward edges of otherwise live oyster reefs (i.e., dead margins) (Garvis et al. 2015).

Canaveral National Seashore (CANA), part of the United States National Park System, occupies approximately 230 km² of Mosquito Lagoon (25.7 km length × 4.3 km maximum width; Hellmann 2013). This equates to approximately 40% of the Park's acreage. CANA was established in 1975 to "preserve and protect the outstanding natural, scenic, scientific, ecologic and historic values of certain lands, shoreline, and waters of the State of Florida, and to provide for public outdoor recreation use and enjoyment of the same" (National Park Service 1975). Estuarine shorelines in CANA are composed primarily of unconsolidated shell, clay, and sand (Hellmann 2013). Housed adjacent to these shorelines are many of CANA's most important historical resources—numerous significant prehistoric shell middens and wooden built structures. Middens in CANA were constructed by Timucuan and Ais Native Americans starting approximately 500 BC (Hellmann 2013).

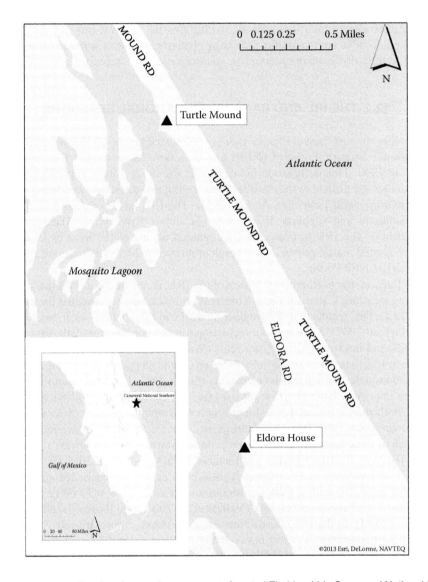

Figure 12.1 Map of stabilization sites on the east coast of central Florida within Canaveral National Seashore.

The community of Eldora sprang up along the eastern shore of Mosquito Lagoon in the 1870s (Hellmann 2013) (Figure 12.1). The original homesteaders and pioneers grew citrus and other crops, which they shipped on steamboats that regularly stopped at Eldora before railroad track completion in 1898. The community swelled to a few hundred individuals, but waned with the combination of winter crop freezes and reduced steamship traffic. The community was replaced by "winter retreats" and small fish camps, with only one original building still standing—the Moulton-Wells House (a.k.a. Eldora State House).

12.3 SEA LEVEL RISE AND LIVING SHORELINES IN CANA

Shoreline hardening was identified as a major threat to marine and estuarine habitats in the Florida Fish and Wildlife Conservation Commission's Comprehensive Wildlife Conservation

Strategy (2012). One alternative to shoreline hardening is living shoreline restoration/stabilization whereby humans deploy appropriate flora and fauna (or substrate for faunal recruitment) to mimic local, natural shoreline communities in areas where losses have occurred through natural (e.g., storms) or anthropogenic (e.g., boat wakes or strikes, trampling) means. Living shorelines are becoming more accepted throughout the United States as cost-effective, long-term alternatives to hard-armored shorelines, especially with expected climate change. For long-term sustainability, conserving and restoring living shorelines is one of the only methods that will be able to both adjust to future climate conditions and preserve essential ecological functions (Borsje et al. 2011; Erwin 2009). Sea level has risen at an average rate of 3.1 mm/year since 1993, and sea levels are predicted to increase over the next century (IPCC 2007). Erosion of shorelines and loss of shoreline habitat are expected to worsen with continued sea level rise and increased extreme weather events in the future (IPCC 2007) and the revised Florida State Wildlife Action Plan (FFWCC 2012) includes climate change planning to address this concern.

Sea level rise in CANA was calculated to be 2.34 mm/year (NPS 2014). Shoreline vegetation can help mediate the effects of sea level rise by trapping sediments, causing accretion to maintain desired elevations with moderate levels of sea level rise and retreat landward if unobstructed (Morris et al. 2002) and retaining shoreline structure and functions as the environment changes. The presence of adjacent oyster beds and emergent flora produces a synergistic effect for buffering waves, providing multiple defenses for erosion (Coen et al. 2007; Manis et al. 2014; Meyer et al. 1997). Manis et al. (2014) documented the dramatic increase in wave attenuation that occurs with living shoreline materials (marshgrass *S. alterniflora* and stabilized oyster shell) between newly deployed and 1-year postdeployment. Using a wave tank, they documented that control (no plants, no shells) trials reduced wave energy by 1%, newly deployed *S. alterniflora* plugs + stabilized shell reduced wave energy by 19%, while 1-year established *S. alterniflora* and live oysters that recruited to the deployed shell reduced the wave energy by 67%.

As mentioned previously, CANA is located within a transition zone between subtropical and temperate climates. For living shorelines, that enables us to use the eastern oyster *C. virginica* (temperate species), smooth cordgrass *S. alterniflora* (temperate species), and some combination of three species of mangroves (subtropical). Combined, this allows us to incorporate a three-tiered strategy to protect CANA shorelines, whereas most stabilization efforts are able to include only one or two of these taxa. These engineered ecosystems, in CANA, in turn then support federally listed species including wood storks and Atlantic salt marsh snakes, and species of special concern, including the American oyster catcher, brown and white pelicans, and numerous wading and shorebirds. Research in Mosquito Lagoon has documented 24 species of wading birds and 149 species of marine flora and fauna on oyster reefs (Barber et al. 2010; Boudreaux et al. 2006; L. Walters, personal communication) and 56 species of birds, fishes, and invertebrates using mixed mangrove and *S. alterniflora* shoreline habitat (M. Donnelly, personal observation). Mosquito Lagoon waters are listed as Fish Habitat Areas of Particular Concern and Essential Fish Habitat by NOAA for snapper, grouper, and bull sharks that come to these shallow, protected waters to forage and give birth.

12.4 GENERAL METHODS

Since 2011, we have deployed stabilized oyster shell and native plants grown from local vegetation sources to preserve local genetic diversity and adaptations at six living shoreline stabilization sites in Mosquito Lagoon and one site in St. Augustine, Florida. Our general site design is as follows: (1) upper intertidal zone: *R. mangle*, *L. racemosa*, and *A. germinans* seedlings planted 2 plants/m; (2) mid-intertidal zone: *S. alterniflora* transplant units (plugs) planted 3 plants/m; and (3) lower intertidal zone: placement of 0.25-m² stabilized shell mats (a.k.a. oyster restoration mats

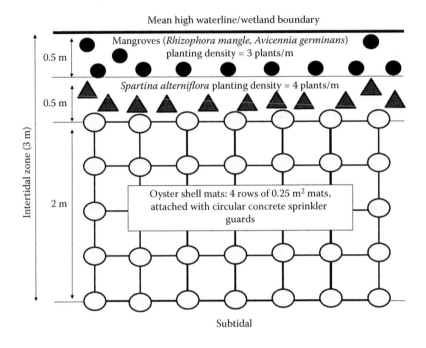

Figure 12.2   Plan view diagrams of stabilization sites in Canaveral National Seashore.

Table 12.1   Summary of Shoreline Parameters Included in Living Shoreline Monitoring

Shoreline Parameters	Monitoring Variables	Sampling Methods	Monitoring Frequency
1. Erosion	Erosion or accretion	Erosion stakes	Monthly
2. Habitat structure	Temperature	Temperature loggers	Monthly
	Soil moisture	Soil moisture meter	Quarterly
	Slope	Laser level on linear transects	Every 6 months
	Elevation	Laser level on linear transects	Every 6 months
	Substrate type	Point-intercept in 0.25-m² quadrats	Quarterly
3. Biodiversity	Vegetation	Point intercept in 0.25-m² quadrats	Quarterly
	Oysters	Total counts in 0.25-m² quadrats	Quarterly
	Fiddler crabs/burrows	Total counts in 0.25-m² quadrats	Quarterly
	Fishes/mobile inverts	Seine net along shoreline	Quarterly
	Birds	Abundance/behavior observations	Quarterly

built from Vexar extruded polyethylene mesh, cable ties, and disarticulated oyster shells drilled near the umbo; Garvis et al. 2015) or 1-m-long oyster shell bags perpendicular to shoreline (3 shell bags/m) seaward of *S. alterniflora* (Figure 12.2). The oyster treatment(s) used would depend on bathymetry, slope, and expected wake intensity. The *S. alterniflora* plugs used in the Turtle Mound case study described below spent approximately 1 month in pots with topsoil and 1 year in pots for mangroves grown from seeds. To improve plant survival rates, we later grew plants for longer durations in pots. For the Eldora State House case study, we established *S. alterniflora* plugs in pots for 6 months predeployment and grew mangroves in pots for 2–4 years, transplanting them at least twice during this time into larger containers.

Retention of deployed materials was monitored weekly for 1 month and then monthly for a minimum of 1 year. Direct measures of plant success (stem growth, number of leaves, and number of flowers/seeds) were recorded after 1, 6, and 12 months. Oyster recruitment was measured on randomly selected mats at 6 and 12 months.

A BACIPS (Before-After-Control-Impact Paired Series) experimental design was used for the abiotic variables (Table 12.1; Underwood 1991). Specifically, we measured the rate of erosion/accretion, characterized habitat structure (slope, relative elevation, temperature, and soil moisture), and documented diversity and abundance of other shoreline plants, mobile and sessile marine invertebrates, fishes, and birds at our living shoreline sites and nearby control areas.

12.5 CASE STUDY: TURTLE MOUND HISTORIC SITE

Turtle Mound is one of Florida's largest and best known archeological sites (Figure 12.1). It was listed on the National Register of Historic Places in 1970 and is currently nominated as a National Historic Landmark. For centuries, generations of Native Americans deposited oyster shell with lesser amounts of conch and clam shells, fish bones, pottery, and ash on Turtle Mound, as well as associated burial mounds (Hellmann 2013). As such, Turtle Mound now contains 1.5 million bushels of shells and extends more than 30 ft above the coastline's flat landscape (Hellmann 2013). Oyster shells excavated from the midden have been radiocarbon dated back to 500 BC through 1565 AD (Hellmann 2013). Radiocarbon dating also suggests that the accumulation of shells is now closer to the lagoon shoreline than it was previously, probably through slumping and movements of shell units over time (Hellmann 2013). For centuries, this midden was an important navigational landmark for Spanish sailors and was included in some of the earliest European maps of Florida.

Before and after national park establishment, the State of Florida and Army Corp of Engineers attempted to stabilize Turtle Mound by placing cement bags in the intertidal zone adjacent to a concrete retaining wall along the northwest face of the midden. Since then, no marine life has attached to these cement bags other than some ephemeral macroalgae (e.g., *Ulva, Enteromorpha*) and the bags are now disintegrating into small chunks that are visible in subtidal waters throughout the area (L. Walters, personal observation). Midden shells now rely on the retaining wall for support. Both the wall and the cement bags remain in place despite their detraction from the scenic beauty of the area. The longer southwest face of Turtle Mound began to show signs of rapid erosion because of storms, high water events, and boat wakes in the late 1990s. In 2009, we received a request from CANA to help stabilize this site. In 2010, we received funding for implementation of a living shoreline demonstration project.

One concern for implementing restoration at Turtle Mound was that it had become an extremely popular fishing spot for individuals without boats and, on weekend days, more than 100 fishers per day walk along the west side of Turtle Mound, often trampling any remaining shoreline vegetation, to get to the favorite sandbar fishing site. CANA and Volusia County had previously attempted to curb erosion by sparsely planting native marsh vegetation, specifically *S. alterniflora* and seedlings of the red mangrove (*R. mangle*). This resulted in no plant survival owing to limited signage, battering by storm surges and boat wakes, complete submersion during the annual fall high water season, and trampling by fishers on these small plants.

This previous stabilization attempt showed us three things. First, before attempting any stabilization at Turtle Mound, we needed to construct a walking path to provide access to the popular fishing spot. We needed to direct visitors away from our planned living shoreline deployments without disturbing the historical resource. To do this, we cleared a path that was then bordered by a rope-and-post fence design for which the cement bases of the posts were aboveground (Figure 12.3a). Because of the historic importance of the site, we were not allowed to bury posts into the ground. The path was completed in a single day in March 2011 and we have received numerous compliments on it as many people had not purposely damaged the shoreline flora over the years, but found no alternative to get to the sandbar. Others were simply happy not to be required to get themselves or their gear wet in getting to the fishing spot. Second, signage to inform visitors of our living shoreline needed to be readily visible at the start of the walking path to engage visitors in the importance of the project. We did not anticipate the time required to obtain approvals for signage, but we did meet our funding deadlines. Third, we learned quickly that our deployed plants needed to be sufficiently tall and sturdy (i.e., woody stem) and to have sufficient root biomass to handle the worst possible lagoon conditions.

On a spring weekend in May of 2011, 620 *S. alterniflora* transplant units (1 month old), 450 *R. mangle* seedlings (1 year old), and 1140 oyster mats were deployed along 200 m of eroding shoreline at Turtle Mound (Figure 12.3b). In 2013, we constructed an 80-m hybrid living shoreline in front of the large sea wall/cement bags on the northern face of Turtle Mound with 165 *S. alterniflora* transplants, 70 mangrove seedlings (*R. mangle* and *A. germinans*), and 800 shell bags. Over the past 4 years, we have documented significant increases in percent cover of vegetation in this intertidal zone, from less than 3% before stabilization efforts to an average percent cover of 70% in the mangrove zone (upper intertidal zone) and 50% in smooth cordgrass zone (middle intertidal zone) (Figure 12.3c). Live oyster densities were less than 5 oysters per square meter before stabilization and have significantly increased to an average of 121 oysters per square meter after 4 years. The increase in plant cover combined with recruitment and growth of oysters in the lower intertidal zone has had a positive effect on sediment trapping and decreased erosion over time (Figure 12.4). Although loss of sediment occurred during the first 2 years after stabilization, especially as a result of Superstorm Sandy, significant accretion has occurred over the past 2 years at the stabilized sites as percent cover of vegetation and oyster density has increased (Figure 12.4), with accretion rates now occurring nearly five times faster than estimated rates of sea level rise in CANA (NPS 2014).

(a)

Figure 12.3 Photographs of Turtle Mound Historic Site: (a) Immediately after installation of walkway, but before stabilization.   (*Continued*)

(b)

(c)

Figure 12.3 (Continued) Photographs of Turtle Mound Historic Site: (b) Community stabilization event. (c) Four years poststabilization.

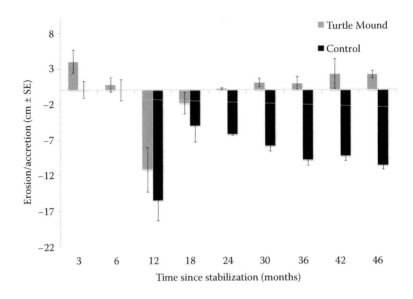

Figure 12.4 Erosion/accretion at Turtle Mound Historic Site. Local sea level rise has been predicted to be 0.23 cm/year.

12.6 CASE STUDY: ELDORA STATE HOUSE

The Moulton-Wells House, also known as the Eldora State House, was built in 1913, and is the only building from the 300+ person community of Eldora still standing (Hellmann 2013) (Figure 12.1). It was listed on the National Register in 2001 and is now used as a history museum for CANA. According to the application to the national register, the house was located 15.2 m east of the shoreline in Mosquito Lagoon on a shell midden site less than 0.4 m above sea level (Hellmann 2013). The shoreline fronting the Eldora State House has been eroding for many years; as of February 2013, the distance from the house to the same location on the shoreline was 13.0 m, a loss of 14.5% of the shoreline from the national register description. Even greater cause for concern is a saltmarsh meadow approximately 100 m south of the Eldora State House that has experienced 6 m of shoreline loss in the same time frame. The difference between the two areas is most likely the result of sediment types: primarily consolidated, crushed shell at the Eldora State House versus less consolidated, organic sediments in the salt marsh.

In early 2013, CANA requested our assistance in deploying a living shoreline at the Eldora State House to prevent any further shoreline loss near this structure. The Park placed rope and post fencing at the ecotone where the intertidal and terrestrial communities merged to lessen the likelihood of trampling. Over 2 days in May 2013, we deployed 756 oyster restoration mats, 340 *S. alterniflora* transplants, and 150 mangrove seedlings (combination of *R. mangle*, *L. racemosa*, and *A. germinans*) along 106 m of eroding shoreline in front of the Eldora State House (Figure 12.5). At this site, we used mangrove seedlings that had been grown in pots for 2 years before planting and *S. alterniflora* that had been potted for 4–6 months. Compared to the plantings at Turtle Mound in 2011, the initial mangrove seedlings were on average 0.25 meter taller and had developed woody stems before planting in the field. After two years, percent cover of vegetation at Eldora significantly increased from less than 5% cover to an average of 71% in the smooth cordgrass zone (middle intertidal zone) and 62% in the mangrove zone (upper intertidal zone). In addition to survival and growth of our planted species, increased plant cover at the Eldora State House has been facilitated by propagule

Figure 12.5   Photograph of the Eldora State House from the water, 2 years poststabilization. What was previously bare sediment in the intertidal zone seaward of the house is now covered with small mangrove trees and the marsh cordgrass *Spartina alterniflora*.

trapping of all three native mangrove species and natural recruitment of native halophytic shrubs. Oyster densities before stabilization were less than 3 oysters per square meter and have increased to an average of 20 oysters per square meter after 2 years. Our stabilization efforts at the Eldora State House had a positive effect on sediment trapping and an average of 5.4 cm of accretion was observed at this site over the past 2 years (2.7 cm annually; Figure 12.6).

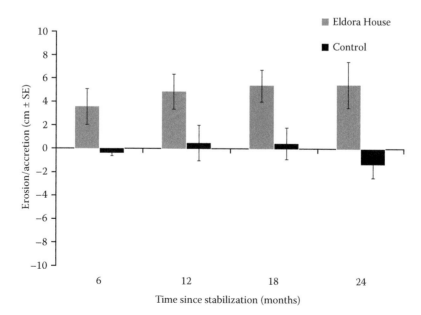

Figure 12.6   Erosion/accretion at the Eldora State House. Local sea level rise has been predicted to be 0.23 cm/year.

12.7 ONGOING ISSUES IN A POPULAR TOURIST AREA— DISTURBANCE, BOAT WAKES, AND TRAMPLING

By all biological and social metrics, our projects have been successful. However, one of the biggest barriers to long-term success remains human actions. One unexpected concern has come from large masses of seagrass wrack that accumulates some fall seasons along IRL shorelines. Although the origins of the wrack are not explicitly known (storm dislodgement vs. propeller scarring), boat wakes and storm wakes push these massive piles of seagrass (primarily *H. wrightii*) into the high intertidal zone where 0.5-m-deep masses completely cover and weigh down deployed *S. alterniflora* and mangroves. Despite our education and outreach efforts, we continue to see individuals moving deployed shell bags to create higher perches to stand on while fishing from shore, as well as placing buckets and gear directly on top of deployed plants that stand less than a meter from our signage. Some people actually pull out the deployed plants to better accommodate their gear. As scientists, we are able to adaptively manage our project objectives for climate change and even anthropogenic damage from boat wakes. What we cannot control are the actions of only a small number of individuals who can decimate years of stabilization efforts in a matter of minutes to hours.

12.8 VOLUNTEER POWER

While some "entitled" or apathetic individuals can quickly destroy a restoration or stabilization project, we rely on the many good people who volunteer to make these efforts happen. Before commencing living shoreline efforts In Mosquito Lagoon, we had been actively involved in community-based oyster reef restoration with numerous community partners since 2005. This provided us with a large pool of volunteers to ask for help when we began our living shoreline efforts, as well as practice on how to run successful community events. For our Turtle Mound project, we engaged 2970 volunteers, and our project was selected as a 2011 Toyota/Field and Stream Heroes of Conservation Project. Combined with Eldora State House shoreline and our other living shorelines, we have a grand total of more than 7250 community volunteers who have contributed more than 18,200 h to living shoreline stabilization in central Florida.

One important lesson learned was that not everyone wants or has the time or physicality to help on the limited number of actual deployment dates. We were able, however, to build a strong base of engaged citizens by providing the community with a myriad of ways to help—from growing mangroves from propagules in elementary school classroom nurseries, to drilling and bagging oyster shell, to sharing our outreach, in the form of hard-copy children's story and activity books, with their families. This has allowed us to engage members of the community from preschool age children to retirees, expanding the overall impact of these volunteer activities. The education and outreach component of our projects has increased awareness about the importance of estuarine systems as well as promoted an environmental ethic supportive of conservation of our natural systems. Although we may not be able to directly measure this change in worldview as easily as documenting increases in plant cover or changes in erosion rates, we believe that it may be one of the most important outcomes of our projects. A young girl once stood up after planting a mangrove along the shoreline and told her parents, "I just saved the planet!" The pride on her face and excitement in her voice showed the impact participation in this project had on her—because, after all, doesn't everyone secretly want to be a superhero?

REFERENCES

Barber, A., L. Walters, and A. Birch. 2010. Potential for restoring biodiversity of macroflora and macrofauna on oyster reefs in Mosquito Lagoon, Florida. *Florida Scientist* 73: 47–62.

Beck, M.W., R.D. Brumbaugh, L. Airoldi, A. Carranza, L.D. Coen, C. Crawford, O. Defeo, G.J. Edgar, B. Hancock, and M.C. Kay. 2011. Oyster reefs at risk and recommendations for conservation, restoration, and management. *Bioscience* 61: 107–116.

Bejder, L., A. Samuels, H. Whitehead, N. Gales, J. Mann, R. Connor, M. Heithaus, J. Watson-Capps, C. Flaherty, and M. Krützen. 2006. Decline in relative abundance of bottlenose dolphins exposed to long-term disturbance. *Conservation Biology* 20: 1791–1798.

Bickel, S.L., J.D. Malloy Hammond, and K.W. Tang. 2011. Boat-generated turbulence as a potential source of mortality among copepods. *Journal of Experimental Marine Biology and Ecology* 401: 105–109.

Bishop, M.J. 2005. Compensatory effects of boat wake and dredge spoil disposal on assemblages of macroinvertebrates. *Estuaries* 28: 510–518.

Bishop, M.J. 2008. Displacement of epifauna from seagrass blades by boat wake. *Journal of Experimental Marine Biology and Ecology* 354: 111–118.

Borsje, B.W., B.K. van Wesenbeeck, F. Dekker, P. Paalvast, T.J. Bouma, M.M. van Katwijk, and M.B. de Vries. 2011. How ecological engineering can serve in coastal protection. *Ecological Engineering* 37: 113–122.

Boudreaux, M.L., J.L. Stiner, and L.J. Walters. 2006. Biodiversity of sessile and motile macrofauna on intertidal oyster reefs in Mosquito Lagoon, Florida. *Journal of Shellfish Research* 25: 1079–1089.

Bowerman, K. and D.E. DeLorme. 2014. Boaters' perceptions of a mobile app for a marine conservation social marketing campaign. *Soc. Mark. Q.* 20: 47–65.

Bozek, C.M. and D.M. Burdick. 2005. Impacts of seawalls on saltmarsh plant communities in the Great Bay Estuary, New Hampshire USA. *Wetlands Ecology and Management* 13: 553–568.

Burgin, S. and N. Hardiman. 2011. The direct physical, chemical and biotic impacts on Australian coastal waters due to recreational boating. *Biodiversity and Conservation* 20: 683–701.

Campbell, D. 2015. Quantifying the effects of boat wakes on intertidal oyster reefs in a shallow estuary. M.S. thesis. Univ. of Central Florida.

Carlton, J.T. 2001. Introduced species in U.S. coastal waters: Environmental impacts and management priorities. Pew Oceans Commission, Arlington, Virginia.

Coen, L.D., R.D. Brumbaugh, D. Bushek, R. Grizzle, M.W. Luckenbach, M.H. Posey, S. Powers, and S.G. Tolley. 2007. Ecosystem services related to oyster reef restoration. *Marine Ecology-Progress Series* 341: 303–307.

Demes, K.W., R.L. Kordas, and J.P. Jorve. 2012. Ferry wakes increase seaweed richness and abundance in a sheltered rocky intertidal habitat. *Hydrobiologia* 693: 1–11.

Donnelly, M. and L. Walters. 2008. Water and boating activity as dispersal vectors for *Schinus terebinthifolius* (Brazilian Pepper) seeds in freshwater and estuarine habitats. *Estuaries and Coasts* 31: 960–968.

Erwin, K.L. 2009. Wetlands and global climate change: The role of wetland restoration in a changing world. *Wetlands Ecology and Management* 17: 71–84.

Fletcher, S.W. and W.W. Fletcher. 1995. Factors affecting changes in seagrass distribution and diversity patterns in the indian River Lagoon complex between 1940 and 1992. *Bulletin Marine Science* 57: 49–58.

Florida Fish and Wildlife Conservation Commission. 2012. Florida's Wildlife Legacy Initiative. Florida's Comprehensive Wildlife Conservation Strategy. Tallahassee, Florida, USA.

Fonseca, M.S., and A. Malhotra. 2012. Boat wakes and their influence on erosion in the Atlantic Intracoastal Waterway, North Carolina. NOAA Technical Memorandum NOS NCCOS # 143.

Gabel, F., X.F. Garcia, I. Schnauder, and M.T. Pusch. 2012. Effects of ship-induced waves on littoral benthic invertebrates. *Freshwater Biology* 57: 2425–2435.

Garvis, S.K., P.E. Sacks, and L.J. Walters. 2015. Assessing the formation, movement and restoration of dead intertidal oyster reefs over time using remote sensing in Canaveral National Seashore and Mosquito Lagoon, Florida. *Journal of Shellfish Research* 34: 251–258.

Grizzle, R.E., J.R. Adams, and L.J. Walters. 2002. Historical changes in intertidal oyster (Crassostrea virginica) reefs in a florida lagoon potentially related to boating activities. *Journal of Shellfish Research* 21: 749–756.

Hellmann, R. 2013. Canaveral National Seashore: Archeological Overview and Assessment. Southeast Archeological Center, National Park Service, Tallahassee, FL.

Indian River Lagoon National Estuary Program. 2008. Indian River Lagoon comprehensive and conservation management plan update, Palm Bay, Florida.

Intergovernmental Panel on Climate Change. 2007. Climate Change 2007: Synthesis Report. Contribution of Working Groups I, II and III to the Fourth Assessment Report of the Intergovernmental Panel on Climate Change. eds. Pachauri, R.K. and A. Reisinger. Geneva, Switzerland.

Kraus, N.C., and W.G. McDougal. 1996. The effects of seawalls on the beach: Part I, an updated literature review. *Journal of Coastal Research* 12: 691–701.

Lorenz, S., F. Gabel, N. Dobra, and M.T. Pusch. 2013. Modelling the effects of recreational boating on selfpurification activity provided by bivalve mollusks in a lowland river. *Freshwater Science* 32: 82–93.

Manis, J., S. Garvis, S. Jachec, and L. Walters. 2014. Wave attenuation experiments over living shorelines over time: A wave tank study to assess recreational boating pressures. *Journal of Coastal Conservation* 1–11.

Meyer, D.L., E.C. Townsend, and G.W. Thayer. 1997. Stabilization and erosion control value of oyster cultch for intertidalmarsh. *Restoration Ecology* 5: 93–99.

Morris, J.T., P.V. Sundareshwar, C.T. Nietch, B. Kjerfve, and D.R. Cahoon. 2002. Response of coastal wetlands to rising sea level. *Ecology* 83: 2869–2877.

National Park Service. 1975. Enabling Legislation for Canaveral National Seashore. Public Law 93-626, 93rd Congress, H.R. 5773.

National Park Service. 2014. Canaveral National Seashore Florida. http://www.nps.gov/cana/index.htm (accessed November 21, 2014).

Pilkey, O.H. and H.L. Wright. 1988. Seawalls versus beaches. In *The Effects of Seawalls on the Beach*, eds. Krauss, N.C. and O.H. Pilkey, *Journal of Coastal Research* Special Issue No. 4: 41–64.

Scheidt, D.M. and C.M. Garreau. 2007. Identification of watercraft use patterns in Canaveral National Seashore. Titusville, Florida: Canaveral National Seashore PIMS Final Report Project No. 1021131.

Schroevers, M., B.J.A. Huisman, M. van der Wal, and J. Terwindt. 2011. Measuring ship induced waves and currents on a tidal flat in the Western Scheldt estuary. *Current, Waves and Turbulence Measurements, IEEE/OES* 10: 123–129.

Sidman, C., T. Fik, R. Swett, B. Sargent, J. Fletcher, S. Fann, D. Fann, and A. Coffin. 2007. A recreational boating characterization of Brevard County. Florida Sea Grant. TP-160.

Smith, N.P. 1987. An introduction to the tides of Florida's Indian River Lagoon. I. Water levels. *Fla. Sci.* 50: 49–56.

Smith, N.P. 1993. Tidal and nontidal flushing of the Florida's Indian River Lagoon. *Estuaries* 16: 739–746.

St. Johns River Water Management District. 2014a. The Indian River Lagoon: An estuary of national significance. http://floridaswater.com/itsyourlagoon/ (accessed November 16, 2014).

St. Johns River Water Management District. 2014b. The Indian River Lagoon: Background and history. http://floridaswater.com/itsyourlagoon/history.html (accessed November 16, 2014).

Underwood, A.J. 1991. Beyond BACI: Experimental designs for detecting human environmental impacts on temporal variations in natural populations. Australian *Journal of Marine Freshwater Research* 42: 569–587.

United States Census Bureau. 2012. USA Counties. http://censtats.census.gov/usa/usa.shtml (accessed June 1, 2012).

United States Coast Guard. 2013. 2012 Recreational boating statistics. Publication COMDTPUB P16754.26.

United States Environmental Protection Agency. 2014. Estuaries and coastal watersheds. http://water.epa.gov/type/oceb/nep/index.cfm#tabs-2 (accessed November 16, 2014).

Valiela, I., J.L. Bowen, J.L., and J.K. York. 2001. Mangrove forests: One of the world's threatened major tropicalenvironments. *Bioscience* 51: 807–815.

Wall, L.M., L.J. Walters, R.E. Grizzle, and P.E. Sacks. 2005. Recreational boating activity and its impact on the recruitment and survival of the oyster Crassostrea virginica on intertidal reefs in Mosquito Lagoon, Florida. *Journal of Shellfish Research* 24: 965–973.

Walters, L.J., A. Roman, J. Stiner, and D. Weeks. 2001. Water Resource Management Plan, Canaveral National Seashore. Titusville, Florida: National Park Service, Canaveral National Seashore Report.

Walters, L.J., P.S. Sacks, M.Y. Bobo, D.L. Richardson, and L.D. Coen. 2007. Impact of hurricanes and boat wakes on intertidal oyster reefs in the Indian River Lagoon: Reef profiles and disease prevalence. *Florida Scientist* 70: 506–521.

Wasson, K., C.J. Zabin, L. Bedinger, M.C. Diaz, M.C., and J.S. Pearse. 2001. Biological invasions of estuaries without international shipping: The importance of intraregional transport. *Biological Conservation* 102: 143–153.

Waycott, M., C.M. Duarte, T.J.B. Carruthers, R.J. Orth, W.C. Dennison, S. Olyarnik, A. Calladine, J.W. Fourqurean, K.L. Heck, A.R. Hughes, G.A. Kendrick, W.J. Kenworthy, F.T. Short, and S.L. Williams. 2009. Accelerating loss of seagrasses across the globe threatens coastal ecosystems. *Proceedings of the National Academy of Sciences* 106: 12377–12381.

Williams, R., A.W. Trites, and D.E. Bain. 2002. Behavioural responses of killer whales (Orcinus orca) to whale-watching boats: Opportunistic observations and experimental approaches. *Journal of Zoology* 256: 255–270.

Zacharias, M.A. and E.J. Gregr. 2005. Sensitivity and vulnerability in marine environments: An approach to identifying vulnerable marine areas. *Conservation Biology* 19: 86–97.

CHAPTER 13

Growing Living Shorelines and Ecological Services via Coastal Bioengineering

Steven G. Hall, Robert Beine, Matthew Campbell, Tyler Ortego, and Jon D. Risinger

CONTENTS

13.1 INTRODUCTION

Coastal bioengineering is a subfield of biological engineering, applied in the coastal zone (Hall 2015) to enhance sustainability and reduce the need for excess material by encouraging growth in coastal structures and techniques. The intersection between coastal bioengineering and living shorelines is significant. Living shoreline projects are coastal restoration techniques that may use sand, plants, submerged reefs, and other biological features to protect the shoreline and the habitat (National Oceanic and Atmospheric Administration [NOAA] 2016). Coastal bioengineering is a method to bring rigorous engineering principles into living shoreline design to enhance success. The use of biology and ecology to address issues at the water/land interface is integral to these techniques, which have been recognized to simultaneously reduce material use, while providing enhanced ecosystem services (Coen et al. 2007; Hall 2009a,b). In fact, the interactions of the biological components on the physical component generally induce combined changes to the environment, which are the essence of bioengineered structures and critical to success of living shorelines. These also introduce challenges in that the final impact of such designs depends on the growth of living organisms.

Traditional living shorelines are normally used in low to moderate wave energy environments where rigorous engineering was not generally conducted to optimize designs for the specific conditions. This is primarily attributed to the relative infancy of the field and the lack of extensive data on relevant engineering parameters that are present in other engineering fields, such as transportation and vertical construction. With the development of these parameters and the aggregation of relevant data and engineering design techniques, which include both physical and biological parameters, coastal bioengineers will be able to enhance the design of living shorelines and their use will be more widely accepted in the development and protection of coastal infrastructure (Hall 2015; Hall et al. 2016).

Many coastal areas can benefit from engineered approaches to living shorelines. Areas such as the more developed east coast of the United States may have significant urbanization and infrastructure, whereas areas like Louisiana and the Gulf of Mexico coast include more wild wetlands. Louisiana has 40% of US coastal wetlands but suffers 90% of the nation's coastal wetland loss (Mendelssohn et al. 2012). It has one of the highest relative sea level rise rates in the world, coupled with a very flat profile (Coleman et al. 1997). Subsidence of the soft sediments is a major issue and predicted land loss is dire (Barras et al. 2003; Blum and Roberts 2009; Penland et al. 2005). In addition, major tropical storms affect many coastal areas on a regular basis (Day et al. 2007), and issues of coastal stability are critical. As such, Louisiana is an ideal laboratory to observe coastal impacts that are beginning to challenge many other areas around the country and the globe.

Many of these diverse areas have historically chosen nonbiological techniques for coastal wetlands and shoreline protection, and in many cases, this has actively affected the biology of the coastal areas in negative ways (Williams et al. 1997). Some of those "hard" techniques are seawalls, bulkheads, and jetties (Phillips and Jones 2005). These structures slowly break down over time, are susceptible to sea level rise, may increase erosion further down shore, and do not provide much habitat (Hall et al. 2012; Scyphers et al. 2011).

In the last few years, more scientists and policy makers are recognizing the impact that channelizing the Mississippi River has had on the coast. They have also been recognizing the effects of hydrocarbon spills, the introduction of salinity changes in the marshes via canals, and other ecosystem stressors. One of the impacts is the loss of aquatic species such as oysters. Globally, 85% of oysters have been lost (Beck et al. 2011). Fisheries such as the Chesapeake have been decimated by a combination of overharvest and environmental degradation, and in the Gulf Coast, large areas of oyster reefs have been lost, although a large fishery remains. Furthermore, scientists are now recognizing the limits of traditional engineering, both in terms of cost and ecological impacts, and are considering other techniques, such as living shorelines, to address coastal challenges.

13.1.1 Coastal Bioengineering

Coastal bioengineering is a subset of biological engineering or bioengineering (Hall 2015) that focuses on understanding the biology and engineering to support biotic components (Hall and Lima 2001). Coastal bioengineering acknowledges and encourages living and growing aspects of shoreline protection. Since engineers design and predict for different environmental and wave conditions, this can be useful to identify locations where living shorelines can be successfully implemented, avoiding some of the challenges introduced when installing or managing living shorelines.

Bioengineering in the coastal environment may be applied to living shorelines and other "soft" approaches, but can also address structural, sedimentary, water quality, navigation, flow, and other parameters that tend to extend beyond the limits of traditional living shorelines. This chapter will address some of the key parameters, present some coastal bioengineering techniques and examples that intersect with living shorelines, and provide considerations for future investigation at this unique interface.

13.1.2 Fundamentals of Coastal Structures and the Environment

Coastal engineers have been placing structures in the environment for centuries. Examples include earthen levees to manage river flooding, gate structures to control flow, jetties and piers to provide access to deeper water and alter nearshore hydrology, seawalls to protect from wave or storm impacts, and breakwaters and groins to alter shoreline morphology (Kamphuis 2010). Other structures are placed in the coastal zone, although not always engineered specifically for coastal applications. Examples include roadways and bridges in the coastal zone, as well as habitable structures and boat-related infrastructure (port, pier, and other infrastructure). When an artificial structure is introduced into the marine environment, it will interact with the environmental components (e.g., water, soil, organisms). If the environmental and structural components are understood, a reasonable expectation of its effects can be predicted (Walker 1988).

The environmental components can be separated into two classes, which include physical and biological. Some of the physical components that would interact with the artificial structure might include waves, currents, water chemistry, turbidity, bathymetric morphology, and complexity of the physical landscape. Some of the biological components could include sessile organisms that would colonize the structure (e.g., oysters, barnacles, bryozoans, and corals), organisms that seek shelter and food in the structure (e.g., juvenile fish, plankton, and predators such as crabs or birds), and organisms that are attracted to the modifications that the artificial structure induces (e.g., plants behind or in sediment accreted in the low-energy areas near or in the structure). These organisms could increase the strength and size of a structure or potentially reduce its strength by boring, predation, or, in the case of plants, excess herbivory. These interactions between biological and physical components both respond to and generally induce changes to the environment. One example is that growth of oysters on bioengineered reefs increases energy losses across the reef systems, increasing sedimentation (see Smith et al. 2014, 2015). Both mass and area are also increased, leading to more stable structures. These interactions and understandings are further discussed in this chapter.

Engineering, ecological, economic, social, and other goals may each have their place in the success of coastal projects. Some have proposed systems that can be financially profitable and therefore more likely to attract private sector investment. Examples include plants or animals with value (e.g., recreational fishing, oysters for food, plants for food, or biofuels). Living shorelines, with input from coastal bioengineers, can provide economically sensible, ecologically friendly solutions to coastal challenges.

13.2 IDENTIFICATION AND ANALYSIS OF KEY PARAMETERS

A living shoreline project is deemed successful if it meets various criteria, such as robust habitat, stable shoreline, and diversity of species. This success can be achieved by considering environmental and engineering parameters during the design stage. Coastal bioengineers utilize these parameters to design living shoreline techniques to achieve the success criteria of the given project. These parameters are defined through a process of defining goals, establishing success criteria, and identifying key parameters that will drive the design of the living shoreline project.

It is important for the coastal bioengineer to understand the environmental, biological, and ecological drivers behind the establishment of target habitat and species associated with the living shoreline project. The target habitat and species will require specific conditions, such as tidal inundation, soil type and chemistry, water quality and chemistry, wave energy, and feature complexity, in order to thrive in the given environment.

The location of the project is also critically important because it will define the existing physical and biologically relevant parameters. These parameters will need to be considered during the design process in order to achieve a successful project. As discussed, wave energy may be higher in some locations or seasons; suspended sediment, salinity, and temperature may drive which species will survive, which would all affect the living shoreline design and implementation. Coastal bioengineers utilize these parameters to design solutions that are capable of resisting expected forces while attempting to provide for the growth of desired plants and other organisms.

Unfortunately, the fields of science and engineering often approach experimental setup from different perspectives. This has resulted in a lack of necessary parameters defined in literature that can be used by coastal bioengineers to design a wide range of living shoreline techniques. Therefore, coastal bioengineers should not only seek existing knowledge from biological and ecological literature but also collect data on relevant parameters necessary for engineering design from field sites. As the physical and biological relationships are established through science, coastal bioengineers will be equipped with more tools to design successful living shoreline projects.

Engineers often use measured or predicted parameters and equations to estimate sustainability and environmental effects of coastal structures or materials (e.g., Hall et al. 2009; Ortego 2006). A prime example of this type of relationship is the tolerance of *Spartina alterniflora* to various wave energy climates in established marsh platforms (Ravens et al. 2009; Roland and Douglass 2005). Roland and Douglass developed a series of curves that represent the wave energy of various conditions of marsh (i.e., healthy, degraded, and nonexistent). Coastal bioengineers use this relationship to effectively and efficiently design coastal structures to reduce wave energy enough to sustain healthy marsh platforms. This kind of information enables coastal bioengineers the ability to optimize designs, which makes them more economical and successful.

13.2.1 Environmental Parameters

As discussed above, environmental parameters are critical to successfully design living shoreline projects. These parameters determine whether a target habitat or species can thrive or even survive the conditions that the project is exposed to. These parameters may include water quality parameters—for example, salinity, temperature, dissolved oxygen (DO), nitrogenous components such as ammonia, nitrite, nitrate, and oil or other toxins (see Mendelssohn et al. 2012). Environmental parameters may also include more physical parameters related to the biology, such as tolerance of a target species to tidal inundation, soil salinity, and wave energy. Habitat patterns are also important for defining environmental parameters, such as growth pattern of sessile organisms on a structure and intertidal vegetation elevation ranges relative to tide. There are other parameters related to the particular location of the project, such as predation pressure and subsidence that may need to be accounted for when designing a living shoreline project.

Environmental parameters can also be identified for the habitat or protective value of the biological feature. The very nature of the living shoreline technique is predicated by the fact that the biological components support both a habitat and a protective function. Examples of these types of biological features are oyster reefs and mangroves, which are considered foundational species, but have also been shown to protect and stabilize shorelines. The quality of the habitat or the density of organisms can also be directly related to the wave dissipation qualities. An example of this was shown in the mangrove forests of Vietnam (Mazda et al. 1997). Therefore, the parameters that optimize growth and density of the habitat will also enhance the protection function of the structure.

A classic example of the habitat and protection function of a living shoreline component is the oyster reef. The environmental parameters are critical for growth and survival of this organism. Oyster growth requires intermediate salinity (usually 10–20 parts per thousand (ppt), as shown in Figure 13.1), moderate temperatures, and sufficient DO, as well as sufficient suspended food for these filter feeders. These parameters ultimately determine the overall health, density, and geometry of the oyster reef.

The growth patterns and rates of the target species are critical information for designing features that can function as both habitat and protection. Using settlement patterns of oyster larvae spatially in the water column, in various water currents, on various materials, and various light penetration scenarios allows the design of structures that enhance the habitat and protection functions of the oyster reef component of the living shoreline project. Some of these types of patterns were investigated by Risinger (2012), who surveyed a series of engineered structures designed to grow oysters. His research showed a distinct pattern of settlement for oysters *Crassostrea virginica*, as

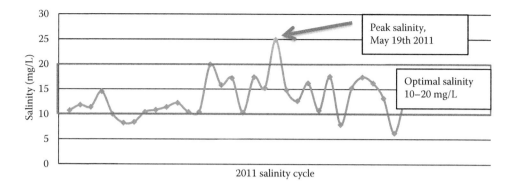

Figure 13.1 Salinity near Rockefeller Refuge (SW Louisiana) spring 2011 shows conditions conducive for oyster spat settlement and growth. (From Risinger, J. 2012. Biologically Dominated Engineered Coastal Breakwaters. PhD Dissertation, LSU. http://etd.lsu.edu/docs/available/etd-07032012-120846/. 128 pp. With permission.)

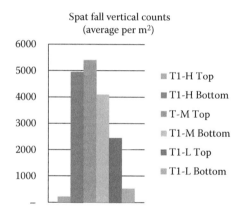

Figure 13.2 Measurement of settlement of spat (immature oysters) at different locations on a structure helps engineers identify where critical components should be deployed in the water column. (From Risinger, J. 2012. Biologically Dominated Engineered Coastal Breakwaters. PhD Dissertation, LSU. http://etd.lsu.edu/docs/available/etd-07032012-120846/. 128 pp. With permission.)

shown in Figures 13.1 and 13.2, and confirmed by growth on bioengineered structures (Figure 13.3). This same type of analysis could be done with other target species.

Learning about optimal growth conditions and enhancing growth is important not only from a biological point of view. From a structural perspective, it is important to engineer systems where the organisms remain healthy, but where the structural strength of the engineered devices is also enhanced by the growth of, in this case, oysters. Thus, both salinity (Figure 13.1, optimal oyster growth at 10–20 ppt) and settlement patterns (Figure 13.2) are critical to design structures that enhance growth in the right areas for sustainable coastal protection.

While the environment affects organisms, the opposite is also true. Many environmental parameters relate to the effect of the biological component on its surrounding environment. This could include the water filtration of a particular organism, such as oysters or barnacles that would improve water clarity and thus allow other organisms to flourish such as seagrass. Oysters and barnacles are filter feeders, filtering approximately 17 L of water per gram of oyster per day under optimal conditions (Ehrich and Harris 2015). This can have a dramatic effect on water quality, reducing sediments, removing carbon and nitrogen, and consequently reducing biochemical oxygen demand, which can

Figure 13.3   Bioengineered rings provide an intertidal wave break that grows horizontally and vertically over time, enhancing sediment deposition and providing ecosystem services. (Courtesy of Hall.)

contribute to hypoxia. Furthermore, the ability of oyster to sequester nitrogen and carbon could have further ecosystem service value (Grabowski et al. 2005; La Peyre et al. 2014; Peterson et al. 2003). Another example would be the protection and enhancement of sediment that would be added to the overall sediment budget of a system, which could reduce shoreline erosion. Figure 13.3 shows an example where the growth of oysters on bioengineered structures has begun to enhance sediment deposition by reducing wave energy and contributing to the local material budget. It should be noted that plants often grow in accreted sediment, sequester carbon, and stabilize shoreline. Combinations of plants, animals, and other organisms can be used in biofriendly engineering applications.

Further, the oyster reefs reduce wave energy, allowing other sediment to be deposited in areas near the reefs for other habitat types to establish (Hall 2015; Piazza et al. 2005). Thus, considering biologically relevant parameters such as salinity (Figure 13.1), specific biological parameters such as spat set (Figure 13.2), and geometric and environmental parameters such as those involved in wave reduction and incorporating them into engineered systems can synergize physical, biological, and ecological parameters, enhancing the sustainability of the system.

13.2.2 Engineering Parameters

The coastal bioengineer must consider not only the environmental parameters discussed above but also the engineering parameters related to the location and the individual components of the living shoreline project. The location-related parameters could include the bearing capacity of the existing soil, which relates to how much weight can be placed on it before the soil fails. These parameters also include the energy environment that the project is located where it could be subject to waves, winds, and water currents. The engineering parameters related to the individual living shoreline components determine their size, weight, geometry, and material type.

The energy-related engineering parameters, such as wind, tide, storm surge, currents, and waves, are important because they allow the coastal bioengineer to assess and design for the long-term stability of the living shoreline features. The energy parameters also determine the energy environment where biological and structural components can survive. These parameters determine the size of the structures needed to reduce waves or to be stable. They determine the slopes of embankments or the elevations of certain plant species and other features. The energy contained in moving water or wind affects the size of particles that can be moved in the water column or sub-aerial environment. Changes in energy can affect erosion (when, e.g., wave energy increases during a storm event) or deposition (e.g., wave energy reduction in the lee of a coastal engineered structure or natural reef) (Hamaguchi et al. 1991; Hall 2015; Meyer et al. 1997). Movement of water can also affect mixing, leading to changes in salinity, which can be critical for estuarine organisms.

These energy-related parameters can be determined through data collection, analytical methods, numerical modeling methods, or a combination. Typically, the existing data at a particular project location will be compiled and used as boundary or forcing conditions of a numerical model. Figure 13.4 shows a typical result from a wind-generated wave model conducted by Louisiana State University Coastal Studies Institute. Similar models are available for tides, currents, and storm surge to determine local conditions at the project site. Further data collection of the project site can be collected to refine the results and increase the accuracy of the numerical models. These models are useful in determining the energy forces that will interact with the living shoreline components in the coastal environment. The better these parameters are understood, the more the coastal bio-engineer can optimize the design to reduce risks and costs.

The structural parameters are related to the individual components of the living shoreline project. Structures are typically described by geometric parameters (e.g., height, diameter, total volume, and surface area), density parameters (mass and density), strength (e.g., yield strength and ultimate failure strength), and other parameters (e.g., flexibility, ability to resist failure). Bioengineered structures will also be described by parameters of interest to encrusting organisms (e.g., surface pH, rugosity or roughness, localized flow parameters, total surface area, and indentations, which may

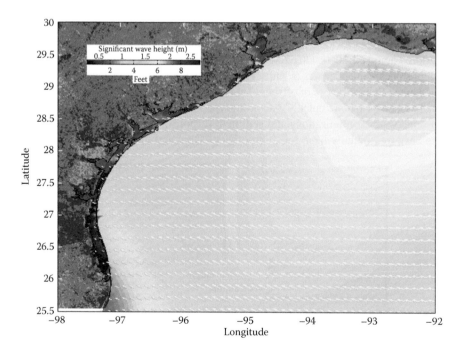

Figure 13.4 Numerical model screenshot from Louisiana State University Coastal Studies Institute numerical model forecast of wave height and direction in the western Gulf of Mexico on April 18, 2016 (with CSI permission) (http://wavcis.csi.lsu.edu/forecasts/forecasts.asp?modelspec=mike21).

provide protection from predators). Coastal bioengineers will address both physical and biological parameters in their designs (O'Beirn et al. 2000; Hall et al. 2015).

The strength of the living shoreline components will need to be analyzed relative to the forces exerted on the structure, which typically originate from the energy parameters described above. Commercial software is available to analyze stress and strain patterns on components under various force-loading schemes. For example, Risinger (2012) used Autodesk Inventor (www.autodesk.com) to run stress, strain, and displacement analyses on bioengineered concrete components (Figure 13.5). The figure shows that the component is likely to fail because of the forces exerted on the structure. This type of analysis is useful in optimizing component design and predicting useful life of components.

Understanding structure–wave interactions and structure–current interactions is necessary when a structure is used as a wave dissipating or current reducing structure. The structure will be required to stay in place during design storm conditions and to reduce the wave or current energy sufficiently to allow other features, such as marsh to survive. These interactions can be determined through analytical and numerical methods. When complex features are included in the design, it may be desired to conduct computational fluid dynamic (CFD) modeling, which uses numerical analysis and algorithms to solve the Navier–Stokes equations and accurately predict fluid flows around objects. With the advances and availability of high-speed computing, this analysis approach has become more prevalent. Figure 13.6 depicts a CFD model of a concrete armor unit structure under hurricane conditions. The model was able to determine the forces on the structure and the reduction of the wave energy. Similar engineered structures may be emplaced, in areas where either riverine or marine water flow may be significant. In either case, understanding interaction between structures and currents is critical, and similar software can be used to address these concerns.

Other engineering parameters are related to how individual components of the living shoreline project affect the morphology of the shoreline being protected and adjacent areas. These have been studied extensively through the field of coastal engineering and can be applied to coastal bioengineering. Typical structures that have been traditionally used in coastal engineering are breakwaters, groins, jetties, sills, reefs, and bulkheads.

Offshore breakwaters provide a physical barrier that dissipates wave energy reaching the shoreline (Benassai 2006; Campbell 2004; CCEZM 1990). Breakwaters are typically constructed from rock, concrete armor stones, sunken barges or ships, or any heavy objects that break up wave action (CCEZM 1990). In the United States, the use of offshore breakwaters as a means of shore protection and passive nourishment has increased more quickly than groin-type structures in the last decades (Benassai 2006).

Typical breakwater configurations and sediment responses are indicated in Figure 13.7. These breakwaters have traditionally been rock, but others have noted that natural or restored shell reefs

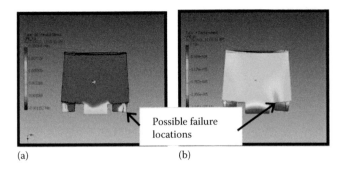

(a)   (b)

Figure 13.5 (a and b) Strain and displacement analyses for thin-walled cylindrical structures intended for stacking in the wave zone to grow artificial reefs show areas of possible failure and can aid in improved structural design. (From Risinger, J. 2012. Biologically Dominated Engineered Coastal Breakwaters. PhD Dissertation, LSU. http://etd.lsu.edu/docs/available/etd-07032012-120846/. 128 pp. With permission.)

Figure 13.6 Commercially available computational fluid dynamics software can be customized to provide predictions of wave energy (or in this case water velocity) near emplaced structures to allow optimal emplacement of properly engineered structures. (From Campbell, M.D. 2004. Analysis and Evaluation of a Bioengineered Submerged Breakwater. http://etd.lsu.edu/docs/available /etd-11052004-152838/. With permission.)

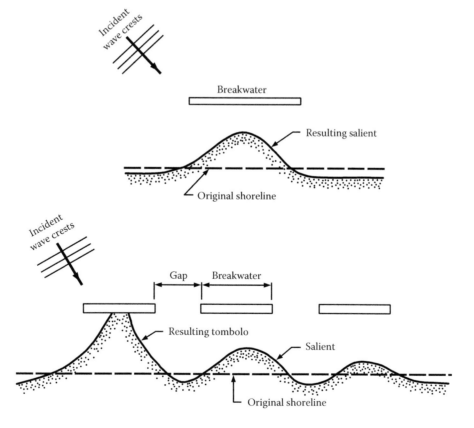

Figure 13.7 Incident wave energy interacts with breakwater to create salient and eventually tombolo (connected with breakwater) landforms. Breakwater and gap distances as well as offshore distance are important engineering parameters.

can dissipate wave energy in a similar manner (Piazza et al. 2005; Scyphers et al. 2011), while also providing ecosystem services and habitat (Coen et al. 2007; Grabowski et al. 2005; La Peyre et al. 2014; Peterson et al. 2003).

Engineering parameters are critical to the success of breakwaters. The relative distance offshore, spacing, length of breakwater, and depth of the breakwater emplacement are important for determining the morphological response in a sandy environment. For positive morphological responses such as salient (i.e., accretion of shoreline toward breakwater) or tombolo (i.e., accretion of shoreline that connects to the breakwater) formation, the key variables are as follows:

- Y, Distance of breakwater from nourished shoreline
- L_s, Length of breakwater structure
- L_g, Gap distance between adjacent breakwater segments
- d_s, Depth (average) at breakwater structure below mean water level

Three dimensionless ratios, Y/d_s, L_s/L_g, and L_s/Y have emerged to separate salient and tombolo response. When the breakwater is long or located close to shore, conditions favor tombolo formation. Pope and Dally (1986) recommend the following:

$$L_s/Y > 1.5 - 2 \text{ for single breakwater} \tag{13.1}$$

$$L_s/Y = 1.5 \text{ for segmented breakwater } (L < L_g < L_s) \tag{13.2}$$

While these are basically geometric parameters that influence environmental parameters such as local wave energy, these provide a starting point for design of engineered structures that account for and enhance growth of biological components.

13.2.3 Integrating Biological and Engineering Parameters

Engineering with the biology is a critical aspect of good coastal bioengineering (Hall and Lima 2001). This is of interest in living shorelines, where appreciation of the needs of coastal plants and animals should guide the development of materials and structures, as well as techniques to emplace plants or encourage the growth or settlement of reef-building organisms. Each of these can help enhance the performance and sustainability of projects over time. After a keen understanding of the biology and ecology of a given target habitat is achieved, the coastal bioengineer is able to develop design criteria and innovative solutions.

The intention of living shoreline techniques is to develop nature-based protective infrastructure. Utilizing native species for a given area and encouraging them to thrive can achieve both objectives. Oysters are an example of nature-based protective infrastructure that live in colonies and can produce an abundance of offspring. As such, they grow reefs, with new generations of oysters often settling year after year on older oysters, building complex and ecologically beneficial reef systems. These provide protection for a variety of other species, offer a massive food source, while enhancing water quality and providing coastal protection.

It is logical that these reef systems have been proposed to replace traditional coastal structures, like breakwaters. These reefs can be engineered into "bioengineered breakwaters" by placing a base structure to allow for oyster colonization. This base structure can be designed to function as a typical rock breakwater, although the base structure should also be designed with the biological parameters of the oyster in mind.

Under reasonable conditions, bioengineered breakwaters can approach or even exceed the performance of a traditional rock structure over time because of their ability to increase in elevation as the oysters grow (Campbell 2004; compared to Ahrens 1987). They can thus be designed using similar coastal engineering criteria as discussed above, but with a time component. However, because they can continue to grow, they should normally maintain or even expand their footprint, whereas rock structures tend to sink below the water level over the course of years, particularly in areas of soft sediments.

One unique aspect of growing systems, such as bioengineered structures or other living shoreline emplacements, is that they can change dramatically over time as growth occurs. Immature systems typically are less stable and underperform, whereas after sufficient growth, systems tend to be more stable, become stronger, reduce more wave energy, and hold sediment much better (Figures 13.3 and 13.7). Growing structures, even though they also subside and sink, can grow back to the surface as plants or reef-building animals grow.

For the purposes of this discussion, two examples are given below that highlight the ways that the biology can enhance the performance of designed structures. These studies were conducted on oysters, but other organisms could be investigated in a similar way. The concept of using the organism's inherent biological properties to accomplish a desired design goal is the essence of these examples.

Bioengineered structures and some structures used with biological components can increase the structural properties of the structure. For example, as encrusting organisms grow on a bioengineered reef, they increase the total mass and volume of the structure (Campbell 2004; Hall et al. 2007). They may also increase total strength (Risinger 2012; Risinger et al. 2012, 2013) and, in many cases, dramatically increase "toughness," a technical term referring to ability to resist total failure, even if partial failure occurs (Risinger et al. 2012). Figure 13.8 shows strength enhancement of bioengineered concrete after oysters grew on the sections over 2 years.

Another example of integrating biological properties with engineering properties is a novel equation developed by Campbell (2004) to include growth, geometric, and wave energy numbers. This number links biological parameters with wave transmission, which is the ratio of wave energy that moves past a structure. Better (lower) wave transmission reduces wave energy on the shoreline

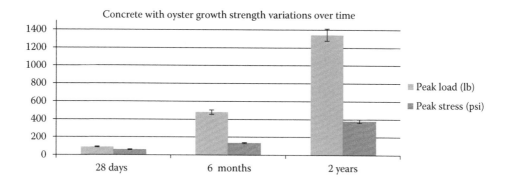

Figure 13.8 Strength was enhanced by 500%–1200% owing to the growth of oysters on sections of bioengineered concrete over a 2-year period. Peak load refers to the maximum force placed on the sample before they broke, while peak stress is this load divided by the cross-sectional area of the original sample. (From Risinger, J. 2012. Biologically Dominated Engineered Coastal Breakwaters. PhD Dissertation, LSU. http://etd.lsu.edu/docs/available/etd-07032012-120846/. 128 pp. With permission.)

and can reduce erosion or protect marsh plants. Campbell called his equation the oysterbreak transmission number, which relates the wave height, wave length, growth stage, and density of the structure. The transmission coefficient is the ratio of how much of an incoming wave passes through the structure. The measured transmission coefficient was plotted with respect to the oysterbreak transmission number (Figure 13.9) to find a relationship. An equation was created, which successfully describes that relationship (Equation 13.3).

$$K_t = \frac{1}{1 + \left(\dfrac{L_w \times r^2 \times \Psi}{H^2} \right)^{0.7721}}$$

(13.3)

It was found that the transmission coefficient, K_t, was related to the wave height (m), H_w, wave length (m), L_w, radial growth on the horizontal bars (m), r, and the number of slats per meter inside the oysterbreak, Ψ.

Using this relationship, Campbell predicted how growth could enhance wave energy reduction. Figure 13.9 shows how transmission coefficient (i.e., the amount of wave energy that propagates through the structure) is reduced as growth occurs on lighter and denser initial structures. What this reveals is that, over time (1–3 years of growth), wave energy can be reduced by up to 70%.

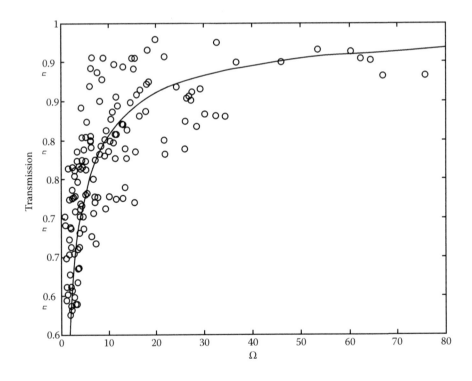

Figure 13.9 Transmission coefficient versus oysterbreak transmission number. Oysterbreak transmission number, $\Omega = \dfrac{H_w^2}{L_w \times (r^2 \times \Psi)}$. (From Campbell, M.D. 2004. Analysis and Evaluation of a Bioengineered Submerged Breakwater. http://etd.lsu.edu/docs/available/etd-11052004-152838/. With permission.)

However, Figure 13.10 reveals that when less dense structures are initially installed, they may have little wave dampening activity. Conversely, although a more dense initial structure achieves excellent wave reduction in approximately 1 year, even less dense structures will achieve this level of wave reduction after 3 years of good growth, with far less use of materials and construction costs (Figure 13.10).

Ultimately, what coastal bioengineers bring to the discussion is the ability to quantify and predict as engineers, with the serious consideration of biotic components and the needs (e.g., water, salinity, temperature) of those components in the design. The design of living shorelines can be used in similar ways that traditional coastal structures, such as bulkheads and breakwaters, have been used to stabilize shorelines and protect infrastructure. Oyster reefs (artificial or natural), for example, may also significantly alter local wave energy, allowing sediment to accrete in low-energy areas near shore (Malveaux and Hall 2015). This, in turn, may allow plant colonization to develop and thrive as a natural buffer of living shorelines and green infrastructure, as seen in Figure 13.11. These natural growth zones can protect other coastal infrastructure (e.g., roads and buildings) from damage during extreme, less frequent storms.

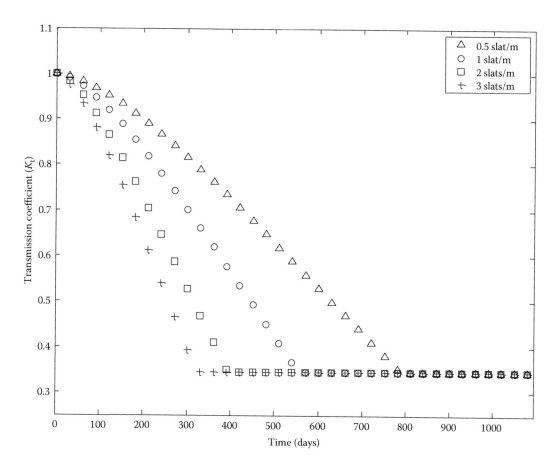

Figure 13.10 Predicted transmission coefficients as density changes in a theoretical bioengineered structure. Growth on the structure should reduce wave energy transmission over time. (From Campbell, M.D. 2004. Analysis and Evaluation of a Bioengineered Submerged Breakwater. http://etd.lsu.edu/docs/available/etd-11052004-152838/. With permission.)

Figure 13.11   2015 photo of LA-08 project in western Louisiana showing growth on structures, sediment accretion, and plant growth on accreting land. (Photo courtesy of C. Malveaux.)

13.3 COASTAL BIOENGINEERING TECHNOLOGIES

Many engineered structures have been built from ancient to modern times, but hard structures are not generally biologically friendly. In fact, seawalls, jetties, and even levees tend to change both water quality and energy parameters in such a way as to reduce available habitat in the coastal environment. However, other structures are able to minimize ecological impact or even provide enhanced ecological options (Hall et al. 2016; Risinger 2012). Such bioengineered structures or materials attempt to enhance the longevity of coastal protection projects, while reducing cost (Dehon 2010; Ortego 2006). For example, rock breakwaters are a traditional technique for reducing wave energy near shorelines. However, in soft deltaic sediments, rocks tend to sink over time, requiring additional maintenance.

Living solutions may be more buoyant but may also be able to grow upward as a structure or sediment sinks. A number of examples of commercially available technologies intended to provide enhanced growth for coastal protection exist. From an ecological point of view, living solutions are far preferable, as living shorelines and biofriendly structures allow or may even enhance ecosystem health in areas otherwise stressed from climate change and other factors. Engineers use mathematical and physical tools to design these technologies more effectively.

For example, tools exist to model hydrodynamic interactions with specific geometries. These tools include physical model tests and CFD software tools. Currently, projects are being designed for the "naked structure," ignoring encrustation and sediment accretion. This is because insufficient knowledge exists to accurately characterize biology-induced changes. This method may result in unduly conservative designs and thus require excessively heavy and expensive emplacements. Thus, one limitation is lack of knowledge of the biological growth and positive impacts of that growth, while another limitation is the need for appropriate water quality parameters for healthy growth of desired biological organisms. Nevertheless, even preliminary steps toward more ecologically friendly and materially efficient designs are steps in the right direction.

13.3.1 Brief Survey of Structural Types

Some of the structures that have been considered or emplaced include artificial reefs; breakwaters; reconstructed coastal barrier islands or dunes; artificial materials to enhance growth of oysters, plants, or other species; and docks or even buildings that may provide structural, biological, and ecological services over the next few decades, with a growing understanding that many of these may become components of future reef systems. Byrum et al. (2014) considered human housing that could enhance biology and ecology over several decades (Figure 13.12) and eventually transition to natural reef systems as subsidence and climate change alter coastal ecologies.

Land growth as depicted in Figure 13.12 could occur for several decades, but with coastal subsidence and sea level rise, eventually the structure would likely submerge, at which time the bio-friendly concrete could provide a good surface for reef-building organisms, and the overall structure could provide habitat and protection for aquatic organisms.

At the more specific level, plants, animals, and even microorganisms have potential to contribute to coastal bioengineering solutions. As an example, living shorelines often utilize plant-based solutions. These technologies have engineering components such as placement, acceptable slopes for soil retention, and an understanding of acceptable or optimal growth conditions. Coastal bio-engineers understand the objectives of growing systems and can specify or design components or systems. Figure 13.13 shows coastal mats with vegetation growing through. These mats must have sufficient integrity to survive wave energy during growth and must also provide root stability and acceptable conditions to allow plant growth.

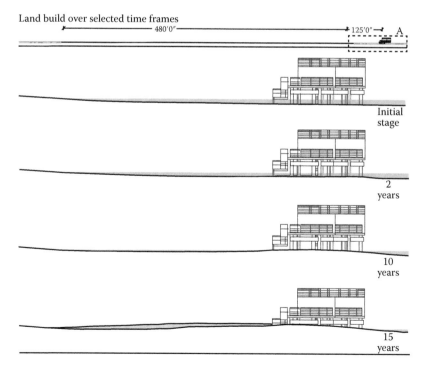

Figure 13.12 A figure depicting land growth that might be accreted under or around a coastal structure over decades. (Byrum, M., S. Hall, J. Erdman, J. Sullivan, L. Harrell, C. Knott, and S. Bertrand. 2014. Culturing Coastal Plants and Animals for Sustainable Housing, ASABE July 2014, Montreal, QC. With permission.)

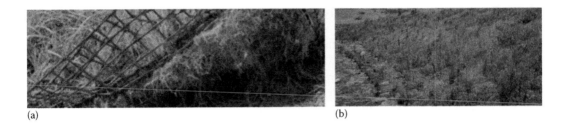

(a)   (b)

Figure 13.13 Coastal mat with vegetation growing through it (a). More mature growth with a slope that should be specified by engineering techniques (b). (Photo adapted from http://www .deltaland-services.com/shore-links.)

While techniques such as coastal mats rely on some amount of land that already exists, other techniques rely on buoyant components that allow plants to float, rising and falling with water levels and eventually enhancing local vegetation while reducing wave energy, even where there is no land permanently above water (Figure 13.14).

It should be noted that most floating plant technologies to date are not designed to survive high energy waves; thus, these tend to be restricted to low-energy areas. Coastal bioengineers may assist with enhancing strength to allow emplacement in moderate wave energy zones.

Another technology generally focused on plant growth is the use of fiber coir logs (Figure 13.15). These logs are typically constructed of coconut fibers or other fibrous materials held together with netting, but allowing plant growth into or adjacent to the material. They reduce sediment loss into streams or coastal zones and eventually biodegrade leaving healthy plant growth under optimal conditions. These are also typically designed for low to moderate energy. Coastal bioengineers can identify, specify, and potentially design with coir logs or redesign the devices themselves to address future needs.

While there are a variety of plant-based technologies, animals can also be used as a component of biologically dominated solutions, and in some cases, multiple types of organisms can be

Figure 13.14 Floating plant technologies allow plants to grow in a buoyant medium. Roots can eventually tap into sediment below and can stabilize or even build land in optimal conditions. (Adapted from Houmaweekly.com.)

Figure 13.15 Coir logs can reduce erosion, hold soil, and allow growth of plants.

used. Risinger (2012) surveyed a variety of materials and structural components that had biologically friendly components or purposes. He focused on bioengineered breakwaters, each of which provided some type of physical and often three-dimensional structural components with biologically friendly surfaces on which native organisms such as the eastern oyster *C. virginica* might thrive. Most of these structures serve as breakwaters, some intertidal and some with components above or below the water surface. Typical breakwater patterns used in living shoreline techniques, whether rock or more biofriendly materials, are low-crested or submerged breakwaters as seen in Figure 13.16, a drawing of biofriendly Oysterbreak rings.

Boyd and Pace (2013) note that, in many areas, coastal zones have been "hardened" with a variety of solid, often concrete or rock structures. For example, 31% of Alabama's coastal shoreline had been armored by 2012, and it was predicted that 45% of Mobile Bay would be armored by 2020 (Boyd and Pace 2013). This not only cuts down on shoreline access for the public but also dramatically reduces and degrades habitat for coastal species, while in some cases it can exacerbate coastal erosion on adjacent land that is not armored. Alternatives such as living shorelines that allow connection between the water and native species, while providing habitat, would be preferable. Coastal bioengineering provides an engineering approach to match materials and structures to wave energy, but with an end goal of maximizing, not minimizing, habitat for native species, while preserving coastal wetlands.

Risinger (2012) surveyed several options of commercial materials and structures intended to encourage biological growth. These include roughly spherical biofriendly concrete components (e.g., Reefballs, Figure 13.17a), triangular steel frames filled with oyster shells (e.g., Reef Blk,

Figure 13.16 Three-dimensional computer-aided design drawings of a multiring configuration of bioengineered coastal reef system. (From Ortego, T.R. 2006. Analysis of Bioengineered Concrete for Use in a Submerged Reef Type Breakwater. Master's Thesis, LSU. http://etd.lsu.edu/docs/available/etd-06122006-223212/. 48 pp.)

(a)   (b)   (c)

Figure 13.17 Reefballs (a) installed in four rows in an intertidal formation are a nearly spherical configuration with openings on top and at other locations on the structures to encourage coral or other organisms to grow. They tend to be relatively low density, meaning they should not sink quickly in areas of soft sediment. ReefBlk (b) are triangular frames into which oyster shell can be poured, allowing growth on the oyster shell. Under some conditions, the shell can be destroyed by burrowing organisms or wave energy, leaving the steel frames. The overall system is medium density, but the steel frames are high density. Eco Disks (c) have rock embedded in a series of vertical disks, making them medium to high density, meaning they may sink in areas of soft sediments. (From Risinger, J. 2012. Biologically Dominated Engineered Coastal Breakwaters. PhD Dissertation, LSU. http://etd.lsu.edu/docs/available/etd-07032012-120846/. 128 pp. With permission.)

Figure 13.17b), disks incorporating calcareous rock and concrete materials (e.g., Eco Disks, Figure 13.17c), cylindrical stackable components (e.g., Oysterbreak, Figure 13.18), and other technologies intending to encourage growth of oysters, corals, coastal plants, or other organisms. These compare to Figures 13.13 through 13.15, examples of materials intended to encourage growth of plants to stabilize shorelines.

Oysterbreak rings (Figure 13.18) are typically cylindrical interlocking rings that can be stacked and emplaced as sub-, inter-, or supratidal wave breaks. They are porous but dramatically reduce wave energy in the lee of the structure, and can, when designed properly, survive in very high energy environments. Smaller or lighter versions can be used economically in low- to medium-energy zones and may sink more slowly in soft sediments.

Figure 13.18 Cylindrical artificial reef components (Oysterbreak) are low to medium density with a high rugosity surface. (Courtesy of Hall.)

In summary, a variety of technologies including engineered structures that are intended to encourage plant or animal growth, thus protecting shorelines and enhancing local ecology, are commercially available. Care must be taken to consider both physical and biological aspects of these structures and materials.

13.4 CHALLENGES AND FUTURE DEVELOPMENTS

Both challenges and opportunities are available for coastal bioengineering applications to living shorelines. Some challenges are that since biology is needed, systems must accommodate and plan for growth, reproduction, and more complex ecological phenomena including predation and ecosystem services. As discussed, a variety of engineered products that may be applied to living shorelines and related nature-based solutions are available. Specific solutions will need further engineering design and development. One area of possible development is the use of living shoreline techniques in higher-energy areas than have traditionally been used.

Opportunities include the production of various services in addition to coastal protection. For example, continued research to enhance multiple positive impacts such as shoreline protection, food production, and ecosystem services is needed. Development of numerical models that incorporate both biological and physical components, including changes over time, will further enhance the ability to design and emplace these types of devices. These specific developments may involve plants for coastal protection, animals such as reef-building organisms, and even human structures, which eventually serve as the basis for future reefs or land building.

We also need to develop overarching engineering solutions that incorporate traditional geometric and physical parameters such as wave energy, geometry, and sediment, with biological (growth and reproduction) or biologically relevant parameters such as water quality and food availability. From an engineering point of view, these solutions would ideally involve a minimal amount of material to maximize growth, providing sustainability. Because of uncertainties about the effects of "mature" systems, often excess material is emplaced to achieve immediate results in terms of wave reduction. However, this is both at the cost of excess material and transportation and emplacement costs, but does not generally maximize growth for ecosystem services. Finding ways to emplace thin-walled but stable devices or to move "matured" units from low-energy areas after they have become biologically dominated can jump-start both ecosystem processes and provide a way to instantaneously provide sufficient wave energy reduction in high-energy areas.

Alternative techniques to combine shoreline protection with valuable food or fiber products could lend additional economic value to these already valuable coastal areas (e.g., Hall 2010; Harding and Mann 2001). Techniques to enhance natural beaches and use of native species such as *S. alterniflora* grasses and black mangrove *Avicennia germinans*, along with animal-dominated reef systems, could each contribute to sustainable long-term solutions (Hall et al. 2007).

Energy-producing systems using wave and wind energy have great potential for suitable coastal locations. In each of these cases, consideration of the biology and environment while engineering structures and systems can enhance long-term sustainability and adaptability in the face of climate or coastal changes.

A number of engineering parameters (physical, including strength and geometry; chemical, including water quality; and biological, including plants and animals) have been discussed in this chapter. Continued development and study of both the parameters themselves and engineering specifications that include the biological and biofriendly components of these parameters are needed.

Analytical and empirical modeling that focuses on wave energy (Campbell 2004) and structural strength (Risinger 2012) has been done, as well as some work on oyster growth and water quality parameters (e.g., Ehrich and Harris 2015). Commercial software is available to predict physical strength of standard strength materials, but development of numerical models that incorporate realistic aspects of biological growth, reproduction, predation, and the interactions between biology and physical parameters—for example, reefs reducing wave energy, plants reducing soil erosion, animals filtering water—is needed.

Full integration of biological growth rates and the effects of that growth (whether animal or plant) on physical and biological parameters are needed. This is a critical area of ongoing research. Typical oyster growth rates are available, but their effect on reef density, sinkage rates, filtration and carbon sequestration rates, and wave energy reduction is still not specified. Further work in these areas is critical.

As discussed, ecosystem services such as habitat (see La Peyre et al. 2014) are critical, and engineered reef systems can provide significantly greater habitat. However, other ecosystem services such as carbon sequestration (Dehon 2010) and nitrogen filtration have not been fully quantified. These synergistic benefits may be extended to combinations of growth and productivity coupled with sustainable harvest.

Oysters are cultured for food, often with on-bottom growth. However, engineered systems could be developed to simultaneously culture oysters and reduce wave energy or provide other desired ecosystem services. One specific technique that has been proposed but not fully implemented is to use long-line systems or similar cage culture with oysters suspended in the water column. This enhances growth potential (Hall 2010) but could also serve to reduce wave energy. Preliminary work suggests that these systems could reduce wave energy by 10%–20% while providing valuable food products.

One of the challenges of living shoreline emplacements is that they are often done piecemeal by small landowners. This not only means that a variety of options are available but also means that industrial equipment may be inappropriate in many cases. Examples of large bioengineered reefs being installed by large equipment have shown that this is possible with minimal impact on the ecosystem, but small-scale emplacements that can be done by hand or small equipment are also important. This suggests a market for small-scale engineered components that could still collectively have a landscape scale impact on natural solutions to coastal challenges.

Coastal bioengineering can contribute to living shorelines by providing a combination of engineering design for materials and structures, with awareness and prediction of growth needs and rates of plants, animals, or other organisms. Further work is needed to develop these opportunities and help enhance living shorelines applications in ways and areas that have not been done to date. Economically viable, environmentally friendly, socially acceptable, and sustainable solutions are needed, and engineers can contribute to these synergistic solutions. These engineering solutions must truly acknowledge the whole range of physics and biology, including environmental conditions, and the impact of growing communities on the local ecosystems. Ultimately, enhanced awareness of biology in engineering design can create more sustainable living shorelines that provide multiple benefits, from food to ecosystem services to coastal protection with long-term sustainability.

REFERENCES

Ahrens, J.P. 1987. Characteristics of reef breakwaters. Coastal Engineering Research Center. Department of the Army, Waterways Experimental Station, Corps of Engineers. Vicksburg, MS.

Barras, J., S. Beville, D. Britsch, S. Hartley, J. Johnston, P. Kemp, Q. Kinler, A. Martucci, J. Porthouse, D. Reed, K. Roy, S. Sapkota, and J. Suhayda. 2003. Historical and projected coastal Louisiana land changes: 1978–2050: USGS Open File Report 03-334, p. 39.

Beck, M. et al. 2011. Oyster reefs at risk and recommendations for conservation, restoration, and management. *Bioscience* 61(2): 107–116.

Benassai, G. 2006. *Introduction to Coastal Dynamics and Shoreline Protection*. WIT Press, Southampton, MD. pp. 249–266.

Blum, M.D. and H.H. Roberts. 2009. Drowning of the Mississippi Delta due to insufficient sediment supply and global sea-level rise. *Nature Geoscience* 2(2009): 488–491.

Boyd, C. and N. Pace. 2013. Coastal Alabama Living Shorelines Policies, Rules and Model Ordinance Manual. NOAA/Mobile Bay National Estuary Program publication, http://www.mobilebaynep.com/images/uploads/library/Coastal-Alabama-Living-Shorelines-Policies-Manual.pdf.

Byrum, M., S. Hall, J. Erdman, J. Sullivan, L. Harrell, C. Knott, and S. Bertrand. 2014. Culturing Coastal Plants and Animals for Sustainable Housing, ASABE July 2014, Montreal, QC.

Campbell, M.D. 2004. Analysis and Evaluation of a Bioengineered Submerged Breakwater. http://etd.lsu.edu/docs/available/etd-11052004-152838/.

Coen, L.D., R.D. Brumbaugh, D. Bushek, R. Grizzle, M.W. Luckenbach, M.H. Posey, S.P. Powers, and G. Tolley. 2007. As we see it: Ecosystem services related to oyster restoration. *Marine Ecology Progress Series* 341: 303–307.

Coleman, J.M., H.H. Roberts, and G.W. Stone. 1997. Mississippi River Delta: An overview. *Journal of Coastal Research* 14(3): 698–716.

Committee on Coastal Erosion Zone Management (CCEZM). 1990. *Managing Coastal Erosion*. National Academy Press, Washington, D.C. pp. 56–62.

Day, J.W. Jr. et al. 2007. Restoration of the Mississippi Delta: Lessons from Hurricanes Katrina and Rita. *Science* 315: 1679.

Dehon, D. 2010. Thesis: Investigating the Use of Bioengineered Oyster Reefs as a Method of Shoreline Protection and Carbon Storage. Master of Science in Biological and Agricultural Engineering. 63 pp.

Ehrich, M. and L. Harris. 2015. A review of existing eastern oyster filtration rate models. *Ecological Modelling* 297: 201–212.

Grabowski, J.H., A.R. Hughes, D.L. Kimbro, and M.A. Dolan. 2005. How habitat setting influences restored oyster reef communities. *Ecology* 86: 1926–1935.

Hall, S. 2009a. Complementary contributions to coastal restoration in Louisiana and Mexico. *Louisiana Agriculture* 53(3): 18–19.

Hall, S.G. 2009b. Considerations for engineering with natural and artificial reefs in oyster and coral dominated environments. In: Thangadurai, N, (Ed.), *Biotechnology in Fisheries and Aquaculture* 35–46.

Hall, S.G. 2010. Culturing eastern oyster, *Crassostrea virginica* in coastal Louisiana for food production, ecological restoration and carbon sequestration. *Proc. AES Issues Forum* 154–165.

Hall, S. 2015. Coastal Bioengineering for Sustainable Cost Effective and Ecological Solutions. Conference Paper 2145044, ASABE 2015, New Orleans, July 2015, 8 pp.

Hall, S.G., R. Beine, M. Campbell, and T. Ortego. 2006. Development of biologically dominated engineered structures for erosion reduction and environmental restoration. In: Xu, Y.J. and V.P. Singh, (Eds.). *Coastal Environment and Water Quality*, Water Resources Publications, pp. 455–464.

Hall, S.G., R. Beine, M. Campbell, and T. Ortego. 2007. Bioengineered wave breaks grow oysters, help restore coast. *Louisiana Agriculture* 51(1): 34–37.

Hall, S.G., R. Beine, T. Ortego, M. Campbell, and M. Turley. 2009. Bioengineered reefs to enhance natural fisheries and culture eastern oyster Crassostrea virginica in the Gulf of Mexico. In: Thangadurai, N., (Ed.), *Biotechnology in Fisheries and Aquaculture* 27–34.

Hall, S.G. and M. Lima. 2001. Problem solving approaches and philosophies in biological engineering: Challenges from technical, social and ethical arenas. *Transactions of the ASAE* 44(4): 1037–1041.

Hall, S.G., J. Risinger, and T. Ortego. 2012. Ecologically Engineered Reefs for Sustainable Coastal Restoration. Abstract, Ecological Engineering Society, Syracuse NY, June 2012.

Hall, S.G., D. Smith, R. Price, M. Thomas, and J. Steyer. 2015. Robots and artificial reefs for water conservation and management. *Louisiana Agriculture*, Fall 2015.

Hall, S., J. Steyer, and M. Thomas. 2016. Coastal bioengineered reefs and engineered cage culture to grow eastern oysters *Crassostrea virginica* for food and coastal protection and restoration. *World Aquaculture Society* Feb 2016, abstr. 457.

Hamaguchi, T., T. Uda, C. Inoue, and A. Igarashi. 1991. Field experiment on wave-dissipating effect of artificial reefs on the Niigata Coast. *Coastal Engineering in Japan, Japan Society for Civil Engineers* 34: 50–65.

Harding, J. and R. Mann. 2001. Oyster reefs as fish habitat: Opportunistic use of restored reefs by transient fishes. *Journal of Shellfish Research* 20(3): 951–959.

Kamphuis, J.W. 2010. *Introduction to Coastal Engineering and Management* (2nd Edition). Advanced Series on Ocean Engineering: V. 30, ISBN 978-981-283-484-3, 564 pp.

La Peyre, M.K., A. Humphries, S. Casas, and J. La Peyre. 2014. Temporal variation in development of oyster reef functional services. *Ecological Engineering* 63: 34–44.

Malveaux, C. and S. Hall. 2015. Use of Semi-Autonomous Multicopters to Assess Coastal Bioengineered Reefs and Habitat. ASABE 2015, New Orleans, Abstract 2189263.

Mazda, Y., M. Magi, M. Kogo, M. Kogo, and P.N. Hong, 1997. Mangroves as a coastal protection from waves in the Tong King delta, Vietnam. *Mangroves and Salt Marshes*. June, 1(2): 127–135. DOI:10.1023/A:1009928003700.

Mendelssohn, I., G. Anderson, D. Baltz, R. Caffey, K. Carman, J. Fleeger, S. Joye, Q. Lin, E. Maltby, E. Overton, and L. Rozas. 2012. Oil impacts on coastal wetlands: Implications for the Mississippi River Delta ecosystem after the Deepwater Horizon oil spill. *BioScience* 62(6): 562–574.

Meyer, D.L., E.C. Townsend, and G.W. Thayer. 1997. Stabilization and erosion control value of oyster cultch for intertidal marsh. *Restoration Ecology* 5: 93–99.

NOAA Habitat Conservation National Marine Fisheries Service. Living Shorelines. Retrieved from: http://www.habitat.noaa.gov/restoration/techniques/livingshorelines.html.

O'Beirn, F., M. Luckenbach, J. Nestlerode, and G. Coates. 2000. Toward design criteria in constructed oyster reefs: Oyster recruitment as a function of substrate type and tidal height. *Journal of Shellfish Research* 19(1): 387–395.

Ortego, T.R. 2006. Analysis of Bioengineered Concrete for Use in a Submerged Reef Type Breakwater. Master's Thesis, LSU. http://etd.lsu.edu/docs/available/etd-06122006-223212/. 48 pp.

Penland, S. et al. 2005. Changes in Louisiana's shoreline: 1855–2002. *Journal of Coastal Research* (Special Issue no. 44): 7–39.

Peterson, C.H., J.H. Grabowski, and S.P. Powers. 2003. Estimated enhancement of fish production resulting from restoring oyster reef habitat: quantitative valuation. *Marine Ecology Progress Series* 264: 249–264.

Phillips, M.R. and A.L. Jones. 2005. Erosion and tourism infrastructure in the coastal zone: Problems, consequences and management. *Tourism Management* 27: 517–524.

Piazza, B.P., P.D. Banks, and M.K. La Peyre. 2005. The potential for created oyster shell reefs as a sustainable shoreline protection strategy in Louisiana. *Restoration Ecology* 13: 499–506.

Pope, J. and W.R. Dally. 1986. Detached Breakwaters for Shore Protection, Technical Report CERC-86-1, U.S. Army Engineer Waterways Experiment Station, Vicksburg, MS.

Ravens, T.M., R.C. Thomas, K.A. Roberts, and P.H. Santschi. 2009. Causes of salt marsh erosion in Galveston Bay, Texas. *Journal of Coastal Research*, 25(2): 265–272. West Palm Beach (Florida), ISSN 0749-0208.

Risinger, J. 2012. Biologically Dominated Engineered Coastal Breakwaters. PhD Dissertation, LSU. http://etd.lsu.edu/docs/available/etd-07032012-120846/. 128 pp.

Risinger, J.D., M. Turley, T. Ortego, and S. Hall. 2013. Bioengineered Oyster Reefs for Sustainable Shoreline Protection & Ecosystem Restoration in the Gulf of Mexico. Presentation, 5th National Conference on Ecosystem Restoration (NCER). Chicago.

Risinger, J.D., M. Turley, T. Ortego, and S. Hall. 2012. Oyster Reefs 101: An Engineer's Perspective. Presentation, Restore America's Estuaries (RAE), Tampa, FL (Invited).

Roland, R.M., and S.L. Douglass. 2005. Estimating wave tolerance of Spartina alterniflora in coastal Alabama. *Journal of Coastal Research* 21(3): 453–463.

Scyphers, S.B., S.P. Powers, K.L. Heck, and D. Byron. 2011. Oyster reefs as natural breakwaters mitigate shoreline loss and facilitate fisheries. *PLoS ONE* 6:e22396.

Smith, D., L. Cross, J. Rivet, and S. Hall. 2014. Design of a semi-autonomous boat for measurements of coastal sedimentation and erosion. Sediment dynamics from the summit to the sea. *Proceedings of the International Association of Hydrological Sciences* 267: 101–112. http://www.proc-iahs.net/367/index.html.

Smith, D., J. Rivet, M. Nalewaik, and S. Hall. 2015. Autonomous Collection of Coastal Bathymetry Data. (Poster) 2189088. (poster 134) ASABE 2015, New Orleans, July 2015.

Steyer, G.D. 2010. Coastwide Reference Monitoring System (CRMS): U.S. Geological Survey Fact Sheet 2010-3018, 2 pp. (Revised August 2010.)

Walker, H.J. 1988 (Ed.). *Artificial Structures and Shorelines*. Boston, Kluwer Publishers.

Williams, S.J., G.W. Stone, and A.E. Burrus. 1997. A perspective on the Louisiana wetland loss and coastal erosion problem. *Journal of Coastal Research* 13(3): 593–594.

CHAPTER 14

Evaluation of Living Shoreline Marshes as a Tool for Reducing Nitrogen Pollution in Coastal Systems

Aaron J. Beck, Randy M. Chambers, Molly M. Mitchell, and Donna Marie Bilkovic

CONTENTS

14.1 INTRODUCTION

The impact of increased nitrogen, phosphorus, and sediment discharge as non–point source pollution into estuarine systems is a growing concern. Nitrogen and phosphorus in part are responsible for the high productivity of estuaries, but overabundance of these nutrients stimulates nuisance algal blooms that can have negative impacts on other species and habitats. For example, epiphytic algae on seagrass blades reduce light availability to the seagrass and ultimately reduce seagrass resilience (e.g., Orth et al. 2006). Algal overproduction in the water column and subsequent algal death and decomposition and expansion of global "dead zones" have been attributed to excess nutrients (Diaz and Rosenberg 2008). Management strategies are needed in many estuarine systems to reduce or eliminate the loading of nitrogen, the nutrient that typically limits algal production.

Wetlands can function as sinks for nitrogen in at least three interrelated ways: short-term retention of N via plant uptake, long-term retention of N via accumulation in soils, and microbial transformation of N via denitrification. Using wetlands to reduce N pollution is a well-established technique dating back to the 1970s. The treatment of wastewater, agricultural runoff, stormwater flow, and highway runoff by constructed wetlands has become a common and accepted practice. Created wetlands can be a cost-effective method for reducing anthropogenic non–point source pollution.

Growing interest in nature-based shoreline protection approaches (termed "living shorelines" henceforth) has led states and localities to consider the value-added potential for living shorelines to reduce nitrogen loads into estuaries. Natural marshes have been shown to be effective filters for nitrogen in both groundwater (Tobias et al. 2001) and tidal waters (Deegan et al. 2007). On the basis of this N removal function by natural marshes, water quality management models (e.g., the Chesapeake Bay Watershed Model, http://www.chesapeakebay.net/about/programs/modeling/53/) have begun to incorporate N removal efficiencies for created marshes. An explicit N removal efficiency has not been determined for living shorelines and their created marshes. These tidal marshes differ from typical created marshes in that they are narrow fringing marshes that are specifically designed to protect shorelines from erosion. This chapter focuses on the potential for living shoreline created marshes to remove nitrogen from groundwater and tidal surface waters, thus improving water quality.

Fringing marshes of living shorelines receive regular flows of surface water via tidal flooding and thus may exchange nitrogen with the open water of the estuary. Owing to their location, fringing marshes may also intercept groundwater discharging from upland sources (Figure 14.1). The submarine groundwater discharge (SGD) of nutrients to Chesapeake Bay has not been systematically

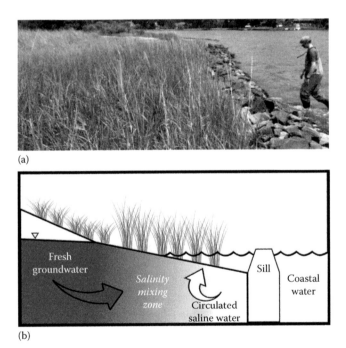

(a)

(b)

Figure 14.1 (a) Photograph showing the living shoreline at the Sturgeon site. (b) Cross-section schematic of a living shoreline site showing fresh (meteoric) groundwater, and circulated seawater N sources to the marsh.

studied and may be as high as 30% of surface inputs (Libelo et al. 1991). SGD and the movement of groundwater through fringing marshes may allow for the interception and processing of nitrogen, but this process has never been quantified for living shoreline marshes.

The goals of this study were to determine the potential for N removal/transformation via three interrelated and not mutually exclusive processes occurring in living shoreline marshes: (1) short-term N retention by plants—annual retention of nitrogen via seasonal aboveground production of marsh grasses, (2) long-term retention in soils—*burial of organic N* that accumulates over years, and (3) N removal from groundwater and tidal waters from the upland to the water's edge. Because the age of a marsh may affect the biogeochemical processes occurring within it (Bilkovic and Mitchell 2016, Chapter 15, this book), a chronosequence of living shorelines was included in the study.

14.2 METHODS

14.2.1 Study Area Description

The Chesapeake Bay is the largest estuarine system in the United States at approximately 11,500 km^2, with some 19,000 km of shoreline (Kemp et al. 2005). The Bay watershed is nearly 15 times that size, and extends into six states (Delaware, Maryland, New York, Pennsylvania, Virginia, and West Virginia) and the District of Columbia. Surface freshwater inputs come primarily from the Susquehanna, Potomac, Rappahannock, York, and James Rivers.

We estimated the range and average rates of nitrogen removal for eight created tidal marshes from living shorelines of the lower Chesapeake Bay. A ninth site was sampled for plant uptake and groundwater and surface water removal of N, but not for burial of N. Living shorelines use natural elements, sometimes in combination with a stabilizing structure, to control erosion, conserve habitat, and maintain coastal processes. Our study focused on hybrid living shoreline projects that included created salt marsh and low, freestanding stone structures (sills or break-waters) placed parallel to and near the marsh shoreline for stabilization; these are more common in higher energy settings. To select marsh-sill sites, all marsh-sill projects in Virginia between 1 and 12 years of age (to ensure at least one full growing season and modern construction standards) were inventoried using Virginia Institute of Marine Science shoreline permit databases (Bilkovic and Mitchell 2013). Because the age of a marsh may influence biogeochemical processes, marshes ranging from newly created to up to 12 years of age were incorporated into the analysis to establish trends of nitrogen removal over time. Besides the age of the project, sites were required to (1) be within the meso-polyhaline salinity regime, (2) be accessible, and (3) meet certain design standards (planted marshes protected by a sill built of oyster shell or rocks not exceeding 300 lb with a crest ≤ 0.3 m above mean high water and regular gaps to allow tidal intrusion). On that basis, the candidate pool was narrowed and permission was obtained for sampling at nine marsh-sill sites (Figure 14.2).

14.2.2 Sampling Design and Methods

Bimonthly sampling of water, plants, and soils was carried out in 2014 (five sampling events: February, April, June, August, and October) to capture the seasonal variability of wetland processes. For each site ($n = 9$), five transects were randomly selected that ran perpendicular from the edge of the marsh to the upland and were at least 2 m apart. Geomorphic characteristics were evaluated onsite and remotely: slope, marsh survey area, area of low and high marsh, and shoreline length. Slope (difference in low to high marsh/distance from the upland border to marsh edge) was calculated in ArcGIS 10.0 along five transects using elevation data obtained with Northwest

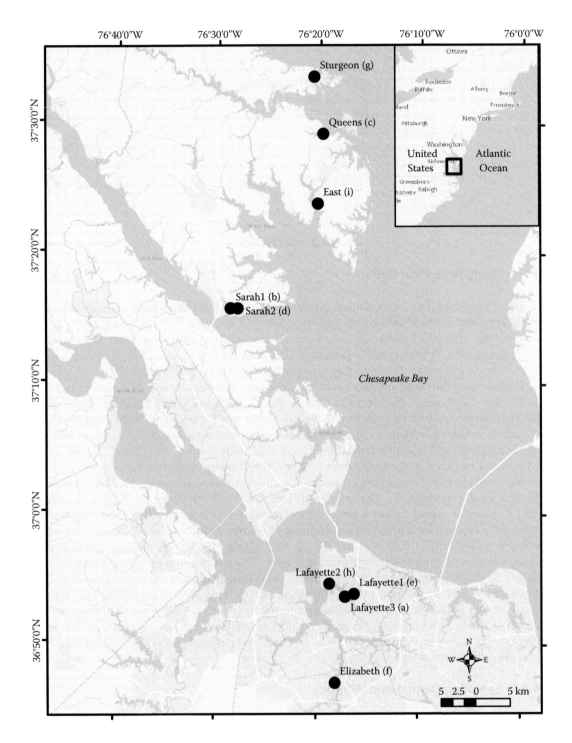

Figure 14.2 Site map showing locations of the nine Living Shoreline marshes in the current study. Associated letters in parenthesis correspond to those used in Figures 14.3 and 14.4.

Instruments NRL802 rotary laser with stadia rod and sensor in combination with a Trimble Geo XH to measure elevations and GPS locations of points along each transect line.

14.2.3 Short-Term N Retention by Plants

Marsh plant biomass and foliar nitrogen content were estimated to determine the potential for short-term N retention. Along each transect, in low (*Spartina alterniflora*) and high (*Spartina patens*) marsh zones, plants were identified and enumerated, and the plant height of the five tallest stems was measured within 0.25 m² quadrats during April–October. Aboveground biomass of *S. alterniflora* and *S. patens* was estimated nondestructively from stem count and average height data using simple linear regression models. Biomass data used in the regressions were based on measurements of vegetation harvested from two marsh-sill sites at the end of the growing season (October). For foliar nitrogen content, both early and late in the growing season, leaf samples were collected from high and low marsh sites along all five transects of all nine marshes. From each site, the third leaf down from the growing shoot was collected from five individual plants and bagged together. Leaf samples were dried and ground in a Wiley mill. Total percent nitrogen by weight was determined on a Perkin-Elmer 2400 Elemental Analyzer and then multiplied by aboveground biomass to calculate short-term N retention per square meter. Net accumulation of biomass N measured during the growing season was assumed to be annual.

14.2.4 Long-Term N Retention in Soils

Total nitrogen in marsh soil was measured to determine the potential for nutrient accumulation (long-term marsh retention) in eight of the living shoreline marshes ranging from 2 to 11 years postcreation. Cores were collected from high marsh sites (dominated by *S. patens*) and low marsh sites (dominated by *S. alterniflora*) from transects 1, 3, and 5 of each marsh. The 7-cm-diameter cores were collected to a depth of 30 cm and then sliced into 10-cm sections (some soils could not be cored to a full 30 cm). Soil percent organic matter was determined as loss on ignition of a subsample of the dried bulk soil. Samples were dried and weighed for determination of bulk density. Macro-organic matter was removed by sieving and separation by hand. The weight percent nitrogen of sieved soil and macro-organic matter was determined separately for each 10-cm section of each core and then combined to calculate total percent N of the soil. Long-term N accumulation per square meter was calculated as the increase in soil N content 0–20 cm relative to the soil N content 20–30 cm.

14.2.5 N Removal from Groundwater and Tidal Waters

Groundwater samples were collected at the shallowest saturated depth (i.e., 0.2–1 m depth) using a stainless steel shielded-screen piezometer (Retract-a-Tip; Charette and Allen 2006). Samples were collected along the longest or most central transect within the experimental plot, from the waterline to the landward side of the living shoreline. Individual groundwater samples were collected at three to five locations along the transect, depending on shoreline width. Surface water samples were collected by hand within 2 m of shore. Approximately 60-mL samples were collected and placed on ice until they could be returned to the laboratory. Samples were centrifuged and filtered (GF/F, 0.7 μm nominal) within 8 h of collection, and stored frozen until analysis. Dissolved NH_4^+ and NO_x $\left(NO_3^- + NO_2^-\right)$ were measured by the manual hypochlorite and spongy cadmium methods, respectively, and analyzed using a Shimadzu UV-1601 spectrophotometer (Grasshoff et al. 1999). Salinity was measured in the field with a handheld YSI556 multiprobe or a handheld refractometer. Salinity mixing diagrams were used to evaluate nonconservative nutrient behavior and resultant influence on N flux through the living shoreline, as described more fully below.

Differences in aboveground biomass and soil and plant tissue characteristics (e.g., g/m^2, % N) by shore position/vegetation type (high marsh/*S. patens* vs. low marsh/*S. alterniflora*) were examined with one-way analyses of variance (ANOVAs) with post hoc comparisons. Plant density, height, and aboveground biomass were regressed against the age of the marsh. Aboveground biomass was log-transformed before analyses to meet the assumptions of parametric statistical tests. Regression analysis was also used to test for significant correlations of marsh age with soil %N and bulk soil N content.

14.3 RESULTS

14.3.1 Short-Term N Retention by Plants

Because of standard planting techniques for living shoreline marshes, diversity was low in all marshes, typically restricted to *S. alterniflora* in the low marsh and *S. patens* in the high marsh. Because other plants were rare, analyses were limited to these two species. In all marshes, *S. alterniflora* was taller but grew less densely than *S. patens*. *S. alterniflora* had an average height of 97 ± 50 cm and an average density of 110 ± 83 stems/m^2. *S. patens* had an average height of 84 ± 29 cm and an average density of 560 ± 479 stems/m^2. There was no relationship between the age of these created marshes and plant density or height (*S. alterniflora* [$F_{1,8} = 0.03, p = 0.9, F_{1,8} = 0.01, p = 0.9$]; *S. patens* [$F_{1,8} = 2.2, p = 0.2, F_{1,8} = 0.7, p = 0.4$]), suggesting that plant productivity did not follow a clear trajectory over the 11-year age time range of the marshes surveyed. Plant aboveground biomass was strongly related to stem height and counts for each plant species (*S. alterniflora:* $R^2 = 85.8$, *S. patens:* $R^2 = 83.2$), allowing development of an allometric relationship for estimating plot biomass based on plot density and plant height. Plant aboveground biomass was calculated for each species using the following equations:

$$Spartina \ alterniflora \ \text{Biomass} = (0.1807e^{0.0332*\text{Mean HT}})* \text{Number of Stems} \qquad (14.1)$$

$$Spartina \ patens \ \text{Biomass} = (0.0381e^{0.04*\text{Mean HT}})* \text{Number of Stems} \qquad (14.2)$$

Aboveground biomass estimates at the end of the growing season (Oct) averaged 601 ± 869 g C/m^2 (*S. alterniflora*) and 298 ± 364 g C/m^2 (*S. patens*) and did not differ with the age of the marsh for *S. alterniflora* ($F_{1,8} = 0.03, p = 0.9$) or *S. patens* ($F_{1,8} = 3.7, p = 0.1$).

Foliar leaf nitrogen content (percent by weight) was higher in the spring than in the fall for both grass species. *S. alterniflora* averaged 2.86% ± 0.33% in the spring and 1.28% ± 0.15% in the fall. *S. patens* averaged 2.58% ± 0.30% in the spring and 1.95% ± 0.25% in the fall. *S. alterniflora* had significantly higher spring foliar leaf nitrogen compared to *S. patens* (one-way ANOVA, $F_{1,66} = 13.67, p = 0.0004$), but fall foliar leaf nitrogen was comparable between the two grass types. Foliar leaf nitrogen content of the grass was related to plot characteristics, with higher leaf nitrogen content in less dense plots of *S. alterniflora* ($R^2 = -0.41$, spring; $R^2 = -0.46$, fall) and denser plots of *S. patens* ($R^2 = 0.50$, fall). For *S. patens*, leaf nitrogen content was also positively related to plot grass height ($R^2 = 0.50$, fall).

Accumulation of N in plant biomass was estimated using the aboveground N inventories of *S. patens* (high marsh) and *S. alterniflora* (low marsh) and N measurements of plant biomass at three locations within each high and low marsh. Plant N ranged between 0.66 and 15.11 g/m^2 for the high marsh and between 1.79 and 20.48 g/m^2 for the low marsh. Averaged across each marsh, plant N ranged between 1.2 and 14.4 g/m^2. Low marsh biomass accounted for 70% ± 17% of the total marsh grass N inventory.

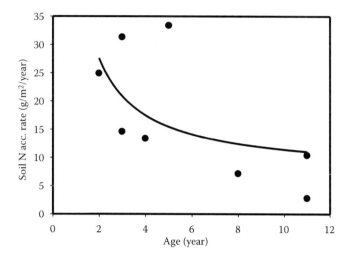

Figure 14.3 Calculated rates of N accumulation in soils as a function of created marsh age.

14.3.2 Long-Term N Retention in Soils

Sediment carbon content by weight decreased slightly (but not significantly, owing to the high variability) with depth (0–10 cm, 2.22% ± 2.31%; 10–20 cm, 1.29% ± 1.11%; 20–30 cm, 0.86% ± 0.77%). Sediments in the top 0–10 cm had low organic carbon content and were similar between low (2.08%C ± 2.41%C) and high marsh plots (2.22%C ± 2.26%C).

Soil macro-organic matter typically comprised less than 1% of the soil weight, and the sieved soil%N contributed >90% of the total bulk soil N. Across all sites and depths, soil N averaged 0.06% ± 0.06% by weight and ranged from 0.01% to 0.16%. For high marsh soils, %N was not significantly correlated with marsh age (*S. patens* $F_{1,22} = 2.2$, $p = 0.15$), but %N increased significantly with age for low marsh soils (*S. alterniflora* $F_{1,22} = 5.31$, $p = 0.03$). Bulk soil N averaged 182 ± 78 g/m^2 in the low marsh and 173 ± 90 g/m^2 in the high marsh. Bulk soil N was not significantly correlated with marsh age (*S. alterniflora* $F_{1,22} = 2.29$, $p = 0.14$; *S. patens* $F_{1,22} = 2.1$, $p = 0.16$). Among marshes, average annual accumulation of N in soils was variable and ranged from 2.8 to 33.4 g/m^2/year (Figure 14.3).

14.3.3 N Removal from Groundwater and Surface Waters

Hydraulic gradients between inland groundwater and bay surface water varied with slope of the living shoreline, from 0.03 to 0.28. These gradients were estimated at mid-to-low tide, and would be expected to vary tidally as observed elsewhere in this region (Beck et al. 2016). Groundwater discharge was calculated according to the Darcy equation ($v = K_h[dh/dl]$), assuming a regional hydraulic conductivity (K_h) of 0.001 cm/s (Gallagher et al. 1996; McFarland and Bruce 2006; Reay et al. 1992; Robinson et al. 1998). A seepage face of 5 m width was estimated. This is a conservatively low estimate, as seepage faces are generally on the order of 20–100 m (Bokuniewicz 1980; Michael et al. 2003; Reay et al. 1992; Shaw and Prepas 1990), but appropriate considering the shallow depth of the living shorelines. Meteoric groundwater discharge was therefore estimated to be between 2.7 and 23.8 cm/day, which corresponds to a volumetric flux of 134–1192 L/day/m shoreline length (Table 14.1). The groundwater flux was lowest at the low gradient Elizabeth site and highest at the steep Lafayette1 site.

Groundwater salinity generally decreased from approximately that of surface water at the waterline to low, but not necessarily fresh salinity at the landward border of the living shoreline (Figure 14.4). Salinity showed little or inconsistent variation along the transect at Lafayette3 and

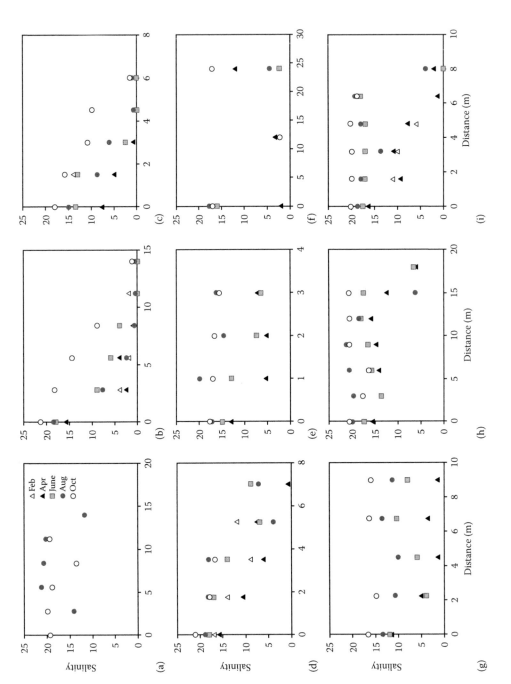

Figure 14.4 Groundwater salinity versus distance across marsh (distance "0" indicates bay surface water) at each site. (a) Lafeyette3, (b) Sarah1, (c) Queens, (d) Sarah2, (e) Lafayette1, (f) Elizabeth, (g) Sturgeon, (h) Lafayette2, and (i) East.

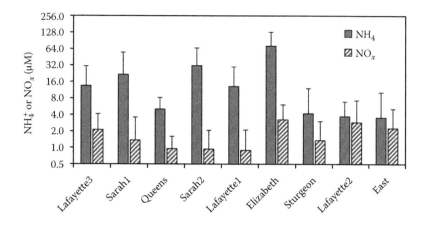

Figure 14.5 Nutrient species (solid bars, NH_4^+; hatched bars, NO_x^-) in all groundwater samples collected at each living shoreline site over the study period. Columns represent mean concentrations and error bars indicate one standard deviation of the mean. Note that the y axis is on a log_2 scale.

Elizabeth sites (Figure 14.4a and f). At the Lafayette2 site, groundwater salinity was high until nearly the upland border of the living shoreline, where it decreased by 10–15 units (Figure 14.4h). Seasonal changes in groundwater salinity were also observed, with lowest salinity in spring, and increasing salinity through fall.

Dissolved total inorganic nitrogen $\left(NO_x^- + NH_4^+; TIN\right)$ average concentrations ranged from 3.5 to 17 μM (0.049–0.24 mg/L) in surface water and from 5.5 to 74 μM (77–1.04 mg/L) in groundwater. Most of the TIN at all sites comprised NH_4^+, with NO_x^- representing an average of 7%–39% of TIN at different sites (Figure 14.5). TIN concentrations were generally enriched in groundwater relative to high-salinity surface water (Figure 14.6), except at the Sturgeon and Lafayette2 sites, which had low TIN concentrations in fresher groundwater (Figure 14.6g and h). No consistent variation of TIN with salinity was observed at the Elizabeth site (Figure 14.6f). No seasonal variation of TIN concentrations was evident at any of the sites.

14.3.4 Groundwater TIN Flux Estimates for Living Shorelines

At most of the sites sampled, TIN concentrations were highest in the low-salinity or fresh groundwater, implying that groundwater is a potential source of fixed-N to Bay waters. However, nonconservative mixing behavior was evident at all sites (Figure 14.6). Trends within single seasons may not have captured full salinity gradients and trends across seasons were generally similar at individual sites. Thus, TIN concentrations in the fresh endmember were assumed to be constant over the year for each site, and data for each site were treated without regard to season. At some sites, the fresh groundwater endmember concentration was taken to be that in the highest observed zero-salinity sample (e.g., 100 μM [1.4 mg/L] at the Sarah1 site, Figure 14.6b). Fresh groundwater was not successfully sampled at all sites; thus, the zero-salinity endmember was estimated by qualitatively extrapolating a linear fit of the concentrations at high salinity (Officer 1979). For example, the highest TIN concentration at the Lafayette1 site (Figure 14.6e) was ~70 μM (0.98 mg/L), at salinity 7. A straight line drawn through the most saline points indicated that the high concentration at salinity 7 was consistent with conservative mixing between the observed surface water and a zero-salinity endmember of 110 μM (1.54 mg/L) for that site.

TIN concentrations across the mixing zone at each marsh were mostly lower than expected for conservative mixing, suggesting removal of fixed-N within the living shoreline. At two sites, Sturgeon and Lafayette2 (Figure 14.6g and h, respectively), the fresh endmember had lower TIN

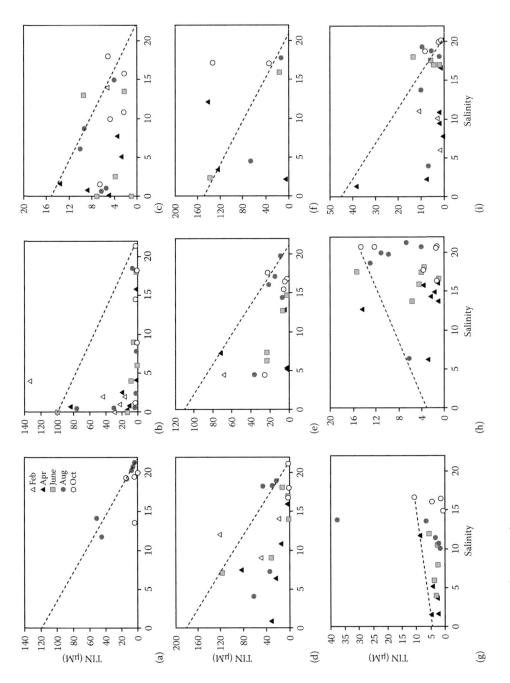

Figure 14.6 Groundwater TIN ($NO_x^- + NH_4^+$) concentrations versus salinity at each site. Dashed lines indicate conservative mixing between surface water and fresh groundwater. (a) Lafayette3, (b) Sarah2, (c) Queens, (d) Sarah1, (e) Lafayette2, (f) Elizabeth, (g) Sturgeon, (h) Lafayette2, and (i) East.

concentrations than surface water, but groundwater samples along the mixing zone still mostly indicated nonconservative removal. The extent of removal for each sample was estimated as the difference between the expected TIN concentration based on the conservative mixing line and that observed in the sample. Negative "removal" rates occurred if the concentration was higher than expected from conservative mixing, and suggested enrichment along the flowpath. Evapotranspiration could also potentially cause apparent enrichment of dissolved nutrients and, if important at these sites, would make our calculations an underestimate of actual removal. Removal rates across all sites ranged between 100% and −63%, and averaged 55% ± 40% (neglecting four samples with greater than 100% removal or enrichment). Approximately 90% of samples showed indication of nonconservative TIN removal along the groundwater mixing flowpath.

The fresh groundwater TIN flux, neglecting removal due to biogeochemical processes occurring within the living shoreline marsh, was estimated as the product of groundwater volumetric discharge (134–1192 L/day/m shoreline length; see above) and TIN concentration in the zero-salinity endmember. These fluxes represent the expected input to Bay surface waters in the absence of the living shoreline buffer zone, and ranged from 0.7 to 131 mmol/day/m shoreline (9.8–1800 mg/day/m shoreline). The TIN removal was estimated at each site using fresh groundwater TIN flux estimates and the average percent removal observed in groundwater samples at the site. Removal of TIN from fresh groundwater within the living shorelines ranged between 0 and 87 mmol/day/m shoreline (0–1200 mg/day/m shoreline; Table 14.1).

14.3.5 Surface Water TIN Flux Estimates for Living Shorelines

The salinity gradients observed across most of the sites in this study (Figure 14.4) indicate substantial recirculation of Bay surface waters through the permeable sands used to construct living shorelines. TIN concentrations in surface waters were generally lower than those in fresh groundwater (Figure 14.6) but were nonetheless often higher than concentrations in brackish groundwater, indicating removal within the living shoreline subsurface.

Because we know of no previous work that has evaluated the relative proportions of meteoric versus recirculated groundwater in living shorelines, the volume of Bay water circulated through these living shoreline marshes cannot be precisely estimated. However, previous studies examining a range of spatial scales and a variety of global sites have shown that recirculated groundwater discharge is usually greater by a factor of 10 or more than the meteoric groundwater discharge (e.g., Burnett et al. 2006; Younger 1996). As a first approximation, Bay water recirculation volume in the living shoreline sites is assumed to be 10-fold higher than the Darcy groundwater flux, although it may be as much as 50-fold higher (Harvey and Odum 1990).

The flux of TIN from Bay surface waters to living shorelines is therefore estimated as the product of recirculated water volume flux, average surface water TIN concentrations at each site, and the average TIN percent removal in groundwater at each site (Table 14.1). Removal fluxes ranged between 0 and 89 mmol/day/m shoreline (Table 14.1) and were generally of comparable magnitude to removal fluxes from the fresh groundwater. However, at the Sturgeon and Lafayette2 sites, low TIN concentrations in the fresh groundwater suggested low groundwater TIN fluxes (0.3 and 0.5 mmol/day/m shoreline [4 and 7 mg/day/m shoreline], respectively), but the recirculated groundwater component drove an estimated TIN removal from Bay water of 6 and 25 mmol/day/m shoreline (84 and 350 mg/day/m shoreline), respectively. The recirculated groundwater component represented 58% ± 25% of the total TIN removal across all sites.

14.3.6 Summary of N Removal in Living Shoreline Marshes

The comparison of N removal/transformation calculations for the eight living shoreline marshes is shown in Table 14.2. Average removal/transformation of TIN from groundwater and surface

Table 14.1 Summary of Groundwater Flows and Calculations of Total Inorganic Nitrogen (TIN) Removal from Groundwater and Surface Water Exchange

Marsh	Age (Years)	Hydraulic Gradient (m/m)	Darcy GW Rate (cm/Day)	Fresh GW Flux (L/m/Day)	Average TIN (µM)	GW TIN Removal (mmol/m/Day)	SW TIN Removal (mmol/m/Day)	Total TIN Removal (mmol/m/Day)
Sarah2	2	0.121	10.5	523	32.4 (35.5)	50	23	73
Queens	3	0.087	7.5	376	5.7 (3.2)	2	6	9
Lafayette1	3	0.276	23.8	1192	13.8 (17.5)	87	89	176
Sarah1	4	0.075	6.5	324	22.8 (34.2)	23	9	32
Elizabeth	5	0.031	2.7	134	74.4 (59.3)	0	0	0
Lafayette2	8	0.071	6.1	307	5.9 (4.8)	0	25	25
Sturgeon	11	0.035	3.0	151	5.5 (8.0)	0	6	6
East	11	0.101	8.7	436	5.7 (7.6)	12	9	21

Note: Groundwater flux across the face of the marsh is in L/m/day. Removal of TIN from groundwater (GW) and surface water (SW) exchange is expressed in millimoles of N per meter of marsh shoreline per day. Total TIN removal is the sum of GW and SW N removal. Standard deviations of the average TIN values are shown in parentheses.

Table 14.2 Comparison of N Dynamics (in g/m²/Year) from Living Shoreline Marshes of Different Ages

Marsh	Age	Short-Term N Retention	Long-Term N Retention	TIN Removal
Sarah2	2	5.0	24.9	55.8
Queens	3	1.2	31.3	7.3
Lafayette1	3	12.7	14.6	231.8
Sarah1	4	3.1	13.4	17.9
Elizabeth	5	9.9	33.4	0.0
Lafayette2	8	14.4	7.1	11.9
East	11	3.4	2.8	9.6
Sturgeon	11	1.9	10.4	4.7

Note: Short-term N retention is the average measured accumulation of N in aboveground plant biomass within a single growing season. Long-term N retention is the average measured accumulation of N in marsh soil from 0–20 cm depth. TIN removal is the calculated average reduction in total dissolved inorganic nitrogen associated with groundwater flow and surface water infiltration and exchange through living shoreline soils.

water was highest (42.4 ± 78.5 g/m²/year), followed by long-term N retention in soils (17.6 ± 10.8 g/m²/year) and short-term N retention in plants (6.4 ± 5.1 g/m²/year). No systematic variation among removals was obvious, in that high rates of TIN removal were not always matched by high rates of soil and plant accumulation; in these marshes, denitrification could be a more important N removal process, but cannot be determined from our results. Likewise, short-term N retention in plants might be the result of translocation of stored nutrients from the soil N pool, or the result of N uptake from incoming groundwater or surface water. Even so, only two sites (East and Lafayette2) had more short-term retention in plants than there was N available in the bulk soil. TIN delivery at those sites, however, was sufficiently large to support annual plant growth.

14.4 DISCUSSION

Living shoreline marshes have a role in nitrogen removal from both groundwater and flooding tidal waters, reducing overall non–point source pollution. Nitrogen removal within living shorelines is expected to occur via plant uptake and accumulation in soils, or via microbe-mediated conversion of fixed nitrogen species to dinitrogen gas (i.e., denitrification and anammox). The potential importance of these different nitrogen sinks was evaluated in the current work by examining aboveground biomass N, soil N, and interpreting mixing plots of dissolved N in porewater. Natural marshes have been found to efficiently remove nitrogen from groundwater (e.g., 90% removal; Tobias et al. 2001) and partially from tidal waters (e.g., 30%–40% removal; Deegan et al. 2007). Likewise, restored marshes from the Gulf of Mexico coast have shown near-complete removal of groundwater nitrogen (Sparks et al. 2015). We saw on average 55% of total dissolved N removal from groundwater and from flooding tidal waters. Living shoreline vegetation should remove nitrogen as effectively as natural marsh vegetation, since the foliar nitrogen content in *S. alterniflora* and *S. patens* from our sites is at least as high as from natural marshes. Because of their relative young age, however, other removal mechanisms (e.g., soil organic storage and denitrification) may be less efficient.

Soil organic nitrogen is the primary pool of nitrogen in a natural salt marsh, comprising approximately 98% of the total plant-soil nitrogen content in a Louisiana salt marsh (Buresh et al. 1980). However, soil nitrogen pools are related to organic matter content, both of which can be much lower in created marshes compared to natural marshes (Langis et al. 1991). Denitrification can be an important removal process for groundwater as it flows out to the estuary or ocean. Tidal salt marshes are considered an effective N sink in coastal systems in part because of their role in denitrification (Merrill and Cornwell 2000), which is controlled by availability of NO_3^- and organic carbon in a reducing environment. Changes in any of these factors have the potential to change an ecosystem's value as a sink for terrestrial nitrogen inputs (Davis et al. 2004). Living shorelines typically are built with clean sand fill, a sediment type that is very different from the organic matrix of natural marshes. Sandy soils tend to be less efficient removers of nitrogen compared to organic soils (Portnoy et al. 1998), because sandy soils can have higher dissolved oxygen content and lower retention time and lower organic matter available for denitrification. In peat soils (e.g., natural marsh soils), maximum denitrification rates are higher and occur closer to the soil surface than in sandy soils (Davidsson et al. 1997). Living shoreline marshes typically are planted in sandy soils, where water circulation is driven by a variety of processes, including wave and tide pumping, density-driven transport, and bioirrigation (Santos et al. 2012). As a result, TIN fluxes within the living shorelines must also include some nitrogen transported from Bay surface water into the marsh. Do N fluxes into these marshes lead to N removal, that is, can living shoreline marshes be considered efficient removers of nitrogen if much of the N is derived from flooding tidal water? Prior research in both natural and created marshes found average ranges of soil N accumulation from 3–8 $g/m^2/$ year up to 14–19 $g/m^2/$year (Craft et al. 2002, 2003). From the current study, the initial rates of N accrual are high—almost 30 $g/m^2/$year, but with age, those rates decline to less than 10 $g/m^2/$year (Figure 14.3). This is still on the high end of N accumulation for created marshes. Living shorelines can represent a major sink for N even in locations where meteoric groundwater contains little TIN and where non–point source TIN flux from groundwater is low.

14.4.1 Groundwater Characteristics of Living Shoreline Sites

Coastal groundwater often plays an important role in nutrient transport and processing in coastal ecosystems, influencing biodiversity, ecosystem structure and function, and benthic and pelagic productivity. Nutrient-enriched SGD can sustain benthic and neritic phytoplankton and seagrasses in coastal regions (Johannes 1980; Rutkowski et al. 1999; Ullman et al. 2003; Valiela et al. 1990). Conversely, SGD-derived chemical input may also be detrimental for its role in promoting harmful algal blooms (Gobler and Sanudo-Wilhelmy 2001; Lee and Kim 2007) or as a source of contaminants such as mercury to benthic communities (Laurier et al. 2007).

Mixing of fresh groundwater and intruding seawater results in a dynamic and highly active zone known as the subterranean estuary (STE; Moore 1999). Groundwater within the saline STE may primarily consist of recycled marine chemical constituents either from surface water or regenerated during microbial remineralization of particulate organic matter (Santos et al. 2009). In the fresh region of the STE, however, terrestrially derived, "new" chemical constituents are transported to the estuary and ultimately the ocean (McLachlan and Illenberger 1986). Geochemical transformations occurring during fresh-saline mixing in the STE can result in nonconservative behavior of dissolved constituents with either land or ocean source (Beck et al. 2007). Because of these transformations, SGD can act as either a source or a sink for different nutrients in the estuary or coastal ocean.

Groundwater salinity variation across the transects indicated that living shorelines effectively represent subterranean estuaries (Moore 1999), where fresh, meteoric groundwater mixes with saline groundwater (i.e., circulated surface seawater) (Figure 14.1). The degree of saline intrusion in living shorelines appears to be related to the hydraulic gradient, with low slope sites (e.g., Lafayette3 and Elizabeth) having a low or unclear salinity gradient. In contrast, high slope sites such as Sarah2 and East have marked salinity gradients, at least during the high-recharge spring season (Figure 14.4). Seasonal variation in recharge and thermosteric sea level anomalies leads to increased saline intrusion and landward retreat of the STE in summer and early fall (Gonneea et al. 2013; Luek and Beck 2014; Michael et al. 2005).

Nutrient concentrations in groundwater at the living shoreline sites (Figure 14.5) were consistent with other observations in the Chesapeake Bay region but were generally lower within the brackish marsh groundwater samples than observed elsewhere (Beck et al. 2016; Gallagher et al. 1996; Reay

2004; Reay et al. 1992). Results of the current study also matched previous work showing low NO_x^- and high NH_4^+ concentrations in suburban/urban land use types (Reay and Simmons 1992 and references therein). The dominance of NH_4^+ in groundwater samples indicates that nutrient cycling processes other than denitrification (e.g., anammox, uptake by plants, or coupled nitrification/denitrification) must be responsible for any observed fixed-N removal.

14.4.2 Removal Mechanisms—Plant Uptake versus N_2 Formation

Nitrogen is the primary limiting nutrient in many salt marshes (Howarth 1988); therefore, there is often a tight relationship between nitrogen availability and plant productivity, biomass, abundance, and leaf nitrogen content (Valiela 1983). Fertilization with nitrogen significantly increases both aboveground plant biomass and plant height in *S. alterniflora* (Buresh et al. 1980). In addition, *S. alterniflora* has a high N:P molar ratio owing to the requirement of nitrogen to created compounds used for osmotic regulation in salt marshes (Smart and Barko 1980). Salt marsh plants are optimized to take up any available nitrogen and have been found to efficiently remove nitrogen from both groundwater (Tobias et al. 2001) and tidal waters (Wolaver et al. 1983).

Neither foliar nitrogen content nor aboveground biomass of marsh plants changed with marsh age. It should be noted that all living shoreline marshes are relatively young (<12 years old) compared to the age of natural marshes (>50 years old) and uncertainty exists regarding how these marshes will mature over time. These marshes may initially have an advantage from fertilization during planting as well as relatively low intra- and interspecific competition that could diminish over time.

Foliar leaf nitrogen was higher in the low marsh (*S. alterniflora*) plants than in the high marsh plants in the spring (when leaf nitrogen was overall higher). The low marsh tends to have higher porewater salinity than the high marsh; thus, *S. alterniflora* may have a higher requirement for nitrogen. However, these differences disappeared in the fall, possibly because of high nitrogen use over the summer months by *S. alterniflora*. Foliar leaf nitrogen was positively related to height and density in *S. patens*, an expected relationship. However, it was negatively correlated with both height and density for *S. alterniflora*. The negative correlation in both cases was less than 50% and may be driven by growth responses to high salinity rather than nitrogen availability. The different patterns in the response of leaf nitrogen content to plant density in the two grass species may be explained by their relative locations in the marsh or as different mechanistic responses. This opens the possibility that they may play different roles in the sequestration of nitrogen in marshes.

Soil and aboveground biomass N represented a removal flux of between 6 and 44 g/m²/year (Table 14.2). This flux is of similar order of magnitude to the estimated removal of TIN during groundwater advection and mixing in the living shoreline. In some marshes, TIN removal exceeded the soil and plant removal flux, suggesting that dinitrogen formation reactions may have occurred at these and probably other sites. In addition to denitrification, the observed co-occurrence of NO_x and NH_4 at most sites indicates potential for anammox (as measured elsewhere in the Chesapeake Bay; Dale and Miller 2007), although anammox is not considered an important pathway in salt marshes (Koop-Jakobsen and Giblin 2009). At sites where organic N accumulation exceeded rates of removal from groundwater, particulate N deposition from suspended particles may have provided a major source.

14.4.3 Summary: Nutrient Removal by Living Shorelines

Living shoreline marshes can intercept nitrogen (N) in both surface waters and groundwater and may be in a position to reduce N loading to estuarine ecosystems. We examined the potential for N removal/transformation in eight living shoreline marshes, considering three interrelated processes: short-term N-removal by plants, long-term N accumulation in soils, and N removal from

groundwater and surface water. From five transects in each of the eight marshes, seasonal accumulation of N in the aboveground portions of plants from the high marsh (*S. patens*) and low marsh
(*S. alterniflora*) ranged from 1 to 15 $g/m^2/year$, and soil N accumulation ranged between 3 and
33 $g/m^2/year$. Analysis of change in dissolved inorganic N in groundwater and surface water flow
through the marshes indicated N removal rates ranging from 0 to 232 $g/m^2/year$. Collectively, living
shoreline marshes exhibit a variable but consistent pattern of N removal at rates comparable to other
created and natural marshes. Although our data do not allow us to quantify the relative importance
of denitrification, plant uptake, geochemical sorption, and burial, the overall indication is that living
shoreline marshes can be effective components of N reduction strategies for coastal systems.

The results of the current work demonstrate that living shorelines can function to remove fixed
nitrogen from both meteoric groundwater and recirculated seawater. It remains unclear whether
the removal is associated with a permanent sink via dinitrogen formation, or whether it is simply
incorporated into biomass that can be subsequently buried or decomposed and remineralized. In
either case, living shorelines appear to provide an effective buffer for non–point source nutrient
inputs. Because biomass formation within these marshes occurs during summer months when surface waters are most strongly impaired by excess nutrients and harmful algae growth, living shorelines may represent an ideal approach for reducing or preventing non–point source nutrient inputs
during the most sensitive seasons.

Even though living shorelines are capable of removing nitrogen from groundwater and tidal
waters, their overall effectiveness depends on both their placement in the watershed and their areal
coverage. Modeling of nutrient removal by freshwater wetlands showed that despite high effectiveness on a unit area basis, they had relatively little impact on the overall river load owing to their
low water residence time and low areal coverage within the watershed (Arheimer and Wittgren
2002). In that case, changes in agricultural management and implementation of Best Management
Practices were far more effective in reducing overall loads (30% vs. 5%, respectively; Arheimer et
al. 2004). In addition, there may be a limit on nitrogen removal by marshes, as removal efficiency
appears to decrease as anthropogenic loads increase (Valiela and Cole 2002). Nonetheless, the current work indicates that living shorelines are capable of removing a major portion of nitrogenous
nutrient species from both groundwater and surface water. Because these living shorelines operate
at the interface between land and sea, they act as a final control on non–point source nutrient inputs
to coastal waters. Living shorelines may therefore be an important component of an effective NPS
pollution reduction scheme, if perhaps not a solitary solution.

ACKNOWLEDGMENTS

This study was supported by the Commonwealth of Virginia. We thank Michele A. Cochran,
Virginia Institute of Marine Science; Tim Russell, The College of William & Mary; Dave Stanhope,
Kory Angstadt, and many other colleagues at the Center for Coastal Resources Management,
Virginia Institute of Marine Science, for field and laboratory support. We are grateful to the anony-
mous reviewer for comments, which helped improve our manuscript. This paper is Contribution No.
3571 of the Virginia Institute of Marine Science, College of William & Mary.

REFERENCES

Arheimer, B. and H.B. Wittgren. 2002. Modelling nitrogen removal in potential wetlands at the catchment
      scale. *Ecological Engineering* 19(1): 63–80.
Arheimer, B., G. Torstensson, and H.B. Wittgren. 2004. Landscape planning to reduce coastal eutrophication:
      Agricultural practices and constructed wetlands. *Landscape and Urban Planning* 67: 205–215.

Beck, A.J., A.A. Kellum, J.L. Luek, and M.A. Cochran. 2016. Chemical flux associated with spatially and temporally variable submarine groundwater discharge, and chemical modification in the subterranean Estuary at Gloucester Point, VA (USA). *Estuaries and Coasts* 39(1): 1–12.

Beck, A.J., Y. Tsukamoto, A. Tovar-Sanchez, M. Huerta-Diaz, H.J. Bokuniewicz, and S.A. Sañudo-Wilhelmy. 2007. Importance of geochemical transformations in determining submarine groundwater discharge-derived trace metal and nutrient fluxes. *Applied Geochemistry* 22: 477–490.

Bilkovic, D.M. and M.M. Mitchell. 2013. Ecological tradeoffs of stabilized salt marshes as a shoreline protection strategy: Effects of artificial structures on macrobenthic assemblages. *Ecological Engineering* 61: 469–481.

Bilkovic, D.M. and M.M. Mitchell. 2016. Designing living shoreline salt marsh ecosystems to promote coastal resilience. In: *Living Shorelines: The Science and Management of Nature-Based Coastal Protection*. Bilkovic, D.M., M.M. Mitchell, M.K. La Peyre, and J.D. Toft (Eds.). CRC Press, Boca Raton, FL.

Bokuniewicz, H. 1980. Groundwater seepage into Great South Bay, New York. *Estuarine and Coastal Marine Science* 10(4): 437–444.

Buresh, R.J., R.D. DeLaune, and W.H. Patrick. 1980. Nitrogen and phosphorus distribution and utilization by *Spartina alterniflora* in a Louisiana Gulf Coast marsh. *Estuaries* 3(2): 111–121.

Burnett, W.C., P.K. Aggarwal, H. Bokuniewicz, J.E. Cable, M.A. Charette, E. Kontar, S. Krupa, K.M. Kulkarni, A. Loveless, W.S. Moore, J.A. Oberdorfer, J. Oliveira, N. Ozyurt, P. Povinec, A.M.G. Privitera, R. Rajar, R.T. Ramessur, J. Scholten, T. Stieglitz, M. Taniguchi, and J.V. Turner. 2006. Quantifying submarine groundwater discharge in the coastal zone via multiple methods. *Science of the Total Environment* 367(2–3): 498–543.

Charette, M.A. and M.C. Allen. 2006. Precision ground water sampling in coastal aquifers using a direct-push, shielded-screen well-point system. *Groundwater Monitoring & Remediation* 26(2): 87–93.

Craft, C., S. Broome, and C. Campbell. 2002. Fifteen years of vegetation and soil development after brackish-water marsh creation. *Restoration Ecology* 10: 248–258.

Craft, C., P. Megonigal, S. Broome, J. Stevenson, R. Freese, J. Cornell, L. Zheng, and J. Sacco. 2003. The pace of eco-system development of constructed *Spartina alterniflora* marshes. *Ecological Applications* 13(5): 1417–1432.

Dale, R.K. and D.C. Miller. 2007. Spatial and temporal patterns of salinity and temperature at an intertidal groundwater seep. *Estuarine, Coastal and Shelf Science* 72(1): 283–298.

Davidsson, T.E., R. Stepanauskas, and L. Leonardson. 1997. Vertical patterns of nitrogen transformations during infiltration in two wetland soils. *Applied and Environmental Microbiology* 63(9): 3648–3656.

Davis, J.L., B. Nowicki, and C. Wigand. 2004. Denitrification in fringing salt marshes of Narragansett Bay, Rhode Island, USA. *Wetlands* 24(4): 870–878.

Deegan, L., J. Bowen, D. Drake, J. Fleeger, C. Friedrichs, K. Galvan, J. Hobbie, C. Hopkins, D. Johnson, J. Johnson, L. LeMay, E. Miller, B. Peterson, C. Picard, S. Sheldon, M. Sutherland, J. Vallino, and R. Warren. 2007. Susceptibility of salt marshes to nutrient enrichment and predator removal. *Ecological Applications* 17(5): S42–S63.

Diaz, R.J. and R. Rosenberg. 2008. Spreading dead zones and consequences for marine ecosystems. *Science* 321(5891): 926–929.

Gallagher, D.L., A.M. Dietrich, W.G. Reay, M.C. Hayes, and G.M. Simmons. 1996. Ground water discharge of agricultural pesticides and nutrients to estuarine surface water. *Groundwater Monitoring & Remediation* 16(1): 118–129.

Gobler, C.J. and S.A. Sañudo-Wilhelmy. 2001. Temporal variability of groundwater seepage and Brown Tide Blooms in a Long Island Embayment. *Marine Ecology Progress Series* 217: 299–309.

Gonneea, M.E., A.E. Mulligan, and M.A. Charette. 2013. Climate-driven sea level anomalies modulate coastal groundwater dynamics and discharge. *Geophysical Research Letters* 40(11): 2701–2706.

Grasshoff, K., K. Kremling, and M. Ehrhardt (Eds.). 1999. *Methods of Sea Water Analysis*. 3rd ed. Wiley-VCH Verlag GmbH, Weinheim.

Harvey, J.W. and W.E. Odum. 1990. The influence of tidal marshes on upland groundwater discharge to estuaries. *Biogeochemistry* 10(3): 217–236.

Johannes, R.E. 1980. The ecological significance of the submarine discharge of groundwater. *Marine Ecology Progress Series* 3: 365–373.

Kemp, W.M., W.R. Boynton, J.E. Adolf, D.F. Boesch, W.C. Boicourt, G. Brush, J.C. Cornwell, T.R. Fisher, P.M. Glibert, J.D. Hagy, L.W. Harding, E.D. Houde, D.G. Kimmel, W.D. Miller, R.I.E. Newell, M.R. Roman, E.M. Smith, and J.C. Stevenson. 2005. Eutrophication of Chesapeake Bay: Historical trends and ecological interactions. *Marine Ecology Progress Series* 303: 1–29.

Koop-Jakobsen, K. and A.E. Giblin. 2009. Anammox in tidal marsh sediments: The role of salinity, nitrogen loading, and marsh vegetation. *Estuaries and Coasts* 32(2): 238–245.

Langis, R., M. Zalejko, and J.B. Zedler. 1991. Nitrogen assessments in a constructed and a natural salt marsh of San Diego Bay. *Ecological Applications* 1(1): 40–51.

Laurier, F.J.G., D. Cossa, C. Beucher, and E. Breviere. 2007. The impact of groundwater discharges on mercury partitioning, speciation, and bioavailability to mussels in a coastal zone. *Marine Chemistry* 104(3–4): 143–155.

Lee, Y.-W. and G. Kim. 2007. Linking groundwater-borne nutrients and dinoflagellate red-tide outbreaks in the southern sea of Korea using a Ra tracer. *Estuarine, Coastal and Shelf Science* 71: 309–317.

Libelo, E.L., W.G. MacIntyre, and G.H. Johnson. 1991. Groundwater nutrient discharge to the Chesapeake Bay: Effects of near-shore land use practices. pp. 613–622. In: *New Perspectives in the Chesapeake System: A Research and Management Partnership. Proceedings of a Conference.* Mihursky, J.A., and A. Chaney (Eds.). CRC Publication 137, Solomons, MD.

Luek, J.L. and A.J. Beck. 2014. Radium budget of the York River estuary (VA, USA) dominated by submarine groundwater discharge with a seasonally variable groundwater end-member. *Marine Chemistry* 165: 55–65.

McFarland, E.R. and T.S. Bruce. 2006. The Virginia coastal plain hydrogeologic framework. US Geological Survey professional paper (1731).

McLachlan, A. and W. Illenberger. 1986. Significance of groundwater nitrogen input to a beach surf zone ecosystem. *Stygologia* 2: 291–296.

Merrill, J. and J. Cornwell. 2000. The role of oligohaline marshes in estuarine nutrient cycling. In: *Concepts and Controversies in Tidal Marsh Ecology.* M. Wienstien and D. Kreeger (Eds.). Springer, Netherlands.

Michael, H.A., A.E. Mulligan, and C.F. Harvey. 2005. Seasonal oscillations in water exchange between aquifers and the coastal ocean. *Nature* 436(7054): 1145–1148.

Michael, H.A., J.S. Lubetsky, and C.F. Harvey. 2003. Characterizing submarine groundwater discharge: A seepage meter study in Waquoit Bay, Massachusetts. *Geophysical Research Letters* 30(6): 1297.

Moore, W.S. 1999. The subterranean estuary: A reaction zone of ground water and sea water. *Marine Chemistry* 65(1): 111–125.

Officer, C.B. 1979. Discussion of the behaviour of nonconservative dissolved constituents in estuaries. *Estuarine and Coastal Marine Science* 9(1): 91–94.

Orth, R.J., T.J. Carruthers, W.C. Dennison, C.M. Duarte, J.W. Fourqurean, K.L. Heck, A.R. Hughes, G.A. Kendrick, W.J. Kenworthy, S. Olyarnik, and F.T. Short. 2006. A global crisis for seagrass ecosystems. *Bioscience* 56(12): 987–996.

Portnoy, J.W., B.L. Nowicki, C.T. Roman, and D.W. Urish. 1998. The discharge of nitrate-contaminated groundwater from developed shoreline to marsh-fringed estuary. *Water Resources Research* 34(11): 3095–3104.

Reay, W.G. 2004. Septic tank impacts on ground water quality and nearshore sediment nutrient flux. *Ground Water* 42(7): 1079–1089.

Reay, W.G., D.L. Gallagher, and G.M. Simmons. 1992. Groundwater discharge and its impact on surface water quality in a Chesapeake Bay inlet. *Water Resources Bulletin* 28: 1121–1134.

Reay, W.G. and G.M. Simmons. 1992. Groundwater discharge in coastal systems: Implications for Chesapeake Bay. Perspectives on Chesapeake Bay, pp. 17–44.

Robinson, M., D. Gallagher, and W. Reay. 1998. Field observations of tidal and seasonal variations in ground water discharge to tidal estuarine surface water. *Groundwater Monitoring & Remediation* 18(1): 83–92.

Rutkowski, C.M., W.C. Burnett, R.L. Iverson, and J.P. Chanton. 1999. The effect of groundwater seepage on nutrient delivery and seagrass distribution in the northeastern Gulf of Mexico. *Estuaries* 22(4): 1033–1040.

Santos, I.R., W.C. Burnett, T. Dittmar, I.G. Suryaputra, and J. Chanton. 2009. Tidal pumping drives nutrient and dissolved organic matter dynamics in a Gulf of Mexico subterranean estuary. *Geochimica et Cosmochimica Acta* 73(5): 1325–1339.

Santos, I.R., B.D. Eyre, and M. Huettel. 2012. The driving forces of porewater and groundwater flow in permeable coastal sediments: A review. *Estuarine, Coastal and Shelf Science* 98: 1–15.

Shaw, R.D. and E.E. Prepas. 1990. Groundwater-lake interactions: I. Accuracy of seepage meter estimates of lake seepage. *Journal of Hydrology* 119(1): 105–120.

Smart, R.M. and J.W. Barko. 1980. Nitrogen nutrition and salinity tolerance of *Distichlis spicata* and *Spartina alterniflora*. *Ecology* 61(3): 630–638.

Sparks, E.L., J. Cebrian, C.R. Tobias, and C.A. May. 2015. Groundwater nitrogen processing in Northern Gulf of Mexico restored marshes. *Journal of Environmental Management* 150: 206–215.

Tobias, C., S. Mako, I. Anderson, E. Canuel, and J. Harvey. 2001. Tracking the fate of a high concentration groundwater nitrate plume through a fringing marsh: A combined groundwater tracer and in situ isotope enrichment study. *Limnology and Oceanography* 46(8): 1977–1989.

Ullman, W.J., B. Chang, D.C. Miller, and J.A. Madsen. 2003. Groundwater mixing, nutrient diagenesis, and discharges across a sandy beachface, Cape Henlopen, Delaware (USA). *Estuarine, Coastal and Shelf Science* 57: 539–552.

Valiela, I. 1983. Nitrogen in salt marsh ecosystems. In: *Nitrogen in the Marine Environment*. E.J. Carpenter and D.G. Capone (Eds.), pp. 649–678. Academic Press, San Diego, California.

Valiela, I. and M.L. Cole. 2002. Comparative evidence that salt marshes and mangroves may protect seagrass meadows from land-derived nitrogen loads. *Ecosystems* 5(1): 92–102.

Valiela, I., J. Costa, K. Foreman, J.M. Teal, B. Howes, and D. Aubrey. 1990. Transport of groundwater-borne nutrients from watersheds and their effects on coastal water. *Biogeochemistry* 10(3): 177–197.

Wolaver, T.G., J.C. Zieman, R. Wetzel, and K.L. Webb. 1983. Tidal exchange of nitrogen and phosphorus between a mesohaline vegetated marsh and the surrounding estuary in the lower Chesapeake Bay. *Estuarine, Coastal and Shelf Science* 16(3): 321–332.

Younger, P.L. 1996. Submarine groundwater discharge. *Nature* 382: 121–122.

PART IV

Synthesis of Living Shoreline Science
Biological Aspects

CHAPTER 15

Designing Living Shoreline Salt Marsh Ecosystems to Promote Coastal Resilience

Donna Marie Bilkovic and Molly M. Mitchell

CONTENTS

15.1 BACKGROUND

As human occupation of coastlines grows and threats from flooding increase because of a changing climate, shoreline protection becomes an increasing priority. Globally, shoreline protection approaches are evolving toward the incorporation of more natural and nature-based features. This is a deliberate movement away from traditional armoring approaches including seawalls, bulkheads, or revetments, which now dominate urban coastal landscapes. Past approaches to shoreline protection have resulted in significant hardening of natural shorelines; for example, 18% of Chesapeake Bay (Bilkovic et al. 2016; Figure 15.1), 17% of New Jersey (Lathrop and Love 2007), 21% of Florida (Florida DEP 1990), and 30% of southern California (Griggs 1998) coastlines were armored, with higher values (>45%–50%) along urban shores in many regions. Shoreline hardening reduces the capacity of the coastline to provide habitat, absorb and reduce floodwaters, and adapt to changing water levels. New understanding of the importance of natural shorelines in promoting coastal resilience and the adverse effects of hardening has led to a reexamination of shoreline management policy in many systems.

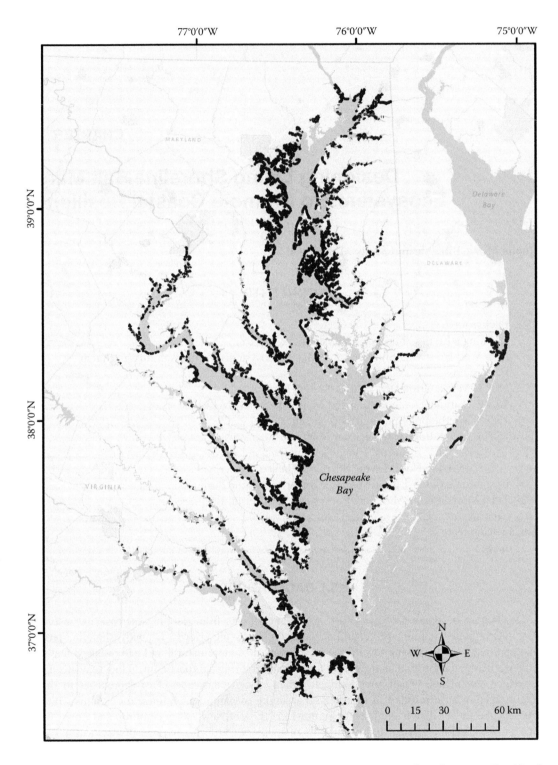

Figure 15.1 Distribution of shoreline armoring (bulkhead, riprap revetment, seawall, and unconventional hard materials) in Chesapeake Bay is represented as black lines along the shore. Approximately 18% of Chesapeake Bay tidal shorelines are armored.

Change in shoreline management policy in response to scientific insights has come in waves. In Chesapeake Bay, the largest estuary in the United States, the predominant form of shoreline protection in the mid to late 20th century was armoring with a bulkhead or seawall (Figure 15.2). Toward the end of the 20th century, a greater scientific understanding about the detrimental impacts of vertical seawalls and bulkheads on coastal intertidal habitats and fauna led to the encouraged use of riprap revetments to lessen those impacts. Even so, the proliferation of armoring has fragmented natural habitats and introduced novel habitat to most coastal environments. Artificial structures can act as ecological barriers, separating upland and wetland functions, increasing the rate of wetland loss, and changing the character of the nearshore ecosystem (Peterson and Lowe 2009 and references within). Fragmentation results in smaller natural habitat patches that are further apart and surrounded by a matrix of artificial or inferior habitat that can affect dispersal of organisms and subsequently species diversity and composition at local and regional scales (Collinge 2009). These changes can result in disrupted connectivity, habitat homogenization, and altered coastal landscapes, with uncertain implications for estuarine and marine faunal community structure and function.

Further innovation designed to better emulate land–water connectivity resulted in nature-based erosion control approaches (*living shorelines* henceforth) that involve offshore stabilizing wave-break structures in combination with natural features such as marshes (Figure 15.3). Innovative living shoreline designs are now being promoted to maximize ecosystem service provision while protecting eroding shores. However, coastal environments are complex, with patterns and processes of the terrestrial landscape (geomorphic and human features) influencing adjacent seascapes (intertidal and submerged marine landscapes). Mimicking the complexity of interactions in a created environment requires a thorough and sophisticated understanding of the connectivity between seascapes and terrestrial landscapes and remains one of the major challenges in coastal ecology.

Some nuanced differences in the definitions of living shorelines are used by various agencies, researchers, or interest groups, but several common elements often exist. Generally, living shorelines have been defined as the use of natural elements, commonly marsh vegetation, sometimes in

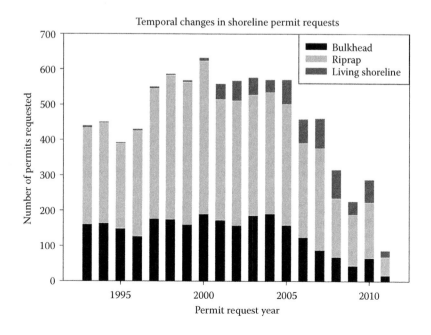

Figure 15.2 Changes in the use of shoreline protection approaches in Virginia, Chesapeake Bay over time.

Figure 15.3 A low-profile stone sill used as a wave break in front of a created salt marsh. (Image by Karen Duhring, CCRM/VIMS.)

combination with a stabilizing structure, to control erosion, restore or conserve habitat, and maintain coastal processes (Bilkovic and Mitchell 2013). A key component of living shorelines is that they provide for both the physical protection and ecological function of coastal habitats. They are engineered shorelines, designed specifically to break wave energy and reduce shoreline erosion while minimizing the adverse effects typically associated with hardened shorelines and therefore require balance between the ecological and engineering design criteria. In this regard, they differ from restored/created marshes that are designed entirely for ecological services. However, the ecological services provided by living shorelines largely have been assumed more than tested, with the idea that in marshes, function follows form: if the engineered shoreline is physically protected and the marsh grass grows, then all other ecological functions eventually will be attained. "Successful" engineering designs, however, may not be a success ecologically if the grass grows but estuarine and marine faunal communities are not supported. A full test of the ecological outcomes of different engineering designs of living shorelines has never been completed.

Here, we present new findings and review case studies and information from both the gray and peer-reviewed literature to synthesize the current understanding on the functioning and ecosystem services provision of a living shoreline marsh ecosystem. We focus on living shoreline approaches that include created salt marsh, a prevalent living shoreline technique found in many US settings but notably along the East and Gulf coasts. In addition, using our synthesis review, we examine the potential for living shoreline ecosystems to improve coastal resilience in urbanized estuaries with threatened coastal habitats.

15.2 DEFINING A LIVING SHORELINE ECOSYSTEM

The hypothetical underpinnings of living shorelines are that (1) natural habitats (including coastal wetlands, oyster reefs, dunes, and sea grasses) reduce the risk of coastal flooding and erosion and enhance resilience defined here as the ability of an ecosystem to absorb disturbance and maintain its structure, function, identity, and feedbacks (Gunderson et al. 2009; Walker et al. 2004); and (2) created habitats will eventually provide some level of ecosystem function in addition to coastal protection that is similar to the natural systems they are designed to emulate.

The first hypothesis has been definitively demonstrated for salt marshes. Marsh vegetation is known to break wave energy (Knutson et al. 1982; Morgan et al. 2009; Shepard et al. 2011) and, in wide enough marshes, can provide significant protection to the upland (Gedan et al. 2011). In areas where wave energies exceed the practical width of living shoreline marshes, wave breaks, built of stone or oyster shell, can be placed immediately channelward of the marsh, enhancing erosion protection.

The second hypothesis is still an area of active research. Because marsh vegetation is both the physical protection element of living shorelines and the foundation species upon which the ecology of the marsh is built, it has become the primary metric of success in evaluation of living shorelines. However, salt marsh ecosystems are a complex web of elements, including the plants, animals, sediments, and microbial communities, all of which are critical for full functionality. As a hybrid human-built ecosystem, a living shoreline marsh with a stabilizing stone structure is not likely to fully realize equivalent structure and function as a salt marsh. Indeed, it is difficult to identify a true natural analog to a living shoreline marsh. The combination of hard structure fronting a marsh may be compared to marsh-oyster reef complexes, but not all structure is designed to emulate reefs and studies suggest that food web dynamics vary (e.g., Bilkovic and Mitchell 2013; Wong et al. 2011). However, by better linking engineering with ecology, new research is seeking to improve designs of novel structures to maximize select ecosystem services (e.g., fish habitat provision, water quality improvement) while accommodating coastal community needs for protection (e.g., Firth et al. 2014).

Restored marshes (frequently extensive marshes built through excavation of uplands or dredge spoil deposition) may require decades to attain equivalence to natural marshes in terms of biodiversity, plant productivity, and biogeochemical properties (e.g., Craft et al. 1999; Morgan and Short 2002; Zedler and Callaway 1999). Primary production and plant species richness are common functional indicators used to assess created marsh performance in relation to natural marshes. Generally, plant structure and productivity develop faster (<10 years) than other properties such as soil organic matter and benthic faunal composition (Craft et al. 1999; Morgan and Short 2002). Soil organic matter is an important functional indicator positively associated with surface accretion, plant growth, and invertebrate density and diversity (Craft et al. 2003; Morgan and Short 2002; Moy and Levin 1991; Sacco et al. 1994). Trajectory models for soil organic matter in created marshes suggest that 15 years or more is often needed to reach near equivalency with reference marshes (Craft et al. 1999; Morgan and Short 2002). Living shorelines may follow a similar maturation trajectory, although the introduction of novel structures or placement of sand fill for hybrid designs may cause living shoreline marshes to deviate from these established trajectories. Because living shorelines are relatively new shoreline protection approaches, existing data on living shoreline marsh properties primarily fall within the early stages of a possible trajectory curve; a much longer monitoring time frame is required to build predictive trajectory models.

Conceptually, we can identify key elements controlling the flow of energy and subsequent ecosystem service provision of a living shoreline salt marsh ecosystem as compared to a natural salt marsh ecosystem (Figure 15.4). This conceptual model can guide the design and implementation of living shorelines, as well as define the parameters of created ecosystem function and service provision.

Salt marshes exist at the interface of terrestrial and marine habitats along protected coastlines in the middle and high latitudes (Mitsch and Gosselink 2015) and range from narrow fringing bands along shorelines to vast complexes with interior tidal creeks. The extent of the salt marsh is largely determined by the slope of the shoreline, tide range, sediment processes, and wave energy climate. The zonation and structure of plants and animals in the marsh are primarily defined by the

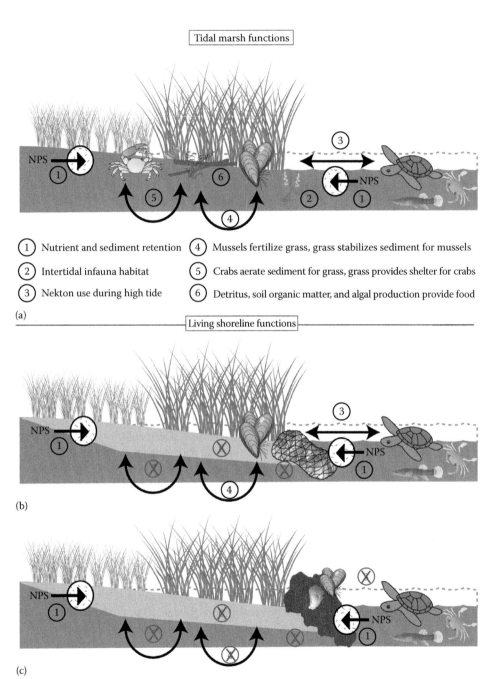

Figure 15.4 Comparative ecosystem conceptual models of (a) natural salt marsh, (b) a living shoreline created salt marsh with an oyster sill structure, and (c) a living shoreline created salt marsh with a stone sill structure. Functions that may be reduced in a given environment are grayed out. Several prominent differences between living shoreline marshes and natural marshes may influence marsh ecosystem development including (1) differences in the resulting plant communities; (2) the use of coarse sand material with low organic matter content as a substrate that may lead to different nutrient pathways; (3) the uniform topography, vegetation, and edge characteristics; and (4) impacts of novel habitat introduction. Oyster sills tend to be low-elevation, intertidal structures that allow full access of fauna to the marsh surface during high tide. Stone sills frequently extend above mean high water, restricting fauna usage and access to the marsh surface.

extent and duration of inundation, fluctuations in salinity, drying and submergence, and extremes in temperature (Mitsch and Gosselink 2015). Biotic factors may further contribute to the regulation of plant distribution and production (e.g., through interspecific competition and grazing/consumer control; Bertness 1985; Silliman and Zieman 2001). Salt marshes predominantly fall within the neap-spring intertidal zone and are often divided into two zones—low marsh (flooded daily) and high marsh (flooded irregularly). Along the East and Gulf coasts, common dominant plants are *Spartina alterniflora* in the low marsh and *Spartina patens* and *Distichlis spicata* in the high marsh.

While the construction of a living shoreline salt marsh is firmly grounded in marsh creation and restoration science, there are several key physical differences from a natural or restored salt marsh that are a result of practical constraints such as project size, availability of upland or subtidal bottom to create proper elevations, and regulatory requirements including sediment fill grain size. In addition, living shorelines are constructed with engineering goals predominating (i.e., shoreline erosion control) that may influence or constrain the water quality and habitat capacity of these ecosystems. A living shoreline salt marsh is most aptly compared to a narrow fringing salt marsh that follows the coastline. It is typically constructed by grading the land or filling the shallow water subtidal zone with clean sand to create suitable elevations for marsh plantings. Living shorelines in higher-energy environments (0.8 km < fetch < 1.6 km) typically include infrastructure that are low-profile (height = 0 to +0.3 m above mean high water) shore-parallel structures to facilitate the establishment of marsh plants and help attenuate waves. These structures are often called sills or breakwaters and may be constructed of quarry stone or native oyster shell contained by bags or may consist of a series of concrete oyster domes or reef balls/castles designed to provide habitat structure for benthic ecosystem engineers. Sills may include gaps, windows, or vents to provide tidal access for fauna. A marsh-sill is often built channelward with conversion of existing intertidal and subtidal lands to sand fill, planted marsh, and sill structure. These designs result in a wider intertidal area and a change in elevation, sediment type, and plant usage but better maintain the upland-water connection compared to riprap revetments and bulkheads. In urban settings, one limitation to project construction is often the availability of land to accommodate the elevation needs for the development of a self-sustaining marsh.

15.3 HOW WELL DO LIVING SHORELINE SALT MARSH ECOSYSTEMS MIMIC NATURAL SALT MARSH ECOSYSTEMS?

The highly dynamic nature of salt marsh systems controls community organization, interaction, and function. Few species have successfully adapted to the widely variable and often harsh physical conditions of an intertidal marsh environment. Plants are limited to the few species highly tolerant of inundation, salinity, and hydrogen sulfide present in the marsh soils. Because of the physical constraints that are placed on fauna (e.g., exposure to heat, drying, and extremes in dissolved oxygen, temperature, and wave energy), trophic interactions are relatively simple with only a few core faunal species groups dominating. Salt marshes along the Atlantic and Gulf coasts of the United States are inhabited by several principal faunal species groups—intertidal bivalves (*Geukensia* spp.), marine crabs (*Uca* spp., *Callinectes* spp.), marine snails (*Littorina* spp.), marsh fish (*Fundulus* spp.), estuarine turtles (*Malaclemys terrapin*), and marsh birds (e.g., *Ammodramus caudacutus* salt-marsh sharp-tailed sparrow). Some species (e.g., ribbed mussels) actively manipulate their habitat, fundamentally influencing the productivity of the marsh. For example, mussels can be considered a secondary foundation species and strongly enhance marsh functions, including decomposition, primary production, facilitation of other marsh species, and water infiltration rates (Angelini et al. 2015). Other species rely on marshes for refuge and feeding either as transitory visitors or resident occupants. Strong trophic interactions among these groups have been documented with implications for ecosystem function and service provision. *S. alterniflora* production has reportedly been enhanced because of ammonium excreted by *Geukensia demissa* (Bertness 1984) and limited by *Littorina* grazing (Silliman and Zieman 2001). The presence of a natural complement of fauna in a living shoreline marsh can be an indication that the ecosystem services provided by the living shorelines are being maximized.

Besides tidal regime (the ultimate driver of salt marsh organization), there are four primary components of a marsh ecosystem tied to its function and, thus, ecosystem service capacity: (1) marsh grass density and growth form, (2) sediment composition and biogeochemical cycling, (3) microtopography and microhabitats, and (4) marsh edge and adjacent habitats. Living shorelines efficiently mimic tidal regime; however, the other four components vary greatly on the basis of design characteristics and, because of the complex ecological interactions in the marsh, may result in greatly decreased function (Figure 15.4).

15.3.1 Marsh Grass Density and Growth Form

Marsh grasses tend to grow efficiently in the sand sediment used for living shorelines. Allowing time for establishment of the marsh, plant structure (stem density and height) in living shoreline marshes may become comparable to natural fringing marshes (Bilkovic and Mitchell 2013; Currin et al. 2008). If proper elevations are established, zonation patterns for marsh plants and colonizing animals mimic natural salt marshes (Priest 2008). However, we recently estimated that, along sheltered coasts in Chesapeake Bay, plant density in living shoreline marshes (number of stems/m^2 ± SE, 108 *S. alterniflora* ± 13; 559 *S. patens* ± 76; $n = 8$; 1–11 years old) was less than that in natural fringing marshes (280 *S. alterniflora* ± 32, 678 *S. patens* ± 163; $n = 10$). Conversely, plant height (cm ± SE) was taller on average in living shoreline marshes (104 ± 6.6, *S. alterniflora*; 87 ± 3.8, *S. patens*) than in natural fringing marshes (63 ± 3.2, *S. alterniflora*; 52 ± 6.6, *S. patens*), possibly because of enhanced nutrient availability in the living shorelines. There was no relationship between the age of these created marshes and plant density or height, suggesting that plant productivity did not follow a clear trajectory over the 11-year age time range of the marshes surveyed. Similarly, soil and plant canopy properties of restored salt marshes in California (2–11 years) did not exhibit strong directional trends for ecosystem development likely because of high interannual variability, and the change toward functional equivalence may take decades (Zedler and Callaway 1999). Relatively low *S. alterniflora* densities in newly created marshes may initially limit the occupation of the new marsh by fauna because of limited canopy cover as predation refuge (e.g., fiddler crabs, Hemmi and Zeil 2003) and lack of suitable habitat or food resources (e.g., mussels tend to aggregate around *Spartina* stems, Nielsen and Franz 1995; the abundance of grazing periwinkles is correlated with stem densities, Kiehn and Morris 2009).

15.3.2 Sediment Composition and Biogeochemical Cycling

In contrast to the organic and fine mineral composition of natural marshes, living shoreline intertidal sediments are composed of predominantly coarse sand with low organic matter content because of the practice of using clean sand fill to create suitable elevations for marsh plantings (Bilkovic and Mitchell 2013; Currin et al. 2008). The implications of using sand as a substrate are not immediately obvious because it has conflicting properties. Sand provides a looser (less stable) substrate and is lacking detrital food, which may limit colonization of infauna (Bilkovic and Mitchell 2013; Levin et al. 1996), invertebrate grazers, and consequently primary and secondary consumers. The coarse sandy soils used to construct marshes differ from the substrate of natural marshes and may not be able to retain enough nitrogen to allow for optimal marsh plant growth (Boyer and Zedler 1998). However, sand fill will be better oxygenated than typical marsh sediments, potentially increasing the availability of some nutrients and improving marsh grass growth. The use of sand substrate also has implications for the filtration capacity of marshes because it may allow water to percolate much faster than through an organic matrix, raising the potential that groundwater nitrogen could move through narrow living shoreline marshes without being removed. However, an assessment of potential nitrogen uptake for marshes with sills in Chesapeake Bay suggests that both young and older more established created marshes were able to remove nitrogen (Beck et al., Chapter 14, this book) despite the sand matrix, indicating that living shorelines can help mitigate eutrophication of tidal systems.

The lack of organic and, in particular, detrital matter may be partially responsible for low grazer populations in living shoreline marshes (Bilkovic and Mitchell, personal observation). Fiddler crabs *Uca pugnax* and *Uca pugilator* and marsh periwinkles *Littorina irrorata* are important to the flow of energy in some East Coast US salt marshes, consuming approximately one-third the net production of a *Spartina* marsh (Cammen et al. 1980). Fiddler crabs are detritivores consuming algae, bacteria, and fungus scraped off of the marsh substrate, as well as plant and animal detritus. Marsh periwinkles are also detritivores predominantly consuming fungi that grow on dead *S. alterniflora* or are induced to grow on live *Spartina* through periwinkle grazing activities that create wounds on the plant (Alexander 1979; Silliman and Newell 2003). In newly created living shorelines, the biomass of dead or decomposing *Spartina* and fungi on the clean sand and plant leaves may by limited and restrict periwinkle and fiddler crab populations. When populations of these grazers are low, the functions they augment in the marsh ecosystem may be limited. While the extent of biological mediation *Uca* spp. contributes to the marsh ecosystem is yet undetermined, this burrowing organism may facilitate salt marsh plant production by oxygenating marsh sediments and increasing the availability of nutrients to the plants (Angelini and Silliman 2012; Bertness 1985; Daleo et al. 2007; McCraith et al. 2003). The importance of *L. irrorata* to the trophic interactions of the marsh has been explored in several systems. When in high densities, *L. irrorata* grazing can reduce the integrity of *S. alterniflora* in already stressed marshes by decreasing the plant's resistance to microbial infection and increasing the rate of marsh die-off (Silliman and Newell 2003; Silliman and Zieman 2001; Silliman et al. 2005). However, new studies suggest that fiddler crabs can mitigate overgrazing by the marsh periwinkle by facilitating *S. alterniflora* growth possibly through bioturbation of anoxic sediments or nutrient redistribution (Gittman and Keller 2013).

Both of these species are important prey items for the economically important blue crab *Callinectes sapidus* (Hamilton 1976; Seed 1980). In the mid-Atlantic, Southeast, and Gulf coast salt marshes of the United States, commonly occurring densities of marsh periwinkles and fiddler crabs are 50–800 ind./m² (Silliman and Bortolus 2003) and 224–480 burrows/m² (Allen and Curran 1974; Bertness and Miller 1984; Katz 1980), respectively. Very limited information exists on the occupation of *L. irrorata* or *Uca* spp. in the created marshes of living shorelines. Within one marsh-sill on the York River, Virginia, that we surveyed (pre- and post-construction; 2011–2013), no *L. irrorata* were present up to 2 years after construction, whereas ~200 ind./m² were present at the adjacent reference marsh during the same time frame. However, *L. irrorata* and *Uca* spp. burrows were present in other surveyed created marshes with sills in Virginia (*n* = 8, 1–11 years old), although largely in relatively low densities (<50 snails/m² and <20 burrows/m²).

15.3.3 Microtopography and Microhabitats

Natural marshes are complex intertidal forms with hummocks of higher elevation, a network of various sized tidal creeks, and crenulated edges. Although dominated by just a few species, salt marshes are host to a diversity of rare plants, which increase the complexity of the habitat. Living shorelines typically lack all of these characteristics. They are frequently planted with only a few species, and this condition can persist many years later (Desrochers et al. 2008). They tend to have lower percentages of high marsh species overall and the low and high marsh plants are strictly segregated.

There are uncertain implications of the lack of microtopography and microhabitats in living shoreline marshes for most marsh species, but in a salt marsh, the existence of microhabitats determines the internal distribution of species and life stages. Differential microhabitat use by resident marsh fish throughout their life histories has been demonstrated in salt marshes. *Fundulus heteroclitus* tend to occupy pool microhabitats near the upland edge of the marsh as larvae and then move to the low marsh as larger juveniles and *Fundulus luciae*, throughout their life history, occupy high marsh habitats (Kneib 1984). Invertebrate prey (e.g., harpacticoid copepods) are also observed in patchy distributions in microhabitats on the marsh surface (Kneib 1986). Very young (2.5 to 14.8 mm carapace width) blue crabs *C. sapidus* have been captured in shallow pool microhabitats that remain on a salt marsh surface on an ebbing tide, suggesting that these vegetated environments are being used for settlement and metamorphosis from the planktonic megalopae to the epibenthic crab stage (Kneib 1997).

The lack of microtopography and microhabitats in created marshes is particularly apparent in relation to their bird populations. Literature on bird use of created marshes typically suggests that created marshes do not attract or support the same bird populations as those using natural marshes. Although some marsh obligate species use is documented, shifts in dominant species are typical (e.g., Darnell and Smith 2004; Desrochers et al. 2008; Havens et al. 1995; Melvin and Webb 1998). Bird use of marsh-sill structures has been documented in the Chesapeake Bay, with ducks, great blue herons, green herons, blue herons, great egrets, lesser cattle egrets, ospreys, Canada geese, mute swans, and bald eagles all being detected (Bosch et al. 2006); however, there is a lack of empirical data on the use of the living shoreline marsh surface by marsh or water birds. The quality of living shorelines, which is most likely responsible for differing bird usage, is their uniformity (Atkinson 2003; Melvin and Webb 1998). Marsh edge, tidal creeks, and hummock topography are cited as the most important microhabitats for bird use in salt marshes (Melvin and Webb 1998). Diversity in vegetative structure also provides important microhabitats; living shorelines tend to have only a narrow band of salt bushes (if any), which reduces their potential use by passerines (Havens et al. 1995).

Living shoreline size may also contribute to differing bird communities. Living shorelines are typically narrow in width, a characteristic that deters birds seeking protective cover (Havens et al. 1995; Seigel et al. 2005). The small size of the living shoreline marsh may result in more birds nesting in less suitable vegetation, resulting in reduced rates of egg hatching and nesting, as has been found for red-wing blackbirds (Desrochers et al. 2008). Even in natural marshes, there is a relationship between the breeding density of salt marsh specialists and marsh size/area (Benoit and Askins 1999; Shriver et al. 2004); thus, it is logical that this relationship would extend to small created marsh habitats.

15.3.4 Marsh Edge and Adjacent Habitats

Living shoreline marsh edges differ from many natural marshes because of their construction, which can incorporate a variety of structures along the seaward edge to provide stability. Oyster sills tend to be low-elevation, intertidal structures that allow full access of nekton to the marsh surface during high tide. Stone sills are primarily installed to break wave energy and therefore frequently extend above mean high water, raising concerns about nekton usage and access. In addition, this reduction in wave energy may result in higher sediment accretion rates in newly created living shoreline marshes with sills than in natural marshes (Currin et al. 2008), possibly diminishing shallow water refuge habitat, increasing marsh elevation, and contributing to high marsh development. Both types of structure can provide habitat for fouling organisms and may create reef-like complex habitats; however, while oyster shell is a native habitat, stone sills are created novel habitat, replacing soft-bottom sediments. Sills can result in a living shoreline with an altered invertebrate community dominated by fouling organisms (e.g., barnacles, oysters) and with fewer subsurface deposit-feeders (Bilkovic and Mitchell 2013; Wong et al. 2011). Moreover, restricted access because of the placement of structures that may act as barriers to the marsh edge and surface, in combination with food and refuge habitat limitations, particularly in the early stages of marsh establishment, may exclude marsh facilitators such as ribbed mussels, fiddler crabs, and periwinkles. This could result in lessened grass vitality, denitrification, biotic filtration of tidal waters, and prey provision for fish and crabs.

15.3.4.1 Estuarine Faunal Recruitment and Access to Marsh Surfaces

Atlantic ribbed mussels (*G. demissa*) are the predominant intertidal bivalve species found throughout salt and brackish marsh systems along the Atlantic Coast and into the Gulf of Mexico. The successful recruitment of ribbed mussels to the marsh depends on a combination of factors including predation pressure, food availability and quality, and larval access to suitable settling substrate on the marsh (Bertness and Grosholz 1985; Nielsen and Franz 1995). During their larval pelagic stage, ribbed mussels settle onto suitable substrate, possibly using settlement cues from substrate or conspecific byssal threads (Ompi 2011) and then transform into their shelled semi-sedentary stage. Postsettlement juvenile mussels are able to adjust their positions within a short distance (<1 m) if needed (Bertness and Grosholz 1985; Nielsen and Franz 1995). They tend to settle on aggregates of adult mussels around the stems of *S. alterniflora* (Nielsen and Franz 1995) and can reach densities (mussels/m^2) of 2000–3000 in New England, and 10,000 in Jamaica Bay, New York (Bertness and Grosholz 1985; Franz 1997, 2001; Kuenzler 1961; Lent 1969; Lin 1989; Stiven and Kuenzler 1979). Similarly, we regularly observed mussel densities of 3000–4000 along the York River, Chesapeake Bay, with a few sites reaching 5000–8000 mussels/m^2 (Bilkovic and Mitchell 2014a). Oysters can also be a significant component of the low intertidal salt marshes; however, their range and numbers have been greatly limited by commercial harvesting, disease, and pollution in Chesapeake Bay, and there are limited areas in the Bay where oysters live intertidally (Nestlerode et al. 2007; Woods et al. 2005). Successful recruitment of these species into the marsh is dependent on access to the marsh surface and the presence of suitable habitat to allow for survival and growth.

Faunal recruitment on the surface of living shoreline marshes may be limited because of restricted access from the high elevations of some stone sills or possibly reduced survival from less-than-optimal habitat conditions. We surveyed 20 sites (9 living shoreline marsh-sills and 11 natural fringing marshes) within lower Chesapeake Bay documenting ribbed mussel and marsh plant abundance during 2012–2014. Within each marsh, we determined mussel and plant density 1-m intervals from the marsh edge representing distances of 0–1 m, 1–2 m, 2–3 m, and 3–4 m from the marsh–estuary edge. We also surveyed sill structures for mussel and oyster density within quadrats placed at low-shore levels equivalent to the marsh surface elevation. Ribbed mussel densities were consistently lower on both the created marsh surface and the rock sill as compared to the natural marsh surfaces (one-way analysis of variance [ANOVA]; $F_{2,26} = 17.9$; $p < 0.0001$; Figure 15.5). In all our surveys, we never found oysters associated with the vegetated portion of the living shorelines, and mussel abundance was dependent on design. Most sills had elevations at least 0.3 m above mean high water, and for these sites, mussel recruitment into the marsh tended to be low (range: 0–15.2 mussels/m^2). The exception to this was one site (Hermitage, Elizabeth River, Virginia) where the sill structure was built low enough for daily overtopping, potentially allowing mussel larvae to wash over the sill structure (406 mussels/m^2). Other considerations are that larvae may be gaining access to the marsh surface but are not surviving. Possible contributors to reduced survival are sediment fill characteristics and dispersed planting configurations. Coarse sand with low organic matter used as fill differs from typical marsh fine sediment with high organic matter and may be less suitable for mussel settlement or survival. Moreover, mussels use high-density clumps of marsh plants as predator or heat stress refuge, and living shoreline marshes tend to be planted in a dispersed manner with single plugs. Silliman et al. (2015) showed that planting in clumps increases *Spartina*

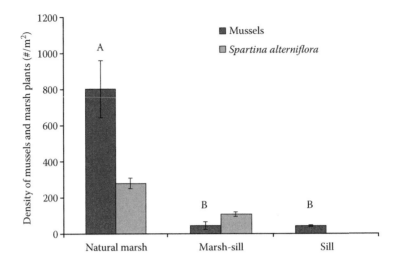

Figure 15.5   Ribbed mussel densities were significantly lower on the surface of the created marsh ($n = 9$) and the sill structure ($n = 8$) than the natural fringing marsh surface ($n = 11$). Likewise, *Spartina* densities were reduced at the marsh-sills compared to the natural marsh (one-way ANOVA; $F = 19.3$, df = 1; $p < 0.001$; $R^2 = 46.6\%$), which may be contributing to the relatively low mussel abundance. Marsh mussel densities represent abundance along the marsh edge (first meter of marsh) and sill mussel densities were measured on the seaward side of the sill at low shore elevation. Created marsh-sills ranged in age from 1 to 11 years and were constructed with clean sand fill.

biomass, density, survivorship, and expansion. This planting approach may facilitate the development of preferred microhabitats to enhance mussel survival and should be investigated further. Mussels are prey for a number of nekton and bird species, so low numbers can impact the trophic web. In addition, the presence of mussel aggregations can increase the number of juvenile crabs in the marsh (Angelini et al. 2015). Therefore, the lack of fiddler crabs in living shorelines may, in part, be related to the low mussel recruitment into the vegetation.

Access is also a concern for the numerous species of fish and crustaceans that use salt marshes for shelter or food to some degree. More than 90% of the commercially important fish and shellfish of the southeastern Atlantic and Gulf coasts reportedly use salt marshes as young (Mitsch and Gosselink 2015). Several species are endemic to salt marshes including mummichog and killifish (*Fundulus* spp.), silversides (*Menidia* spp.), shrimp (*Palaemonetes* spp.), and blue crabs (*C. sapidus*). Nekton play an essential role in the transfer of energy from the marsh to the greater estuary and coastal food webs (Beck et al. 2001; Kneib 2000). There remain limited data on the value of living shorelines to nekton populations; however, enhanced invertebrate and fish diversity and abundance along living shorelines compared to armored shorelines as well as documented fish utilization of living shoreline habitats suggest that they are providing refuge and forage habitat (Currin et al. 2008; Davis et al. 2006; Gittman et al. 2015a; Hardaway et al. 2007; Scyphers et al. 2011). Nekton community composition in created marshes behind sills can be similar to natural control marshes after marsh establishment (typically ≥2 years) and inclusive of marsh-resident species *Fundulus, Menidia, Cyprinodon variegatus, Palaemonetes* spp., and prevalent marsh transient species blue crab *C. sapidus*, mullets *Mugil* spp., spot *Leiostomus xanthurus*, and pinfish *Lagodon rhomboides* (e.g., Davis et al. 2006; Gittman et al. 2015a). Further characterization of fish utilization in different living shoreline habitats (e.g., marsh, shallows, oyster, or stone sill) over time using empirical measures of performance (growth, condition, and diet) is needed to more fully understand the magnitude of the value of living shorelines as nursery habitat.

The estuarine turtle, diamondback terrapin (*Malaclemys terrapin*) relies on salt marshes for food and shelter but must lay its eggs on dry land (Brennessel 2006). Disruptions to the aquatic–terrestrial connection (e.g., shoreline armoring) will prevent terrapin from reaching nesting habitat above the tideline (Roosenburg 1991). Terrapin in Chesapeake Bay are strongly associated with areas with >10% marsh within 750 m and limited by the presence of armoring (>17%); as such, the preferential use of living shorelines in place of armoring to stabilize a shore may act to enhance or maintain terrestrial–aquatic connectivity for terrapin (Isdell et al. 2015). The usefulness of living shoreline marsh habitat will be dictated by the afforded access to the marsh surface for the terrapin. As an upper-level predator, the implications of the exclusion of terrapin from marsh habitat could resonate throughout a marsh food web.

15.3.4.2 Novel Living Shoreline Habitats

In estuaries, such as Chesapeake Bay, where soft-bottom habitat dominates and rocky shorelines are rare, the introduction of novel artificial rocky structure, such as a stabilizing stone sill or breakwater, may enhance recruitment of species that are limited by the availability of hard substrate, including native and introduced species (Bilkovic and Mitchell 2013). There is a significant lack of empirical data on the types of epibiotic assemblages that colonize artificial structures, including information on seasonal changes in species composition and abundance and the prevalence of non-native species. These structures alter the hydrodynamic and physical conditions around them, likely affecting the distribution of planktonic larvae that rely on currents to transport them to suitable substrate for settlement. Species recruitment to sills or breakwaters is expected to be highly variable because it is regulated by many factors such as species distribution, life history, distance from source populations, relative dominance of pioneer species, estuarine circulation patterns, physical conditions, and wave energy (Barnes et al. 2010; Bushek 1988; Sutherland and Karlson 1977). In addition, the placement of a sill requires the conversion of existing shallow subtidal bottom, which often supports high macrobenthic functional diversity (Bilkovic and Mitchell 2013). Because of the loss of shallow bottom under the sill footprint and the filling to create new intertidal habitat, infaunal diversity can be reduced for several years (Bilkovic and Mitchell 2013). Further, slight changes in living shoreline design may have significant implications on the biological complement of a site. Even within a single breakwater project (York River, Chesapeake Bay, Virginia, USA), we observed differing fouling assemblages between individual breakwater structures related to differing created elevations behind those structures that affected inundation patterns and thus recruitment of bivalves (Bilkovic and Mitchell 2014b). There were low densities of ribbed mussels *G. demissa* and oysters *Crassostrea virginica* on the landward side of breakwaters with sand fill built to elevations around the high tide mark (24 and 43 ind./m^2, respectively), while those that experienced regular inundation behind the breakwater during at least half of the tide cycle possessed densities (110 and 86 ind./m^2, respectively) similar to the seaside of the breakwater (153 and 93 ind./m^2, respectively). There was also elevated water temperature (1°C–2°C higher) behind the breakwaters with high elevations in comparison to the seaside. These differences illustrate the sensitivity of the biology to nuances in the design of shoreline protection approaches.

Colonization of living shoreline structural components by fouling organisms, including bivalves, in many cases occurs rapidly. The immediate offshore structure (stone sill, oyster-shell sill) can be colonized within the first year of construction if near a larval supply. For example, after only 1 month, we detected increases in colonizing species diversity and average size of individual animals collected from an offshore sill constructed with bags of oyster shell (120–193 half-shells per bag) in Ware River, Chesapeake Bay. After oyster sill construction (late April 2014), six sacrificial experimental oyster bags were removed monthly for 5 months (June to October 2014) and 1 year later (November 2015) to document faunal colonization. Within 5 months, average individual oyster surface area (shell length × width) per bag increased from 14.5 to 548.3 mm^2 (one-way ANOVA; $F_{1,23} = 137.2$; $p < 0.0001$) and the number of spat was nearly 100 times greater ($F_{1,23} = 20.7$; $p = 0.0001$; Figure 15.6). There was also evidence of a second spat set during the late summer 2014.

During the first year, average individual ribbed mussel surface area more than tripled from 13.8 to 54.1 mm^2 ($F_{1,23}$ = 12.8; p = 0.002), while the number of mussels increased 10-fold from 8 to 95 animals ($F_{1,23}$ = 24.8; p < 0.0001). Total number of colonizing organisms rose from 129 to 1253 ($F_{1,23}$ = 10.56; p = 0.0035), and the average number of species found in each bag increased from 3 to 5. In the following year (November 2015), settlement continued to increase for a total of 1638 colonizing organisms (1236 oysters and 247 mussels) with an average of 7 species in each bag. Oyster and mussel growth also continued with average individual sizes reaching 881.4 mm^2 for oysters and 133.4 mm^2 for mussels. On average, oysters were 1.3 inches (33 mm) in length after two growing seasons and 17% were 2 inches or larger, attaining the minimum harvest size for many East Coast states. There were a total of 14 species observed to be associated with the oyster sill throughout the sampling period including oysters, ribbed mussels, hooked mussels, barnacles (*Balanus* spp.), naked goby *Gobiosoma bosci* eggs, limpets, shrimp, anemone, tunicates, and clams (*Tagelus plebeius*, *Mya arenaria*). Many of these species are commonly observed in association with oyster reefs, which suggests that the oyster sill may be serving as oyster reef habitat.

However, the extent and abundance of the colonizing organisms are strongly related to the material and design of the structures; predominantly submerged sills constructed of bagged oyster shell or dome concrete structures designed to recruit oysters can be expected to have higher species diversity and abundance than the slightly higher-profile (~0.3 m above MSL) stone sills or breakwaters constructed with larger quarry stone. On an ~3-year-old breakwater constructed of exterior stone between 225 and 1150 kg each with smaller interior core stone (York River, Chesapeake Bay, Virginia), we observed only barnacles, ribbed mussels, and oysters, and large areas of the rocks were bare, whereas, as described above, we detected 14 species on a new oyster sill that occupied increasing space over time (Bilkovic and Mitchell 2014b; Bilkovic et al. 2014; Figure 15.6). Within the low shore zone (bottom edge of 0.25 m^2 quadrat placed at low tide elevation), breakwater rocks were 84% bare (landward side) and 55% bare (seaside) and animals were predominantly observed within rock niches. In contrast, 42% of the bagged shell was colonized by epibiota within a few months of placement. Scyphers et al. (2014) observed similar patterns with higher densities of ribbed mussels on submerged bagged oyster breakwaters (sills) in comparison to Reef Ball breakwaters possibly because of differing levels of complexity. In addition, they demonstrated that juvenile and resident fish species diversity was higher for both types of breakwaters in comparison to control mudflat sites. Still, further study is needed to determine the

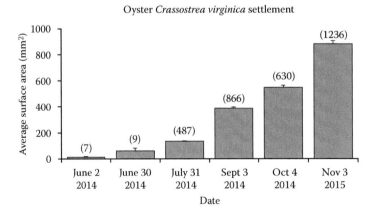

Figure 15.6   The number and size (surface area: shell length × width) of oyster spat on an offshore sill constructed with bags of oyster shell in Ware River, Chesapeake Bay, Virginia, USA, increased dramatically in the immediate months after sill placement in April 2014. Total number of spat is shown in parentheses.

extent and relative difference in faunal recruitment and postsettlement mortality among various types of structures.

15.4 DESIGNING A BETTER LIVING SHORELINE
FOR CONSERVATION PURPOSES

Designing a living shoreline with natural infrastructure to not only prevent erosion but also increase the habitat capacity of a system is a complicated endeavor requiring input from both coastal engineers and ecologists. As with any wetlands creation project, attention should first be given to siting criteria mainly driven by fixed physical features. Generally, shoreline orientation that allows ~6 h of direct sunlight during the growing season, predictable salinity regimes and flooding, and low to moderate wave climate (including boat wake activity) are necessary shoreline components for living shoreline marsh establishment (Burke et al. 2005; Duhring et al. 2006; Garbisch and Garbisch 1994). Once landscape/reach level dynamics are determined to be appropriate for living shoreline placement, site-level components can be adapted, within reason, to meet project goals and can be extremely influential on resulting shoreline function. As such, during the planning stages, the expected benefits from the living shoreline must be measured against the loss or diminishment of existing subtidal and riparian resources.

Living shoreline design elements that determine ecosystem service capacity include vegetation buffer width, plant species composition, density, and configuration, soil composition, and habitat access. Marsh vegetative species composition and arrangement affect sedimentation rates and organic content production and, through these, marsh elevation (Pasternack et al. 2000). A disruption of one of these processes can cause a cascade of changes to the others. Therefore, vegetation selection and placement could have significant impacts on the sustainability of the living shoreline. Changes in plant composition and marsh elevation are linked to changes in local topography, which affects flooding duration (Rozas and Reed 1993), benthic invertebrate community composition (Yozzo and Diaz 1999), and nekton use of the marsh (Yozzo and Smith 1998). Bosch et al. (2006) observed that the presence of both high and low marsh may help stabilize eroding banks and provide wildlife habitat. Additionally, year-round erosion protection was noted in marshes with limited winter die-back of plants (e.g., *Spartina* and *Juncus* species in their southern range of the United States).

Living shorelines paired with offshore sill structures may introduce habitat diversity and could subsidize secondary productivity particularly in areas where loss of complex biogenic habitat (e.g., oyster reefs) has occurred. Sill structures can support relatively high epibenthic community production compared to habitats without hard structures (Bilkovic and Mitchell 2013; Wong et al. 2011). As previously discussed, differing material and dimensions of the sills can result in varying abundance and diversity of the colonizing assemblages. Borrowing from research on the habitat enhancement of hardened coastal defense structures (e.g., seawalls, revetments), attention to the amount and configurations of niches, available interstitial space, and microhabitats in and around stone sills can greatly enhance the abundance and diversity of colonizing organisms (Chapman and Underwood 2011; Lundholm and Richardson 2010). Firth et al. (2016) suggest that to increase abundance and diversity of native species, structures should be built within the low intertidal zone with ample shaded surfaces for refuge from heat and predation. However, low-shore assemblages on artificial breakwater structures are often composed of few common taxa, suggesting that not all species are able to establish populations on these structures even in areas where larval supply is not in question (Bacchiocchi and Airoldi 2003; Gacia et al. 2007; Moschella et al. 2005; Vaselli et al. 2008). Moreover, caution is warranted because the introduction of artificial substrates, even those used in combination with natural elements, to protect shorelines has the potential to act as an anthropogenic dispersal mechanism for opportunistic native, nonnative, or invasive species potentially altering species abundance and distributions at local and regional scales (Bulleri and Airoldi 2005; Glasby and Connell 1999; Vaselli et al. 2008). Additional larger questions to consider are whether the increasing amounts of artificial stone structures may influence the source/sink dynamics of shellfish, either positively or negatively, and whether differences in oyster assemblages on rock structures and its natural analog, the oyster reef, can be overcome with improved ecological engineering.

To ensure the maximum provision of ecosystem services, faunal colonization of the marsh surface is of equal importance to the faunal colonization of hard surfaces. A trade-off exists between protection of the shore from wave climate and access to the marsh via tidal gaps, which may allow pockets of shoreline erosion. Certain design elements, such as lower-profile sills with tidal gaps, vents, or lowered sections of stone, allow greater access to the marsh habitat by aquatic organisms and encourage faunal colonization. Insufficient tidal gaps will restrict flushing, possibly causing elevated temperatures and reduced access by mobile aquatic species (Burke et al. 2005). Regrettably, there is an absence of information on the efficacy of varying distances and configurations of tidal gaps to allow optimal marsh inundation and access for marine fauna. However, Hardaway et al. (2007) observed that in addition to tidal gaps, marsh access for fish was maintained through macropores in the sill stone and by overtopping of the sill by tidal waters. Innovative vent designs (e.g., staggered system), porosity in sill stone, and overtopping by tidal waters are potential balancing mechanisms to ensure the coexistence of erosion control with desired ecological functions. Marsh-resident species often have relatively small home ranges, which can provide some guidance on the maximum distances at which an access point becomes necessary. For example, the resident marsh fish mummichog (*F. heteroclitus*) has been reported to have a small adult home range of 36–38 linear meters along a tidal creek bank (Lotrich 1975), although greater ranges of 110–640 m have been reported in other systems (Skinner et al. 2005; Sweeney et al. 1998; Teo and Able 2003). Described home ranges of other marsh species include the tendency for juvenile spot (*L. xanthurus*) to remain within a single marsh creek (Weinstein and O'Neil 1986) and distances of 9 m for pinfish (*L. rhomboides*) (Potthoff and Allen 2003) and 30 m for juvenile black sea bass (*Centropristis striata*) (Able and Hales 1997). Nekton access to a marsh has also been linked to marsh edge geomorphology (i.e., slope, irregularity, and elevation) (La Peyre and Birdsong 2008), which influences the duration and frequency of flooding on the adjacent marsh. Ensuring marsh use by nekton may also be the key to ensuring shorebird use as appropriate prey populations are necessary to attract shorebirds (Atkinson 2003).

15.5 TIDAL MARSH LIVING SHORELINE ECOSYSTEMS FOR COASTAL RESILIENCE

Living shorelines were conceived as a means to remedy the ecologically disruptive nature of coastal defenses. Undeniably, these nature-based approaches to shoreline protection are an improvement over traditional armoring. Armoring has been shown to adversely affect local populations (e.g., Bilkovic and Roggero 2008; Bilkovic et al. 2006; Chapman 2003; Moschella et al. 2005; Toft et al. 2007) and converts coastal landscapes from a heterogeneous mosaic of natural habitats to human-dominated landscapes with homogenized, fragmented habitats (Figure 15.7). Hardened shorelines make up approximately 14% of US coastal shorelines and the majority of the armoring has occurred along sheltered shorelines (Gittman et al. 2015b), which are prime candidates for nature-based protection. Rates of armoring in the United States have been estimated to be 200 km/year, with the expectation that the percentage of armored shoreline will be doubled by 2100 (Gittman et al. 2105b). In Virginia, more than 80% of shorelines are appropriate for living shorelines based on fetch conditions (Center for Coastal Resources Management 2013). Further, living shorelines were estimated to be able to enhance ecosystem services capacity by more than 90%, compared to their previous status (Rodríguez-Calderón 2014). This suggests that if living shorelines are implemented at appropriate sites as opposed to armoring, they can be a viable alternative for habitat conservation, particularly in urbanized or developing coastal localities. Living shorelines that maintain or restore

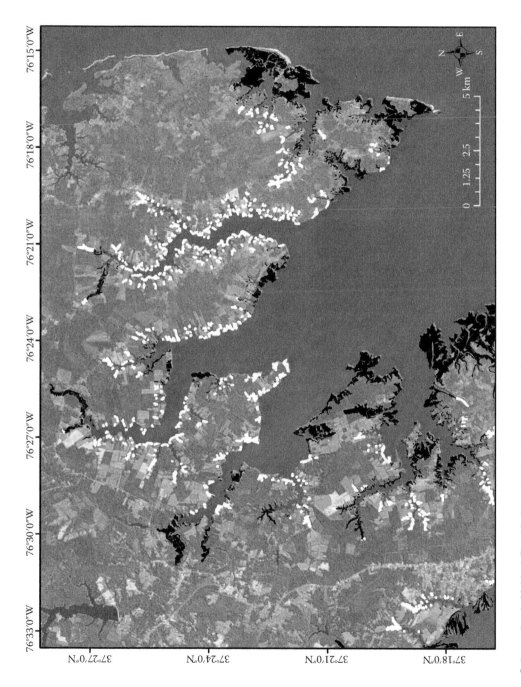

Figure 15.7 Salt marshes in Mobjack Bay, Virginia, which contain varying levels of development, were grouped by proximity using a density-based algorithm in GIS into two broad categories: clustered (black) and dispersed (white). In areas of developed land use or hardened shorelines, 71% of marshes were dispersed.

coastal habitats and processes may have a significant role in mitigating further coastal habitat fragmentation. However, the field of shoreline restoration is nascent and many questions remain about the evolution, persistence, and maximum achievable functionality (e.g., habitat capacity and ecosystem service provision) of living shoreline habitats.

In addition to preserving existing ecosystem services, living shorelines are of particular interest as an adaptation to sea level rise. Hardened shorelines, combined with sea level rise, can increase flooding and erosion of adjacent shorelines (Holleman and Stacey 2014), while natural and living shorelines have been credited with absorbing floodwaters, helping attenuate waves (Barbier et al. 2011), and providing erosion protection during storms (Gittman et al. 2014). Living shorelines are more resilient than hardened shorelines in the face of sea level rise (Vandenbruwaene et al. 2011) because they include salt marshes that have some ability to capture and accumulate sediment, building their elevation (Hardaway et al. 1984). Marshes enhance sedimentation through the reduction of wave energy (e.g., Barbier et al. 2013), which is dependent on the width of the marsh and the height and density of the vegetation (Hardaway et al. 1984). In addition, marsh sill structures may enhance the capacity of the marshes to accrete sediment by breaking waves (e.g., Currin et al. 2008). However, their resilient nature is dependent on an adequate sediment supply and sufficient reduction of wave energy to enhance sediment capture. Therefore, living shorelines will be most successful as long-term adaptations if they are situated to allow landward migration of the wetlands with sea level rise rather than solely depending on marsh accretion. Despite their resilient nature, living shorelines are not the dominant choice of protection in most locales. For example, only approximately 20% of Virginia shoreline permit requests are for living shorelines, compared to traditional armoring (Bilkovic et al. 2016; Figure 15.2). This suggests that there is progress to be made through policy and public education to truly make living shorelines a successful form of adaptation.

15.6 CONCLUSIONS

Living shorelines show great promise as a combined ecological/engineering adaptation to changing coastal pressures, which can promote both human and ecological coastal resilience. However, to fully reach their potential, living shorelines must be designed to replicate the complex ecosystem interactions present in natural marshes. The proposed conceptual framework delineating theoretical differences in living shoreline salt marshes from natural marshes can be used to identify research priorities as well as engineering challenges to improve the capacity of these created habitats. Prominent differences potentially influencing the ecosystem development and contribution to coastal resilience include (1) differences in the resulting plant communities; (2) the use of coarse sand material as a substrate; (3) the uniform topography, vegetation, and edge characteristics; and (4) impacts of novel habitat introduction. Some of these differences may change over time (e.g., soil organic matter content is likely to increase) and some may be lessened with design modifications. Constraints on the accessibility of the marsh surface by marine fauna resulting from some engineering designs should be remedied when possible. Research with the intent to inform the design and implementation of nature-based projects should involve the collaboration of both ecologists and coastal engineers. For example, such a partnership is essential to truly understand how to best maximize tidal access (i.e., frequency and configuration of faunal access features on sills or breakwaters) without compromising shore protection. To minimize uncertainty of restoration success and maximize living shoreline resilience, several research avenues should be pursued:

- Species abundance and distribution within living shoreline habitats in relation to surrounding natural habitat complexes
- Long-term evolution of faunal populations and living shoreline habitat structure, diversity, and complexity (e.g., plant integrity, sediment composition, microbial community, and microhabitats)

- Sill structural characteristics that simultaneously maximize protection and aquatic access
- Living shoreline design characteristics to maintain or enhance sedimentation and accretion; promoting increased ecosystem function longevity in the face of sea level rise

In addition, on a larger spatial scale, the cumulative effect of living shorelines on coastal resilience, particularly after major storm events, should be quantified. An essential priority in the near term is the development of systematic and standardized monitoring at both regional and national scales to fully evaluate living shoreline effectiveness as a strategy for both coastal defense and habitat conservation. These data are essential to inform project designs, improve siting, and shorten the time frame until a living shoreline is functioning as designed.

Moving living shoreline science forward as a cross-disciplinary study will maximize benefits to both the human and natural systems, promoting coastal areas that are truly resilient into the next century.

ACKNOWLEDGMENTS

This study was supported by the National Science Foundation, Women in Science and Engineering, the National Fish and Wildlife Foundation, and the Commonwealth of Virginia. We thank our colleagues at the Center for Coastal Resources Management, Virginia Institute of Marine Science, for field and laboratory support. Special thanks to J. Herman for assistance with geospatial analyses to derive marsh clustering and fragmentation patterns. We are grateful to P. Mason and anonymous reviewers for comments on our manuscript. This paper is Contribution No. 3524 of the Virginia Institute of Marine Science, College of William & Mary.

REFERENCES

Able, K.W. and L.S. Hales, Jr. 1997. Movements of juvenile black sea bass *Centropristis striata* (Linnaeus) in a southern New Jersey estuary. *Journal of Experimental Marine Biology and Ecology* 213: 153–167.

Alexander, S.K. 1979. Diet of the Periwinkle *Littorina irrorata* in a Louisiana Salt Marsh. *Gulf Research Reports* 6(3): 293–295.

Allen, E.A. and H.A. Curran. 1974. Biogenic sedimentary structures produced by crabs in lagoon margin and salt marsh environments near Beaufort, North Carolina. *Journal of Sedimentary Petrology* 44(2): 538–548.

Angelini, C. and B.R. Silliman. 2012. Patch size-dependent community recovery after massive disturbance. *Ecology* 93(1): 101–110.

Angelini, C., T. van der Heide, J.N. Griffin, J.P. Morton, M. Derksen-Hooijberg, L.P.M. Lamers, A.J.P. Smolders, and B.R. Silliman. 2015. Foundation species' overlap enhances biodiversity and multifunctionality from the patch to landscape scale in southeastern United States salt marshes. *Proceedings of the Royal Society B: Biological Sciences* 282: 20150421.

Atkinson, P.W. 2003. Can we recreate or restore intertidal habitats for shorebirds? *Bulletin-Wader Study Group* 100: 67–72.

Bacchiocchi, F. and L. Airoldi. 2003. Distribution and dynamics of epibiota on hard structures for coastal protection. *Estuarine, Coastal and Shelf Science* 56(5): 1157–1166.

Barbier, E.B., I.Y. Georgiou, B. Enchelmeyer, and D.J. Reed. 2013. The value of wetlands in protecting southeast Louisiana from hurricane storm surges. *PLoS ONE* 8(3): e58715.

Barbier, E.B., S.D. Hacker, C. Kennedy, E.W. Koch, A.C. Stier, and B.R. Silliman. 2011. The value of estuarine and coastal ecosystem services. *Ecological Monographs* 81: 169–193.

Barnes, B.B., M.W. Luckenbach, and P.R. Kingsley-Smith. 2010. Oyster reef community interactions: The effect of resident fauna on oyster (*Crassostrea* spp.) larval recruitment. *Journal of Experimental Marine Biology and Ecology* 391: 169–177.

Beck, A.J., R.M. Chambers, M. Mitchell, and D.M. Bilkovic. In review. In: *Living Shorelines: The Science and Management of Nature-Based Coastal Protection*, CRC Press.

Beck, M.W., K.L. Heck Jr., K.W. Able, D.L. Childers, D.B. Eggleston, B.M Gillanders, ..., and M.P. Weinstein. 2001. The identification, conservation, and management of estuarine and marine nurseries for fish and invertebrates: A better understanding of the habitats that serve as nurseries for marine species and the factors that create site-specific variability in nursery quality will improve conservation and management of these areas. *Bioscience* 51(8): 633–641.

Benoit, L.K. and R.A. Askins. 1999. Impact of the spread of *Phragmites* on the distribution of birds in Connecticut tidal marshes. *Wetlands* 19: 194–208.

Bertness, M. 1984. Ribbed mussels and *Spartina alterniflora* production in a New England salt marsh. *Ecology* 65(6): 1794–1807.

Bertness, M.D. 1985. Fiddler crab regulation of *Spartina alterniflora* production on a New England salt marsh. *Ecology* 66(3): 1042–1055.

Bertness, M. and E. Grosholz. 1985. Population dynamics of the ribbed mussel, *Geukensia demissa*: The costs and benefits of an aggregated distribution. *Oecologia* 67: 192–204.

Bertness, M.D. and T. Miller. 1984. The distribution and dynamics of *Uca pugnax* (Smith) burrows in a New England salt marsh. *Journal of Experimental Marine Biology and Ecology* 83(3): 211–237.

Bilkovic, D.M. and M.M. Mitchell. 2013. Ecological tradeoffs of stabilized salt marshes as a shoreline protection strategy: Effects of artificial structures on macrobenthic assemblages. *Ecological Engineering* 61: 469–481.

Bilkovic, D.M. and M.M. Mitchell. 2014a. Biofiltration potential of ribbed mussel populations. Final Report to the Women in Science and Engineering (WISE), National Science Foundation, College of William & Mary. Virginia Institute of Marine Science, Center for Coastal Resources Management, Gloucester Point, Virginia. 7 pp. http://ccrm.vims.edu/publications/NSF_WISE_mussels.pdf

Bilkovic, D.M. and M. Mitchell. 2014b. Composition, distribution, and dynamics of intertidal epibiota on coastal defense structures. Final Report to the Women in Science and Engineering (WISE), National Science Foundation, College of William & Mary. Virginia Institute of Marine Science, Center for Coastal Resources Management, Gloucester Point, Virginia. 7 pp. http://ccrm.vims.edu/publications/NSF_WISE_epibiota.pdf

Bilkovic, D.M., M.M. Mitchell, and R. Isdell. 2014. Johns Point Landing Living Shoreline Ecological Monitoring. Final Report to Gloucester County and NFWF. Virginia Institute of Marine Science, Center for Coastal Resources Management, Gloucester Point, Virginia. 7 pp. http://ccrm.vims.edu/publications/JohnsPt.pdf

Bilkovic, D.M., M. Mitchell, P. Mason, and K. Duhring. 2016. The Role of living shorelines as estuarine habitat conservation strategies. *Coastal Management*. DOI:10.1080/08920753.2016.1160201.

Bilkovic, D.M. and M. Roggero. 2008. Effects of coastal development on nearshore estuarine nekton communities. *Marine Ecology Progress Series* 358: 27–39.

Bilkovic, D.M., M.M. Roggero, C.H. Hershner, and K. Havens. 2006. Influence of land use on macrobenthic communities in nearshore estuarine habitats. *Estuaries and Coasts* 29(6): 1185–1195.

Bosch, J., C. Foley, L. Lipinsky, C. McCarthy, J. McNamara, A. Naimaster, A. Raphael, A. Yang, and A. Baldwin. 2006. Constructed wetlands for shoreline erosion control: Field assessment and data management. Prepared for Maryland Department of the Environment for submittal to the U.S. Environmental Protection Agency.

Boyer, K.E. and J.B. Zedler. 1998. Effects of nitrogen additions on the vertical structure of a constructed cordgrass marsh. *Ecological Applications* 8(3): 692–705.

Brennessel, B. 2006. Diamonds in the marsh: a natural history of the diamondback terrapin. UPNE, Lebanon, NH.

Bulleri, F. and L. Airoldi. 2005. Artificial marine structures facilitate the spread of a non-indigenous green alga, *Codium fragile* ssp. *tomentosoides*, in the north Adriatic Sea. *Journal of Applied Ecology* 42(6): 1063–1072.

Burke, D.G., E.W. Koch, and J.C. Stevenson. 2005. Assessment of Hybrid Type Shore Erosion Control Projects in Maryland's Chesapeake Bay—Phases I & II. Final Report for Chesapeake Bay Trust, Annapolis, MD, p. 112.

Bushek, D. 1988. Settlement as a major determinant of intertidal oyster and barnacle distributions along a horizontal gradient. *Journal of Experimental Marine Biology and Ecology* 122: 1–18.

Cammen, L.M., E.D. Seneca, and L.M. Stroud. 1980. Energy flow through the fiddler crabs *Uca pugnax* and *U. minax* and the marsh periwinkle *Littorina irrorata* in a North Carolina salt marsh. *American Midland Naturalist* 103: 238–250.

CCRM (Center for Coastal Resources Management). 2013. Comprehensive Coastal Inventory. http://ccrm .vims.edu/gis_data_maps/shoreline_inventories/index.html (accessed September 2015).

Chapman, M.G. 2003. Paucity of mobile species on constructed seawalls: Effects of urbanization on biodiversity. *Marine Ecology Progress Series* 264: 21–29.

Chapman, M.G. and A.J. Underwood. 2011. Evaluation of ecological engineering of "armoured" shorelines to improve their value as habitat. *Journal of Experimental Marine Biology and Ecology* 400(1): 302–313.

Collinge, S.K. 2009. *Ecology of Fragmented Landscapes*. The Johns Hopkins University Press. 360 pp.

Craft, C., P. Megonigal, S. Broome, J. Stevenson, R. Freese, J. Cornell, L. Zheng, and J. Sacco. 2003. The pace of ecosystem development of constructed *Spartina alterniflora* marshes. *Ecological Applications* 13(5): 1417–1432.

Craft, C.B., J. Reader, J. Sacco, and S.W. Broome. 1999. Twenty-five years of ecosystem development of constructed *Spartina alterniflora* (Loisel) marshes. *Ecological Applications* 9: 1405–1419.

Currin, C.A., P.C. Delano, and L.M. Valdes-Weaver. 2008. Utilization of a citizen monitoring protocol to assess the structure and function of natural and stabilized fringing salt marshes in North Carolina. *Wetlands Ecology Management* 16: 97–118.

Daleo, P., E. Fanjul, A.M. Casariego, B.R. Silliman, M.D. Bertness, and O. Iribarne. 2007. Ecosystem engineers activate mycorrhizal mutualism in salt marshes. *Ecology Letters* 10(10): 902–908.

Darnell, T.M. and E.H. Smith. 2004. Avian use of natural and created salt marsh in Texas, USA. *Waterbirds* 27: 355–361.

Davis, J.L.D., R.L. Takacs, and R. Schnabel. 2006. Evaluating ecological impacts of living shorelines and shoreline habitat elements: An example from the upper western Chesapeake Bay. In: Erdle, S.Y., J.L.D. Davis, and K.G. Sellner (Eds.), *Management, Policy, Science, and Engineering of Nonstructural Erosion Control in the Chesapeake Bay*, pp. 55–61. CRC Publication No. 08-164, Gloucester Point, VA.

Desrochers, D.W., J.C. Keagy, and D.A. Cristol. 2008. Created versus natural wetlands: Avian communities in Virginia salt marshes. *Ecoscience* 15(1): 36–43.

Duhring, K.A., T.A. Barnard, Jr., and C.S. Hardaway, Jr., 2006. A survey of the effectiveness of existing marsh toe protection structures in Virginia. Virginia Institute of Marine Science, Gloucester Point, Virginia.

Firth, L.B., R.C. Thompson, K. Bohn, M. Abbiati, L. Airoldi, T.J. Bouma, F. Bozzeda, V.U. Ceccherelli, M.A. Colangelo, A. Evans, and F. Ferrario. 2014. Between a rock and a hard place: Environmental and engineering considerations when designing coastal defence structures. *Coastal Engineering* 87: 122–135.

Firth, L., F.J. White, M. Schofield, M.E. Hanley, M. Burrows, R.C. Thompson, M.W. Skov, A.J. Evans, P.J. Moore, and S.J. Hawkins. 2016. Facing the future: The importance of substratum features for ecological engineering of artificial habitats in the rocky intertidal. *Marine and Freshwater Research* 67(1): 131–143.

Florida DEP, Department of Environmental Protection Bureau of Beaches and Coastal Systems. 1990. Coastal armoring in Florida. Final Status Report. Tallahassee, Florida.

Franz, D. 1997. Resource allocation in the intertidal salt-marsh mussel *Geukensia demissa* in relation to shore level. *Estuaries* 20: 134–148.

Franz, D. 2001. Recruitment, survivorship, and age structure of a New York Ribbed Mussel population (*Geukensia demissa*) in relation to shore level—A nine year study. *Estuaries* 24: 319–327.

Gacia, E., M.P. Satta, and D. Martin. 2007. Low crested coastal defence structures on the Catalan coast of the Mediterranean Sea: How they compare with natural rocky shores. *Scientia Marina* 71(2): 259–267.

Garbisch, E.W. and J.L. Garbisch. 1994. Control of upland bank erosion through tidal marsh construction on restored shores: Application in the Maryland portion of Chesapeake Bay. *Environmental Management* 18(5): 677–691.

Gedan, K.B., M.L. Kirwan, E. Wolanski, E.B. Barbier, and B.R. Silliman. 2011. The present and future role of coastal wetland vegetation in protecting shorelines: Answering recent challenges to the paradigm. *Climatic Change* 106(1): 7–29.

Gittman, R.K., F.J. Fodrie, A.M. Popowich, D.A. Keller, J.F. Bruno, C.A. Currin, C.H. Peterson, and M.F. Piehler. 2015b. Engineering away our natural defenses: An analysis of shoreline hardening in the US. *Frontiers in Ecology and the Environment* 13(6): 301–307.

Gittman, R.K. and D.A. Keller. 2013. Fiddler crabs facilitate *Spartina alterniflora* growth, mitigating periwinkle overgrazing of marsh habitat. *Ecology* 94(12): 2709–2718.

Gittman, R.K., C.H. Peterson, C.A. Currin, F.J. Fodrie, M.F. Piehler, and J.F. Bruno. 2015a. Living shore-lines can enhance the nursery role of threatened estuarine habitats. *Ecological Applications*. http://dx.doi.org/10.1890/14-0716.1

Gittman, R.K., A.M. Popowich, J.F. Bruno, and C.H. Peterson. 2014. Marshes with and without sills protect estuarine shorelines from erosion better than bulkheads during a Category 1 hurricane. *Ocean & Coastal Management* 102: 94–102.

Glasby, T.M. and S.D. Connell. 1999. Urban structures as marine habitats. *Ambio* 28: 595–598.

Griggs, G.B. 1998. The armoring of California's coast. In: Magoon, O.T., H. Converse, B. Baird, and M. Miller-Henson (Eds.), *California and the World Ocean '97, Conference Proceedings*, pp. 515–526. American Society of Civil Engineers, Reston, VA.

Gunderson, L.H., C. Allen., and C.S. Holling (Eds.). 2009. *Foundations of Ecological Resilience*. Island Press, Washington, DC.

Hamilton, P.V. 1976. Predation on *Littorina irrorata* (Mollusca: Gastropoda) by *Callinectes sapidus* (Crustacea: Portunidae). *Bulletin of Marine Science* 26(3): 403–409.

Hardaway, C.S., J. Shen, D. Milligan, C. Wilcox, K. O'Brien, W. Reay, and S. Lerberg. 2007. Performance of Sill, St. Mary's City, St. Mary's River, Maryland. Technical Report prepared by Virginia Institute of Marine Science, Gloucester Point, Virginia.

Hardaway, C.S., G.R. Thomas, A.W. Zacherle, and B.K. Fowler. 1984. Vegetative Erosion Control Project: Final Report. Prepared for Virginia Division of Soil and Water. Technical Report. Virginia Institute of Marine Science, Gloucester Point, VA, 275 pp.

Havens, K.J., L.M. Varnell, and J.G. Bradshaw. 1995. An assessment of ecological conditions in a constructed tidal marsh and two natural reference tidal marshes in coastal Virginia. *Ecological Engineering* 4: 117–141.

Hemmi, J.M. and J. Zeil. 2003. Burrow surveillance in fiddler crabs I. Description of behaviour. *Journal of Experimental Biology* 206(22): 3935–3950.

Holleman, R.C. and M.T. Stacey. 2014. Coupling of sea level rise, tidal amplification, and inundation. *Journal of Physical Oceanography* 44(5): 1439–1455.

Isdell, R.E., R.M. Chambers, D.M. Bilkovic, and M. Leu. 2015. Effects of terrestrial–aquatic connectivity on an estuarine turtle. *Diversity and Distributions* 1–11.

Katz, L.C. 1980. Effects of burrowing by the fiddler crab *Uca pugnax*. *Estuarine and Coastal Marine Science* 11: 233–237.

Kiehn, W.M. and J.T. Morris. 2009. Relationships between *Spartina alterniflora* and *Littoraria irrorata* in a South Carolina salt marsh. *Wetlands* 29(3): 818–825.

Kneib, R.T. 1984. Patterns in the utilization of the intertidal salt marsh by larvae and juveniles of *Fundulus heteroclitus* (Linnaeus) and *Fundulus luciae* (Baird). *Journal of Experimental Marine Biology and Ecology* 83(1): 41–51.

Kneib, R.T. 1986. The role of *Fundulus heteroclitus* in salt marsh trophic dynamics. *American Zoologist* 26(1): 259–269.

Kneib, R.T. 1997. Early life stages of resident nekton in intertidal marshes. *Estuaries* 20(1): 214–230.

Kneib, R.T. 2000. Salt marsh ecoscapes and production transfers by estuarine nekton in the southeastern United States. In: M.P. Weinstein and D.A. Kreeger (Eds.), *Concepts and Controversies in Tidal Marsh Ecology*, pp. 267–291. Kluwer Academic, Boston.

Knutson, P.L., W.N. Seeling, and M.R. Inskeep. 1982. Wave dampening in *Spartina alterniflora* marshes. *Wetlands* 2: 87–104.

Kuenzler, E. 1961. Structure and energy flow of a mussel population in a Georgia salt marsh. *Limnology and Oceanography* 6: 191–204.

La Peyre, M.K. and T. Birdsong. 2008. Physical variation of non-vegetated marsh edge habitats and nekton habitat use patterns in Barataria Bay, LA, USA. *Marine Ecology Progress Series* 356: 51–61.

Lathrop, R.G., Jr. and A. Love. 2007. Vulnerability of New Jersey's coastal habitats to sea level rise. Grant F. Walton Center for Remote Sensing & Spatial Analysis Rutgers University and the American Littoral Society. Highlands, NJ. 17 pp.

Lent, C. 1969. Adaptations of the ribbed mussel, *Modiolis demissus* (Dillwyn), to the intertidal habitat. *American Zoologist* 9: 283–292.

Levin, L.A., D. Talley, and G. Thayer. 1996. Succession of macrobenthos in a created salt marsh. *Marine Ecology Progress Series* 141: 67–82.

Lin, J. 1989. Influence of location in a salt marsh on survivorship of ribbed mussels. *Marine Ecology Progress Series* 56: 105–110.

Lotrich, V.A. 1975. Summer home range and movements of *Fundulus hereroclitus* (Pisces: Cyprinodontidae) in a tidal creek. *Ecology* 56(1): 191–198.

Lundholm, J.T. and P.J. Richardson. 2010. Habitat analogues for reconciliation ecology in urban and industrial environments. *Journal of Applied Ecology* 47: 966–975.

McCraith, B.J., L.R. Gardner, D.S. Wethey, and W.S. Moore. 2003. The effect of fiddler crab burrowing on sediment mixing and radionuclide profiles along a topographic gradient in a southeastern salt marsh. *Journal of Marine Research* 61(3): 359–390.

Melvin, S.L. and J.W. Webb. 1998. Differences in the avian communities of natural and created *Spartina alterniflora* salt marshes. *Wetlands* 18(1): 59–69.

Mitsch, W.J. and J.G. Gosselink. 2015. *Wetlands*, 5th ed., John Wiley & Sons, Inc., New York.

Morgan, P.A., D.M. Burdick, and F.T. Short. 2009. The functions and values of fringing salt marshes in Northern New England, USA. *Estuaries and Coasts* 32: 483–495.

Morgan, P.A. and F.T. Short. 2002. Using functional trajectories to track constructed salt marsh development in the Great Bay Estuary, Maine/New Hampshire, U.S.A. *Restoration Ecology* 10(3): 461–473.

Moschella, P.S., M. Abbiati, P. Åberg, L. Airoldi, J.M. Anderson, F. Bacchiocchi, F. Bulleri, G.E. Dinesen, M. Frost, E. Gacia, L. Granhag, P.R. Jonsson, M.P. Satta, A. Sundelöf, R.C. Thompson, and S.J. Hawkins. 2005. Low-crested coastal defence structures as artificial habitats for marine life: Using ecological criteria in design. *Coastal Engineering* 52: 1053–1071.

Moy, L.D. and L.A. Levin. 1991. Are *Spartina* marshes a replaceable resource? A functional approach to evaluation of marsh creation efforts. *Estuaries* 14: 1–16.

Nestlerode, J.A., M.W. Luckenbach, and F.X. O'Beirn. 2007. Settlement and survival of the oyster *Crassostrea virginica* on created oyster reef habitats in Chesapeake Bay. *Restoration Ecology* 15(2): 273–283.

Nielsen, K.J. and D.R. Franz. 1995. The influence of adult conspecifics and shore level on recruitment of the ribbed mussel *Geukensia demissa* (Dillwyn). *Journal of Experimental Marine Biology and Ecology* 188(1): 89–98.

Ompi, M. 2011. Settlement behaviour and size of mussel larvae from the family Mytilidae *Brachidontes erosus* (Lamarck, 1819), *Brachidontes rostratus* (Dunker, 1857), *Trichomya hirsutus* (Lamarck, 1819), and *Mytilus galloprovincialis* (Lamarck, 1819). *Journal of Coastal Development* 13(3): 215–227.

Pasternack, G.B., W.B. Hilgartner, and G.S. Brush. 2000. Biogeomorphology of an upper Chesapeake Bay river-mouth tidal freshwater marsh. *Wetlands* 20(3): 520–537.

Peterson, M.S. and M.R. Lowe. 2009. Implications of cumulative impacts to estuarine and marine habitat quality for fish and invertebrate resources. *Reviews in Fisheries Science* 17: 505–523.

Potthoff, M.T. and D.M. Allen. 2003. Site fidelity, home range, and tidal migrations of juvenile pinfish, *Lagodon rhomboides*, in salt marsh creeks. *Environmental Biology of Fishes* 67: 231–240.

Priest, W.I. 2008. Design Criteria for Tidal Wetlands. In: Erdle, S.Y., J.L.D. Davis, and K.G. Sellner (Eds.), *Management, Policy, Science and Engineering of Nonstructural Erosion Control in the Chesapeake Bay: Proceedings of the 2006 Living Shoreline Summit*, 136 pp. CRC Publ. No. 08-164, Gloucester Point, VA.

Rodríguez-Calderón, C. 2014. Forecasting ecosystem service capacity: Effects from sea level rise and management practices, Chesapeake Bay, Virginia. Dissertation. College of William & Mary.

Roosenburg, W.M. 1991. The diamondback terrapin: population dynamics, habitat requirements, and opportunities for conservation. *New Perspectives in the Chesapeake System: A Research and Management Partnership. Proceedings of a Conference* 227–234.

Rozas, L.P. and D.J. Reed. 1993. Nekton use of marsh-surface habitats in Louisiana (USA) deltaic salt marshes undergoing submergence. *Marine Ecology Progress Series* 96: 147–157.

Sacco, J.N., E.D. Seneca, and T.R. Wentworth. 1994. Infaunal community development of artificially established salt marshes in North Carolina. *Estuaries* 17: 489–500.

Scyphers, S.B., S.P. Powers, K.L. Heck Jr., and D. Byron. 2011. Oyster reefs as natural breakwaters mitigate shoreline loss and facilitate fisheries. *PLoS ONE* 6(8): e22396.

Scyphers, S.B., S.P. Powers, and K.L. Heck. 2015. Ecological value of submerged breakwaters for habitat enhancement on a residential scale. *Environmental Management* 55 (2): 383–391

Seed, R. 1980. Predator–prey relationships between the mud crab *Panopeus herbstii*, the blue crab, *Callinectes sapidus* and the Atlantic ribbed mussel *Geukensia* (= *Modiolus*) *demissa*. *Estuarine and Coastal Marine Science* 11(4): 445–458.

Seigel, A., C. Hatfield, and J.M. Hartman. 2005. Avian response to restoration of urban tidal marshes in the Hackensack Meadowlands, New Jersey. *Urban Habitats* 3(1): 87–116.

Shepard, C.C., C.M. Crain, and M.W. Beck. 2011. The protective role of coastal marshes: A systematic review and meta-analysis. *PLoS ONE* 6: 27374.

Shriver, W.G., T.P. Hodgman, J.P. Gibbs, and P.D. Vickery. 2004. Landscape context influences salt marsh bird diversity and area requirements in New England. *Biological Conservation* 119(4): 545–553.

Silliman, B.R. and A. Bortolus. 2003. Underestimation of *Spartina* productivity in western Atlantic marshes: Marsh invertebrates eat more than just detritus. *Oikos* 101: 549–554.

Silliman, B.R. and S. Newell. 2003. Fungal farming in a snail. *Proceedings of the National Academy of Sciences USA* 100: 15643–15648.

Silliman, B.R., E. Schrack, Q. He, R. Cope, A. Santoni, T. Van Der Heide, R. Jacobi, M. Jacobi, and J. Van De Koppel. 2015. Facilitation shifts paradigms and can amplify coastal restoration efforts. *Proceedings of the National Academy of Sciences* 112(46): 14295–14300.

Silliman, B.R., J. van de Koppel, M.D. Bertness, L.E. Stanton, and I.A. Mendelssohn. 2005. Drought, snails, and large-scale die-off of southern U.S. salt marshes. *Science* 310: 1803–1806.

Silliman, B.R. and J.C. Zieman. 2001. Top-down control of *Spartina alterniflora* production by periwinkle grazing in a Virginia salt marsh. *Ecology* 82(10): 2830–2845.

Skinner, M.A., S.C. Courtenay, W.R. Parker, and R.A. Curry. 2005. Site fidelity of mummichogs (*Fundulus heteroclitus*) in an Atlantic Canadian estuary. *Water Quality Research Journal of Canada* 40(3): 288–298.

Stiven, A. and E. Kuenzler. 1979. The response of two salt marsh molluscs, *Littorina irrorata* and *Geukensia demissa*, to field manipulations of density and *Spartina* litter. *Ecological Monographs* 49: 151–171.

Sutherland, J.P. and R.H. Karlson. 1977. Development and stability of the fouling community at Beaufort, North Carolina. *Ecological Monographs* 47: 425–446.

Sweeney, J., L. Deegan, and R. Garritt. 1998. Population size and site fidelity of *Fundulus heteroclitus* in a macrotidal saltmarsh creek. *Biological Bulletin* 195: 238–239.

Teo, S.L.H. and K.W. Able. 2003. Habitat use and movement of the mummichog (*Fundulus heteroclitus*) in a restored salt marsh. *Estuaries* 26(3): 720–730.

Toft, J.D., J.R. Cordell, C.A. Simenstad, and L.A. Stamatiou. 2007. Fish distribution, abundance, and behavior along city shoreline types in Puget Sound. *North American Journal of Fisheries Management* 27(2): 465–480.

Vandenbruwaene, W., T. Maris, T.J.S. Cox, D.R. Cahoon, P. Meire, and S. Temmerman. 2011. Sedimentation and response to sea-level rise of a restored marsh with reduced tidal exchange: Comparison with a natural tidal marsh. *Geomorphology* 130: 115–126.

Vaselli, S., F. Bulleri, and L. Benedetti-Cecchi. 2008. Hard coastal-defence structures as habitats for native and exotic rocky-bottom species. *Marine Environmental Research* 66: 395–403.

Walker, B., C.S. Holling, S.R. Carpenter, and A. Kinzig. 2004. Resilience, adaptability and transformability in social–ecological systems. *Ecology and Society* 9(2): 5.

Weinstein, M.P. and S.P. O'Neil. 1986. Exchange of marked juvenile spots between adjacent tidal creeks in the York River Estuary, Virginia. *Transactions of the American Fisheries Society* 115: 93–97.

Wong, M., C. Peterson, and M. Piehler. 2011. Evaluating estuarine habitats using secondary production as a proxy for food web support. *Marine Ecology Progress Series* 440: 11–25.

Woods, H., W.J. Hargis Jr., C.H. Hershner, and P. Mason. 2005. Disappearance of the natural emergent 3-dimensional oyster reef system of the James River, Virginia, 1871–1948. *Journal of Shellfish Research* 24(1): 139–142.

Yozzo, D.J. and R.J. Diaz. 1999. Tidal freshwater wetlands. Invertebrate diversity, ecology, and functional significance. In: Batzer, D.P., R.B. Rader, and S.A. Wissinger (Eds.), *Invertebrates in Freshwater Wetlands of North America: Ecology and Management*, pp. 889–918. Wiley, Hoboken, NJ.

Yozzo, D.J. and D.E. Smith. 1998. Composition and abundance of resident marsh-surface nekton: Comparison between tidal freshwater and salt marshes in Virginia, USA. *Hydrobiologia* 362(1–3): 9–19.

Zedler, J.B. and J.C. Callaway. 1999. Tracking wetland restoration: Do mitigation sites follow desired trajectories? *Restoration Ecology* 7(1): 69–73.

CHAPTER 16

Ecological Performance of Hudson River Shore Zones
What We Know and What We Need to Know

David L. Strayer and Stuart E.G. Findlay

CONTENTS

16.1 INTRODUCTION

The shore zones along New York's Hudson River have been highly modified to allow for use as sites for railroads and roads, shipping, housing, commercial and industrial activity, disposal of dredge spoils and waste, recreation, and other activities. It is likely that such modifications will continue and even intensify into the future as sea levels rise, damage from recent storms (e.g., Superstorm Sandy) and failing infrastructure is repaired, and people continue to be drawn to the river's edge. In the past, such modifications were usually done with little consideration for their effects on the biodiversity and ecological processes of shore zones, resulting in widespread, severe ecological damage. More recently, as the ecological value of shore zones (and the extent of past damage) has been recognized, there has been growing interest in retaining or restoring as much ecological value as possible while still meeting human needs in the shore zone. This chapter reviews what we have learned about the ecological performance of Hudson River shore zones, both natural and engineered, and lays out some challenges for understanding, managing, and enhancing the ecological structure and function of Hudson River shore zones. We use the term "ecological performance" broadly to encompass ecological structure (e.g., biodiversity), functions (e.g., decomposition rates), and services (e.g., recreational potential). Sites that vary in any of these characteristics have different ecological performance.

For the purposes of this chapter, we follow our earlier definition of the shore zone as "the region closely adjoining the shoreline in which strong and direct interactions tightly link the terrestrial ecosystem to the aquatic ecosystem, and vice versa" (Strayer and Findlay 2010, p. 128). In practice in our field studies, this has usually meant the zone extending from 1 m vertically below the mean lower low water mark up to 1 (or 1.25) m vertically above the mean higher high water mark. Because shore zones along the Hudson range from nearly horizontal to nearly vertical, shore zones defined in this way range from being <1 m wide to >100 m wide, and encompass a wide range of habitats from steel walls and steep bedrock to sandy beaches and marshes.

Our focus has been on the freshwater tidal Hudson River in eastern New York (Figure 16.1), extending from RKM 100 (i.e., river kilometer 100 as measured from The Battery at the southern end of the river in New York City) to the head of tide at RKM 248 in Troy. This section of the Hudson is a large (mean depth and width are ~8 m and 900 m, respectively), warmwater, turbid (transparency usually 1–1.5 m) river that almost never contains detectable sea salt. Nevertheless, the entire study area is tidal; tidal range (0.8–1.6 m) does not diminish upriver and is as large at

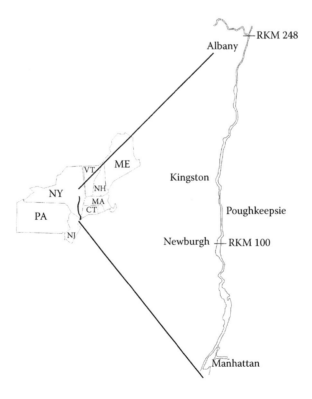

Figure 16.1 Location of the Hudson River. (Modified from Strayer, D.L., S.E.G. Findlay, D. Miller, H.M. Malcom, D.T. Fischer, and T. Coote. 2012. Biodiversity in Hudson River shore zones: Influence of shoreline type and physical structure. *Aquatic Sciences* 74: 597–610.)

Troy as at New York City (Geyer and Chant 2006). Water levels and currents along the Hudson's shores are determined chiefly by tides rather than by freshwater flows (Strayer and Findlay 2010), although downriver currents can be strong during floods. In *addition* to tidal and flood currents, the Hudson's shores are subject to strong forces from wakes, wind-driven waves, and ice (Georgas et al. 2015; LaPann-Johannessen et al. 2015; Miller and Georgas 2015). The Hudson's water is moderately hard, turbid, and fertile (Caraco et al. 1997; Simpson et al. 2006). The macrophytes *Vallisneria americana* (submersed) and *Trapa natans* (floating-leaved) are common (Nieder et al. 2004), but their coverage varies considerably from year to year. The fish and invertebrate fauna is dominated by species typical of warm freshwaters, although diadromous fishes (e.g., *Alosa* spp.) and several typically brackish-water fishes and invertebrates are common (e.g., Simpson et al. 1985). Approximately 50% of the shoreline of the freshwater tidal Hudson has been engineered (Miller 2005) to stabilize shorelines and protect them from erosion, provide deepwater access to boats, or contain dredge spoils. Riprapped revetments are the most common modification, especially along the rail lines that border both sides of the river, but cribbing, pilings, sheet pile, and other structures also occur.

The Hudson River Sustainable Shorelines Project (https://www.hrnerr.org/hudson-river-sustainable-shorelines/) was set up in 2008 by the Hudson River National Estuarine Research Reserve as a partnership with the Cary Institute of Ecosystem Studies, the Consensus Building Institute, the New York State Department of Environmental Conservation, the Stevens Institute of Technology, and other partners "to provide science-based information about the best shoreline management options for preserving important natural functions of the Hudson River Estuary's shore zone...." As part of this project, we and other ecologists have been investigating the ecological characteristics and functions of different kinds of shore zones, both natural and engineered, and testing ways to improve the ecological functioning of engineered shore zones. Shore zones provide as wide a range in ecological functions as other ecosystems, including production of plant and animal biomass, nutrient cycling, provision of habitats, and so on. We consider several (but not all) of these and provide examples and overviews in this chapter. Notable omissions from our Hudson studies are amphibians, reptiles, birds, and mammals, which have been considered in depth for other shore zones (e.g., Isdell et al. 2015; Kaufman et al. 2014; Meager et al. 2012; Tracy-Smith et al. 2012). This chapter reviews our major findings, lists some important challenges, and offers some suggestions for moving forward.

16.2 WHAT WE KNOW

16.2.1 There Is Wide Variation in Ecological Performance across Sites

Our group has measured organic matter accumulation and decay (Harris et al. 2014), plant composition and cover (Strayer et al. 2012, 2016), terrestrial and aquatic invertebrate abundance and composition (Coote 2015; Strayer et al. 2012), and fish density and species composition (Strayer et al. 2012 and unpublished) along the Hudson's shores. Regardless of the ecological variable that we measured, we found wide variation in ecological structure and function among sites, often ranging from near zero to very high values (Figure 16.2). It appears that many ecological variables reach higher values in natural shore zones than in engineered shore zones (Figure 16.2b; Strayer et al. 2012), although variation is very high within each of these classes.

As a result of this high variation, it is difficult to make generalizations about the ecology of the Hudson's shore zones that apply to all shore types. Furthermore, this high variation means that the replacement of one shore type with another, or the restoration or degradation of a shore zone, can have very large ecological consequences.

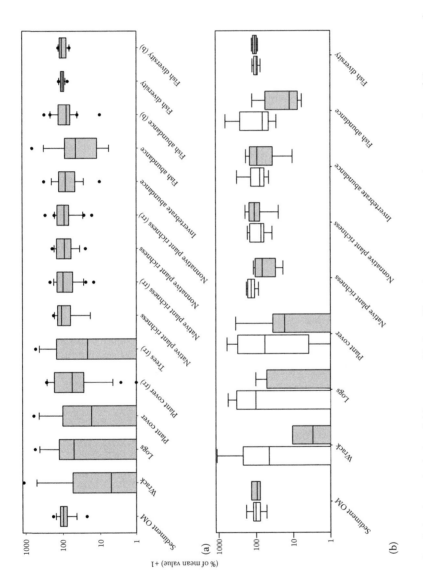

Figure 16.2 Wide range of variation in ecological variables across Hudson River shore zones, expressed as percentage of the mean value observed across all sites (note the log scale), for (a) all study sites and (b) comparison of natural (white) versus engineered (gray) shore zones. The box extends from the 25th to the 75th percentile, the horizontal line shows the median, the whiskers show the 10th and 90th percentiles, and dots show outliers. Most studies included 18–21 sites. Unless indicated otherwise, the studies covered randomly selected sites from a wide range of shore zone types, both natural and engineered; on the x-tic labels in (a), (b) indicates data from built shorelines only, and (rr) indicates data from riprapped revetments only. From left to right, variables are the % organic matter content of shallow subtidal sediments; the wrack index (a measure of organic matter accumulation, calculated as the product of the thickness and width of the wrack line); the number of large logs on the shore; plant cover (modified Braun–Blanquet scale in quadrats for all sites; % cover on a vertical cover board for rr) along the shore; the number of trees (stems >3 m tall) along the shore; taxonomic richness of native and nonnative plants; abundance of aquatic invertebrates in kick-net samples; abundance of fish in electrofishing surveys; and Gini diversity of fish in electrofishing surveys. See Strayer et al. (2012, 2016) and Harris (2014) for more details.

16.2.2 Several Natural and Anthropogenic Factors Contribute to Cross-Site Ecological Variation

Several factors contribute to the large variation across the Hudson's shore zones. Here, we briefly describe the influence of a few of the most important factors: slope, physical complexity, exposure to physical forces, species invasions, and management by humans.

16.2.2.1 Slope

Slope influences the ecological performance of shore zones in two ways: by affecting the area of the shore zone and by affecting habitat quality. If we define the shore zone as the area lying between two topographic contours, it is obvious that increasing the slope of the shore zone will decrease its areal extent and therefore decrease the overall magnitude of any ecological variable or process that is areally based (e.g., population size of a riparian plant, primary production). In the extreme case of a vertical wall, the shore zone nearly disappears, and many of its ecological values are minimized.

Slope affects many of the ecological variables that we studied along the Hudson's shores, often leading to diminished performance on steeper slopes (Figure 16.3). Specifically, steep shore zones collect

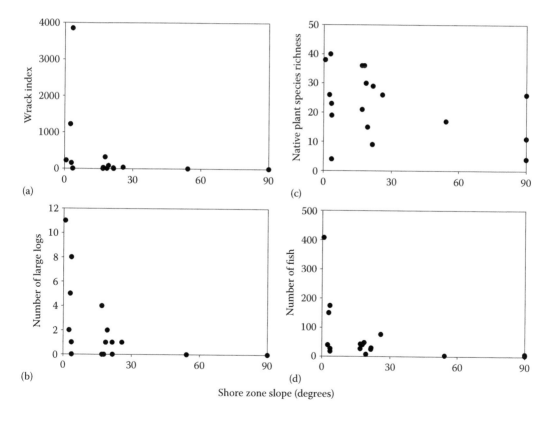

Figure 16.3 Examples of diminished ecological performance of steep shore zones along the freshwater tidal Hudson River: (a) wrack index (a measure of organic matter accumulation, calculated as the product of the thickness and width of the wrack line); (b) the number of large logs on the shore; (c) species richness of terrestrial vascular plants; (d) number of fish collected during three 5-min electrofishing samples. Sites include both natural and engineered shore zones. (From Strayer, D.L., S.E.G. Findlay, D. Miller, H.M. Malcom, D.T. Fischer, and T. Coote. 2012. Biodiversity in Hudson River shore zones: Influence of shoreline type and physical structure. *Aquatic Sciences* 74: 597–610.)

fewer pieces of large driftwood and less organic matter (wrack) than flatter shore zones, have faster rates of organic matter loss (Harris et al. 2014), and often support lower densities and fewer species of plants and animals (Strayer et al. 2012, 2016). Because engineered shore zones are often intentionally steeper than "natural" (unengineered) shore zones (Strayer and Findlay 2010; Strayer et al. 2012), steepness may therefore contribute to the diminished ecological function often observed in engineered shore zones.

Few studies have investigated the reasons behind the ecological poverty of steep shore zones. Steep shores are often exposed to strong wave, current, and ice forces, and gravitational forces on steep slopes will tend to move small or light particles downslope rather than remaining in place. The paucity of fine particles and organic matter, along with strong physical forces (especially from ice and floating debris), may discourage the development of vegetation, which could in turn discourage animal use of steep shore zones. It is even possible that there is a positive feedback between slope development and vegetational development, such that steep slopes would shed fine particles and organic matter, discouraging the growth of vegetation that could accumulate fine particles and prevent further erosion and steepening of the slope. Conversely, initially flatter slopes or finer sediments could allow dense vegetation to develop, which could then allow fine particles and organic matter to accumulate, further encouraging vegetation. Such a feedback could also imply the existence of a threshold initial slope or particle size distribution that will determine whether the shore will develop into a barren, coarse, steepening shore or a vegetated, fine, flattening shore. It is possible that steep engineered shores, which often are deliberately smooth to shed debris, reduce friction, and so on, would discourage this dynamic between vegetational development, roughness, and the accumulation of fine or organic particles.

16.2.2.2 Exposure

The shore zones along the Hudson vary widely in their exposure to physical forces such as waves, wind, wakes, currents, ice, and floating debris. Although it has proved difficult to summarize or measure "exposure" satisfactorily (see discussion of this point below), it is nevertheless clear even from imperfect characterization of the exposure regime that exposure to physical forces strongly influences the structure and function of the Hudson's shore zones, including the abundance and composition of the biota (Figure 16.4).

Exposure could have multiple effects on ecological variables. Strong physical forces, whether more or less continuous (e.g., currents, wind-driven waves) or infrequent (e.g., heavy ice driven onto a shore, hurricane-force winds and the waves they produce), could make the shore zone inhospitable to some organisms (e.g., small fishes whose swimming speeds are inadequate to cope with strong currents or waves—Wolter and Arlinghaus 2003) or create strong disturbances that pave the way for colonization by disturbance-dependent species, such as many nonnative plants (Davis et al. 2000). In addition, exposure must affect the physical character of the substratum (particle size, organic content) and water movement near and within the substratum, both of which should strongly affect biotic distributions, redox conditions, and biogeochemical processes.

It is worth noting that engineered shore zones along the Hudson tend to be more exposed than unengineered sites (see Table 2 of Strayer et al. 2012). Whether this is because they were built out into the channel where they were exposed to stronger physical forces or because protection was needed on the sites that were naturally most exposed, this high exposure may contribute to the distinctive ecological character of engineered sites.

16.2.2.3 Physical Complexity

Ecologists have long known that physical complexity or heterogeneity tends to favor high biodiversity and high values of ecological processes (e.g., Lovett et al. 2005; Stein et al. 2014). In keeping with this general pattern, ecologists working on all kinds of shore zones have reported that physically complex or heterogeneous shore zones support denser and more diverse communities (reviewed by Strayer and Findlay 2010). Measures of physical complexity and heterogeneity often include surface roughness (at multiple scales), variation in particle size, extent of holes or penetrations, and whether shore zone materials are homogeneous (the same rock type for instance) or an intermingling of several different materials (rocks, wood, etc.). We have found such effects in the Hudson's shore zones (Figure 16.5), although these effects are generally rarer and weaker than the effects of slope and exposure (Strayer et al. 2012; Villamagna et al. 2009). The apparently

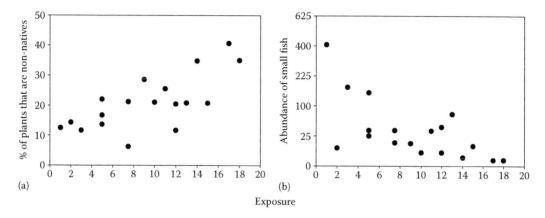

Figure 16.4   Examples of the influence of exposure to physical forces on shore zone ecology along the Hudson: (a) fraction of the shore zone plant species that are not native ($r^2 = 0.50$, $p = 0.001$); (b) abundance of small fish (number of fish <200 mm long caught in 15 min of electrofishing; $r^2 = 0.41$, $p = 0.004$). Note the square-root scale on the y axis of (b). "Exposure" is an index based on readings of dynamometers, mass loss from clod cards, and sediment grain size (see Strayer et al. 2012 for details); higher values indicate exposure to greater physical forces. (Modified from Strayer, D.L., S.E.G. Findlay, D. Miller, H.M. Malcom, D.T. Fischer, and T. Coote. 2012. Biodiversity in Hudson River shore zones: Influence of shoreline type and physical structure. *Aquatic Sciences* 74: 597–610.)

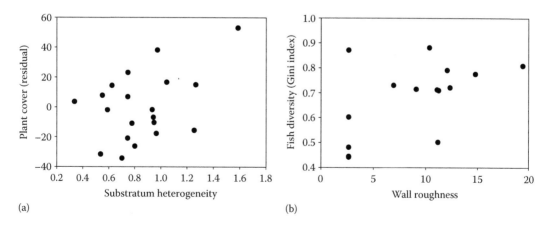

Figure 16.5   Examples of the effects of physical complexity on density and diversity of shore zone communities: (a) plant cover (% coverage of a vertical cover board) on riprapped revetments (residuals after the effects of shore zone slope have been removed, and excluding one site where the vegetation was mowed), as a function of particle heterogeneity (SD of the pebble count; Strayer et al. 2016) ($r^2 = 0.17$, $p = 0.07$); (b) species diversity of fish communities along walls, as a function of the physical complexity of the wall ($r^2 = 0.29$, $p = 0.02$). (Modified from Strayer, D.L., E. Kiviat, S.E.G. Findlay, and N. Slowik. 2016. Vegetation of rip-rapped revetments along the freshwater tidal Hudson River, New York. *Aquatic Sciences*. 78: 605–614; and Strayer, D.L., D. Miller, and S.E.G. Findlay, unpublished.)

subordinate role of physical complexity along the Hudson may simply reflect a lack of study, but may be a result of the particular physical conditions along the Hudson (very wide range in slope and exposure, including many steep, exposed shores resulting from human construction [Miller 2005] and glacial scouring [Sirkin and Bokuniewicz 2006]).

16.2.2.4 Species Invasions

Nonnative species are abundant and conspicuous in the Hudson's shore zones, perhaps reflecting the general success of invaders in heavily disturbed habitats (e.g., Davis et al. 2000) and the accessibility of the Hudson's shore zones to vectors such as commercial shipping, boat launches, canals, fish stocking, bait bucket releases, and water dispersal of plants (hydrochory) and animals (Mills et al. 1996, 1997). Many of the most abundant species in the aquatic or terrestrial part of the Hudson's shore zones are nonnatives (Table 16.1), and 29%–36% of the terrestrial plant species and 30%–36% of the fish species found in various studies along the Hudson's shores were not native (Strayer et al. 2012, 2016). Although only a few of these nonnative species have received serious study, we know

Table 16.1 Nonnative Species That Were Abundant along the Shore Zone of the Freshwater Tidal Hudson River in the Early 21st Century

Scientific Name	Common Name
Plants	
Ailanthus altissima	Tree of heaven
Amorpha fruticosa	False indigo
Artemisia vulgaris	Mugwort
Celastrus orbiculatus	Oriental bittersweet
Centaurea stoebe	Spotted knapweed
Cynanchum louiseae	Black swallowwort
Gallium mollugo	Stockport weed
Lonicera X bella	Bell's honeysuckle
Lythrum salicaria	Purple loosestrife
Melilotus officinalis	Sweet-clover
Phragmites australis	Common reed
Trapa natans	Water chestnut
Verbascum thapsus	Common mullein
Invertebrates	
Bithynia tentaculata	Faucet snail
Dreissena polymorpha	Zebra mussel
Vertebrates	
Cygnus olor	Mute swan
Cyprinus carpio	Common carp
Ictalurus punctatus	Channel catfish
Micropterus dolomieui	Smallmouth bass
Micropterus salmoides	Largemouth bass

Source: From Nieder, W.C., E. Barnaba, S.E.G. Findlay, S. Hoskins, N. Holochuck, and E.A. Blair. 2004. Distribution and abundance of submerged aquatic vegetation and *Trapa natans* in the Hudson River estuary. *Journal of Coastal Research* 45: 150–161; Strayer, D.L., S.E.G. Findlay, D. Miller, H.M. Malcom, D.T. Fischer, and T. Coote. 2012. Biodiversity in Hudson River shore zones: Influence of shoreline type and physical structure. *Aquatic Sciences* 74: 597–610; Strayer, D.L., E. Kiviat, S.E.G. Findlay, and N. Slowik. 2016. Vegetation of rip-rapped revetments along the freshwater tidal Hudson River, New York. *Aquatic Sciences.* 78: 605–614; Strayer, D.L., D. Miller, and S.E.G. Findlay, unpublished, and other unpublished observations.

that nonnative species have had strong and far-reaching effects on biodiversity and biogeochemistry in the Hudson (e.g., Caraco et al. 2006; Strayer et al. 1999, 2003; Tall et al. 2011) and in shore zones elsewhere (e.g., Airoldi et al. 2005; Crooks 1998; Thompson et al. 2002), and almost certainly have affected people's perception, use, and valuation of the shore zone.

It is likely that new species invasions will continue to strongly affect the Hudson's shore zones. Many species capable of strong ecological and economic effects in the shore zone have either arrived recently in the Hudson basin or are likely to appear in the near future (e.g., mile-a-minute vine, hydrilla, grass carp, round gobies, the emerald ash borer, and the New Zealand mudsnail). It is an open question whether the many established invaders in the Hudson's shore zone will make it more susceptible ("invasional meltdown"—Jeschke et al. 2012; Simberloff and Von Holle 1999) or less susceptible (e.g., because of competition from aggressive plants such as false indigo and oriental bittersweet, or predation from abundant predatory fish such as channel catfish and smallmouth bass) to these future invasions and their effects.

16.2.2.5 Management

The deliberate management of shore zones also strongly affects their structure and function. In addition to construction and maintenance of structures such as bulkheads and revetments, the Hudson's shore zones are often graded, mowed, herbicided, or planted with ornamental plants. Not surprisingly, these activities can affect the biodiversity and ecological functioning of shore zones (e.g., Strayer et al. 2016).

Despite the undoubted importance of the factors just discussed, they do not account for all of the observed variation in the Hudson's shore zones, and all of our studies have found a large amount of unexplained variation. This unexplained variation is partly a result of our inability to accurately quantify factors such as "exposure," but presumably also reflects the influence of other, unstudied, factors. Some factors that probably affect the Hudson's shore zone ecosystems, and which may be worth studying or managing, include the specific disturbance history of the site, the specific site history of engineered sites, the local climate, the chemical and physical properties of construction materials such as riprap, and the nature and use of the surrounding area.

16.3 WHAT WE NEED TO KNOW

16.3.1 What Drives Shoreline Modification and Design Choices?

Shorelines are modified by human activity for several reasons: to preserve or enhance a current use, allow a future use, or avoid a future risk. Decisions about moving forward with a modification generally hinge on the perceived need balanced against the cost and difficulty of actually completing a project. The need to take action is often triggered by evidence that the shoreline is either not now supporting its intended use or will not be in the near future. The intended use might be as specific and demanding as allowing berthing of large ships or as simple as being a visually appealing swath along the waterfront or serving as a physical buffer or protection from waves and currents.

Many uses depend on knowing the various physical forces imposed on the shore; for instance, use as a swimming area assumes relatively calm water. However, simple yet reliable field measures of these physical forces have been difficult to develop (Strayer et al. 2012). The Hudson River Sustainable Shorelines Project addressed this problem by developing tools showing the best estimate of forces (waves, wakes, and ice) on Hudson River shorelines (Georgas et al. 2015; LaPann-Johannessen et al. 2015; Miller and Georgas 2015). While undoubtedly useful for initial screening of sites, such tools often will be inadequate. They generally do not provide sufficient detail to evaluate a specific project site (too coarse spatial scale, inadequate characterization of long-term variability, including extreme events). Furthermore, it may be difficult to translate physical forces expressed as wave heights, shear stresses, and so on into the variables that affect site use and maintenance (e.g., rates of erosional loss, and especially measures of ecological performance such as survival of plants, habitat suitability for animals, rates of nutrient transformation, and the like). Until we have better measures to quickly and quantitatively estimate the functional exposure regime at a site, it seems likely that we will overbuild shore defenses and be unable to adequately predict ecological performance (including the performance of green infrastructure) at project sites. Adding uncertainty to the decision process is the obvious stochastic element in preparing for future events. While it seems likely that storm frequency and severity will increase for some shorelines (e.g., Lin et al. 2012), our inability to accurately predict storm frequency and severity will again lead to a tendency to overbuild and modify shore zones too frequently.

Human land use in shore zones also drives decisions to take action. Some land uses have clear and specific requirements for protection (rail lines, fuel storage, wastewater treatment plants, etc.); hence, designing to meet these requirements is reasonably straightforward. In contrast, some land uses can tolerate occasional flooding, moderate erosion, or inputs of debris (e.g., some parts of parks), or it may be unclear what essential attribute of the shore zone is at risk from flooding or erosion. For instance, many humans prefer shore zones that are "tidy" and easy to move around (Dalton 2012), and floods may make the shore zone untidy. However, tidy shore zones often provide low habitat heterogeneity and cover for animals. Rather than insisting that flooding must be prevented because it makes the shore zone messy, it may be useful to ask the question about how much "neatness" will suffice for human use while balancing the benefits to other organisms, and design shore defenses accordingly. As before, better information on what is driving decisions about shoreline modification will allow for better choices of actions. Last, it remains difficult to balance human preferences (or ecological performance) against the costs of construction and maintenance. In many instances, it is obvious what a landowner or decision-maker is or is not willing to pay and this provides an estimate of value that could be assigned to either a conventional or ecologically enhanced shore zone treatment. On the other hand, if there are multiple users of public land who may have different desired uses, it becomes much harder to determine how much additional cost (either in dollars or risk of failure) is justified for a nontraditional treatment (cf. Hostmann et al. 2005).

16.3.2 What Are Gaps to Picking and Implementing the "Best" Solution?

Once a decision is made to undertake some active shoreline modification, there are several gaps between desire and a perfect project. First, while the general shoreline attributes that often lead to better ecological function are reasonably clear (e.g., low slopes, high heterogeneity), design, permitting, and construction require detailed drawings, material lists, and so on, rather than general notions. Exactly what grade should a "low slope" be? What range of stone sizes should be used to build a "heterogeneous" shoreline? Today, good "off-the-shelf" designs are not available for most designs in most locations. This is a dilemma because until there are field-tested designs available for a shoreline treatment in a given setting, there will not be much construction of that type, and until there is construction, there won't be robust designs. It also seems possible that this scarcity of specific designs may actually lead to reduced innovation because once a few designs receive a "stamp of approval," they may become the default. Below, we briefly discuss experimental approaches, but it is important to recognize that innovation may not easily be put into practice.

Another significant difficulty is quantifying cumulative benefits or damages. While any specific shoreline treatment may have small positive or negative effects on the larger ecosystem, there is rarely an easy way to measure the consequences of an additional segment or to determine what total amount is allowable (for negative effects) or desirable (for positive effects). It can therefore be difficult to promote (or prohibit) small projects without some mechanism to assess their whole-system benefits or damages. Adding to this difficulty is the possibility that small amounts of very good (or very bad) shore treatments could have disproportionate consequences for organisms and ecological processes. One could imagine that a short segment of river with vertical sheet-pile shores on both sides might represent an effective barrier to up- or downstream movements of small fishes that require continuous shelter. Similarly, a short reach of high turbulence could generate a large downstream plume of resuspended sediment; thus, the consequences of a particular treatment could be out of scale with their physical extent. Conversely, one might imagine that a small area of very high quality habitat in the right place could represent a significant "oasis" for organisms in transit. Hence, there is an important knowledge gap in identifying how much of a shoreline treatment, placed at what location, will lead to whole-system benefits or damages.

Several of the issues raised above are best addressed through experimental "demonstration" sites for treatments that might be novel or at least have not been tested under local conditions. These sites are a valuable proving ground for accelerating adoption of new approaches. Without these examples, property owners are unlikely to suggest innovation, regulators are less likely to grant permits, and engineers will be reluctant to sign off. While it may be true that a particular shoreline treatment has been tried somewhere, there is (for good reason) wariness for extrapolating findings from other systems to a new location. Forces on the shoreline might differ between the two systems (e.g., ice vs. no ice); thus, there will be questions about physical performance, local materials and contractors may not be suited to the approach, and the local ecosystem may be very different from the system in which the technique was developed (e.g., marine vs. freshwater). Some capacity for local demonstration seems essential, but there will be real constraints in funding, permits, and willingness to experiment.

16.3.3 Consequences of Choices

There is an assumption that an unavoidable trade-off exists between ecologically beneficial and physically robust shoreline treatments. At the extreme, this may seem sensible since a sandy beach will be more susceptible to erosion than a well-anchored sheet-pile bulkhead. In actuality, the trade-off diminishes if one considers (1) the nature of forces on a shore and (2) the tolerance or consequences of failure (Figure 16.6). Stretches of shore experiencing relatively weak forces are amenable to a much broader range of treatments (with a corresponding range of ecological

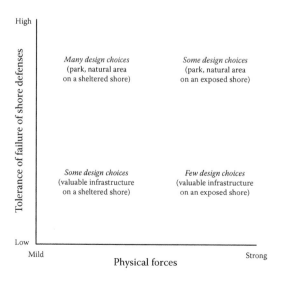

Figure 16.6 Risk of failure and site exposure constrain design choices.

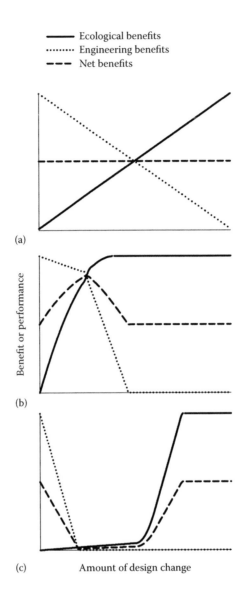

Figure 16.7 Schematic diagram of some possible trade-offs between ecological (solid line) and engineering (dotted line) benefits of shorelines that are modified to provide ecological benefits. In the diagram, the net benefit (dashed line) of the shore treatment is assumed to be an average of the ecological and engineering benefits. In practice, the net benefit will depend on the values held by stakeholders—how they value the various uses of the site and the costs of failure (cf. Hostmann et al. 2005). In (a), both ecological and engineering benefits vary linearly and inversely with the amount of change to the design; thus, all designs provide the same net benefit. In (b), ecological benefits rise rapidly to an asymptote, whereas engineering performance does not much begin to degrade until a threshold is reached. Wrack accumulation on a shore might provide such an example. The greatest net benefits are provided when the design is a little modified. In (c), even a small amount of design change greatly degrades engineering performance, while little ecological benefit is provided until radical changes are made. Structural modifications to a bulkhead face might offer an example. In this case, the optimal treatment that provides the greatest net benefits is either no change or radical change to the design, with partially modified designs being highly suboptimal. See text for further discussion.

performance) than shores exposed to extreme erosive forces. Likewise, shores that are allowed to "fail" occasionally will again have a broader spectrum of options than shores where failure is absolutely intolerable. As above, having better information on forces (exposure), uses, and tolerance to failure will at least open the door to more ecologically beneficial options.

In weighing consequences of various options, it is generally possible to identify the expected direction of change in ecological and engineering performance in response to some design change, but we often have imperfect information about the precise amount of these changes. It is therefore difficult to balance X amount of change in ecological performance against Y amount of change in engineering performance and Z amount of change in cost. For instance, hanging structures from a vertical sheet-pile bulkhead to serve as refuges for small organisms will probably lead to some habitat benefit and some structural detriment, as well as additional cost. The actual magnitude of these changes (especially the ecological changes) is probably not "knowable" at any sort of meaningful scale, and even if it were, there would be questions about how to value these changes. There are many possible relationships among cost, engineering performance, and ecological performance; hence, the best solution is not generally obvious without careful analysis (Figure 16.7). Local demonstration sites with a diversity of well-quantified attributes can address this issue, but in reality, no set of demonstration sites will ever cover the full range of possible treatments. As argued above, lack of quantitative information will lead to perpetuation of current designs and a tendency to overbuild. In addition, more precise information about engineering performance than ecological performance may favor the former in design decisions.

16.4 SUGGESTIONS FOR MOVING FORWARD

There are many ways by which the science and management of shore zones might be advanced. We will highlight just two ways forward that we think might be promising on the Hudson and elsewhere. First, the evaluation and adoption of innovative designs and management of shore zones may be inhibited by a perverse positive feedback. Developers, engineers, architects, and regulators may be reluctant to adopt or approve new designs or management approaches until they know that they have been evaluated adequately and demonstrated to be effective (durable, reasonably priced, provide their intended engineering and ecological benefits) in the field (Dalton 2011; Roberts 2015), while scientists and engineers cannot adequately evaluate new approaches until they are built at scale in nature. Consequently, construction and management tend to use conservative approaches, while scientists are restricted to evaluating innovative approaches using small-scale experiments (e.g., Browne and Chapman 2014; Goff 2010; Villamagna et al. 2009) that may fail to reflect the actual performance of these approaches. Alternatively, scientists can conduct field surveys of existing shore zones, but these are plagued by weak inference and a poor match between the existing study sites and the innovative approach that is being evaluated.

Possible ways forward here might include developing a network of demonstration sites showing the use of innovative approaches (see above), encouraging scientists and developers to work together when new approaches are being tried, and developing an experimental permit that can be issued for projects that use innovative but not fully tested approaches. Demonstration sites are thought to be helpful in influencing developers, architects, and engineers to use new approaches (Dalton 2011); hence, the Hudson River Sustainable Shorelines Project developed a list of demonstration projects (https://www.hrnerr.org/hudson-river-sustainable-shorelines/demonstration-site-network/) that can be visited and observed. When new approaches are tried at large scales, it would be very helpful if their performance could be rigorously evaluated. Thus, it may be helpful to involve scientists in design discussions to ensure that such evaluations can easily occur. As a reward for using new, ecologically enhanced designs and allowing scientists to evaluate them, it may be useful to offer developers an expedited experimental permit that allows the use of new approaches, as long as they have *promised* to significantly improve ecological functioning, will be rigorously evaluated, and are tried on a site where failure would not have catastrophic consequences. It may even be helpful to offer a companion program of small grants to scientists and developers to support postconstruction monitoring of innovative projects.

Second, it may be helpful to work more with shore zone owners and users to assess and guide their preferences for shore zones, especially with regard to ecological trade-offs associated with different shore zone treatments. People will tend to shape shore zones to a particular purpose or visual appearance and are most likely to use shore zones that share these attributes. It seems therefore useful for shore zone managers and scientists to understand the motivations behind these choices so that we design shore zones that will actually be used and valued by people. In cases where current preferences are for structures or practices with poor ecological performance, it may be that these choices arise from an ignorance of their high ecological costs. It may therefore be helpful to educate people about these ecological costs, and perhaps try to shift decisions that result in poor ecological performance and do not have well-founded design benefits (e.g., shore zones that are designed to be tidy purely for reflexive reasons). For instance, people who use the Hudson's shore zones not only appreciate the ecological value of shore zones but also prefer tidy landscapes that have poor ecological performance (Dalton 2012). If we could get shore zone users to see tidy shore zones as ecological deserts, they might learn to better appreciate less tidy and more ecological valuable shore zones, ultimately shifting practices along the Hudson.

ACKNOWLEDGMENTS

Our work on Hudson River shorelines has been supported by the National Oceanic and Atmospheric Administration, the Hudson River Foundation, the New York State Department of Environmental Conservation, and the New York–New Jersey Harbor Estuary Program. We thank our colleagues in the Hudson River Sustainable Shorelines Project and the Cary Institute of Ecosystem Studies for many helpful ideas and discussions, and two reviewers for helpful comments.

REFERENCES

Airoldi, L., M. Abbiati, M.W. Beck, S.J. Hawkins, P.R. Jonsson, D. Martin, P.S. Moschella, A. Sundelof, R.C. Thompson, and P. Aberg. 2005. An ecological perspective on the deployment and design of low-crested and other hard coastal defence structures. *Coastal Engineering* 52: 1073–1087.

Browne, M.A. and M.G. Chapman. 2014. Mitigating against the loss of species by adding artificial intertidal pools to existing seawalls. *Marine Ecology Progress Series* 497: 119–129.

Caraco, N.F., J.J. Cole, P.A. Raymond, D.L. Strayer, M.L. Pace, S.E.G. Findlay, and D.T. Fischer. 1997. Zebra mussel invasion in a large, turbid river: Phytoplankton response to increased grazing. *Ecology* 78: 588–602.

Caraco, N.F., J.J. Cole, and D.L. Strayer. 2006. Top down control from the bottom: Regulation of eutrophication in a large river by benthic grazing. *Limnology and Oceanography* 51: 664–670.

Coote, T.W. 2015. New gastropod records for the Hudson River, New York. *American Malacological Bulletin* 33: 114–117.

Crooks, J.A. 1998. Habitat alteration and community-level effects of an exotic mussel, *Musculista senhousia*. *Marine Ecology Progress Series* 162: 137–152.

Dalton, S. 2011. Hudson River Sustainable Shorelines Project Report: Decision-Making Regarding Shoreline Design and Management. 18 pp. https://www.hrnerr.org/doc/?doc=240186112.

Dalton, S. 2012. Shoreline Use and Perceptions Survey Report. Report to the Hudson River Sustainable Shorelines Project. 35 pp. https://www.hrnerr.org/doc/?doc=240189572.

Davis, M.A., J.P. Grime, and K. Thompson. 2000. Fluctuating resources in plant communities: A general theory of invasibility. *Journal of Ecology* 88: 528–534.

Georgas, N., J.K. Miller, Y. Wang, Y. Jiang, Y., and D. D'Agostino. 2015. Tidal Hudson River Ice Cover Climatology. Stevens Institute of Technology, TR-2949; in association with and published by the Hudson River Sustainable Shorelines Project, Staatsburg, NY 12580. https://www.hrnerr.org/doc/?doc =274314925.

Geyer, W.R. and R. Chant. 2006. The physical oceanography processes in the Hudson River Estuary. In: J.S. Levinton and J.R. Waldman (Eds.). *The Hudson River Estuary*, pp. 24–38. Cambridge University Press, New York.

Goff, M. 2010. Evaluating habitat enhancements of an urban intertidal seawall: Ecological responses and management implications. M.S. Thesis, University of Washington. Available at https://docs.google.com /viewer?a=v&pid=sites&srcid=dXcuZWR1fHNlYXR0bGUtc2Vhd2FsbC1wcm9qZWN0fGd4OjMzYj BkNzI3ZWM1YWFjNTM.

Harris, C., D.L. Strayer, and S. Findlay. 2014. The ecology of freshwater wrack along natural and engineered Hudson River shorelines. *Hydrobiologia* 722: 233–245.

Hostmann, M., M. Borsuk, P. Reichert, and B. Truffer. 2005. Stakeholder values in decision support for river rehabilitation. *Archiv für Hydrobiologie Supplementband* 155: 491–505.

Isdell, R.E., R.M. Chambers, D.M. Bilkovic, and M. Leu. 2015. Effects of terrestrial–aquatic connectivity on an estuarine turtle. *Diversity and Distributions* 21: 643–653.

Jeschke, J.M., L.G. Aparicio, S. Haider, T. Heger, C.J. Lortie, P. Pyšek, and D.L. Strayer. 2012. Support for major hypotheses in invasion biology is uneven and declining. *Neobiota* 14: 1–20.

Kaufman, P.R., R.M. Hughes, T.R. Whittier, S.A. Bryce, and S.G. Paulsen. 2014. Relevance of lake physical habitat indices to fish and riparian birds. *Lake and River Management* 30: 177–191.

LaPann-Johannessen, C., J.K. Miller, A. Rella, and E. Rodriguez. 2015. Hudson River Wake Study. Stevens Institute of Technology, TR-2947; in association with and published by the Hudson River Sustainable Shorelines Project, Staatsburg, NY 12580. https://www.hrnerr.org/doc/?doc=274514894.

Lin, N., Emanuel, K., Oppenheimer, M., and E. Vanmarcke. 2012. Physically based assessment of hurricane surge effect under climate change. *Nature Climate Change* 2: 462–467.

Lovett, G.M., C.G. Jones, M.G. Turner, and K.C. Weathers (Eds.). 2005. *Ecosystem Function in Heterogeneous Landscapes*. Springer, New York.

Meager, J.J., T.A. Schlacher, and T. Nielsen. 2012. Humans alter habitat selection of birds on ocean-exposed sandy beaches. *Diversity and Distributions* 18: 294–306.

Miller, D. 2005. Shoreline inventory of the Hudson River. Hudson River National Estuarine Research Reserve, New York State Department of Environmental Conservation. Available at http://gis.ny.gov/gisdata/inventories /details.cfm?DSID=1136.

Miller, J.K. and N. Georgas, 2015. Hudson River Physical Forces Analysis: Data Sources and Methods. Stevens Institute of Technology, TR-2946; in association with and published by the Hudson River Sustainable Shorelines Project, Staatsburg, NY 12580. https://www.hrnerr.org/hudson-river-sustainable-shorelines /shorelines-engineering/physical-forces-statistics/.

Mills, E.L., J.T. Carlton, M.D. Scheuerell, and D.L. Strayer. 1997. Biological invasions in the Hudson River: An inventory and historical analysis. *New York State Museum Circular* 57: 1–51.

Mills, E.L., D.L. Strayer, M.D. Scheuerell, and J.T. Carlton. 1996. Exotic species in the Hudson River basin— A history of invasions and introductions. *Estuaries* 19: 814–823.

Nieder, W.C., E. Barnaba, S.E.G. Findlay, S. Hoskins, N. Holochuck, and E.A. Blair. 2004. Distribution and abundance of submerged aquatic vegetation and *Trapa natans* in the Hudson River estuary. *Journal of Coastal Research* 45: 150–161.

Roberts, E. 2015. Outreach and Informal Conversations with Regulators and Permit Staff: A Summary for the Hudson River Sustainable Shorelines Project. 10 pp. https://www.hrnerr.org/doc/?doc=274524177.

Simberloff, D. and B. Von Holle. 1999. Positive interactions of nonindigenous species: Invasional meltdown? *Biological Invasions* 1: 21–32.

Simpson, H.J., S.N. Chillrud, R.F. Bopp, E. Schuster, and D.A. Chaky. 2006. Major ion geochemistry and drinking water supply issues in the Hudson River Basin. In: J.S. Levinton and J.R. Waldman (Eds.), *The Hudson River Estuary*, pp. 79–96. Cambridge University Press, New York.

Simpson, K.W., J.P. Fagnani, D.M. DeNicola, and R.W. Bode. 1985. Widespread distribution of some estuarine crustaceans (*Cyathura polita, Chiridotea almyra, Almyracuma proximoculi*) in the limnetic zone of the lower Hudson River, New York. *Estuaries* 8: 373–380.

Sirkin, L. and H. Bokuniewicz. 2006. The Hudson River Valley: Geological history, landforms, and resources. In: J.S. Levinton and J.R. Waldman (Eds.), *The Hudson River Estuary*, pp. 13–23. Cambridge University Press, New York.

Stein, A., K. Gerstner, and H. Kreft. 2014. Environmental heterogeneity as a universal driver of species richness across taxa, biomes and spatial scales. *Ecology Letters* 17: 866–880.

Strayer, D.L., N.F. Caraco, J.J. Cole, S. Findlay, and M.L. Pace. 1999. Transformation of freshwater ecosystems by bivalves: A case study of zebra mussels in the Hudson River. *BioScience* 49: 19–27.

Strayer, D.L. and S.E.G. Findlay. 2010. The ecology of freshwater shore zones. *Aquatic Sciences* 72: 127–163.

Strayer, D.L., S.E.G. Findlay, D. Miller, H.M. Malcom, D.T. Fischer, and T. Coote. 2012. Biodiversity in Hudson River shore zones: Influence of shoreline type and physical structure. *Aquatic Sciences* 74: 597–610.

Strayer, D.L., E. Kiviat, S.E.G. Findlay, and N. Slowik. 2016. Vegetation of rip-rapped revetments along the freshwater tidal Hudson River, New York. *Aquatic Sciences* 78: 605–614.

Strayer, D.L., C. Lutz, H.M. Malcom, K. Munger, and W.H. Shaw. 2003. Invertebrate communities associated with a native (*Vallisneria americana*) and an alien (*Trapa natans*) macrophyte in a large river. *Freshwater Biology* 48: 1938–1949.

Tall, L., Caraco, N, and R. Maranger. 2011. Denitrification host spots: Dominant role of invasive macrophyte *Trapa natans* in removing nitrogen from a tidal river. *Ecological Applications* 21: 3104–3114.

Thompson, R.C., T.P. Crowe, and S.J. Hawkins. 2002. Rocky intertidal communities: Past environmental changes, present status and predictions for the next 25 years. *Environmental Conservation* 29: 168–191.

Tracy-Smith, E., D.L. Galat, and R.B. Jacobson. 2012. Effects of flow dynamics on the aquatic-terrestrial transition zone (ATTZ) of Lower Missouri River sandbars with implications for selected biota. *River Research and Applications* 28: 793–813.

Villamagna, A., D. Strayer, and S. Findlay. 2009. Effects of surface roughness on ecological function: Implications for engineered structures in the Hudson River shore zone. Section III: 25 pp. In: Fernald, S.H., D. Yozzo and H. Andreyko (Eds.), *Final Reports of the Tibor T. Polgar Fellowship Program, 2008.* Hudson River Foundation.

Wolter, C. and R. Arlinghaus. 2003. Navigation impacts on freshwater fish assemblages: The ecological relevance of swimming performance. *Reviews in Fish Biology and Fisheries* 13: 63–89.

CHAPTER 17

San Francisco Bay Living Shorelines
Restoring Eelgrass and Olympia Oysters
for Habitat and Shore Protection

**Katharyn Boyer, Chela Zabin, Susan De La Cruz, Edwin Grosholz,
Michelle Orr, Jeremy Lowe, Marilyn Latta, Jen Miller, Stephanie Kiriakopolos,
Cassie Pinnell, Damien Kunz, Julien Moderan, Kevin Stockmann,
Geana Ayala, Robert Abbott, and Rena Obernolte**

CONTENTS

17.1 INTRODUCTION

Living shorelines projects utilize a suite of sediment stabilization and habitat restoration techniques to maintain or build the shoreline, while creating habitat for a variety of species, including invertebrates, fish, and birds (see National Oceanic and Atmospheric Administration [NOAA] 2015 for an overview). The term "living shorelines" denotes provision of living space and support for estuarine and coastal organisms through the strategic placement of native vegetation and natural materials. This green coastal infrastructure can serve as an alternative to bulkheads and other engineering solutions that provide little to no habitat in comparison (Arkema et al. 2013; Gittman et al. 2014; Scyphers et al. 2011). In the United States, the living shorelines approach has been implemented primarily on the East and Gulf Coasts, where it has been shown to enhance habitat values and increase connectivity between wetlands, mudflats, and subtidal lands, while reducing shoreline erosion during storms and even hurricanes (Currin et al. 2015; Gittman et al. 2014, 2015).

There have been fewer living shorelines projects along the US West Coast, with most occurring on small private parcels along Puget Sound in Washington state; however, recognition of the many potential benefits of this approach is growing in the region, in part because of increasing concerns about sea level rise and storm surge and the need to protect valuable residential, commercial, and industrial assets (Gallien et al. 2011; Heberger et al. 2011; McGranahan et al. 2007). In developing the California State Resources Agency Climate Change Adaptation Strategy (Natural Resources Agency 2015), California state agencies recommended the use of living shorelines as a climate change adaptation strategy to reduce the need for engineered hard shoreline protection while enhancing habitat functions as sea level rises. The California State Coastal Conservancy Climate Change Policy (State Coastal Conservancy 2011) and the California Coastal Commission Sea Level Rise Guidance (California Coastal Commission 2015) also recommended implementation of living shorelines because of their potential to reduce erosion and trap sediment while providing intertidal and subtidal habitat and helping to maintain and protect adjacent tidal wetlands. Further, the San Francisco Bay Subtidal Habitat Goals Project proposed piloting of living shorelines projects that test the roles and potential synergy of integrating restoration of multiple species for both habitat and shoreline protection benefits (State Coastal Conservancy 2010). In addition, a 2015 climate change update to the Baylands Ecosystem Habitat Goals Report (Goals Project 2015) recommended multihabitat, multiobjective approaches and living shorelines in order to increase resiliency of San Francisco Bay tidal wetlands and associated habitats to climate changes such as sea level rise.

Concordant with these recommendations, the San Francisco Bay Living Shorelines: Near-shore Linkages Project was implemented in 2012 by the State Coastal Conservancy and an interdisciplinary team of biological and physical scientists. In this chapter, we review our objectives and project design, and evaluate outcomes 3 years after installation, concluding with an assessment of early lessons learned and design criteria for future projects in San Francisco Bay and elsewhere.

17.2 FOCUS ON EELGRASS AND OLYMPIA OYSTERS

Although there are numerous options for species and materials to be utilized in living shorelines designs, this first living shorelines project in San Francisco Bay focused on restoration of two native species, eelgrass (*Zostera marina*) and Olympia oysters (*Ostrea lurida*). We selected these two species for several reasons. First, worldwide declines in both seagrasses and native shellfish species have made their restoration a major priority (Beck et al. 2009; Cunha et al. 2012; Kirby 2004; NOAA Fisheries National Shellfish Initiative 2011; Orth et al. 2006, 2010; Waycott et al. 2009), in part to recover the many associated species that utilize them as primary or critically important habitat (Coen et al. 2007; Hughes et al. 2009; Luckenbach et al. 1995; Ramsey 2012; Scyphers et al. 2011). Second, both seagrasses and shellfish have been shown to attenuate waves and accrete sediments, making them desirable for use in shoreline protection (Fonseca et al. 1982; La Peyre et al. 2015; Lenihan 1999; Meyer 1977; Piazza et al. 2005; Scyphers et al. 2011). Third, within San Francisco Bay, *Z. marina* and *O. lurida* have been identified as major targets for restoration, with increases of 3200 ha of each proposed over 50 years (State Coastal Conservancy 2010). Finally, incorporation of these two species together in a living shorelines design was of interest because of the potential for positive interactions that could enhance establishment or growth of either species or increase the variety of organisms attracted to the complex habitat structure (e.g., Kimbro and Grosholz 2007; Wall et al. 2008).

Eelgrass provides valued ecological functions and services in San Francisco Bay (De La Cruz et al. 2014; Hanson 1998; Kitting 1993; Kitting and Wyllie-Echeverria 1992; Spratt 1981) but covers only ~1200 ha, or approximately 1% of submerged lands (Merkel and Associates 2004, 2009, 2015). Historic coverage and distribution are not well known (a few locations were noted by Setchell 1922, 1927, 1929), but many shallow areas that were likely to have been suitable for eelgrass growth were filled or dredged as commercial shipping and infrastructure around the bay developed. Although submarine light levels in the bay are relatively low and consequently limiting for eelgrass growth (Zimmerman et al. 1991), biophysical modeling indicates that 9490 ha of bottom area may be suitable habitat (Merkel and Associates 2005). Recent studies on restoration methodologies and donor source selection (Boyer et al. 2010), genetic diversity (Ort et al. 2012, 2014), invertebrate usage (Carr et al. 2011), trophic dynamics (Carr and Boyer 2014; Kiriakopolos 2013; Lewis and Boyer 2014; Reynolds et al. 2012), and abiotic effects on eelgrass (Santos 2013) have contributed to an understanding of the opportunities for eelgrass restoration within the bay (reviewed in Boyer and Wyllie-Echeverria 2010). Further, declines in suspended sediment concentrations measured in the last decade indicate improving water clarity (Schoellhamer 2011); restoration measures could proactively advance population expansion in San Francisco Bay, taking advantage of improvements in water quality conditions.

Olympia oysters were historically an abundant part of the fauna in West Coast estuaries (Baker 1995); however, the popularity of the fishery that began in the 1850s as well as other impacts resulted in a collapse of native oyster populations in the region by the early 20th century (Baker 1995; Barnett 1963; Kirby 2004; Zu Ermgassen 2012). Little is known about the pre-European contact distribution and abundance of oysters in San Francisco Bay, much less the ecosystem services they provided; however, aggregations of native oysters were likely to have been habitat for numerous sessile and mobile animals (Ramsey 2012); they are known today to increase invertebrate species richness even at small scales (Kimbro and Grosholz 2007). Because it has not been an important fishery since Gold Rush days, the Olympia oyster has been poorly studied compared to its larger cousins, the Atlantic (*Crassostrea virginica*) and Pacific oyster (*Crassostrea gigas*). Restoration of Olympia oysters, which began in Puget Sound in 1999, is still relatively new compared with efforts in the Atlantic and Gulf coasts and much remains to be learned about effective restoration for these oysters. Lessons learned from restoration on the East and Gulf Coasts are not directly transferrable for several reasons, including differences (1) between the species in terms of life history and ecology; (2) in key limiting factors (such as disease, which is a major issue in many East Coast systems, but not on the West Coast); (3) in restoration goals, which, on the East and Gulf Coasts, frequently include restoring the commercial and recreational fishery as well as habitat, while West Coast restoration efforts have focused solely on oyster population and habitat enhancement; and (4) in the use of hatchery-reared oysters for population enhancement, which has not been used widely in West Coast projects to date.

Monitoring of oysters in SF Bay has resulted in detailed population data for more than 20 intertidal sites (presence/absence data for more than 80 sites), and an increased understanding of the factors that limit oyster populations today (e.g., A. Chang, unpublished data; Deck 2011; Grosholz et al. 2008; Harris 2004; Polson and Zacherl 2009; Wasson et al. 2014; Zabin et al. 2010). This research, along with earlier recruitment studies and small-scale restoration projects, indicates the potential to restore oysters in many areas of the bay through the placement of hard substrate at appropriate tidal elevations, relying entirely on naturally occurring recruitment (Abbott et al. 2012; Grosholz et al. 2008; Wasson et al. 2014; Welaratna 2008; Zabin et al. 2010), although enhancement with hatchery-reared oysters may improve success at some sites.

With these advances in our understanding of the dynamics of eelgrass and Olympia oyster populations and their restoration in San Francisco Bay, the timing was appropriate to increase the scale of restoration of both of these species to acreages large enough to permit evaluation of their effects on physical processes as well as habitat usage by highly mobile bird and fish species. The San Francisco Bay Living Shorelines: Near-shore Linkages Project further tests restoration techniques, restores critical eelgrass and oyster habitat, examines the individual and interactive effects of restoration techniques on habitat values, and tests alternatives to hard/structural stabilization in a multiobjective pilot climate adaptation and restoration project.

17.3 PROJECT GOAL AND OBJECTIVES

The overarching goal of the project is to create biologically rich and diverse subtidal and low intertidal habitats, including eelgrass and oyster reefs, as part of a self-sustaining estuary system that restores ecological function and is resilient to changing environmental conditions.

The objectives of the project are as follows:

1. Use a pilot-scale, experimental approach to establish native oysters and eelgrass at multiple locations in San Francisco Bay.
2. Compare the effectiveness of different restoration treatments in establishing these habitat-forming species.

3. Determine the extent to which restoration treatments enhance habitat for invertebrates, fish, and birds, relative to areas lacking structure and pretreatment conditions.
4. Determine if the type of treatment (e.g., oyster reefs, eelgrass plantings, or combinations of oyster reefs and eelgrass) influences habitat values differently.
5. Begin to evaluate potential for subtidal restoration to enhance functioning of nearby intertidal mudflat, creek, and marsh habitats, for example, by providing food resources to species that move among habitats.
6. Evaluate potential for living subtidal features intended for habitat to also reduce water flow velocities, attenuate waves, and increase sedimentation, and assess whether different restoration treatments influence physical processes differently.
7. Determine if position in the Bay, and the specific environmental context at that location, influences foundational species establishment, habitat provision, and physical processes conferred by restoration treatments.
8. Where possible, compare the ability to establish restoration treatments, habitat functions, and physical changes along mudflats/wetlands versus armored shores.

17.4 SITING AND DESIGN

The two locations for the project (Figure 17.1) were the San Rafael shoreline (parcel owned by The Nature Conservancy) and the Eden Landing Ecological Reserve in Hayward (owned by the California Department of Fish and Wildlife). The San Rafael site included a larger-scale and a small-scale study, while the Hayward site included only a small-scale study, as described below. Oyster treatments were constructed and eelgrass plantings were installed in late July through early August 2012.

17.4.1 Larger-Scale Experiment to Test Both Biological and Physical Effects (San Rafael Only)

This portion of the project included a larger-scale experimental design with four 32×10 m treatment plots situated parallel to the shore, approximately 200 m from shore. The scale of these four plots allowed for evaluation of the effects of native oyster substrate (mounds of bagged Pacific oyster shell), eelgrass, and both together, in comparison to a control plot of the same size (Figures 17.1 and 17.2). The experiment was designed to be large enough in scale to compare effects on physical factors such as wave attenuation and sediment accretion, as well as effects on biological properties that operate at larger scales (e.g., highly mobile invertebrate, bird, and fish utilization).

The Pacific oyster shell mound treatment plot, described in detail below, had a footprint of 1×1 m per element. These were laid out in sets of four elements to make larger units of 4 m^2 (Figures 17.2 and 17.3). To minimize scour, the design included spaces of the same size (4 m^2) between these oyster shell mound units. There were 3 rows of 8 units, for a total of 24 units per plot (96 elements).

Eelgrass was planted and seeded in the eelgrass treatment plot with the same spacing as the oyster reef units. The central 1.5×1.5 m (2.25 m^2) space within every other 4-m^2 space was planted with clusters of shoots and also seeded. The planting technique entailed using a bamboo stake to anchor each shoot in place until rooted (Figure 17.3). Two donor beds were used for transplant material at each site: Point San Pablo and Point Molate (both on the Richmond shoreline) were the sources at San Rafael, while Eden Landing Ecological Reserve in Hayward (small patches offshore) and Bay Farm Island near Alameda were the sources planted at the Hayward small-scale project site (Figure 17.1). Flowering shoots were only available from Point San Pablo at the time of project implementation in late summer 2012 and were collected for use in buoy-deployed seeding (Pickerell et al. 2005) at the San Rafael site only, with a mesh bag of is flowering shoots anchored by a PVC pipe at the center of each unit.

The combined oyster and eelgrass plot was based on an additive design, with eelgrass placed into the central 2.25 m^2 of the 4-m^2 spaces between oyster substrate features (Figure 17.2). This

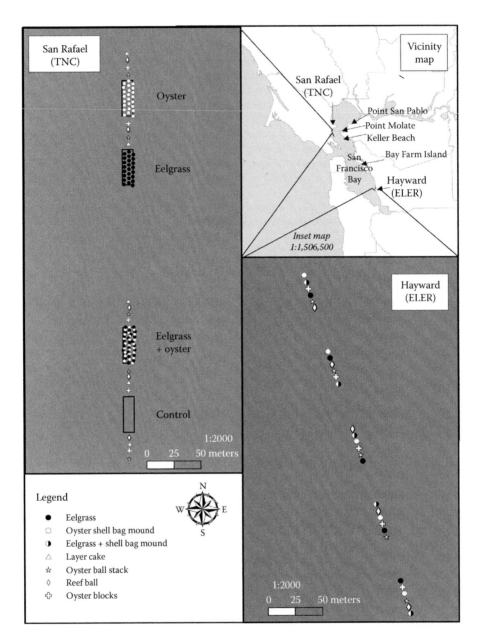

Figure 17.1 Maps showing the location and configuration of (left) the larger-scale and small-scale experiment designs at San Rafael (property of The Nature Conservancy [TNC]) and (right) the small-scale design at Hayward (offshore of Eden Landing Ecological Reserve [ELER]). Space was left at the center of the San Rafael project for preexisting test plots of eelgrass. Eelgrass transplants were collected from Point San Pablo and Point Molate for the San Rafael site and from Bay Farm Island and offshore of ELER for the Hayward site (top right map). Point Molate and Keller Beach eelgrass beds were used as reference sites for epibenthic invertebrate community development at San Rafael.

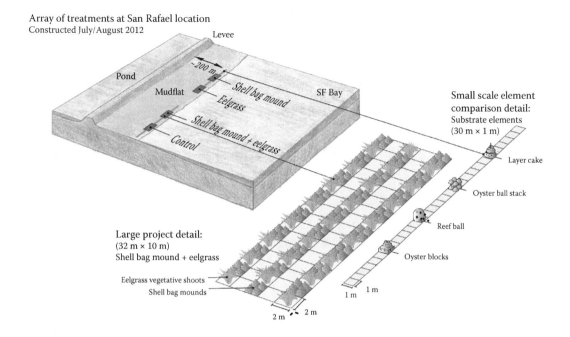

Array of treatments at San Rafael location
Constructed July/August 2012

Array of treatments at Eden Landing Ecological Reserve North
Constructed July/August 2012

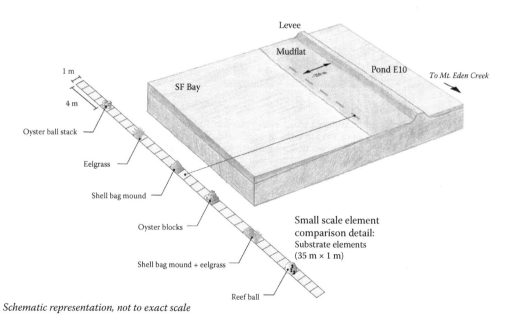

Schematic representation, not to exact scale

Figure 17.2 Schematic of the Living Shorelines: Nearshore Linkages project. Top: the larger-scale project design, as placed at the San Rafael site, with the four types of baycrete elements (the small-scale substrate design) in rows between the four large plots. Bottom: the small-scale substrate design used at the Hayward site. Shell bag mounds were placed as single elements for comparison to baycrete at the Hayward site, and small eelgrass plots, alone and adjacent to oyster elements, were included. (Drawings courtesy Environmental Science Associates.)

Figure 17.3 Top: Photos of treatments used in the project. Bottom: Eelgrass planting using bamboo stake technique, including, on the right, a schematic of planting design within an eelgrass unit at San Rafael and Hayward. Two donors were used to plant each site, as indicated by shading in the schematic. For San Rafael, the donor in the center alternated in each patch.

design permitted us to maintain a spacing of oyster substrate that would minimize scour, while providing enough space around eelgrass plantings to permit access for sampling.

A treatment control plot of the same size was also included (Figures 17.1 and 17.2). The four treatments were arranged randomly in the four possible positions, with 30 m between each plot. Adjacent to the overall treatment area, a large project control area of equal size to the four plots was monitored throughout the project period for certain measures (e.g., bird use of completely unstructured habitat relative to the whole treatment area containing structure).

17.4.2 "Substrate Element" Experiment to Examine Small-Scale Biological Effects (San Rafael and Hayward)

This smaller-scale experiment consisted of five replicate elements of different substrate (surface) types, intended to compare native oyster recruitment, growth, and survival to inform future restoration projects. At the San Rafael site, this experiment was situated in the 30-m spaces between and on either side of the line of larger-scale plots described above (Figures 17.1 through 17.3). At San Rafael, the elements included reef balls, oyster ball stacks, oyster blocks, and a layer cake design all made of "baycrete," a mixture of roughly 20% marine-grade cement and a high proportion of materials (roughly 80%) derived from the Bay including dredged sand and shell (Figure 17.3). These substrate types were replicated five times, for a total of 20 elements placed in groups (blocks), with each of the four substrate types represented in each block.

The Hayward site also included 1-m^2 substrate elements made of baycrete, replicated in five blocks and aligned parallel with the shoreline at ~200 m from shore (Figures 17.1 through 17.3). However, there were five treatments (substrate types): reef balls, oyster ball stacks, oyster blocks, Pacific oyster shell mounds alone, and the latter placed along with adjacent eelgrass plantings. The layer cakes were not included at this site because of concerns about structural integrity under higher wave action, and the oyster shell mounds were added since there was no large-scale project to test their effectiveness at this site as at San Rafael.

17.5 BRIEF PERMITTING REVIEW

The State Coastal Conservancy coordinated with permit agencies before permit application submittals to discuss draft designs and regulatory mechanisms. Permitting discussions focused on project methods and resulting effects on bay species, seasonal windows for the work, and issues regarding the placement of clean Pacific oyster shell and baycrete structures as beneficial fill to create habitat. Permit applications were submitted in February 2012, and numerous follow-up meetings and correspondence occurred on particular aspects of each agency's requirements. Final permits were secured in July 2012, just before construction in late July and August 2012. Permit applications and approvals included the following:

- US Army Corps of Engineers: Nationwide Permit 27 (Aquatic Habitat Restoration, Establishment, and Enhancement Activities).
- NOAA Fisheries consultation with US Army Corps of Engineers: Section 7 consultation relative to the Endangered Species Act, Essential Fish Habitat consultation relative to the Magnuson Stevens Fishery Conservation and Management Act and Fish and Wildlife Conservation Act.
- San Francisco Bay Conservation and Development Commission (BCDC): Administrative permit.
- California Department of Fish and Wildlife consultation with BCDC: Consultation to limit any impacts and maximize benefits to state-listed fish and wildlife; Scientific Collecting Permit for eelgrass donor collections; Letter of Authorization for transplanting eelgrass to restoration sites.
- San Francisco Bay Regional Water Quality Control Board: Section 404 Water quality certification.
- California State Lands Commission: Coordination to confirm that the project is not on state-leased lands.
- California Environmental Quality Act: The project was categorically exempt under Guidelines Section 15333 (14 Cal. Code Regs. §15333) as a small habitat restoration project, not exceeding 5 acres, to restore and enhance habitat for fish, plants, or wildlife and with no significant adverse impact on endangered, rare, or threatened species or their habitat, no known hazardous materials at or around the project site and, given the scale and methodology, no potential for cumulatively significant effects.

In addition to permits, agreements and letters of permission with the landowners (The Nature Conservancy for the San Rafael site and the California Department of Fish and Wildlife for the Hayward site) and local government (City of San Rafael) were obtained.

17.6 KEY FINDINGS, 3 YEARS AFTER INSTALLATION (THROUGH SUMMER 2015)

17.6.1 San Rafael Site

17.6.1.1 Eelgrass

After replanting eelgrass in April 2013 (as the original late-summer planting in 2012 did not succeed), plants at the larger-scale San Rafael project site performed well, reaching 50% of planted densities on average by summer 2013 and 124% by summer 2014 (Figure 17.4). By summer 2015, vegetative shoot counts had reached more than 200% of planted densities in the eelgrass-only plot and just more than 100% in the eelgrass + oyster plot. Although we did not detect seedlings from of buoy-deployed seeding effort in 2012, flowering shoots developed in the plots by summer each year (data not shown), suggesting the possibility of additional recruitment from seed. Maximum plant heights typically reached 160 cm or more during spring–fall, with a marked decrease in height during winter (Figure 17.5). Vegetative shoot density was significantly higher in the eelgrass-only plot starting in spring 2014 (Kruskal-Wallis, $p < 0.05$ from then on). Vegetative shoot heights also tended to be shorter in the eelgrass + oyster plot (Kruskal-Wallis, $p < 0.05$ from July 2014 on). The trend of lower overall densities and heights in the eelgrass + oyster plot compared to the eelgrass-only plot may be attributed to abrasion of plants against the oyster shells, limited space for spread within the matrix of the mixed habitat plot, or somewhat higher epiphytic algal loads on leaves (unpublished data). During the period when the two donors could still be tracked (through summer 2014), plants originating from Point Molate produced significantly higher numbers of shoots than those from Point San Pablo (Kruskal-Wallis, $\chi^2 = 18.21$, df = 1, $p < 0.0001$), perhaps owing to better matching of site conditions between the Point Molate and San Rafael sites (finer sediments than Point San Pablo; Boyer and Wyllie-Echeverria 2010).

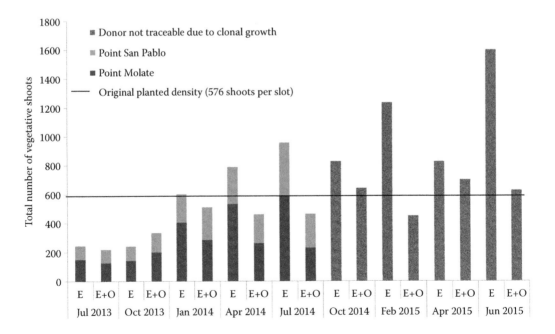

Figure 17.4 Total number of vegetative eelgrass shoots present, per donor and treatment plot at the San Rafael site, quarterly through summer 2015. E = eelgrass plot, E+O = eelgrass and oyster plot. Plants originating from the Point Molate and Point San Pablo donor sites could only be distinguished through July 2014 and were pooled thereafter.

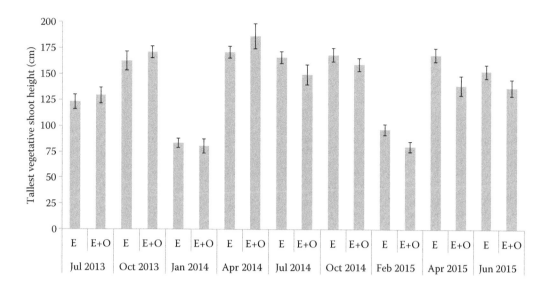

Figure 17.5 Mean height of the tallest vegetative eelgrass shoot in each unit (*n* = 24; ±95% CI), by treatment at the San Rafael site for each quarterly monitoring effort through summer 2015. E = eelgrass plot, E+O = eelgrass + oyster plot. The Point Molate and Point San Pablo donors did not differ in height on any date and were pooled here.

17.6.1.2 *Olympia Oysters*

Olympia oysters quickly recruited to the shell mound structures (by the first fall), with an estimate of more than 2 million present in the first year (Figure 17.6). To be conservative, the population estimates included only the top layer of the oyster shell mounds (the upper third of the 1-m-tall structures), as the lower layers have accumulated sediment and may not support living oysters. The total population reached an estimated peak of 3 million in spring 2013, but has declined since fall that year, with the current population (as of summer 2015) estimated at 750,000 (Figure 17.6). This decline does not appear to be attributable to space competition among growing oysters, but may be the result of (expected) mortality of some of the oysters that settled in the first 2 years as the oysters increased in size, combined with lower recruitment of oysters to the site in 2014 and 2015, as determined by recruitment tiles placed along the shoreline (unpublished data). No differences in oyster numbers or sizes were obvious between the oyster only and eelgrass + oyster treatment.

Oysters also recruited readily to the small "baycrete" structures. Measures of these structures in small quadrats (100 cm^2) early in the project indicated that twice as many oysters were present at lower and mid-level elevations (approximately −20 cm and 0 cm MLLW, respectively) than at the high elevation (~+50 cm MLLW) and on vertical than on horizontal faces; north sides of the elements also typically had 50% more oysters than did south sides. Elevational and directional differences in densities decreased over time, however. There were no differences in oyster sizes across these various surfaces or element types.

There were no differences in oyster densities between the various baycrete element structure types, with the exception of the layer cake configuration, which has more horizontal surface area, on which there were fewer oysters (Figure 17.7). In addition, the stacked small oyster balls tended to collapse; hence, the larger reef balls and oyster blocks have performed best overall among the baycrete structures. Overall, baycrete structures did not support as many oysters as the shell bag elements (Figure 17.7), attributed at least in part to the greater surface area provided by the shells and perhaps also to the lower tidal elevation of the shell bags (the tops of which are at ~+25 cm MLLW).

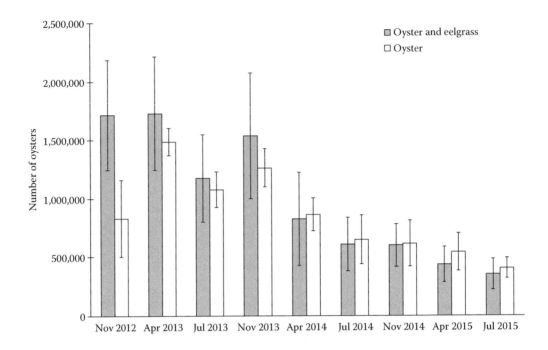

Figure 17.6   Estimated total number of native oysters on shell bag mounds at the San Rafael site over time
in the oyster-only plot and oyster + eelgrass plot. To be conservative, only the upper portion of
the mounds is included here. Means (±95% CI) were calculated from five replicate shell bags
removed from the mounds for oyster counts on each date, which were then scaled up to estimate
oyster numbers at the plot level.

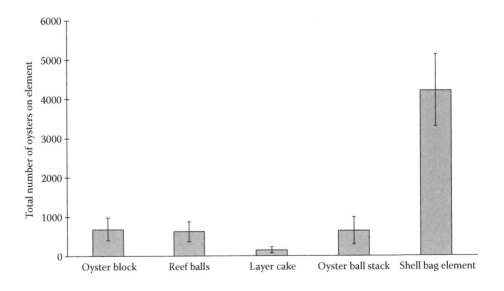

Figure 17.7   Estimated native oyster abundance per baycrete or shell bag element, July 2015. Means (±95%
CI) were generated by scaling up from 10 small replicate shell bags (five each from oyster-only
and oyster–eelgrass treatment plots) or from six 100-cm² quadrats placed on each of five replicate baycrete elements at the San Rafael site.

17.6.1.3 Epibenthic Invertebrate Response

Epibenthic invertebrates were assessed quarterly using baited minnow and oval traps, suction sampling, and shoot collection (for detailed methods, see Pinnell 2016). Trapping with minnow and oval traps for 24 h each quarter indicated an early response of species reliant on physical structure, including shrimp (bay shrimp *Crangon franciscorum* and oriental shrimp *Palaemon macrodactylus*), seen in higher abundance in all treatment plots compared to pretreatment (Kruskal-Wallis $\chi^2 = 24.85$, df = 4, $p < 0.0001$), and Pacific rock crab (*Romaleon antennarium*), which was significantly more abundant in the oyster plots than pretreatment levels (Kruskal-Wallis $\chi^2 = 26.51$, df = 4, $p < 0.0001$). Additional species known to be attracted to physical structure have been trapped in plots with oyster reef or eelgrass present, including native red rock crabs (*Cancer productus*) and northern kelp crabs (*Pugettia producta*), as well as a few nonnative green crabs (*Carcinus maenas*). Suction sampling of epibenthic invertebrates (using a battery-powered aquarium pump on each type of structure or the sediment in the control or pretreatment sampling) showed that community composition was distinct in the plots with oyster reefs present, relative to the control plot and preconstruction conditions (PERMANOVA [Bray Curtis], $p < 0.001$), with the eelgrass-only assemblage in between (Figure 17.8a; Appendix). Further, the invertebrate assemblage in the eelgrass + oyster plot was intermediate between that in the eelgrass-only and oyster-only plots (although more similar to the oyster-only plot). Similarly, freshwater dips of eelgrass shoots to assess epifauna communities (Carr et al. 2011) showed slight differences if oyster reef was present along with eelgrass (Figure 17.8b). Epifauna assemblages on eelgrass at the San Rafael site have not converged with those at Point Molate and Keller Beach, two natural beds just across the bay (Figure 17.8b). Notably, two native species known to remove epiphytes from eelgrass leaves to the benefit of eelgrass growth (Lewis and Boyer 2014) continue to be absent (the isopod *Pendidotea resecata*) or very rare (the sea hare *Phyllaplysia taylori*) at the restored site (only two individuals found during July 2014).

17.6.1.4 Fish Response

Trapping of fish (the same oval and minnow traps described above for invertebrates, with deployment for 24 h once each quarter) showed much overlap in species composition among the treatments; however, a pattern of black surfperch (*Embiotoca jacksoni*) and bay pipefish (*Syngnathus leptorhynchus*) having a greater association with eelgrass habitat emerged. Seining results indicated early recruitment to eelgrass by bay pipefish (within 1 month of the April 2013 replant) and that eelgrass presence increased the occurrence of certain fish species among oyster reef structures, including bay pipefish, shiner surfperch (*Cymatogaster aggregata*) and saddleback gunnel (*Pholis ornata*). Acoustic monitoring using an array of 69-kHz receivers to detect tagged fish showed that individuals of several species visited the site, including two white sturgeon (*Acipenser transmontanus*), a green sturgeon (*Acipenser medirostris*, a threatened species), a leopard shark (*Triakis semifasciata*), a steelhead (*Oncorhynchus mykiss*) smolt, and a striped bass (*Morone saxatilis*). Positional analysis, currently underway, will help determine the degree to which the fish were lingering at the site.

17.6.1.5 Bird and Infaunal Invertebrate Response

To evaluate bird and infaunal invertebrate responses, the treatment area at San Rafael was subdivided into a zone encompassing the eelgrass and oyster treatment plots (zone B) as well as 150-m zones immediately inshore (zone A) and offshore (zone C) of the plots, and a nearby control (unmanipulated) area was divided in the same way; here, we focus on zone B. Avian density and behavior were surveyed at high tide (>0.8 m MLLW) and low tide (<0.25 m) from shore two times a month during the fall (September, October, and November), winter (December, January, and February), and spring (March, April, and May). Benthic cores were collected (10 cm diameter)

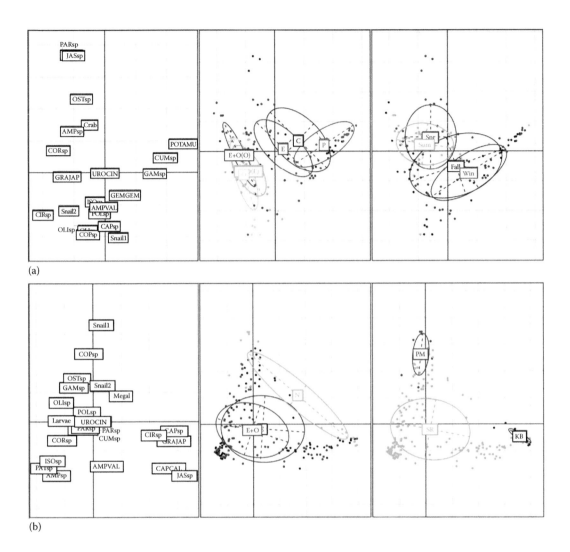

Figure 17.8 Correspondence analysis of epiphytic invertebrates: (a) San Rafael suction sampling patterns by taxa, treatment, and season, fall 2013 through summer 2014 (Year 2 of the project), in comparison to pretreatment (P) samples. C = control, E = eelgrass, O = oyster, E+O(E) = eelgrass from E+O plot, and E+O(O) = oyster from E+O plot. (b) Eelgrass shoot collection patterns in spring 2014 comparing assemblages at the San Rafael (SR) plots from the E or E+O plots to that of two natural (N) beds at Keller Beach (KB) and Point Molate (PM). Two species, *Phyllaplysia taylori* (Taylor's sea hare) and *Pentidotea resecata* (an isopod), were absent or rare at San Rafael and were removed from b owing to their presence obscuring differences produced by other parts of the assemblage. Taxa abbreviations as in Appendix.

during September and May of each year to sample infaunal invertebrates along transects that bisected each zone. Densities of American black oystercatcher (*Haematopus bachmani*) increased in the treatment area in comparison to preinstallation and control densities, and Forster's terns (*Sterna forsteri*) and wading birds (herons and egrets) began using the treatment area after installation (Figure 17.9). Comparing behavior of all bird species during low tide, the treatment area was used more for foraging than was the control area (Figure 17.10); nonforaging (resting, preening, etc.) behaviors were predominant at high tide. Overall benthic invertebrate densities and biomass increased from preinstallation (spring 2012) to year 2 postinstallation (spring 2014) in oyster,

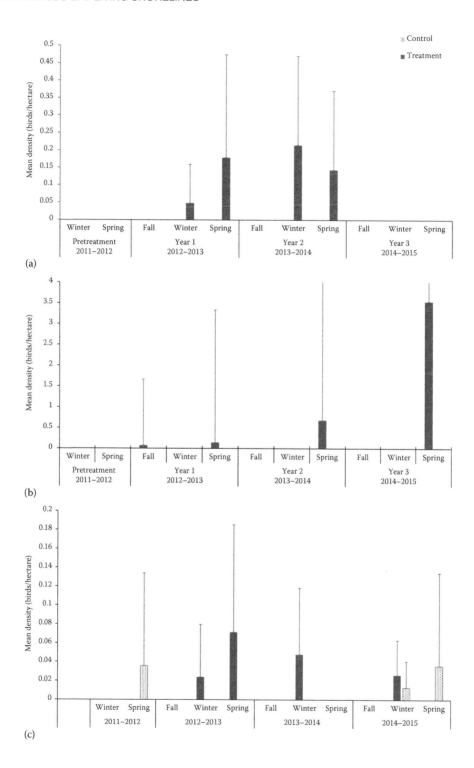

Figure 17.9 Mean seasonal density (with 95% CIs) of (a) black oystercatchers, (b) Forster's terns, and (c) wading birds during low tide at the San Rafael site among pretreatment (2011–2012) and post-treatment years (2012–2015), in the control (gray) and treatment (black) areas. No surveys were conducted during fall of the pretreatment year. Note: *y* axis differs among graphs.

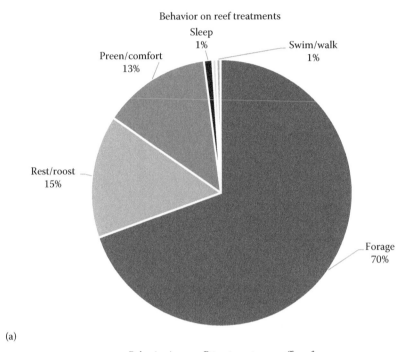

(a)

(b)

Figure 17.10   Percentage of birds (all species) engaged in different behaviors based on low tide scan surveys in Zone B (oyster and eelgrass treatment plots) at San Rafael. (a) Bird behaviors in the treatment plots ("reefs") only, and (b) bird behaviors excluding individuals directly in treatment plots ("off-reef").

eelgrass-only, and eelgrass + oyster treatments (unpublished data). While amphipods were the densest invertebrate, polychaetes comprised the majority of ash free dry weight at San Rafael, and 83% of total polychaete biomass is attributed to a single species, the bamboo worm, *Sabaco elongatus*.

17.6.1.6 Physical Effects

Our measurements show localized sedimentation adjacent to the reefs, with sedimentation over the larger mudflat area less pronounced. Hydrographic surveys of the mudflat surface within 100 m of the reefs at San Rafael in May 2012 and June 2014 show a pattern of erosion bayward (east) of the plots and sedimentation (approximately 0.07 m) shoreward of the plots. Patterns of erosion and deposition are similar for the treatment and control plots. The mudflat surveys also show a north–south trend of increasing erosion to the north, which may be related to the proximity of San Rafael Creek to the north. These results are for only one repeat survey; future surveys are needed to identify longer-term trends. Localized sedimentation has occurred adjacent to both the baycrete structures and the shell mound units and, to a greater extent, inside the shell mound elements comprising the shell mound units. After an initial pulse of sedimentation adjacent to the shell mound units (average of 0.17 m in the first year), sedimentation rates slowed, and in some areas, a net loss of sediment has been observed since construction. The reefs subsided approximately 10 cm in the first 5 months, followed by largely stable conditions (Figure 17.11). The combination of shell bag settling, sediment accumulation around the reefs, and subsidence means that not all of the surface area of the individual elements is available to support oysters (Figure 17.11).

Wave heights show different patterns in the lee (shoreward) of the oyster–eelgrass plot and the control plot, with fewer waves in the lee of the oyster–eelgrass plot. Waves measured over a 2-month period in February to April 2013 ranged in height from 0.06 m (the minimum analyzed) to 0.26 m for both plots. However, there were far fewer waves above 0.06 m shoreward of the oyster–eelgrass compared to the control (21 and 45, respectively) (Figure 17.12). According to wave modeling conducted for the project, for waves immediately offshore of the plots, the oyster–eelgrass plot dissipates approximately 30% more wave energy than the control at mean tide level (MTL). This reduction adds to the wave attenuation benefits of the broad offshore mudflat, which extracts substantial energy before waves reach the plots.

17.6.2 Hayward Site

17.6.2.1 Eelgrass

Eelgrass at this smaller-scale project site reached 75% of planted densities by July 2013 (after a May 2013 replant) and survived through the fall months; however, major declines occurred during the next winter and only two shoots remained by summer 2014 across the 10 small plots. Eelgrass was always shorter at Hayward (~80 cm) than San Rafael, perhaps owing to shallower site conditions at the former. Plants at this site had high densities of the Eastern mud snail, *Ilyanassa obsoleta* (both adults and eggs) on their leaves and also appeared to experience substantial sediment movement and burial; either or both could have contributed to the observed eelgrass mortality.

17.6.2.2 Olympia Oysters

Oyster recruitment at Hayward did not occur until spring 2013 and at a much lower rate than at San Rafael. At its peak in summer 2013, the population on the restoration substrates was estimated at ~2000 oysters on our test elements there; even this relatively modest effort increased the population of native oysters at that site by one order of magnitude. Currently, it appears that there are few oysters on the restoration substrates at this site. Oyster blocks and higher tidal elevations appeared

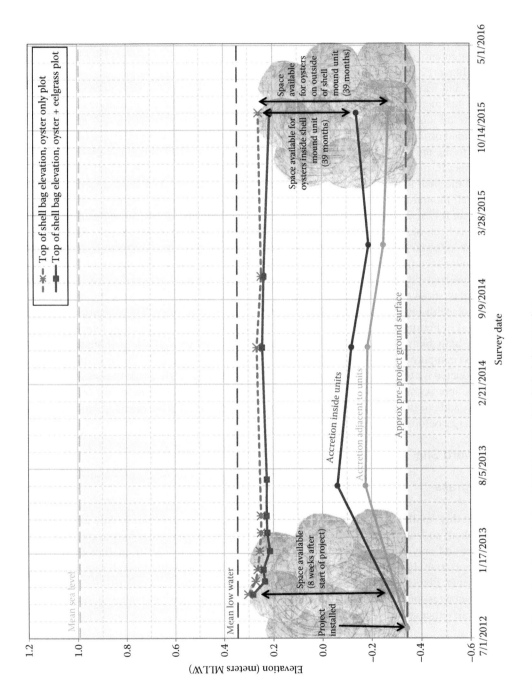

Figure 17.11 Sedimentation and oyster space for shell bags at the San Rafael site over time.

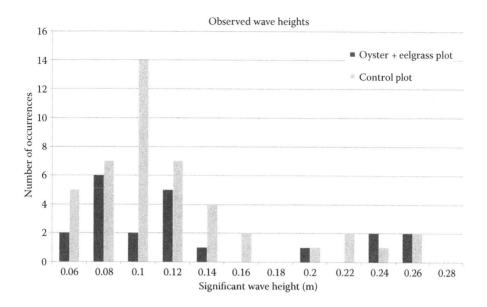

Figure 17.12   Wave heights measured on the shore side of the oyster + eelgrass and control plots at the San Rafael site, February 26, 2013, to April 15, 2013. There were a total of 45 significant waves measured in-shore of the control plot and 21 significant waves measured in-shore of the oyster + eelgrass plot for the sampling duration, indicating that the latter limits significant wave occurrences.

to be the best at supporting oysters for the longest at this site, in contrast to the oyster shell bags and lower tidal elevations performing best at San Rafael. This difference is likely attributable to predation by the Atlantic oyster drill, *Urosalpinx cinerea*, which is more abundant at lower tidal elevations, where it exerts greater predation pressure (as confirmed by field experiments at the site, in which >50% of oysters were killed on substrates placed at +7 cm MLLW within a month, but no predation occurred on substrates placed at +37 cm). The drill is not present at the San Rafael site.

17.6.2.3 Epibenthic Invertebrate Response

Trapping results at Hayward showed that shore crab (*Hemigrapsus oregonensis*) abundances increased within the treatment area relative to the control area and preproject conditions. Eastern mud snails (*I. obsoleta*) were by far the most common invertebrates in traps, with hundreds found per trap in some seasons but no difference with added structure relative to the control area. Suction sampling of epibenthic invertebrates on the oyster shell mounds and eelgrass plots indicated that the mounds developed a distinct community relative to eelgrass when the eelgrass was still present, but in general, there was much overlap in assemblage characteristics with the control area and preproject conditions, perhaps because of the small footprint of the added structure at this site (Pinnell 2016).

17.6.2.4 Fish Response

Only trapping was conducted to assess fish use of this site, in the treatment area versus control (unmanipulated) area. Besides leopard sharks (*T. semifasciata*), which were commonly caught in both control and treatment areas, only one to three individuals of other species were caught (barred surfperch [*Amphistichus argenteus*], Pacific staghorn sculpin [*Leptocottus armatus*], topsmelt [*Atherinops affinis*], jacksmelt [*Atherinopsis californiensis*], Pacific sand dab [*Citharichthys sordidus*], and sevengill shark [*Notorynchus cepedianus*]) over the course of the project to date, making it impossible to discern patterns relative to the addition of reef structure (and eelgrass before the end of 2013).

17.6.2.5 Bird and Infaunal Invertebrate Response

Although the footprint of the treatment area was substantially smaller at Hayward than at San Rafael, the same zone arrangement was used to assess bird and infauna responses to treatments and for consistency between the two sites. While avian diversity and richness were higher at San Rafael, both pre- and postinstallation avian densities were higher at the Hayward treatment and control sites, where small shorebirds predominated. Even with the small project footprint, wader species increased substantially (ANCOVA, $F_{1,117} = 3.52$, $p = 0.063$) postinstallation in the treatment area at Hayward. As at San Rafael, the Hayward treatment area was used primarily for foraging at low tide and nonforaging (resting, preening, etc.) behaviors at high tide. We observed a substantial increase in bivalves in the first posttreatment installation sampling period. Several years of monitoring at this site have established a baseline of avian and infaunal invertebrate characteristics that will be very useful if larger-scale restoration projects go forward in the future.

17.6.2.6 Physical Effects

Subsidence of the individual elements at Hayward was similar to San Rafael and was not found to differ by substrate type. The small-scale treatments did not allow for physical monitoring of wave attenuation and sediment accretion.

17.7 PROGRESS IN ADDRESSING THE PROJECT'S OBJECTIVES

17.7.1 Objective 1: Use a Pilot-Scale, Experimental Approach to Establish Native Oysters and Eelgrass at Multiple Locations in San Francisco Bay

As this project is the first living shorelines design carried out in San Francisco Bay and one of few focused on native oyster and eelgrass habitats on the West Coast, it was important to start small to gain acceptance for such projects among regulators and the public. However, we recognized the need for the project to be large enough to allow assessment of physical effects along shorelines and to attract species that require a larger habitat area for food or refuge services. Thus, at the San Rafael site, we chose a size deemed large enough to meet our science goals but small enough to still be a reasonable pilot project to install and permit.

An experimental approach was important to the project team, as we wished to understand the successes and shortcomings of the restoration project in a rigorous way. However, we settled on only one replicate of each treatment type at the San Rafael site because of space limitation on the San Rafael Shoreline parcel (owned by The Nature Conservancy). Also, current regulatory policies limit the amount of fill (including oyster shell) that can be placed in the estuary; thus, our project team worked thoughtfully to limit the overall size of the installation to meet current permit requirements, while carefully experimenting with methods and techniques to construct the largest reefs in San Francisco Bay to date. The goal of this pilot project is to learn what materials, designs, and approaches work best, ideally leading to additional pilot projects at more sites and also larger-scale projects of this type in the future. From the standpoint of statistical analysis, having only one plot per treatment type means that replicate samples within a plot are not true replicates, as they are not interspersed with other treatment types across the space of the San Rafael property. The risk in interpreting data with only the four large plots spread across the site is that there could be other differences across that space that are not related to the treatments (e.g., sedimentation), thus confounding interpretation of differences by treatment. Still, with care in interpretation, we can say quite a bit about how the treatments evolved habitat and physical functioning characteristics over time and relative to each other. For the smaller-scale comparison of oyster substrates, we were able to achieve true replication at both the San Rafael and Hayward sites, making a rigorous comparison of treatments possible statistically for a number of measures.

We intended to repeat the same design in multiple locations around the bay so that we could determine how environmental context influenced our results; however, we found it difficult to identify locations that met our site selection criteria (e.g., simple bathymetry, relative ease of access, appropriate depths for eelgrass and oysters, willing landowners, etc.) and thus began with just one larger-scale project in this first phase of the work. At Hayward, many of our site selection criteria were met; however, we felt we did not have enough information about the site to be confident that we could establish both oysters and eelgrass and were unwilling to scale up to a larger project until that was achieved.

The project team is assessing seven candidates sites in SF Bay for a next-phase living shorelines project, to actively enhance four native foundation species: eelgrass and Olympia oysters as in the current project, as well as the tidal marsh plants Pacific cordgrass (*Spartina foliosa*) and marsh gumplant (*Grindelia stricta*). Our integrated approach involves restoring these habitats as a linked gradient from marsh to intertidal reefs and subtidal aquatic beds, to increase habitat connectivity and structure and promote both restoration goals and physical goals such as wave attenuation.

17.7.2 Objective 2: Compare the Effectiveness of Different Restoration Treatments in Establishing These Habitat-Forming Species

We have used five approaches to address the effectiveness of different restoration treatments in establishing native oysters and eelgrass. First, our project explicitly aimed to test whether restoring oysters and eelgrass together versus each organism alone would improve outcomes for either species. This test entails evaluating eelgrass growth patterns (densities, heights, etc.) when eelgrass is grown alone versus in proximity to oyster shell reef, and similarly by assessing oyster growth patterns (densities and sizes) when oyster shell reef is restored alone versus in proximity to eelgrass. Second, we tested five types of oyster settlement substrates to determine which would perform the best. In the ideal, a substrate would promote native oyster recruitment, growth, and survival, while discouraging the growth of nonnative species; would not be prone to sinking into soft sediment substrates; and would not cause significant scour, or accumulate large amounts of sediment. Obviously, restoration substrates also need to maintain their structural integrity over time or until biogenic species can add or maintain physical structure independently. Third, we tested transplants versus seeding of eelgrass at the San Rafael site. Fourth, we tested whether the donor (the natural bed collected from) mattered to the outcomes achieved for eelgrass establishment and development of functional attributes of the restored eelgrass. Fifth, we assessed whether the position on oyster elements or the placement of whole oyster settlement substrates at different elevations would influence the effectiveness of native oyster success.

For the first approach, several lines of evidence suggest that there is a benefit to restoring native oysters and eelgrass together. Although trapping has caught a limited number of individuals, a few species of fish were found among oyster reefs at San Rafael only when eelgrass was also present. In addition, suction sampling of epibenthic invertebrates showed that the eelgrass in the combined eelgrass + oyster treatment at San Rafael supported additional species found in the oyster-only plots as well as those found in the eelgrass-only plot. On the other hand, we have not found benefits of oyster reef presence to eelgrass growth characteristics (and in fact eelgrass spread is likely to be limited by the surrounding oyster reefs in our checkerboard design), nor have we seen oyster abundance or size increase in the presence of eelgrass. At Hayward, eelgrass was present for a limited time; thus, we are unable to assess this effect there. We are also collecting stable isotope samples from the common producer and consumer species at the San Rafael site, and these may prove useful in indicating how the food web may differ in either habitat with the presence of the other species and associated species in that habitat. Further, stable isotope analysis should allow us to disentangle trophic links within and among those different treatments, to assess the level of connectivity with adjacent habitats (bare mudflat, marsh) and to identify the main sources of organic matter fueling the food webs and supporting target restoration species' growth. In order to adequately test for effects of dual restoration, we need additional sites where oysters and eelgrass are restored both together and separately, although we suggest greater spacing between oyster reefs and eelgrass in future projects.

For our second approach, we found that oysters performed equally well across the various types of baycrete structures at San Rafael, with one exception—there were far fewer oysters on layer cakes. This was likely because oysters generally did better on vertical versus horizontal surfaces, and layer cake surface area is primarily horizontal. Shell bag mounds outperformed all baycrete structures in terms of number of oysters on a per-element basis. Two element types appear to have less structural integrity than the others: layer cakes and small reef ball stacks, both of which are beginning to shift or break down. Very little sediment accumulated on the surfaces of baycrete elements (generally <2 mm). While shell bag mounds did trap significant amounts of sediment on the lower portions, they still outperformed the baycrete elements. We have not formally analyzed the cover of nonnative species, but the sponges, tunicates, and large arborescent bryozoans found particularly at lower tidal elevations on the elements are not present inside the shell bags.

At Hayward, oysters recruited initially to shell bags only, but currently longer-term survival appears to be best on the oyster blocks, with the other baycrete structures doing less well (layer cakes were not included at this site because of the expectation that they would not hold up under high wave action). This may be because the oyster block elements at Hayward have more vertical surface area at higher tidal elevations than the other structures, which appears to discourage oyster drills.

For our third approach, we were only able to use buoy-deployed seeding at the San Rafael site and flowering shoots only from the Point San Pablo donor site, as flowering shoots were not available at the time of our late summer project start for the other three populations used as donors for transplant material. At San Rafael, we did not detect seedling recruitment in the spring of 2013 after buoy-deployed seeding, and we did not repeat seeding after we conducted the second transplant that April; we would not have had flowering shoots available until summer and did not want to risk damaging transplants by adding the seed buoys into the plots afterward. Thus, in comparing the two methods of eelgrass establishment, we conclude that transplanting whole shoots was the more effective technique overall, in terms of both availability of propagules and success of establishment. However, we still recommend seeding when possible because sexual reproduction can increase the genetic diversity of restored stock and may therefore increase the resiliency of eelgrass to perturbations at restoration sites over time.

In our fourth approach, the Point Molate donor bed initially showed a trend of greater transplant success at San Rafael, with higher overall densities than the Point San Pablo donor. This trend continued and became magnified over time, especially in the eelgrass-only plot. We suggest that Point Molate eelgrass may be better adapted to the sediment conditions found at San Rafael, as both sites have a higher proportion of fine sediments than at Point San Pablo (Boyer and Wyllie-Echeverria 2010). Although we found no difference in growth characteristics between the two donors used at the Hayward site in the limited time we had to assess the eelgrass, the trend of differential success among donors at San Rafael, and similar evidence from previous projects (Lewis and Boyer 2014), lends support to our hypothesis that donor choice can matter to restoration success.

In our fifth and final approach to assessing restoration techniques, we found tidal height, surface orientation, and direction to have strong effects on oyster density at the San Rafael site, although these effects decreased over time. Across all element types at San Rafael for the first several sampling periods, more oysters were present at the lower and mid-level elevations than at the high elevation. More oysters were present on the north side than on the south side and on vertical versus horizontal faces. While longer immersion times could explain greater abundance at lower tidal elevations, the north–south and surface orientation differences strongly suggest that heat or desiccation stress was a factor in determining initial oyster abundance at San Rafael. Oyster abundances at the mid- and low tidal elevations began to decline in spring 2014, while those at the highest elevation remained unchanged, and as of July 2015, densities at all tidal elevations were similar. This decrease is likely concurrent with our observed increase in fouling species, particularly bryozoans, sponges, and algae at these lower tidal elevations, which may compete with oyster spat for settlement space or overgrow adult oysters. At Hayward, while oysters recruited initially to shell bags and then to the interior surfaces of the large oyster balls, two structure types that would be expected to be the best in mitigating heat and desiccation stress, more oysters are currently found on the higher elevations of oysters blocks and large reef balls. As mentioned above, this can likely be attributed to predation by the Atlantic oyster drill *U. cinerea*, which is more abundant at the lower elevations. Results from this work and elsewhere (e.g., Trimble et al. 2009) indicate that oysters generally settle in higher numbers and grow faster at lower tidal elevations. At Hayward, this nonnative predator may thus be restricting oysters to a nonoptimal tidal elevation.

17.7.3 Objective 3: Determine the Extent to Which Restoration Treatments Enhance Habitat for Invertebrates, Fish, and Birds, Relative to Areas Lacking Structure and Pretreatment Conditions

We have accumulated evidence that providing the physical structure of our project design attracted mobile invertebrates that benefit from such structure. Preliminary data suggest that several fish species of concern visited the project site at San Rafael, although additional analysis is necessary to evaluate these patterns. At both San Rafael and Hayward, wading bird presence increased after the placement of reef structures, and at San Rafael, black oystercatchers and Forster's terns are utilizing the reefs for foraging and roosting. Additional monitoring is necessary to determine how the strengths of these relationships develop over time.

17.7.4 Objective 4: Determine if the Type of Treatment (e.g., Oyster Reefs, Eelgrass Plantings, or Combinations of Oyster Reefs and Eelgrass) Influences Habitat Values Differently

Preliminarily, we can conclude from the San Rafael experiment that certain species are benefited more by one substrate than the other. Black oystercatchers and wading birds increased in the presence of the oyster reef structures. Black surfperch and bay pipefish were shown to have a greater association with eelgrass habitat than with oyster-only or control plots, and epibenthic invertebrate assemblages are beginning to become differentiated between the eelgrass and oyster reef habitats. Eelgrass presence increased the occurrence of certain fish species among oyster reef structures (bay pipefish, shiner surfperch, and saddleback gunnel), suggesting that restoring the two habitats in proximity to each other can increase the richness of species present.

17.7.5 Objective 5: Begin to Evaluate Potential for Subtidal Restoration to Enhance Functioning of Nearby Intertidal Mudflat, Creek, and Marsh Habitats (e.g., by Providing Food Resources to Species That Move among Habitats)

As we do not have marsh or creek habitat in proximity to the San Rafael site, we are not able to determine the degree to which our added structures influence functioning or provide subsidies to these habitats. We are able to say that increasing physical structure enhances functions relative to mudflats, at least for species that benefit from the refuge and food resources that are provided by our project. An increase in wading birds and in black oystercatchers through the addition of our project is a good indication that certain guilds of birds are benefiting.

17.7.6 Objective 6: Evaluate Potential for Living Subtidal Features to Reduce Water Flow Velocities, Attenuate Waves, and Increase Sedimentation, and Assess whether Different Restoration Treatments Influence Physical Processes Differently

Our measurements of physical processes have shown accumulation of sediment adjacent to the reefs, but only a small impact on accretion across the whole area of the project; additional measurements are needed over time to assess this trend. We observed less and shorter-term subsidence of the reefs in soft sediment than we expected. Our data showing only a 10-cm subsidence into the sediments, which ended after 5 months, suggest that even in the very soft sediments of the San Rafael site, sinking of reef structures is not a great concern. Sediment accumulating around the oyster shell bags means that these are unlikely to support oyster survival at the lower elevations. This led us to include only the upper portions of the reefs in our estimates of oyster abundance and also suggests that future projects should consider this issue when predicting habitat availability on the reefs. Since, with the exception of the layer cakes and small reef ball stacks, the different element types appear to have performed similarly in terms of stability, the choice for the construction of future reefs should be made based on their performance in oyster habitat terms, which may point to the use of shell bags, reef balls, or perhaps oyster blocks (based on the Hayward results). Future deployments should allow for the loss of available space for oysters owing to subsidence and sedimentation. Larger elements, if used in the future, will tend to subside more.

Our reefs achieved a reduction in wave energy (30%) more so than the broad mudflat alone at MTL; however, we are cautious in our interpretation of this result considering we measured only a limited combination of waves and water levels. Ideally, we would have similar reefs located in multiple locations with different slopes and wave regimes to permit further assessment of such structures in attenuating wave energy along San Francisco Bay shorelines.

17.7.7 Objective 7: Determine If Position in the Bay, and the Specific Environmental Context at That Location, Influences Foundational Species Establishment, Habitat Provision, and Physical Processes Conferred by Restoration Treatments

Although we currently have just two project sites to compare, and only the small substrate comparison that can be made at the Hayward site, there are a number of preliminary conclusions we can draw about the effects of environmental context. For example, eelgrass persistence and spread was far superior at San Rafael, perhaps because of much less exposure on the low tides in this deeper site or because of the Eastern mud snails at Hayward (not present at San Rafael) weighing down the plants or blocking light to the leaves with their egg masses. In addition, oyster shell bags easily outperformed other substrates in terms of oyster recruitment at San Rafael, but at Hayward, oyster blocks appeared to be the best. A shell bag element offers more surface area than any of the baycrete elements and greater protection from heat or desiccation stress attributed to more shading and water retention and perhaps the somewhat lower tidal elevation relative to the baycrete structures. However, at Hayward, where predation pressure is strong and greater at lower elevations, taller structures with more exposed surfaces have ultimately outperformed shell bags. Thus, it appears that selection of optimal substrate needs to be guided by an understanding of the key stressors for eelgrass and oysters at each site. Having additional sites at which to deploy test substrates and measure potential stressors would be useful to further refine site-specific design criteria.

17.7.8 Objective 8: Where Possible, Compare the Ability to Establish Restoration Treatments, Habitat Functions, and Physical Changes along Mudflats/Wetlands versus Armored Shores

At this point, our project does not include a comparison of a soft shoreline versus hardened shoreline environment. A future project at Hayward could accomplish this by comparing areas north (riprap) and south (marsh) of Mount Eden Creek. We are working to identify additional areas where such a comparison could be made in future phases of the work, which we also intend to include active restoration of foundational marsh plant species in an integrated design with eelgrass and oyster reefs, as described earlier.

17.8 FUTURE DESIGN CRITERIA

So far, we are able to draw the following conclusions toward future designs:

- This project and several others (Boyer, unpublished data) suggest that eelgrass should be restored early in the growing season; we did not have success in establishing eelgrass at either site in late July and early August 2012. Our second planting in April and early May 2013 was much more successful at both sites (although the Hayward site failed to support eelgrass by fall/winter 2013).
- We can eliminate two of the baycrete element designs: layer cakes and small reef ball stacks. Neither stands up well structurally over time, and layer cakes have fewer oysters compared with other configurations.
- Key stressors for oysters vary with location within San Francisco Bay and may also shift over the life of a restoration project. It is unlikely that there is a single best design that can be used across estuaries or even within the Bay. Identifying potential stressors and taking these into account in project design may increase project success. For example, shell bags potentially offer protection from heat and desiccation stress and provide a lot of complex surface area for oysters and other organisms to attach to and live in, and greater recruitment and faster growth may occur at lower tidal elevations, but surfaces and tidal elevations that are more stressful in terms of exposure may provide oysters with some measure of protection from marine predators and nonnative fouling species where these species are a concern, especially over the longer term.
- Where possible, pre–site selection surveys and experimental deployments should evaluate longer-term survival as well as recruitment of oysters over several tidal elevations. This might help us identify the "sweet spot" for oysters that provides the best balance between the biotic and abiotic stresses associated with different tidal elevations.
- Additional protection from oyster predators and cover of fouling species might be gained by encouraging larger mobile predators (such as cancrid crabs) and mesograzers to settle on restoration substrates. Future designs might include developing substrate types and configurations that attract large crabs and fish.
- We tentatively suggest that restoration projects incorporating both oyster reef and eelgrass together should be considered; although neither species appears to be benefiting from the other so far, the preliminary evidence that differences in the two habitats encourage a greater number of invertebrate and fish species suggests that their co-location will maximize habitat value. A different configuration for integrating oysters and eelgrass, including spacing them farther apart, might reduce the negative impacts on eelgrass noted in this project.
- Oyster reef designs should consider the fact that the lower portion of elements will experience sediment burial. Future designs could be elevated on materials (such as oyster blocks made of baycrete) that are less difficult to source than bags of Pacific oyster shell, which will be less available in the future.
- Wave energy reduction measured in our San Rafael project is encouraging, but we recommend additional sites be used for similar projects and measurements in order to determine optimal designs and the need for site-specific differences in reef configuration.

ACKNOWLEDGMENTS

The California State Coastal Conservancy has provided funding and leadership in this effort. We appreciate our other funding partners, including the California Wildlife Conservation Board, the Environmental Protection Agency through the San Francisco Estuary Partnership, NOAA Fisheries, and the Golden Gate Bridge and Highway Transportation District. We are also grateful to our land-owner partners, The Nature Conservancy and the California Department of Fish and Wildlife, for supporting the project and permitting access. Construction support was provided by the California Wildlife Foundation, Reef Innovations, Drakes Bay Oyster Company, and Dixon Marine Services. The use of trade, product, or firm names in this publication is for descriptive purposes only and does not imply endorsement by the U.S. Government.

REFERENCES

Abbott, R.R., R. Obernolte, K.E. Boyer, and B. Mulvey. 2012. San Francisco Estuary Habitat Restoration for Salmonids Project. Final Programmatic Report to the National Fish and Wildlife Foundation.

Arkema, K.K., G. Guannel, G. Verutes, S.A. Wood, A. Guerry, M. Ruckelshaus, P. Kareiva, M. Lacayo, and J.M. Silver. 2013. Coastal habitats shield people and property from sea-level rise and storms. *Nature Climate Change*, published online July 14, 2013.

Baker, P. 1995. Review of ecology and fishery of the Olympia oyster, *Ostrea lurida*, with annotated bibliography. *Journal of Shellfish Research* 14: 501–518.

Barnett, E.M. 1963. The California oyster industry. California Fish and Game Bulletin 123. 103 pp.

Beck, M.W., R.D. Brumbaugh, L. Airoldi, A. Carranza, L.D. Coen, C. Crawford, O. Defeo, G.J. Edgar, B. Hancock, M. Kay, H. Lenihan, M.W. Luckenbach, C.L. Toropova, and G. Zhang. 2009. *Shellfish Reefs at Risk: A Global Analysis of Problems and Solutions*. The Nature Conservancy, Arlington VA. 55 pp.

Boyer, K.E. and S. Wyllie-Echeverria. 2010. Eelgrass Conservation and Restoration in San Francisco Bay: Opportunities and Constraints. Appendix 8-1, San Francisco Bay Subtidal Habitat Goals Report. 84 pp. http://www.sfbaysubtidal.org/report.html.

Boyer, K.E., S. Wyllie-Echeverria, L.K. Reynolds, L.A. Carr, and S.L. Kiriakopolos. 2010. Planning for Eelgrass Restoration in San Francisco Bay. Final Report Prepared for the California State Coastal Conservancy, Interagency Agreement No. 05-103.

California Coastal Commission. 2015. Sea Level Rise Policy Guidance. www.coastal.ca.gov.

Carr, L.A. and K.E. Boyer. 2014. Variation at multiple trophic levels mediates a novel seagrass-grazer interaction. *Marine Ecology Progress Series* 508: 117–128.

Carr, L.A., K.E. Boyer, and A. Brooks. 2011. Spatial patterns in epifaunal community structure in San Francisco Bay eelgrass (*Zostera marina*) beds. *Marine Ecology* 32: 88–103.

Coen, L.D., R.D. Brumbaugh, D. Bushek, R. Grizzle, M.W. Luckenbach, M.H. Posey, S.P. Powers, and S.G. Tolley. 2007. Ecosystem services related to oyster restoration. *Marine Ecology Progress Series* 341: 303–307.

Cunha, A.H., N.N. Marbá, M.M. van Katwijk, C. Pickerell, M. Henriques, G. Bernard, M.A. Ferreira, S. Garcia, J.M. Garmendia, and P. Manent. 2012. Changing paradigms in seagrass restoration. *Restoration Ecology* 20: 427–430.

Currin, C.A., J. Davis, L.C. Baron, A. Malhotra, and M. Fonseca. 2015. Shoreline change in the New River Estuary, NC: Rates and consequences. *Journal of Coastal Research* 31: 1069–1077.

Deck, A.K. 2011. Effects of interspecific competition and coastal oceanography on population dynamics of the Olympic oyster, *Ostrea lurida*, along estuarine gradients. Master's thesis. University of California, Davis. 84 pp.

De La Cruz, S.E.W., J.M. Eadie, A.K. Miles, J. Yee, K.A. Spragens, E.C. Palm, and J.Y. Takekawa. 2014. Resource selection and space use by sea ducks during the non-breeding season: Implications for habitat conservation planning in urbanized estuaries. *Biological Conservation* 169: 68–78.

Fonseca, M.S., J.S. Fisher, J.C. Zieman, and G.W. Thayer. 1982. Influence of the seagrass *Zostera marina* (L.) on current flow. *Estuarine, Coastal, and Shelf Science* 15: 351–364.

Gallien, T.W., J.E. Schubert, and B.F. Sanders 2011. Predicting tidal flooding of urbanized embayments: A modeling framework and data requirements. *Coastal Engineering* 58: 567–577.

Gittman, R.K., F.J. Fodrie, A.M. Popowich, D.A. Keller, J.F. Bruno, C.A. Currin, C.H. Peterson, and M.F. Peihler. 2015. Engineering away our natural defenses: An analysis of shoreline hardening in the US. *Frontiers in Ecology and the Environment* 13: 301–307.

Gittman, R.K., A.M. Popowich, J.F. Bruno, and C.H. Peterson. 2014. Marshes with and without sill protect estuarine shorelines from erosion better than bulkheads during a Category 1 hurricane. *Ocean & Coastal Management* 102: 94–102.

Goals Project. 2015. The Baylands and Climate Change: What We Can Do. Baylands Ecosystem Habitat Goals Science Update 2015 prepared by the San Francisco Bay Area Wetlands Ecosystem Goals Project. California State Coastal Conservancy, Oakland, CA. www.baylandsgoals.org.

Grosholz, E., J. Moore, C. Zabin, S. Attoe, and R. Obernolte. 2008. Planning services for native oyster restoration in San Francisco Bay. Final report to the California Coastal Conservancy, Agreement #05-134. 41 pp.

Hanson, L.A. 1998. Effects of suspended sediment on animals in a San Francisco Bay eelgrass habitat. Master's thesis. California State University, Hayward, CA.

Harris, H.E. 2004. Distribution and limiting factors of *Ostrea conchaphila* in San Francisco Bay. Master's thesis. San Francisco State University, San Francisco. 76 pp.

Heberger, M., H. Cooley, P. Herrera, P.H. Gleick, and E. Moore. 2011. Potential impacts of increased coastal flooding in California due to sea-level rise. *Climatic Change* 109: 229–249.

Hughes, A.R., S.L. Williams, C.M. Duarte, K.L. Heck, Jr., and M. Waycott. 2009. Associations of concern: Declining seagrasses and threatened dependent species. *Frontiers in Ecology and the Environment* 7: 242–246.

Kimbro, D.L. and E.D. Grosholz. 2007. Disturbance influences oyster community richness and evenness, but not diversity. *Ecology* 87: 2278–2388.

Kirby, M.X. 2004. Fishing down the coast: Historical expansion and collapse of oyster fisheries along continental margins. *Proceedings of the National Academy of Sciences* 101: 13096–13099.

Kiriakopolos, S.L. 2013. Herbivore-driven semelparity in a typically iteroparous plant, *Zostera marina*. Master's thesis, San Francisco State University.

Kitting, C.K. 1993. Investigation of San Francisco Bay shallow-water habitats adjacent to the Bay Farm Island underwater excavation. A report for the U. S. Department of Commerce/NOAA, National Marine Fisheries Service, Long Beach and Santa Rosa. CA. 41 pp.

Kitting, C.L. and S. Wyllie-Echeverria. 1992. Seagrasses of San Francisco Bay: Status Management and Conservation. pp. 388–393. Natural Areas Global Symposium. National Park Service. NPS D-374. 667 pp.

La Peyre, M.K., K. Serra, T.A. Joyner, and A. Humphries. 2015. Assessing shoreline exposure and oyster habitat suitability maximizes potential success for sustainable shoreline protection using restored oyster reefs. *PeerJ* 3: e1317.

Lenihan, H.S. 1999. Physical–biological coupling on oyster reefs: How habitat structure influences individual performance. *Ecological Monographs* 69: 251–275.

Lewis, J.T. and K.E. Boyer. 2014. Grazer functional roles, induced defenses, and indirect interactions: Implications for eelgrass restoration in San Francisco Bay. *Diversity* 6: 751–770.

Luckenbach, M.W., R. Mann, and J.A. Wesson, eds. 1995. Oyster reef habitat restoration: A synopsis and synthesis of approaches: Proceedings from the Symposium, Williamsburg, VA, April 1995.

McGranahan, G., D. Balk, and B. Anderson. 2007. The rising tide: Assessing the risks of climate change and human settlements in low elevation coastal zones. *Environment and Urbanization* 19: 17–37.

Merkel, K.W., and Associates. 2004. Baywide eelgrass (*Zostera marina*) inventory in San Francisco Bay: Eelgrass atlas. Prepared for the California Department of Transportation and NOAA Fisheries. Available at www.biomitigation.org.

Merkel, K.W. and Associates. 2005. Baywide eelgrass (*Zostera marina* L.) inventory in San Francisco Bay: Eelgrass bed characteristics and predictive eelgrass model. Report prepared for the State of California Department of Transportation in cooperation with NOAA Fisheries. www.biomitigation.org.

Merkel, K.W. and Associates. 2009. San Francisco Bay eelgrass atlas, October–November 2009. Submitted to California Department of Transportation and National Marine Fisheries Service.

Merkel, K.W. and Associates. 2015. San Francisco Bay eelgrass atlas, 2014. Submitted to California Department of Transportation and National Marine Fisheries Service.

Meyer, B.L. 1997. Stabilization and erosion control value of oyster cultch for intertidal marsh. *Restoration Ecology* 5: 93–99.

Natural Resources Agency. 2015. Safeguarding California: Reducing Climate Risk: An update to the 2009 California Climate Adaptation Strategy. http://resources.ca.gov/climate/safeguarding/.

NOAA. 2015. Guidance for considering the use of living shorelines. National Oceanic and Atmospheric Administration (NOAA) Living Shorelines Workgroup. 36 pp.

NOAA Fisheries National Shellfish Initiative. 2011. nmfs.noaa.gov/aquaculture/docs/policy/natl_shellfish _init_factsheet_summer_2013.pdf.

Ort, B.S., C.S. Cohen, K.E. Boyer, L.K. Reynolds, S.M. Tam, and S. Wyllie-Echeverria. 2014. Conservation of eelgrass (*Zostera marina*) genetic diversity in a mesocosm-based restoration experiment. *PLoS ONE*. DOI: 10.1371/journal.pone.0089316.

Ort, B.S., C.S. Cohen, K.E. Boyer, and S. Wyllie-Echeverria. 2012. Population structure and genetic diversity among eelgrass (*Zostera marina*) beds and depths in San Francisco Bay. *Journal of Heredity* 103: 533–546.

Orth, R.J., T.J.B. Carruthers, W.C. Dennison, C.M. Duarte, J.W. Fourqurean, K.L. Heck, A.R. Hughes, G.A. Kendrick, W.J. Kenworthy, S. Olyarnik, F.T. Short, M. Waycott, and S.L. Williams. 2006. A global crisis for seagrass ecosystems. *BioScience* 56: 987–996.

Orth, R.J., S.R. Marion, K.A. Moore, and D.J. Wilcox. 2010. Eelgrass (*Zostera marina* L.) in the Chesapeake Bay region of Mid-Atlantic Coast of the USA: Challenges in conservation and restoration. *Estuaries and Coasts* 33: 139–150.

Piazza, B.P., P.D. Banks, and M.K. La Peyre. 2005. The potential for created oyster shell reefs as a sustainable shoreline protection strategy in Louisiana. *Restoration Ecology* 13: 499–506.

Pickerell, C.H., S. Schott, and S. Wyllie-Echeverria. 2005. Buoy deployed seeding: A new approach to restoring seagrass. *Ecological Engineering* 25: 127–136.

Pinnell, C.M. 2016. Invertebrate response to eelgrass and oyster restoration in San Francisco Estuary. Master's thesis, San Francisco State University.

Polson, M.P. and D.C. Zacherl. 2009. Geographic distribution and intertidal population status for the Olympia oyster, *Ostrea lurida* Carpenter 1864, from Alaska to Baja. *Journal of Shellfish Research* 28: 69–77.

Ramsey, J. 2012. Ecosystem services provided by Olympia oyster (*Ostrea lurida*) habitat and Pacific oyster (*Crassostrea gigas*) habitat: Dungeness crab (*Metacarcinus magister*) production in Willapa Bay, WA. Final report submitted to Oregon State University. 63 pp.

Reynolds, L.K., L.A. Carr, and K.E. Boyer. 2012. A non-native amphipod consumes eelgrass inflorescences in San Francisco Bay. *Marine Ecology Progress Series* 451: 107–118.

Santos, G. 2013. Nutrient dynamics and production in San Francisco Bay eelgrass (*Zostera marina*) beds. Master's thesis, San Francisco State University.

Schoellhamer, D.H. 2011. Sudden clearing of estuarine waters upon crossing the threshold from transport to supply regulation of sediment transport as an erodible sediment pool is depleted: San Francisco Bay, 1999. *Estuaries and Coasts* 34: 885–899.

Scyphers, S.B., S.P. Powers, K.L. Heck, Jr., and D. Byron. 2011. Oyster reefs as natural breakwaters mitigate shoreline loss and facilitate fisheries. *PLoS ONE* 6(8).

Setchell, W.A. 1922. *Zostera marina* in its relation to temperature. *Science* 56: 575–577.

Setchell, W.A. 1927. *Zoster marina latifolia*: Ecad or ecotype? *Bulletin of the Torrey Botanical Club* 54: 1–6.

Setchell, W.A. 1929. Morphological and phenological notes on *Zostera marina* L. *University of California Publications in Botany* 14: 389–452.

Spratt, J.D. 1981. The evolution of California's herring roe fishery: Catch allocation, limited entry and conflict resolution. *California Fish and Game* 78: 20–44.

State Coastal Conservancy. 2010. San Francisco Bay Subtidal Habitat Goals Report. http://www.sfbaysubtidal .org/report.html.

State Coastal Conservancy. 2011. Climate Change Policy and Project Selection Criteria. www.scc.ca.gov.

Trimble, A.C., J.L. Ruesink, and B.R. Dumbauld. 2009. Factors preventing the recovery of a historically overexploited shellfish species, *Ostrea lurida* Carpenter 1864. *Journal of Shellfish Research* 28: 97–106.

Wall, C.C., B.J. Peterson, and C.J. Gobler. 2008. Facilitation of seagrass *Zostera marina* productivity by suspension-feeding bivalves. *Marine Ecology Progress Series* 357: 165–174.

Wasson, K., C. Zabin, J. Bible, E. Ceballos, A. Chang, B. Cheng, A. Deck, T. Grosholz, M. Latta, and M. Ferner. 2014. A guide to Olympia oyster restoration and conservation: Environmental conditions and sites that support sustainable populations in Central California. San Francisco Bay National Estuarine Research Reserve. 43 pp.

Waycott, M., C.M. Duarte, T.J.B. Carruthers, R.J. Orth, W.C. Dennison, S. Olyarnik, A. Calladine, J.W. Fourqurean, K.L. Heck, A.R. Hughes, G.A. Kendrick, W.J. Kenworthy, F.T. Short, and S.L. Williams. 2009. Accelerating loss of seagrasses across the globe threatens coastal ecosystems. *Proceedings of the National Academy of Sciences* 106: 19761–19764.

Welaratna, S. 2008. The native oyster recruitment study in Central and South San Francisco Bay 2006–07. Master's thesis, San Jose State University.

Zabin, C.J., S. Attoe, E.D. Grosholz, and C. Coleman-Hulbert. 2010. Shellfish Conservation and Restoration in San Francisco Bay: Opportunities and Constraints. Appendix 7-1, San Francisco Bay Subtidal Habitat Goals Report. 107 pp. http://www.sfbaysubtidal.org/report.html.

Zimmerman, R.C., J.L. Reguzzoni, S. Wyllie-Echeverria, M. Josselyn, and R.S. Alberte. 1991. Assessment of environmental suitability for growth of *Zostera marina* L. (eelgrass) in San Francisco Bay. *Aquatic Botany* 39: 353–366.

Zu Ermgassen, P.S.E., M.D. Spalding, B. Blake, L.D. Coen, B. Dumbauld, S. Geiger, J.H. Grabowski, R. Grizzle, M. Luckenbach, K. McGraw, W. Rodney, J.L. Ruesink, S.P. Powers, and R. Brumbaugh. 2012. Historical ecology with real numbers: Past and present extent and biomass of an imperiled estuarine habitat. *Proceedings of the Royal Society B* 279: 393–400.

APPENDIX

Taxon	Abbreviation	Site	Survey
Annelids			
Oligochaete	OLIsp	SR, KB, PM, H	su, sh
Polychaete	POLsp	SR, KB, PM, H	su, sh
Crustaceans			
Crabs			
Cancer maenas	CANMAE	SR, H	t
Cancer productus	CANPRO	SR	t
Hemigrapsus oregonensis	HEMORE	SR, H	t
Megalopae	Megal	SR, KB, PM	sh
Metacarcinus magister	METMAG	SR, H	t
Pugettia productus	PUGPRO	SR	t
Romaleon antennarium	ROMANT	SR	t
Amphipods			
Ampelisca sp.	AMPsp	SR	su, sh
Ampithoe valida	AMPVAL	SR, KB, PM, H	su, sh
Caprella californica	CAPCAL	SR	sh
Caprella sp. (incl. juveniles)	CAPsp	SR, KB, PM, H	su, sh
Corophidae (incl. *Monocorophium* sp.)	CORsp	SR, KB, PM, H	su, sh
Gammarus sp.	GAMsp	SR, PM, H	su, sh
Grandidierella japonica	GRAJAP	SR, KB, PM, H	su, sh
Jassa sp.	JASsp	SR, KB	su, sh
Paradexamine sp.	PARsp	SR, KB, PM, H	su, sh
Isopods			
Isopod	ISOsp	SR, PM, H	su, sh
Pentidotea resecata	PENRES	KB, PM	sh
Shrimp			
Cumacean	CUMsp	SR, H	su, sh
Shrimp (incl. *Crangon franciscorum* and *Palaemon macrodactylus*)	Shrimp	SR	t

(Continued)

Taxon	Abbreviation	Site	Survey
Other crustaceans			
Cirripedia	CIRsp	SR, H	su, sh
Copepod	COPsp	SR, KB, PM, H	su, sh
Ostracod	OSTsp	SR, H	su, sh
Bivalves			
Gemma gemma	GEMGEM	SR, H	su
Potamocorbula amurensis	POTAMU	SR, H	su
Siliqua patula	SILPAT	H	su
Gastropods			
Ilyanassa obsoleta	ILYOBS	H	t
Patella sp.	PATsp	SR	sh
Phyllaplysia taylori	PHYTAY	SR, PM	sh
Urosalpinx cinerea	UROCIN	H	su
Snail (round)	Snail 1	SR, KB, PM, H	su, sh
Snail (cork)	Snail 2	SR, H	su, sh

CHAPTER **18**

Comparison of Oyster Populations, Shoreline Protection Service, and Site Characteristics at Seven Created Fringing Reefs in Louisiana
Key Parameters and Responses to Consider

Megan K. La Peyre, Lindsay Schwarting Miller, Shea Miller, and Earl Melancon

CONTENTS

18.1 BACKGROUND

Coastal erosion threatens many low-lying areas around the globe. Rising sea levels from climate change are expected to increase coastal erosion and exacerbate flooding and storm surges. This is particularly true in low-lying coastal Louisiana, which developed as the Mississippi River changed course (delta switching) over the past 7000 years. Periods of land loss and gain resulted in an intricate coastal environment composed of shallow water areas with wetlands, swamps, barrier islands, and ridges (Day et al. 2007). This complex habitat sustains high economic and biological productivity, supporting the largest commercial fishery in the lower 48 states, providing habitat for important species of fish and wildlife, mitigating storm surge, and delivering protection for oil and gas production facilities, including five of the nation's largest ports. Because of past and ongoing geological and physical processes, such as subsidence, sea level rise, tropical cyclonic activity, and direct human activities (Barras 2009; Chmura et al. 1992; Georgiou et al. 2005), coastal Louisiana is estimated to have lost an area almost the size of Delaware (4877 km^2) between 1932 and 2010, with recent analyses indicating losses averaging 42.9 km^2/year (Couvillion et al. 2011).

In response, coastal planners have identified multiple approaches to reduce this loss, resulting in more than \$50 billion worth of restoration and management projects proposed for coastal Louisiana (Coastal Protection and Restoration Authority [CPRA] 2012). These projects include significant reengineering of the rivers, sediment diversions, marsh and barrier island restoration, and shoreline protection projects, including the creation of "oyster barrier reefs" (CPRA 2012; Peyronnin et al. 2013). The use of these oyster barrier reefs to mitigate coastal erosion is an effort to capitalize on the natural benefits of a living shoreline.

If appropriately located, living shorelines offer long-term sustainability through natural building, as well as direct economic and ecosystem benefits including shoreline stabilization, resilience to subsidence and sea level rise, and the enhancement of fisheries and water quality (Grabowski et al. 2012; Humphries and La Peyre 2015; La Peyre et al. 2014). However, in coastal Louisiana, reef living shorelines are dependent on recruitment, growth, and survival of the reef building eastern oyster, *Crassostrea virginica*. These oysters require habitat with immediate and long-term acceptable water quality (i.e., salinity, food resources) and are particularly vulnerable to large-scale anthropogenic changes (Soniat et al. 2013). Understanding how local conditions and discrete events affect sustainability of oyster reef living shorelines is critical to maximizing their benefits.

Though *C. virginica* is a prolific species with a broad range of habitat in which it can survive, the conditions for eastern oysters to prosper as a reef can be limited by a number of factors. *C. virginica* reef sustainability depends on locations conducive to high production of shell substrate through settlement and growth, a necessary requirement for reef longevity (Powell et al. 2006; Walles et al. 2015b). Ultimately, *C. virginica* reef sustainability depends on the complex interaction of a number of environmental factors that affect their population dynamics, growth, reproduction, recruitment, and survival. These multiple factors include salinity, temperature, food, water circulation, sedimentation, disease, predation, and bottom type (La Peyre et al. 2009; Powell et al. 2003; Soniat et al. 1988; Wang et al. 2008).

Of the many factors affecting *C. virginica* populations and reef longevity, salinity is a key variable, as it affects many different aspects of the oyster's life including growth, mortality, reproduction, predation, and disease infection levels (Dekshenieks et al. 1993; Shumway 1996). Oysters are well known for their wide tolerance to salinity, ranging from 5 to 40 psu (Shumway 1996); however, within that range, salinity can affect basic physiological rates, affecting overall population dynamics in different ways (Newell and Langdon 1996). In Louisiana, most oyster production is limited above 15 psu because of excessive mortality owing to *Perkinsus marinus* infections (Powell et al. 2012; Soniat et al. 2012) and predation from oyster drills (Brown and Richardson 1987; Brown et al. 2008; Mackenzie 1970); however, some self-sustaining populations have been documented in areas with salinities below 3.5 psu for five consecutive months of the year (Butler 1954).

In addition to salinity, other factors interacting with reef design likely influence reef sustainability and function. Reef exposure, the percentage of time within the study period the water level is below the reef-top elevation, may affect oyster growth rates, survival, and biofouling on the reef (e.g., Bahr 1976; Byers et al. 2015; Fodrie et al. 2014; Littlewood et al. 1992; Ridge et al. 2015). Similarly, water quality variables (e.g., dissolved oxygen) and sedimentation rates can affect oyster recruitment, growth, and survival (Shumway 1996). At the same time, other variables may affect the provision of services beyond reef sustainability. For example, shoreline exposure may influence the effectiveness of the fringing reef in providing erosion protection (La Peyre et al. 2015) while adjacent habitat types may influence reef habitat value (Gregalis et al. 2009). Many of these variables interact with reef design parameters and need to be integrated into site selection and reef design decisions.

Identifying the correct site location for creating an oyster reef as a living shoreline is critical (Beseres Pollack et al. 2012; Coen and Luckenbach 2000). One essential requirement is the selection of suitable habitat for sustainable oyster populations (Cake 1983; Melancon et al. 1998; Soniat et al. 2013). Habitat suitability indices (HSIs) were developed for environmental impact assessments initially (Cake 1983) and more recently used for aquaculture, conservation, and restoration applications (Beseres Pollack et al. 2012; Soniat et al. 2013). These models all differ slightly in the parameters and thresholds used but essentially use a combination of salinity descriptors, substrate availability, and historic conditions to identify good sites for oyster growth or reef restoration. Despite numerous modeling and habitat suitability approaches available, results of many reef creation projects vary enormously across the Louisiana coast, possibly reflecting local site variability (Casas et al. 2015) along with rapidly changing conditions across estuaries experiencing significant subsidence, sea level rise, and large-scale river management affecting freshwater inflows into the estuaries (Soniat et al. 2013).

18.2 BIOENGINEERED EASTERN OYSTER LIVING SHORELINE PROJECTS IN COASTAL LOUISIANA

Over the last decade, a number of living shoreline projects based on *C. virginica* reefs have been developed in coastal Louisiana (Figure 18.1). These projects range from experimental oyster reefs

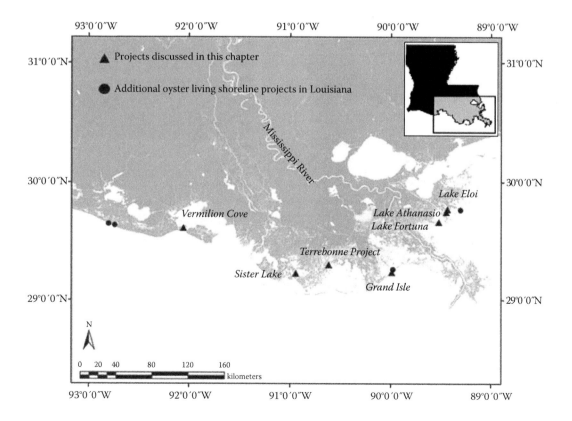

Figure 18.1 Location of eastern oyster living shoreline projects across coastal Louisiana.

using loose shell cultch (Casas et al. 2015; La Peyre et al. 2014), to more bioengineered reefs using a variety of techniques in demonstration projects (Melancon et al. 2013), to large-scale on-the-ground shoreline protection bioengineered projects (La Peyre et al. 2013b,c). These bioengineered reefs have used a variety of engineered base structures. These reef bases all have the common property of installing immediate vertical structure to the nearshore environment, either with concrete (i.e., A-Jack blocks, OysterBreak) or with other materials including mesh cages filled with oyster shell (i.e., ReefBlk) or mesh mats filled with limestone (i.e., Gabion Mats) (Figure 18.2).

The advantages of using oyster reefs as living shorelines include enhancing coastal Louisiana's important oyster population, reducing marsh edge retreat, and providing a potentially sustainable framework for this erosion protection through sustainable reefs. Through shell growth and the continued recruitment of new individuals, oyster reefs will physically expand and become self-sustaining over time. Specifically, in the right location, an oyster reef used for shoreline protection can respond to changing conditions including subsidence and sea level rise (Casas et al. 2015; Mann and Powell 2007; Walles et al. 2015b). While the primary goal of these projects is to help stabilize

(a)

(b)

(c)

(d)

Figure 18.2 Bioengineered reef designs along coastal Louisiana: OysterBreak ([a] Vermilion Cove), Reefblk ([b] Lake Fortuna), A-Jacks ([c] Terrebonne Bay), and Gabion Mats ([d] Terrebonne Bay]).

Table 18.1 Descriptions of Basic Parameters of the Seven Projects with Data Reported in This Work

Location	Material	Year Built	Length (m)	Monitoring Period	No. of Segments	Cost ($/linear m)[a]
Terrebonne	A-Jacks; Gabion Mats; ReefBlk	2007	915	2007–2012	9	1509; 1759; 1309
Sister Lake	Loose shell cultch	2009	225	2009–2012	9	168
Grand Isle	ReefBlk	2010	1400	2010–2014	3	653
Vermilion Cove	OysterBreak rings	2011	480	2011–2014	8	676
Lake Eloi	ReefBlk	2012	1300	2012–2014	3	653
Lake Fortuna	ReefBlk	2012	2400	2012–2014	3	653
Lake Athanasio	OysterBreak rings	2014	700	2014	6	1007

[a] Cost per linear foot depends on site location (mobilization, distance to site, and demobilization) and amount of material ordered (price per linear meter usually diminishes with bulk orders of more material). Cost includes manufacture and installation of reef, not long-term monitoring.

shoreline edges and reduce shoreline erosion, most projects promise delivery of other ecosystem services, including fisheries habitat and water quality enhancement, based on literature from other areas that quantify the contributions of healthy shellfish reefs (Grabowski et al. 2012).

Recent surge and wave modeling for the state of Louisiana's coastal restoration master plan found that waves were significantly reduced near oyster reefs (Cobell et al. 2013), which has increased interest in developing more living shorelines using the eastern oyster. More specifically, recent analyses have demonstrated that reef-based living shorelines along Louisiana's marsh edges are most effective in higher-energy locations (La Peyre et al. 2015). Initial preliminary reports of a number of the constructed living shoreline projects across coastal Louisiana indicate ambivalent results (La Peyre et al. 2013b,c; Melancon et al. 2013); however, much of this uncertainty may be resolved with longer-term data and the use of data from across multiple sites and years. Here, we present an overview of results from seven different oyster reef living shoreline projects distributed across coastal Louisiana (Figure 18.1; Table 18.1), focusing on reef sustainability, location data, and shoreline impact data.

18.3 PROJECT DESCRIPTIONS

Results from seven independent oyster reef restoration living shoreline projects were analyzed across the coast of Louisiana. Sites were spread across four different estuaries, with multiple locations in several estuaries, and included (1) Vermilion, Vermilion Cove (29°36_39.99_N, 92°3_19.70_W); (2) Terrebonne, Sister (Caillou) Lake (29°12_50.70_N, 90°56_3.12_W), and Terrebonne Bay (29°17_1.59_N, 90°37_1.32_W; 29°17_11.41_N, 90°37_9.26_W; 29°18_19.897 N, 90°34_03.958 W); (3) Barataria, Grand Isle (29°13_48.22_ N, 90°0_56.96_ W); and (4) Breton Sound/Biloxi Marsh, Lake Eloi (29°45_47.4_N, 89°26_39.30_W), Lake Fortuna (29°40_47.9_N, 89°31_63.5_W), and Lake Athanasio (29°44_47.04_N, 89°26_46.73_W; Figures 18.1 and 18.2). All sites had fringing bioengineered reefs constructed between 2007 and 2011 with the primary objective of enhancing shoreline protection and secondary goals of increasing provision of ecosystem services such as fisheries habitat and water quality enhancement. Reefs were similar in that they were all located adjacent to eroding marsh (<50 m from the eroding marsh edge); reefs differed in terms of reef length, adjacent habitat, site water quality characteristics, and shoreline orientation (Table 18.1).

Vermilion Bay is a shallow, relatively fresh bay located in Iberia and Vermilion Parishes. It is separated from the saltier waters of the Gulf of Mexico by Marsh Island. On the west side of Marsh Island, the narrow, deep (>25 m in some places) Southwest Pass connects Vermilion Bay to the Gulf of Mexico. The study area in Vermilion Bay faces south, and is characterized by shallow mean water depths (<1 m). It is adjacent to one of Louisiana's historic public oyster seed grounds; however, increased freshwater input from the Atchafalaya River has greatly diminished oyster production in recent years (Louisiana Department of Wildlife and Fisheries 2011). Intertidal reefs were constructed in Vermilion Bay beginning in 2009 in a series of experimental reef segments located less than 25 m from the marsh edge (La Peyre et al. 2013c). The segments were composed of bioengineered Oysterbreak structures; concrete rings, ranging from 50 to 61 cm in height, placed adjacent to one another in varying sub- and intertidal formations.

Sites in Terrebonne Bay include that of Sister (Caillou) Lake (Casas et al. 2015; La Peyre et al. 2014) and the Terrebonne demonstration project (Melancon et al. 2013) located east of Sister Lake. Sister Lake has been a designated Public Oyster Seed Reservation since 1940 (Louisiana Department of Wildlife and Fisheries 2011) and approximately 30% of the area consists of subtidal reefs. Sister Lake is primarily an open body of water 1–3 m in depth, surrounded by brackish marsh. Reefs in Sister Lake were constructed in 2009 at three locations across the lake using piles of shell cultch to form fringing oyster reefs for experimental purposes and measurement of reef development and ecosystem service provision (Casas et al. 2015; La Peyre et al. 2014).

The Terrebonne project site is located along the northeast shore of Terrebonne Bay, a large open body of water with 1–3 m depth similar to that of Sister Lake. There are numerous private oyster leases near the Terrebonne site that are harvested as a subtidal fishery. Fringing intertidal oyster reefs along the northern shore of Terrebonne Bay site are scarce because of the unstable soils attributed to high shoreline erosion rates (Melancon et al. 2013). Soils along the Terrebonne Bay shoreline are composed of a Timbalier–Muck association. This soil is a very poorly drained organic soil that is found in saline marsh habitats. The organic layer extends approximately 1.5 m below the ground surface. Below this layer lies a very fluid clay substratum (USDA 2007). The Terrebonne project site is composed of three bioengineered reef features, all constructed during 2007: ReefBlk (foreshore), A-Jacks (onshore), and the Gabion Mat (onshore). The ReefBlk structures were constructed of triangular rebar frames fitted with mesh bags and filled with clean oyster shell, and placed on top of a crushed stone foundation and anchored. The A-Jack treatment consists of concrete "jack"-shaped structures, each 0.6 m (2 ft) tall, lashed together with steel cables, and also built on top of a crushed stone foundation. The Gabion Mat was fabricated of mattress-shaped mesh frames, each filled with crushed stone, and laid partially submerged on the marsh edge.

Bioengineered reefs were constructed along eroding shorelines at Grand Isle (bay side of island), with the goal of reducing marsh retreat. Grand Isle is a barrier island located at the mouth of Barataria Bay, where it meets the Gulf of Mexico. The Gulf side of the island is dominated by sand beaches; however, the "back" bay side is fringed with marsh habitat. Despite supporting a year-round population, as well as a substantial recreational fishing community, Grand Isle experiences heavy coastal erosion and marsh loss (La Peyre et al. 2013b). This project was constructed near the Grand Isle Oyster Hatchery, which is located just east of the reefs, and is composed of a 1.4-km-long ReefBlk segment. The reef structure was constructed in 2011, approximately 25–50 m away from the shoreline.

Two similar ReefBlk reef extents were constructed in Breton Sound (Lake Fortuna and Lake Eloi) for similar purposes and using similar design (La Peyre et al. 2013b). Both areas, located on either side of the former Mississippi River Gulf Outlet shipping channel, are considerably more exposed than Grand Isle in terms of wave energy and fetch. Bioengineered ReefBlk segments, ranging from 1.3 to 2.4 km in length, run north–south along the shoreline and were built 5–10 m away from the shoreline. In 2014, the Lake Athanasio project was added in Breton Sound and is composed of a 700-m-long Oysterbreak segment. All three sites (Lakes Fortuna, Eloi, and Athanasio) presently support extensive oyster production on both private and public leases.

18.4 METRICS

These seven projects were compared and contrasted using a set of common parameters collected at each of the projects through independent monitoring programs. Specifically, we present and discuss data on (1) environmental site conditions, (2) eastern oyster recruitment and population dynamics, (3) biotic interactions (competitors, biofouling), and (4) adjacent marsh retreat. The Terrebonne site is an 8-year project, and the data presented in this report are based on 4-year, preliminary postconstruction metrics. The Terrebonne project has three structure types with different configurations, and the data presented here are a composite of all three. The goal here is comparison of locations as opposed to comparison of engineered material or reef configuration. The assumption is that site environmental characteristics are the dominant factors controlling reef development and sustainability.

18.4.1 Environmental Data

Daily salinity, temperature, and water levels from continuous data recorders located adjacent to or near each project site were downloaded for calendar years 2008–2014. All sites also had discrete site sampling measuring water turbidity (NTU; Hach 2100P, Hach 2100Q, Hach, CO), dissolved oxygen (mg L^{-1}; YSI-85, YSI Incorporated, OH), and chlorophyll a (ug L^{-1}; EPA Method 456.0). A survey of reef top elevation using a TOPCON GTS-226 electronic total station was conducted once, approximately 1 year postconstruction at all sites, except at Terrebonne. The Terrebonne site elevations were determined immediately postconstruction (February 2008) and 3 years postconstruction (February 2011) using traditional cross-sectional transects and real time kinematic survey methods (Melancon et al. 2013). These surveys established elevations on the upper surface of the structures to document structure heights and settlement over time. All survey data were established using or adjusted to the tie-in with the Louisiana Coastal Zone GPS Network. Elevation, along with daily water levels, was used to calculate the percentage of time that reef tops were above the water line and exposed (exposure time).

18.4.2 Oyster Population

Oyster populations were measured (ind m^{-2}, shell height [mm]) annually during winter (November–February) periods to access the sites during low water periods, because of low water clarity. Sampling approach varied based on bioengineered reef material, but in all cases, we used a random sampling design, stratified by windward (bay-facing) and leeward (marsh-facing) faces of the reef. Sampling for oyster populations resulted in comparable measures of oyster density (ind m^{-2}) and population demographics (shell height [mm]).

Reefs created with Oysterbreak rings (Vermilion Bay, Lake Athanasio) with smooth cement sides were sampled visually using a 0.1-m^2 quadrat. At each location, three reef sample sites (10 m linear stretch of the reef) were selected, and five replicates were taken per site (three sites × five replicates per year). Data were converted to ind m^{-2} and shell heights were recorded for all oysters found within quadrats.

ReefBlk reefs (Lake Eloi, Lake Fortuna, Grand Isle) were sampled by removing approximately 10 shells/clusters to generate density (ind m^{-2}) and record shell heights of a random sample. For each sample period, three randomly selected sites were sampled by collecting five samples of approximately 10 shells/clusters, which were removed from the top half of the reef and placed in a mesh bag. Samples were taken back to the laboratory where oyster size (shell height [mm]) and density (ind m^{-2}) were measured and recorded.

Shell cultch reefs (Sister Lake) were sampled at three random sites per reef (6 reefs × 3 sites = 18) using quadrats to remove 0.25 m^2 of shell, excavated to 10 cm depth. All contents were taken to the laboratory where shell height (mm) was measured for all live oysters, and density was converted to ind m^{-2}.

At Terrebonne sites, with the three different reef structures, data were collected on oyster density (ind m^{-2}), shell height (mm), and shell loss (Melancon et al. 2013). Gabion Mats and A-Jacks were sampled using random stratified (by reef side) quadrat samples (n = 45/material). ReefBlk were sampled by taking 10 stratified (by side of reef) samples at 3 reef locations (n = 30). Each of the 30 samples consisted of excavating the middle shell bag to a depth of 0.3 m (half the bag). For all sites, oyster size (shell height [mm]) and density data were recorded, and density was converted to ind m^{-2}.

18.4.3 Biotic Interactions

At the Terrebonne project sites, densities on the competing and fouling organism, the hooked mussel, *Ischadium recurvum*, were collected using the same winter sampling periods and within the same quadrats and methods as detailed above for oyster populations.

18.4.4 Shoreline Stabilization

All projects, except the Terrebonne, measured shoreline movement using similar methods. Briefly, shoreline position change was measured using techniques similar to Meyer et al. (1997) and Piazza et al. (2005). A minimum of five sites at each project location, with nearby reference shoreline sites, was established with permanent base stakes located in the marsh and in the water. For each sample, a tape measure was stretched level between base stakes and read at the shoreline edge along the same compass heading each time. Shoreline edge is defined as the farthest waterward extent of the emergent wetland macrophytes. Change in shoreline position was calculated as the difference (cm) between measurements. Positive values indicate accretion, and negative values indicate erosion. Shoreline change for each location and observation period is reported in m year^{-1}. For the Terrebonne project, shoreline position was determined using aerial photographs and the Digital Shoreline Analysis System (DSAS version 2.1.1) extension of ArcView GIS (Thieler et al. 2003). Shoreline positions were determined by digitizing aerial photographs at a 1:800 scale following Steyer et al. (1995), which defines shoreline position as the edge of the live emergent vegetation (as above). Numerous periods were analyzed, but we present here only the shoreline change from the immediate postconstruction period (September 16, 2007) to 5 years postconstruction (October 28, 2012). Additional information on how shoreline change was determined can be found in Melancon et al. (2013).

For all sites, across all locations, we combined the measurements and focus on the relative difference between control and reef within each site, rather than across site comparisons.

18.5 RESULTS

18.5.1 Site Environmental Characteristics

Temperatures across the shallow coastal waters were similar between all sites. Mean salinity differed significantly between sites, but was within the range for development of sustainable oyster populations (9–21 psu). Dissolved oxygen, turbidity, and chlorophyll a varied between sites but were all within the same range (Table 18.2). Site characteristics varied in terms of mean, range, and timing of low and high salinities across the seven sites (Table 18.2). To compare site characteristics within similar years, daily salinities at all sites were examined between 2010 and 2014, a period when most sites had the fringing reefs in place (Figure 18.3). Interestingly, mid-salinity sites

Table 18.2 Water Quality Parameters (Mean ± SE; Range) of Reported Projects

Location	Salinity	Temperature (°C)	Dissolved Oxygen (mg L⁻¹)	Chl. A (µg L⁻¹)	Exposure Time (%)
Terrebonne	16.2 ± 0.01 (3.7–27.0)	22.2 ± 0.03 (1.4–36.0)	7.8 ± 0.1 (3.0–14.6)	18.9 ± 0.6 (7.3–31.4)	5.5, 2.4, 0.1[a]
Sister Lake	10.9 ± 0.0 (0.3–29.8)	22.7 ± 0.1 (2.2–34.4)	7.8 ± 0.2 (0.4–17.3)	14.6 ± 0.4 (1.8–43.6)	14
Grand Isle	16.1 ± 0.02 (0.7–31.5)	22.6 ± 0.03 (1.0–35.4)	8.9 ± 1.5 (3.7–76.0)	23.8 ± 2.9 (1.3–182.9)	n/a
Vermilion Cove	9.2 ± 0.03 (0.3–39.1)	22.1 ± 0.04 (0.3–35.3)	6.4 ± 0.2 (0.3–9.2)	15.6 ± 0.6 (0.8–37.3)	66
Lake Eloi	14.2 ± 0.02 (2.3–27.9)	22.5 ± 0.03 (0.1–37.5)	6.1 ± 0.4 (0.4–8.8)	17.8 ± 2.1 (4.6–115.6)	<0.1
Lake Fortuna	8.8 ± 0.02 (0.4–25.1)	22.3 ± 0.04 (−0.6 to 36.7)	5.8 ± 0.4 (0.4–8.4)	12.0 ± 0.7 (3.6–36.3)	2.2
Lake Athanasio	14.2 ± 0.02 (2.3–27.9)	22.5 ± 0.03 (0.1–37.5)	4.7 ± 0.1 (4.4–5.0)	8.9 ± 0.5 (6.1–11.1)	7.1

Note: Ranges reported for exposure time represent differences along the multiple sections of living reefs. Temperature and salinity data originate from Louisiana's Coastwide Reference Monitoring System stations within Terrebone (CRMS TE45H01 and TE45H02), Grand Isle (CRMS0178), Vermilion Cove (CRMS05401), Lake Eloi (CRMS1024), Lake Fortuna (CRMS0147), and Lake Athanasio (CRMS1024). For Sister Lake, data were obtained from the United States Geological Survey (USGS07381349).

[a] Terrebonne site exposure times are for Gabion Mats, A-Jacks, and ReefBlks, respectively.

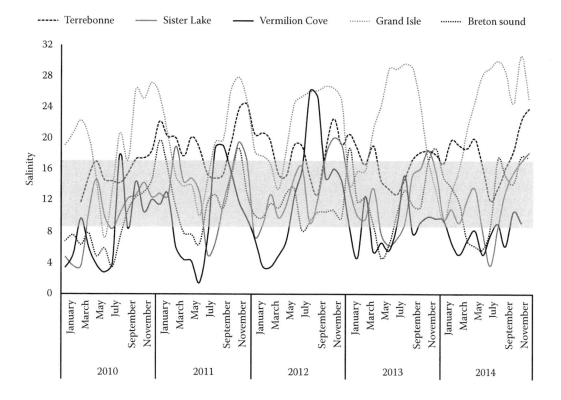

Figure 18.3 Mean monthly salinity at all study sites from 2010 to 2014. Breton Sound data represents three individual sites from this study: Lake Eloi, Fortuna, and Athanasio.

(Sister Lake, Terrebonne, Lake Eloi, Lake Fortuna, and Lake Athanasio) had smaller ranges of salinity (0.4–27.5 psu). These areas are all adjacent to productive subtidal state oyster-producing grounds. In contrast, both low (Vermilion) and high (Grand Isle) mean salinity sites (9.2 and 21.0 psu, respectively) had much larger ranges of salinity over the 5 years examined (from 0.3 to 39.1 psu). These sites are located adjacent to and near areas where subtidal oyster production has histori-cally been extremely low, or only viable when protected by predator cages (i.e., Grand Isle).

For sites where we had elevation data, reef exposure periods ranged from less than 1% exposure to more than 50% exposure periods. As this region is microtidal, exposure events were not regular and occurred more during fall and winter months from storm passage than during other times of the year (Table 18.2).

18.5.2 Oyster Populations

On-reef density and population size distribution differed between reef sites and by age of reef (Figure 18.4). Specifically, two of the mid-salinity sites (Sister Lake, Terrebonne) had the highest densities of oysters, exceeding more than 500 ind m^{-2}, 2 years postconstruction. The other three mid-salinity sites (Lake Eloi, Lake Fortuna, and Lake Athanasio), located within the same coastal

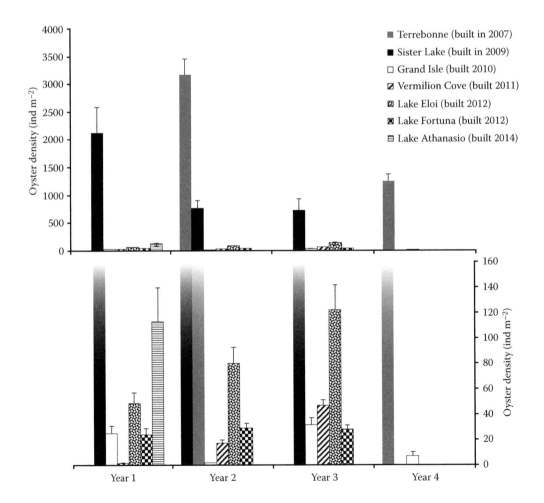

Figure 18.4 Comparison of oyster density (mean ± SE) by living shoreline reef age.

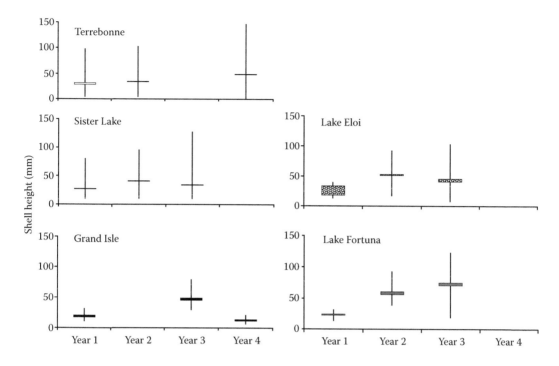

Figure 18.5 Shell height (mm) by age of reef. Collecting shell heights on OysterBreak reefs was not possible, so no data from Vermilion or Lake Athanasio are reported.

area, also recruited oysters, but densities remained low after their first year of construction (range, 50–150 ind m^{-2}). In contrast, the low-salinity (Vermilion) and high-salinity (Grand Isle) sites maintained low or no oyster density on reefs.

Oyster size class information further informs these results with Sister Lake and Terrebonne populations showing slowly increasing mean sizes and increasing ranges of oyster class sizes over time. This indicates continued recruitment and survival of different age oysters over time (Figure 18.5). Lake Eloi and Lake Fortuna show similar trends, but on a much slower time scale. In contrast, the high-salinity site, Grand Isle, indicates oyster recruitment but no long-term survival, as the size range does not increase over time.

18.5.3 Biotic Interactions

The dominant competitor for space and food with the oyster was the hooked mussel (Figure 18.6). Mussels were three times more abundant than oysters on the three Terrebonne structure types in winter surveys (Figure 18.7), causing significant concern about the long-term sustainability of these living shoreline projects. Specifically, in Gulf Coast estuaries, only the eastern oyster builds true three-dimensional reefs; if prevented from doing so by a competitor such as the hooked mussel, which does not cement into reefs and has comparatively fragile shell that fragments easily, the living shoreline will not ultimately be sustainable and provide shoreline protection. Under certain circumstances and specific restoration goals, the presence of multiple foundation species has been argued to be a benefit to a restoration project, such as enhanced filtration capacity and valuable structured habitat (Coen and Luckenbach 2000; Coen et al. 2007; Crain and Bertness 2006; Gedan et al. 2014). For example, Gedan et al. (2014) found that hooked mussels may in fact complement oyster filtration services by more effectively filtering smaller plankton (1.5–3 μm) and, except with larger size

Figure 18.6 Hooked mussels on A-Jacks structure embedded with oysters, observed during the winter 2011 survey at Terrebonne Bay.

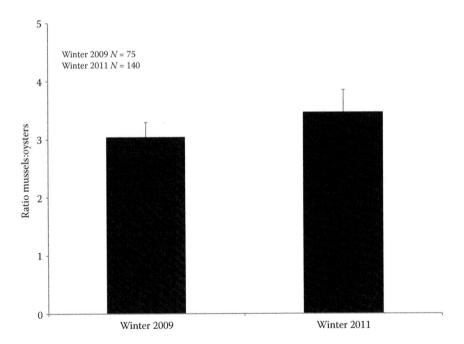

Figure 18.7 Mussel-to-oyster ratio (bars indicate standard error) at Terrebonne project site comparing winter 2009 and winter 2011 surveys. The hooked mussel (*Ischadium recurvum*) is a major competitor for reef space and resources.

class mussels, this filtration was complementary and not competitive with the eastern oyster. The presence of multiple foundation species (i.e., hooked mussel and eastern oyster) could also have the added benefit of expanding the environmental conditions under which the reef may provide eco-system services (Crain and Bertness 2006) and help provide resiliency to reefs particularly in areas with changing conditions. However, if the competing species does not function as a true foundation species (i.e., building reef) and excludes the targeted reef-building organism, this will result in a failure of three-dimensional reef building and maintenance. In Gulf Coast estuaries, where the creation of living shorelines with eastern oyster reefs has a primary goal of shoreline stabilization, projects are not likely to be resilient long term if the eastern oyster is outcompeted by hooked mus-sels, and any enhancement from increased filtering capacity and other ecosystem service provision will remain secondary to that of shoreline protection.

Further interacting with the competition at some sites from hooked mussel, one site in particu-lar, the Terrebonne Bay project, also began to experience shell loss within the ReefBlk structures

(a)

(b)

Figure 18.8 Failure of ReefBlk 4 years postconstruction in Winter 2011. Failure (shell loss within bags) can be seen across multiple blocks (a) and within a single block (b). (From Melancon EJ, Jr., Curole GP, Ledet AM and Fontenot QC. 2013. 2013 Operations, maintenance and monitoring report for Terrebonne bay shore protection demonstration (TE-45), Coastal Protection and Restoration Authority of Louisiana, Thibodaux, Louisiana, 75 pp. and Appendices.)

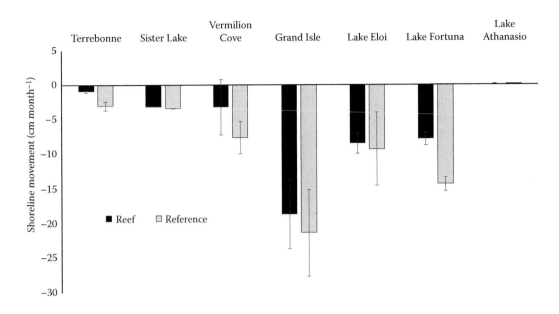

Figure 18.9 Shoreline erosion rates by site, comparing reef and adjacent mud-bottom reference sites. Sites report the overall mean over different periods (Table 18.1). During this time, four hurricanes (Gustav, 2008; Ike, 2008; Ida, 2009; and Isaac, 2012) and four named storms (Edouard, 2008; Bonnie, 2010; Lee, 2011; and Karen, 2013) affected the Louisiana coastline.

(Figure 18.8) and showed evidence of extensive colonization by boring sponges, polychaetes, and Gulf stone crabs, *Menippe adina*. By winter 2011, 4 years postconstruction, the ReefBlk structures at one of the three experimental sites had experienced greater than 50% shell loss (Melancon et al. 2013). Such a large quantity of shell loss equates to structure failure in its ability to support oyster populations and reef building. Observations at the other two experimental sites for the Terrebonne Bay ReefBlk indicate that some shell loss is beginning to occur there as well (Melancon, personal observation).

18.5.4 Shoreline Stabilization

All sites continued to show marsh retreat at both reef and reference sites, with the exception of Lake Athanasio, which had a very short study duration (Figure 18.9). Marsh retreat rates were lower at most reef sites compared to their paired mud edge reference sites, although differences were not consistent across reef sites. Marsh edge retreat rates ranged from 0.5 to 23 cm month^{-1} at all sites, except Lake Athanasio where marsh edges (reef and reference) appeared to be relatively stable over the short period of data collection (4 months). While not presented in this work, there was no evidence that shoreline protection effectiveness increased with reef age.

18.6 LESSONS LEARNED

The use of oyster reefs as a living shoreline in estuarine environments requires development of a sustainable oyster population where high production of shell over time, through settlement and growth, are a necessity (Powell et al. 2012; Walles et al., 2015a,b). Within the required temperature conditions, and when bioengineered substrate is provided for a starting base, salinity drives oyster population development on reefs. There is an extensive literature documenting the effects of salinity on oyster populations (i.e., Shumway 1996) that supports this observation. It is clear that salinity regimes conducive to recruitment, growth, and shell sustainability are critical determinants of success in oyster reef creation. In this region, sites with moderate mean salinities (11–16 psu), and not experiencing extreme salinity (<5 or >25 psu) for extended periods, seemed more conducive to the rapid population development and high production necessary to build sustainable oyster populations and maintain reefs over the long term.

In addition to oyster population responses to environmental factors, biotic interactions need to be considered in site selection. The Terrebonne Bay project demonstrated that, under the right conditions, organisms, such as hooked mussels, may be a significant competitor with oysters in terms of space and food. This is not a new phenomenon, as hooked mussels associated with oysters have been documented from Chesapeake Bay (Lipcius and Burke 2006) to Texas, with Abbot (1974) reporting that it was first described from the mouth of the Mississippi River in 1820. The impact of hooked mussel abundance on long-term reef development and provision of ecosystem services requires further investigation.

Understanding the factors driving shell loss at the Terrebonne site appears to be as important as understanding oyster population dynamics for predicting long-term reef sustainability. The shell loss at the Terrebonne site appears to result from a combination of biofouling by shell pests such as boring sponges and shell brittleness created by polychaete worms. The abundance of sponges and worms was probably stimulated by two factors: salinity and reef exposure time. The Terrebonne site where shell loss occurred was near the upper range of ideal salinities (Figure 18.3) and inundated more than 99% of the time (Table 18.2). This combination of salinity and inundation likely enabled sponge and polychaete colonization and also allowed for Gulf stone crabs, *M. adina*, to enter the shell bags and feed on the colonized shell pest and encrusting organisms such as bryozoans and in the process crush shell.

Although the relationships between inundation duration and oyster reef dynamics have not been clearly delineated for this region, in general, given the right salinity conditions, inundated reefs experience higher rates of predation and biofouling than exposed reefs (Fodrie et al. 2014). While high inundation times have been associated with increased growth owing to more access to food resources and lower stress from decreased exposure (e.g., Bahr 1976; Bartol et al. 1999; Byers et al. 2015), other studies have suggested that regularly exposed oysters have accelerated growth and improved survival, reduced poychaete infestations, and lowered disease and predation (Littlewood et al. 1992; Moroney and Walker 1999; O'Beirn et al. 1994). In coastal North Carolina, intertidal reefs with 20%–40% regular exposure rates were found to have the greatest growth rates, with lower and upper boundaries at 10% and 55% exposure for reef growth (Ridge et al. 2015). With one exception (Sister Lake), reefs studied here fall outside this exposure range. This suggests that the relationship between exposure and reef growth dynamics may be slightly more complex in a region such as the Gulf Coast where exposure is not tidal, but rather more influenced by seasonal storm and weather patterns. The relatively higher salinities near and above the ideal range, combined with high inundation times at this site likely enabled sponge and polychaete colonization (Table 18.2). Oyster population dynamics and biotic interactions in relation to inundation and salinity regimes should be explored to help inform the design of living shoreline reefs as the synergism of these factors seems to be a significant factor in reef creation success.

The development of highly productive reefs for long-term sustainability and shoreline protection effectiveness is particularly important within coastal Louisiana given the high subsidence rates and high relative sea level rise (1–35 mm year^{-1}; CPRA 2012). For the reef to remain viable over the long term, shell accretion would need, at a minimum, to keep pace with site-specific relative sea level rise. Recent work on the Sister Lake reef documented shell accretion rates, which kept pace with local subsidence rates (Casas et al. 2015); this is important because Terrebonne Basin is one of the areas experiencing highest subsidence rates along the coast. These findings, however, suggest that the two best-performing reefs in this study, based on oyster recruitment and survival

(Sister Lake and Terrebonne), may be the only reefs of the seven producing at a rate sufficient to keep pace with relative sea level rise over the long term. However, this shell accretion at both sites could be jeopardized by biotic interactions resulting in decreased oyster shell accretion (through competition for space) and shell loss. Understanding how exposure time may influence boring sponges and crab predation rates on these living shoreline reefs could be useful in designing reefs with targeted elevations.

Sustainable oyster populations on these fringing oyster reefs hold the key to the bioengineered reef functioning as living shorelines and providing associated ecosystem services. Both hurricanes (Morton and Barras 2010; Stone et al. 1997) and cold fronts (Watzke 2004) have been found to erode coastal marshes, and these narrow fringing shoreline reefs have been suggested as an approach to help reduce these effects (i.e., Piazza et al. 2005; Scyphers et al. 2011, 2015). Since the end of 2007 when the first project reported above was constructed, a multitude of tropical storms and hurricanes have crisscrossed and skirted the Louisiana coast, causing high wind and wave activity resulting in coastal erosion: Tropical Storm Edouard (August 2008), Hurricane Gustav (September 2008), Hurricane Ike (September 2008), Hurricane Ida (November 2009), Tropical Storm Bonnie (July 2010), Tropical Storm Lee (September 2011), Hurricane Isaac (August 2012), and Tropical Storm Karen (October 2013) (NOAA National Hurricane Center).

Despite significant storm activity across all the sites, marsh retreat was generally lower at shorelines adjacent to reefs as compared to mud-bottoms, confirming previous findings that fringing oyster reefs may provide marsh stabilization services (La Peyre et al. 2014; Meyer et al. 1997; Piazza et al. 2005; Scyphers et al. 2011). All comparisons, however, had large error bars, indicating that other factors, such as shoreline exposure, adjacent marsh characteristics, or local subsidence may be critical to identifying the most likely sites for successful shoreline protection. A recent analysis involving several of these sites indicated that shoreline exposure (wave energy + fetch) explained a lot of the variation in fringing oyster reef impacts on shoreline retreat (La Peyre et al. 2015). Success in developing eastern oyster–based living shorelines will require consideration of not only oyster habitat suitability but also predator and competitor habitat suitability, as well as energy exposure. While some factors (energy exposure and oyster population dynamics) are dependent on local conditions, other factors (e.g., initial inundation regime and exposure to competitors and predators) may be manipulated through bioengineering of the reef bases.

18.7 FUTURE CONSIDERATIONS

Given the importance of site selection in the success of living shorelines based on oyster reefs, developing models for site selection that incorporate not only present conditions at sites but also projected future conditions is critical. Data from monitoring of the seven reefs examined in this work illustrate the importance of salinity and inundation time, which should inform both site selection and reef design criteria. In coastal Louisiana, rapid land loss and subsidence alter the sediment and marsh properties along shorelines, changing site conditions across the coast at rapid rates (Couvillion et al. 2013). Similarly, significant coastal restoration activities, including river and sediment management, alter salinity regimes and sediment loads in the water, further affecting conditions important to oyster production (La Peyre et al. 2013a; Soniat et al. 2013). As a result, selecting sites for sustainable oyster populations needs to consider not only current habitat and water conditions but also future scenarios.

In coastal Louisiana, modeling efforts to predict future estuarine conditions and to understand impacts of proposed restoration activities have been developed (Peyronnin et al. 2013). These efforts provide an opportunity to examine potential living shoreline reef locations under future conditions. For oyster production, Soniat et al. (2013) applied an HSI to Breton Sound, Louisiana, under three potential future conditions of average, low, and high river inflow rates. The results of this exercise demonstrated dramatic changes in locations deemed suitable for oyster growth. Selecting which scenarios to use for restoration decision making can be tricky. However, consideration of a broad range of potential conditions can provide planners with the ability to make informed management decisions, including those that might affect the selection of sites for oyster reefs as living shorelines.

ACKNOWLEDGMENTS

These data were made possible by funding and support from the Louisiana Department of Wildlife and Fisheries; The Nature Conservancy, Louisiana Chapter; the Coastal Wetlands Planning, Protection, and Restoration Act (CWPPRA); and the Louisiana Coastal Protection and Restoration Authority (CPRA). We thank the many students and colleagues who helped with field work and laboratory processing through these projects. Thank you to Dr. Lesley Baggett for a critical review that improved this manuscript, as well as comments from two anonymous reviewers. Any use of trade, firm, or product names is for descriptive purposes only and does not imply endorsement by the US Government.

REFERENCES

Abbot, R.T. 1974. *American Seashells, the Marine Mollusca of the Atlantic and Pacific Coasts of North America*. Second Edition. New York: Van Nostrand Reinhold Company. 663 pp.

Bahr, L. 1976. Energetic aspects of intertidal oyster reef community at Sapelo Island, Georgia (USA). *Ecology* 57: 121–131.

Barras, J.A. 2009. Land area change and overview of major hurricane impacts in coastal Louisiana, 2004–2008. Scientific Investigations Map 3080. Reston, Virginia: U.S. Geological Survey, scale 1:250,000, 6 sheets.

Bartol, I.K., R. Mann, and M. Luckenbach. 1999. Growth and mortality of oysters (*Crassostrea virginica*) on constructed intertidal reefs: Effects of tidal height and substrate level. *Journal of Experimental Marine Biology and Ecology* 237: 157–184.

Beseres Pollack, J., A. Cleveland, T.A. Palmer, A.S. Reisinger, and P.A. Montagna. 2012. A restoration suitability index model for the eastern oyster (*Crassostrea virginica*) in the Mission-Aransas estuary, Texas, USA. *PLoS ONE* 7:e40839. DOI:10.1371/journal.pone.0040839.

Brown, K.M., G.J. George, G.W. Peterson, B.A. Thompson, BA, and J.H. Cowan Jr. 2008. Oyster predation by black drum varies spatially and seasonally. *Estuaries and Coasts* 31: 597–604.

Brown, K.M. and T.D. Richardson. 1987. Foraging ecology of the southern oyster drill *Thais haemastoma* (Gray): Constraints on prey choice. *Journal of Experimental Marine Biology and Ecology* 114: 123–141.

Butler, P.A. 1954. Summary of our knowledge of the oyster in the Gulf of Mexico. *Fishery Bulletin of the Fish and Wildlife Service* 55: 479–489.

Byers, J.E., J.H. Grabowski, M.F. Piehler, A.R. Randall Hughes, H.W. Wiskel, J.C. Malek, and D.L. Kimbro. 2015. Geographic variation in intertidal oyster reef properties and the influence of tidal prism. *Limnology and Oceanography* 50: 1051–1063.

Cake, E.W. 1983. Habitat suitability index models: Gulf of Mexico American oyster. FWS/OBS-82/10.57. U.S. Department of Interior, Fish and Wildlife Service. 37 pp.

Casas, S.M., J.F. La Peyre, and M.K. La Peyre. 2015. Restoration of oyster reefs in an estuarine lake: Population dynamics and shell accretion. *Marine Ecology Progress Series* 524: 171–184.

Chmura, G.L., R. Costanza, and E.C. Kosters. 1992. Modeling coastal marsh stability in response to sea level rise: A case study in coastal Louisiana, USA. *Ecological Modelling* 64: 47–64.

Cobell, Z., H. Zhao, H.J. Roberts, F.R. Clark, and S. Zou. 2013. Surge and wave modeling for the Louisiana 2012 Coastal Master Plan. *Journal of Coastal Research* 67: 88–108.

Coen, L.D., D. Brumbaugh, D. Bushek, R. Grizzle, M.W. Luckenbach, H. Posey, S.P. Powers, and S. Tolley. 2007. Ecosystem services related to oyster restoration. *Marine Ecology Progress Series* 341: 303–307.

Coen, L.D. and M.W. Luckenbach. 2000. Developing success criteria and goals for evaluating oyster reef res-
    toration: Ecological function or resource exploitation? *Ecological Engineering* 15: 323–343.
Couvillion, B.R., J.A. Barras, G.D. Steyer, W. Sleavin, M. Fischer, H. Beck, N. Trahan, B. Griffin, and D.
    Heckman. 2011. Land area change in coastal Louisiana from 1932 to 2010: US Geological Survey
    Scientific Investigations Map 3164. Scale 1:265,000, 12 p. pamphlet.
Couvillion, B.R., G.D. Steyer, H. Wang, H. Beck, and J.M. Rybczyk. 2013. Forecasting the effects of coastal
    protection and restoration projects on wetland morphology in coastal Louisiana under multiple environ-
    mental uncertainty scenarios. *Journal of Coastal Research* 67: 29–50.
CPRA (Coastal Protection and Restoration Authority). 2012. Louisiana's comprehensive plan for a sustainable
    coast. Baton Rouge, LA: Coastal Protection and Restoration Authority of Louisiana. 189 pp.
Crain, C.M. and M.D. Bertness. 2006. Ecosystem engineering across environmental gradients: Implications
    for conservation and management. *BioScience* 56: 211–218.
Day, J.W. Jr., D.F. Boesch, E.J. Clairain, G.P. Kemp, S.B. Laska, W.J. Mitsch, K., Orth, H. Mashriqui, D.J.
    Reed, L. Shabman, C.A. Simenstad, B.J. Streever, R.R. Twilley, C.C. Watson, and J.T. Wells. 2007.
    Restoration of the Mississippi Delta: Lessons from Hurricanes Katrina and Rita. *Science* 315(5819):
    1679–1684.
Dekshenieks, M.M., E.E. Hofmann, and E.N. Powell. 1993. Environmental-effects on the growth and devel-
    opment of eastern oyster, *Crassostrea virginica* (Gmelin, 1791), larvae—A modeling study. *Journal of
    Shellfish Research* 12: 241–254.
Fodrie, F.J., B. Antonio, C.J. Rodriguez, M.C. Baillie, S.E. Brodeur, R.K. Coleman, D.A. Gittman, M.D.
    Keller, A.K. Kenworthy, J.T. Poray, E.J. Ridge, J. Theuerkauf, and N.L. Lindquist. 2014 Classic para-
    digms in a novel environment: Inserting food web and productivity lessons from rocky shores and salt-
    marshes into biogenic reef restoration. *Journal of Applied Ecology* 51: 1314–1325.
Gedan, K.B., L. Kellogg, and D.L. Breitburg. 2014. Accounting for multiple foundation species in oyster reef
    restoration benefits. *Restoration Ecology* 22:517–524.
Georgiou, I.Y., D.M. Fitzgerald, and G.W. Stone. 2005. The impact of physical processes along the Louisiana
    coast. In: Finkl C.W. and S.M. Khalil (Eds.), Saving America's Wetland: Strategies for Restoration of
    Louisiana's Coastal Wetlands and Barrier Islands. *Journal of Coastal Research* Special Issue No. 44,
    pp. 72–89.
Grabowski, J.H., R.D. Brumbaugh, R.F. Conrad, A.G. Keeler, J.J. Opaluch, C.H. Peterson, M.F. Piehler, S.P.
    Powers, and A.R. Smyth. 2012. Economic valuation of ecosystem services provided by oyster reefs.
    BioScience 62: 900–909.
Gregalis, K.C., M.W. Johnson, and S.P. Powers. 2009. Restored oyster reef location and design affect responses
    of resident and transient fish, crab, and shellfish species in Mobile Bay, Alabama. *Transactions of the
    American Fisheries Society* 138: 314–327.
Humphries, A.T. and M.K. La Peyre. 2015. Oyster reef restoration supports increased nekton biomass and
    potential commercial fishery value. *PeerJ* 3: e1111; DOI:10.7717/peerj.1111.
La Peyre, M.K., B.S. Eberline, T.M. Soniat, and J.F. La Peyre. 2013a. Differences in extreme low salinity
    timing and duration differentially affect eastern oyster (*Crassostrea virginica*) size class growth and
    mortality in Breton Sound, LA. *Estuarine, Coastal and Shelf Science* 135: 146–157.
La Peyre, M.K., B. Gossman, and J.F. La Peyre. 2009. Defining optimal freshwater flow for oyster production:
    Effects of freshet rate and magnitude of change and duration on eastern oysters and *Perkinsus marinus*
    infection. *Estuaries and Coasts* 32: 522–534.
La Peyre MK, Humphries AT, Casas SM and La Peyre JF. 2014. Temporal variation in development of ecosys-
    tem services from oyster reef restoration. *Ecological Engineering* 63: 34–44.
La Peyre, M.K., L. Schwarting, and S. Miller. 2013b. Preliminary assessment of bioengineered fringing
    shoreline reefs in Grand Isle and Breton Sound, Louisiana: U.S. Geological Survey Open-File Report
    2013–1040, 34 pp.
La Peyre, M.K., L. Schwarting, and S. Miller. 2013c. Baseline data for evaluating development trajectory
    and provision of ecosystem services of created fringing oyster reefs in Vermilion Bay, Louisiana: U.S.
    Geological Survey Open-File Report 2013–1053, 43 pp.
La Peyre, M.K., K. Serra, T.A. Joyner, and A. Humphries. 2015. Assessing shoreline exposure and oyster
    habitat suitability maximizes potential success for sustainable shoreline protection using restored oyster
    reefs. *PeerJ* 3: e1317; DOI:10.771/peerj.1317.

Littlewood, D.T.J., R.N. Wargo, J.N. Kraueter, and R.H. Watson. 1992. The influence of intertidal height on growth, mortality and *Haplosporidium nelsoni* infection in MSX mortality resistant eastern oysters, *Crassostrea virginica* (Gmelin, 1791). *Journal of Shellfish Research* 11: 59–64.

Lipcius, R.N. and R.P. Burke. 2006. Abundance, biomass and size structure of eastern oyster and hooked mussel on a modular artificial reef in the Rappahannock River, Chesapeake Bay. Gloucester Point, VA: Virginia Institute of Marine Science, College of William and Mary; Special Report in Applied Marine Science and Ocean Engineering No. 390.

LDWF (Louisiana Department of Wildlife and Fisheries). 2011. 2011 Oyster stock assessment. http://www.wlf.louisiana.gov/fishing/oyster-program; accessed 10/1/2015.

Mann, R. and E.N. Powell. 2007. Why oyster restoration goals in the Chesapeake Bay are not and probably cannot be achieved. *Journal of Shellfish Research* 26(4): 905–917.

Mackenzie, C.L., Jr. 1970. Causes of oyster spat mortality, conditions of oyster setting beds, and recommendations for oyster bed management. *Proceedings of the National Shellfish Association* 60: 59–67.

Melancon, E.J., Jr., G.P. Curole, A.M. Ledet, and Q.C. Fontenot. 2013. 2013 Operations, maintenance and monitoring report for Terrebonne bay shore protection demonstration (TE-45), Coastal Protection and Restoration Authority of Louisiana, Thibodaux, Louisiana, 75 pp. and Appendices.

Melancon, E., T. Soniat, V. Cheramie, R. Dugas, J. Barras, and M. Lagarde. 1998. Oyster resource zones of the Barataria and Terrebonne estuaries of Louisiana. *Journal of Shellfish Research* 17: 1143–1148.

Meyer, D.L., E.C. Townsend, and G.W. Thayer. 1997. Stabilization and erosion control value of oyster cultch for intertidal marsh. *Restoration Ecology* 5: 93–99.

Moroney, D.A. and R.L. Walker. 1999. The effects of tidal and bottom placement on the growth, survival and fouling of the eastern oyster *Crassostrea virginica*. *Journal of the World Aquaculture Society* 30: 433–441.

Morton, R.A. and J.A. Barras. 2010. Hurricane impacts on coastal wetlands: A half-century record of storm-generated features from southern Louisiana. *Journal of Coastal Research* 27: 27–43.

Newell, R.I.E. and C.J. Langdon. 1996. Mechanisms and Physiology of Larval and Adult Feeding. In: Kennedy V.S., R.I.E. Newell, and A.F. Eble (Eds.), *The Eastern Oyster, Crassostrea virginica*. College Park, MD: Maryland Sea Grant College. pp. 185–230.

O'Beirn, F.X., C.C. Dean, and R.L. Walker. 1994. Prevalence of *Perkinsus marinus* in the eastern oyster, *Crassostrea virginica* in relation to tidal placement in a Georgia tidal creek. *Northeast Gulf Science* 13: 79–87.

Peyronnin, N., M. Green, C. Parsons-Richards, A. Owens, D. Reed, J. Chamberlain, D.G. Groves, W.K. Rhinehart, and K. Belhadjali. 2013. Louisiana's 2012 Coastal Master Plan: Overview of a science-based and publicly informed decision-making process. *Journal of Coastal Research* 67: 1–15.

Piazza, B.P., P.D. Banks, and M.K. La Peyre. 2005. The potential for created oyster shell reefs as a sustainable shoreline protection strategy in Louisiana. *Restoration Ecology* 13: 499–506.

Powell, E.N., J.M. Klinck, E.E. Hofmann, and M.A. McManus. 2003. Influence of water allocation and freshwater inflow on oyster production: A hydrodynamic-oyster population model for Galveston Bay, Texas, USA. *Environmental Management* 31: 100–121.

Powell, E.N., J.N. Kraeuter, and K.A. Ashton-Alcox. 2006. How long does oyster shell last on an oyster reef? *Estuarine Coastal and Shelf Science* 69: 531–542.

Ridge, J.T., A.B. Rodriquez, F.J. Fodrie, N.L. Lindquist, M.C. Brodeur, S.E. Coleman, J.H. Grabowski, and E.J.Theuerkauf. 2015. Maximizing oyster-reef growth supports green infrastructure with accelerating sea-level rise. *Scientific Reports* 5, 14785; DOI:10.1038/srep14785.

Scyphers, S.B., S.P. Powers, and K.H. Heck Jr. 2015. Ecological value of submerged breakwaters for habitat enhancement on a residential scale. *Environmental Management* 55: 383–391.

Scyphers, S.B., S.P. Powers, K.L. Heck, and D. Byron. 2011. Oyster reefs as natural breakwaters mitigate shoreline loss and facilitate fisheries. *PLoS ONE* 6(8): e22396.

Shumway, S. 1996. Natural environmental factors. In: Kennedy, V.S., R.I.E. Newell, and A.F. Eble (Eds.), *The Eastern Oyster: Crassostrea virginica*. College Park, MD: Maryland Sea Grant College, University of Maryland. pp. 467–513.

Soniat, T.M., C.P. Conzelmann, J.D. Byrd, D.P. Roszell, J.L. Bridevaux, K.J. Suir, and S.B. Colley. 2013. Predicting the effects of proposed Mississippi river diversions on oyster habitat quality: Application of an oyster habitat suitability index model. *Journal of Shellfish Research* 32: 629–638.

Soniat, T.M., J.M. Klinck, E.N. Powell, and E.E. Hofmann. 2012. Understanding the success and failure of oyster populations: Periodicities of *Perkinsus marinus*, and oyster recruitment, mortality, and size. *Journal of Shellfish Research* 31: 635–646.

Soniat, T.M. and M.S. Brody. 1988. Field validation of a habitat suitability index model for the American oyster. *Estuaries* 11: 87–95.

Steyer, G.D., R.C. Raynie, D.L. Stellar, D. Fuller, and E. Swenson. 1995. Quality management plan for Coastal Wetlands Planning, Protection, and Restoration Act Monitoring Program. Open-file series no. 95-01 (Revised June 2000). Baton Rouge: Louisiana Department of Natural Resources, Coastal Restoration Division. 97 pp.

Stone, G.W., J.M. Grymes III, J.R. Dingler, and D.A. Pepper. 1997. Overview and significance of hurricanes on the Louisiana coast, USA. *Journal of Coastal Research* 13: 656–669.

Thieler, E.R., D. Martin, and A. Ergul. 2003. The Digital Shoreline Analysis System, Version 2.0: Shoreline Change Measurement Software Extension for ArcView: USGS U.S. Geological Survey Open-File Report 03-076.

USDA. United States Department of Agriculture. 2007. Natural Resources Conservation Service (NRCS). Soil Survey of Terrebonne Parish, Louisiana. 318 pp.

Walles, B., R. Mann, T. Ysebaert, K. Troost, P.M.J. Herman, and A.C. Smaal. 2015a. Demography of the ecosystem engineer *Crassostrea gigas*, related to vertical reef accretion and reef persistence. *Estuarine Coastal and Shelf Science* 154: 224–233.

Walles, B., J. Salvador de Paiva, B. van Prooijen, T. Ysebaert, and A. Smaal. 2015b. The ecosystem engineer *Crassostrea gigas* affects tidal flat morphology beyond the boundary of their reef structures. *Estuaries and Coasts* 1: 1–10.

Wang, H., M.A. Harwell, L. Edmiston, E. Johnson, P. Hsieh, K. Milla, J. Christensen, J. Steward, and X. Liu. 2008. Modeling oyster growth rate by coupling oyster population and hydrodynamic models for Apalachicola Bay, Florida, USA. *Ecological Modeling* 211: 77–89.

Watzke, D.A. 2004. Short-term evolution of a marsh island system and the mportance of cold front forcing, Terrebonne Bay, Louisiana. MS Thesis. Louisiana State University, Baton Rouge, LA.

CHAPTER 19

Species Richness and Functional Feeding Group Patterns in Small, Patchy, Natural and Constructed Intertidal Fringe Oyster Reefs

Mark S. Peterson, Kevin S. Dillon, and Christopher A. May

CONTENTS

19.1 INTRODUCTION

The continual degradation of coastal environments and decline of species diversity has led to an increased need to understand and manage coastal regions (La Peyre et al. 2014a; Needles et al. 2015; Sanger et al. 2015) worldwide. The ongoing worldwide loss and modification of biodiversity is the result of anthropogenic activities, including, but not limited to, habitat alteration, invasive species, and climate change (Munsch et al. 2015; Stork 2010; Verdell-Cubedo et al. 2012; Worm et al. 2006). Oyster reefs, which support diverse faunal assemblages within estuarine ecosystems, are a widespread native habitat and therefore aid in sustaining biodiversity and maintaining ecosystem services (Beck et al. 2011; Grabowski et al. 2012) and are often used in living shoreline designs (Gittman et al. 2016; La Peyre et al. 2014a; Scyphers et al. 2011).

As ecosystem engineers, oysters create reefs with multiple age classes and build complex three-dimensional habitat (Jones et al. 1994; Micheli and Peterson 1999) that drive a multitude of ecosystem services (Beck et al. 2011; Coen et al. 2007). For example, oysters/oyster reefs build and stabilize shorelines (Coen et al. 2007; Meyer et al. 1997; Piazza et al. 2005; Scyphers et al. 2011; Stricklin et al. 2010), maintain refuge and nursery habitat for macrofauna that generate organic materials for benthic and pelagic consumers (Abeels et al. 2012; Carroll et al. 2015; Gittman et al. 2016; Hadley et al. 2010; Humphreys et al. 2011a,b; Kingsley-Smith et al. 2012; La Peyre et al. 2014a; Stunz et al. 2010), filter suspended matter (Gedan et al. 2014; Grizzle et al. 2008; Newell 2004), and are a classic example of a foundation species (Angelini et al. 2011; Bracken et al. 2007; Bruno et al. 2003; Yakovis et al. 2008). By restoring oyster reefs, the immediate increase in physical habitat quickly attracts and supports fauna and promotes elevated levels of faunal richness and density (Carroll et al. 2015; Coen et al. 2007; Dillon et al. 2015; Hadley et al. 2010; Luckenbach et al. 2005), and can lead to habitat cascades with other foundation species (Angelini et al. 2011).

Further, oyster reef communities provide valuable ecosystem services that outweigh the direct economic harvest benefit (Grabowski et al. 2012; Peterson et al. 2003). For example, the patchy, intertidal reefs at Grand Bay National Estuarine Research Reserve (GBNERR) provide nekton habitat and trophic transfer potential as well as potential water filtration and nitrogen removal, which may reduce localized eutrophication and anoxic conditions (Dillon et al. 2015). In Louisiana, La Peyre et al. (2014a) found that oysters and resident nekton abundances increased quickly (<1 year) on constructed reefs, but larger transient nekton did not increase and no increases in abundance were noted afterward (up to 3 years), suggesting that consumer populations reach their apex rapidly. Given this limited information on the faunal use, richness/diversity, and community structure of intertidal reefs in the Eastern oyster's (*Crassostrea virginica*) range (Carroll et al. 2015; Dillon et al. 2015; Glancy et al. 2003; Hadley et al. 2010; Kingsley-Smith et al. 2012; Luckenbach et al. 2005; Meyer and Townsend 2000), there is an obvious need to investigate the biodiversity of less well-studied patchy intertidal fringe reefs of the northern GOM. As oyster reefs are restored, ecosystem services may be regenerated to such an extent that a number of services are equivalent to or exceed those of natural reefs (Dillon et al. 2015; Lenihan et al. 2001). Thus, building oyster reefs alone and incorporating oysters as structure into various living shoreline designs can facilitate reductions in climate change impacts (Borsje et al. 2011; Scyphers et al. 2015; Temmerman et al. 2013).

Species richness/diversity and density estimates of macrofauna are thought to be useful measures of the successful restoration of a community (Dillon et al. 2015; Hadley et al. 2010; Kingsley-Smith et al. 2012; La Peyre et al. 2014a; Luckenbach et al. 2005), with resident fauna being particularly important (Walters and Coen 2006). However, production of fauna of lower trophic levels clearly enhances the production and sustainability at upper trophic levels (Harding and Mann 1999); thus, diversity and density alone do not necessarily describe trophic interactions of restored reef systems (Walters and Coen 2006). Stable isotopes have been successfully used to evaluate trophic structure in many aquatic ecosystems including patchy intertidal oyster reefs in the northern GOM. Dillon et al. (2015) found that $\delta^{13}C$ and $\delta^{15}N$ values were not significantly different between paired constructed and natural reef sites in the GBNERR marsh, suggesting that food web structure after 22 months was equivalent. Consumer $\delta^{13}C$ values showed greater spatial and temporal differences than $\delta^{15}N$ values, and these differences were coincident with changes in salinity along a gradient of sample sites (Dillon et al. 2015).

Furthermore, functional feeding group (FFG) approaches have been used to elucidate food webs in marine ecosystem (Kürten et al. 2013) and, in particular, they have been used to indicate an oyster reef's ability to support ecological functions and reflect food web relationships (Rodney and Paynter 2006). Classifying all organisms into FFGs allows for smaller groups of various sized and numerically abundant organisms to be collectively assessed, as many are good estuarine ecosystem indicators (particularly meiofauna and nematodes; Alves et al. 2015), and to gain insight as to the trophic interactions, energy transfer potential, and trophic networks of oyster reef habitat, which drive sustainability (Beck et al. 2011; Rodney and Paynter 2006; Tolley and Volety 2005). To our knowledge, no studies have compared stable carbon and nitrogen isotope values (i.e., carbon sources and trophic position) of different organisms from intertidal oyster reefs that are categorized into FFGs.

We hypothesized that constructed intertidal Eastern oyster reefs are successful functionally if they have equivalent oyster density, faunal assemblage structure, and trophic levels compared to adjacent natural reefs. Specifically, we evaluated oyster reef assemblages at the completion of a 22-month study by (1) quantifying the similarities in species/taxa richness (called species richness hereafter) and density on constructed and natural intertidal fringe reefs, (2) determining which animals were responsible for any differences in that structure, (3) quantifying density and species composition of FFGs on constructed and natural intertidal fringe reefs, and (4) elucidating carbon sources and trophic structure of the FFGs by comparing the $\delta^{13}C$ and $\delta^{15}N$ values from Dillon et al. (2015) of consumers within these feeding guilds. We argue that established constructed intertidal oyster reefs alone, although spatially small in area, provide some of the same ecosystem functions as other living shorelines designs (Currin et al. 2007; Georgia DNR 2013; Gittman et al. 2016; La Peyre et al. 2014a; Munsch et al. 2015; Scyphers et al. 2011).

19.2 METHODS

19.2.1 Site Description

The GBNERR is an ecologically valuable estuary on the eastern Mississippi Gulf Coast with a high diversity of animal and plant species (Peterson et al. 2007). The GBNERR is a primarily marine dominated ecosystem that has been retrograding since being abandoned by its natal river. Tides are seasonally diurnal or semidiurnal, microtidal (~0.5 m; Seim et al. 1987) in nature, and are primarily wind driven (Ennis et al. 2014). Freshwater input is mainly from precipitation, subsequent runoff, and perhaps groundwater seepage (Peterson et al. 2007).

Replicate sites in shallow water were selected based on nearby available oyster habitat, water flow, salinity, substrate, and slope suitable for natural seeding and development of self-sustaining reefs (Cake 1983; Stanley and Sellers 1986) coupled with the need for shoreline stabilization (Stricklin et al. 2010) in three sub-bayous of the GBNERR (30°23′N, 88°24′W): Bayou Cumbest, Crooked Bayou, and North Rigolets (see Figure 1 in Dillon et al. 2015). Bayou Cumbest is the farthest inland of the three sub-bayous with a well-consolidated clay and sand shore adjacent to a steep, upland erosional *Spartina alterniflora* (smooth cordgrass) and *Juncus roemerianus* (black needlerush) marsh edge. Crooked Bayou is the middle sub-bayou with a poorly consolidated muddy bottom and is adjacent to *S. alterniflora* marsh. North Rigolets is located between Point aux Chenes Bay and Middle Bay, composed of unconsolidated mud, adjacent to emergent *S. alterniflora* marsh grass. Mean salinity during the 22-month study at Bayou Cumbest, Crooked Bayou, and North Rigolets was 21.1, 24.2, and 25.1, respectively; however, large fluctuations over short periods are common (Dillon et al. 2015). For example, 1 month before this sampling, salinity at the GBNERR system-wide monitoring program station located in Bayou Cumbest dropped from a salinity of 20 to less than 2 and then rose to near 18 during the final sampling effort when salinity at all three sites was between 18.1 and 18.9.

Small, fringing intertidal Eastern oyster reefs in GBNERR are scattered and sporadic with a maximum coverage of 30%–35% (Dillon et al. 2015; Wieland 2007); these reef types are typical of the region (La Peyre et al. 2014a). In Mississippi Sound, Eastern oysters spawn in all months except December through February, with peak spawning between May and early June and again in September (Cake 1983). The density of live oysters within GBNERR reefs has not been completely evaluated but appears to be highly variable spatially and temporally (Dillon et al. 2015; Wieland 2007).

19.2.2 Reef Construction

Details of reef construction and associated habitat conditions can be found in Dillon et al. (2015). Briefly, each of three constructed Eastern oyster reefs was placed at least 92 m away from a natural reef within the intertidal zone of three sub-basins of the GBNERR: Bayou Cumbest (BC), Crooked Bayou (CB), and North Rigolets (NR) during August 2006. Each natural and constructed reef area (30.5 m × 1.8 m = 55.8 m²) was adjacent to salt marshes dominated by *J. roemerianus* with *S. alterniflora* along the tidal creek edge and divided into three equal sections. Each reef was gridded into equal smaller areas to facilitate random placement of 7 sample trays (dimensions: 48.3 cm L × 30.5 cm W × 11.4 cm H; area = 0.15 m²) within each section (21 trays total per reef), which were lined with 3.12 mm (0.125 in) window screen and filled with one bag (~1 ft³ or 0.028 m³) of recycled oyster shell. Trays were seated approximately 5–8 cm into the sediment within each reef section at low tide. Same-sized bags of shell were added to the remaining constructed reef area to simulate the 30%–35% coverage of the natural reefs (Dillon et al. 2015). After deployment, all shell bags were cut open to mimic natural habitat and there was no noticeable loss of shell material attributed to sinking. Daily collection times were between 9 a.m. and 2 p.m., which typically corresponded to mid- to high tide during the study (see Figure 2 in Dillon et al. 2015).

19.2.3 Sampling Procedures

Constructed and natural reefs were sampled for density and overall taxa richness (defined as number of species per tray) quarterly for 2 years starting in November 2006 and ending in June 2008 (22 months after construction). However, for this study, only samples from June 6, 2008, were used. Salinity, temperature (°C), and dissolved oxygen (mg L⁻¹) were measured during each sampling event with a YSI model 85 handheld meter at each site. Natural reefs were sampled by placing a bottomless plastic tray into the sediment at random locations within the three equal-sized sections. Live oysters, shells, and fauna down to 5 cm within each tray were removed with a small spade and dip nets and placed into buckets for transport to the laboratory. If water covered the sampling trays, a tall, bottomless wooden frame (53.34 cm × 36.83 cm) was pressed into the sediment before the sample collection to reduce sample material loss for both natural and constructed reefs. For constructed reefs, each sample tray was covered with cloth and then removed from each reef section (three trays per reef) and placed into a labeled plastic tray for transport to the laboratory. Only two trays were collected at the CB constructed site during June 2008 because of an apparent boat traffic disturbance before sampling. All samples were individually sieved through a 0.5-mm sieve, fixed in 10% formalin for 24 h, stained with Rose Bengal, and then sorted to the lowest possible taxa and enumerated in the laboratory.

For this study we pooled the three life stages of Eastern oyster (spat [<2.54 cm], seed [2.55–7.61 cm], and adults [>7.62 cm]) for analysis. Nematodes, Platyhelminthes, and Nemertea were identified to phylum, whereas annelids were identified to the level of family, and arthropods, mollusks, and fishes were identified to genus and species whenever possible. Keys used to identify the fauna were Heard (1982), Hoese and Moore (1998), Brusca and Brusca (2003), Abele and Kim (1986), Bousfield and Hoover (1997), Pennak (1978), Foster et al. (2004), and LeCroy (2000a,b, 2004, 2007). Taxonomy of some species follows Williams (1983) and Sarver et al. (1992).

We initially estimated density of all organisms on the oyster reefs on a per 0.15 m² area (area of the sampling tray) because of the extremely patchy and low-density nature of intertidal reefs in the GBNERR. The data, however, are provided as individuals per square meter basis so that comparisons can be easily made to other studies, but extrapolating our estimates in this system to a square meter basis should be approached cautiously since it could easily over- or underestimate these values.

Invertebrates and small fish samples were processed for analysis while subsamples of muscle tissue were collected from larger fish, such as *Opsanus beta* (Gulf toadfish) and *Ctenogobius boleosoma* (darter goby). Small individual organisms of the same taxa (i.e., amphipods, polycheates, and small crabs) for each collection were combined to achieve enough mass for stable isotope analysis and thus reflect population-level stable isotope values. All samples were rinsed thoroughly with deionized water and then oven dried (60°C) to a constant weight and ground to a fine powder with a Crescent Wig-L-Bug grinding mill. After grinding, all samples were acid washed (10% HCl) to remove carbonates and rinsed thrice with deionized water. After the final rinse, samples were dried at 60°C and then homogenized with a Crescent Wig-L-Bug grinding mill. All ground samples were placed into clean scintillation vials and stored in desiccators. Subsamples were packed in tin capsules and analyzed for $\delta^{13}C$ and $\delta^{15}N$ at the UC-Davis Stable Isotope Laboratory. Most (72%) of the $\delta^{13}C$ and $\delta^{15}N$ samples were analyzed in triplicate while some could only be analyzed in duplicate (5%) or a single analysis (23%) because of limited sample material.

19.2.4 Data Analysis

Faunal assemblage composition and density (# m^{-2}) were analyzed using PRIMER-E (version 6.1.6; Clarke 1993, Clarke and Gorley 2006) on fauna collected at the end of the study (using only June 6, 2008, data) based on habitat type (natural, constructed) in all three sub-bayous. Data were square root transformed to reduce the importance of the most abundant taxa. A Bray–Curtis similarity matrix was examined using nonmetric multidimensional scaling (NMDS) (Clarke 1993), which attempts to create groupings of samples through a generated similarity matrix. We also conducted a similarity profile (SIMPROF) test on a group-average cluster analysis to search for meaningful structure within clusters and we superimposed SIMPROF groups onto the ordination to search for mutual consistency between techniques (Clarke and Gorley 2006). NMDS stress is the measure of how well the three-dimensional plot is represented in a two-dimensional plot and thus the distances between the data points (Clarke 1993).

A two-way analysis of similarity (ANOSIM; Clarke 1993) was conducted using the Bray–Curtis index to test for differences of species density (# m^{-2}) of each assemblage by habitat and sub-bayou. The global *R* statistic and *p* value were used to determine differences. Overall and in a pairwise comparison of assemblages, *R* statistic values close to 1 indicate very different assemblages, whereas values close to 0 indicate very similar assemblages. A similarity of percentages (SIMPER) analysis was used to assess which species had the greatest contribution to any dissimilarity in assemblage structure among sub-bayous or habitat types. Levels of significance were set to $p < 0.05$ for all PRIMER analyses.

FFGs were assessed by first categorizing each identified taxa/species into groups based mainly on Fauchald and Jumars (1979) and Ranasinghe et al. (1993). All species were classified as grazers, deposit feeders, filter feeders, omnivores, or carnivores. Grazers are those organisms that feed on periphyton on submerged underwater surfaces; deposit feeders are organisms that feed on biodeposits; filter feeders filter plankton from the water column; omnivores are those generalist fauna that feed on dead and living organic matter; and carnivores are those organisms that feed on other organisms. We compared species richness (taxa/species m^{-2}) of each of the five FFGs separately with a two-way analysis of variance (ANOVA) with sub-bayou ($n = 3$) and habitat type ($n = 2$) as main effects. ANOVA assumptions of normality and homogeneity of variance were tested before analysis using a Kolmogorov–Smirnov one-sample test and Levene's test, respectively. If a significant *F* value was found, mean values were separated with a post hoc Sidak test. Significance was determined when $p \leq 0.05$, and all parametric tests were conducted with SPSS (ver. 20). To assess the composition of each FFG by sub-bayou and habitat type, the percent mean density of each feeding group (n = 5) was calculated and plotted.

19.3 RESULTS

19.3.1 Faunal Assemblages

The SIMPROF test indicated significant structure based on cluster analysis at the 40% and 60% similarity levels; these were superimposed onto the NMDS plot. However, the NMDS analysis revealed considerable overlap and no clear separation between habitat types or among sub-bayous, indicating similar faunal assemblages in natural and constructed oyster reefs within Bayou Cumbest, Crooked Bayou, and North Rigolets (stress = 0.07) by 22 months (Figure 19.1). Assembly structure (Table 19.1) differed by habitat type (R = 0.424, p = 0.005) but not by sub-bayou (R = 0.389, p = 0.133). Thus, a one-way SIMPER analysis by habitat type showed a strong similarity of assemblages on constructed reefs (similarity = 52.79%), while the natural reefs had a lower similarity of 39.92%, supporting the ordination patterns. Analysis between habitat types indicated that the assemblages were quite dissimilar (61.72%). The overall dissimilarity between constructed and natural oyster reefs was caused by the differences in density of spionid, syllid, and nereid polychaetes, the arthropods *Melita* spp. (*nitida* complex; Family Melitidae), *Apocorophium louisianum* (Family Corophiidae), *Panopeus simpsoni* (mud crab), Panopeid juveniles, *Hargeria rapax* (estuarine tanaid), the mollusk *Ischadium recurvum* (hooked mussel), *Cerithidea pliculosa* (plicate horn shell), and the Eastern oyster (Table 19.1).

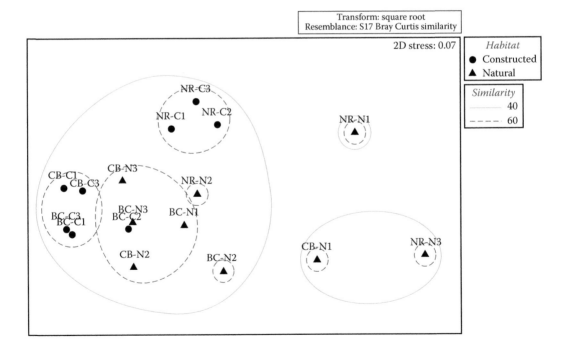

Figure 19.1 Nonmetric multidimensional scaling plot of intertidal reef faunal assemblage density (# m^{-2}), with each point labeled by sub-bayou (BC = Bayou Cumbest, CB = Crooked Bayou, NR = North Rigolets), habitat type (C = constructed, N = natural reefs), and replicate (1, 2, 3). For example, NR-N1 = North Rigolets, natural reef, replicate 1. The ordination uses a Bray–Curtis similarity index based on square root transformed data. Note that there are only two samples from the constructed reef in Crooked Bayou.

Table 19.1 Summary of the SIMPER Analysis of the Mean Density (# m⁻²) of Species after 22 Months Collected on Constructed and Natural Intertidal Oyster Reefs and the Percent Contribution to the Overall Dissimilarity

	Constructed Reefs Mean Density	Natural Reefs Mean Density	% Contribution
Spionidae	39.20	18.21	12.63
Melita spp. (*nitida* complex)	39.20	11.31	12.37
Crassostrea virginica	36.50	10.38	11.40
Apocorophium louisianum	13.91	10.48	5.16
Syllidae	13.89	8.88	5.04
Panopeus simpsoni	14.62	6.06	4.06
Panopeid juveniles	14.77	6.42	3.94
Ischadium recurvum	11.95	1.72	3.85
Nereidae	15.04	9.66	3.73
Hargeria rapax	6.40	7.75	3.27
Cerithidea pliculosa	12.39	6.66	3.24
Capitellidae	3.34	7.63	2.73
Nematoda	4.62	2.04	2.69
Grandidierella bonneroides	7.50	3.74	2.55
Crab megalopae	7.67	2.75	2.35
Glyceridae	0.32	4.39	2.07
Apocorophium lacustre	4.01	4.15	1.96
Tellidtrid amphipods	3.92	1.50	1.72
Balanus spp.	4.46	1.80	1.62
Platyhelminthes	2.73	0.98	1.43
Eunicidae	2.54	0.81	1.30
Geukensia granosissima	3.00	0.57	1.30
	52.79	39.92	90.40%

19.3.2 Species Richness and FFGs

There were significant sub-bayou ($F_{2,16} = 40.95$, $p < 0.0001$) and habitat type ($F_{1,16} = 94.60$, $p < 0.0001$) terms, but there was no habitat type × sub-bayou interaction ($F_{2,16} = 2.47$, $p = 0.130$) for species richness. Species richness was greater in Bayou Cumbest than in Crooked Bayou or North Rigolets (Sidak: $p < 0.05$), but Crooked Bayou and North Rigolets were not different (Sidak: $p > 0.05$). Species richness was always greater in constructed compared to natural sites (Sidak: $p < 0.05$) regardless of sub-bayou.

Analysis of FFGs revealed that the constructed reefs had a complexity comparable to natural reefs (see Table 19.2) found on each sub-bayou regardless of habitat type (Figure 19.2 and Table 19.3), though the predominant FFG and species richness were variable between habitat types. There were no significant interaction terms for any FFG analysis (all df = 2,16; all $p > 0.05$); thus, main effects could be directly discussed. There were no differences between natural and constructed reefs (ANOVA: all df = 1,16; all $p > 0.05$) or among sub-bayous (ANOVA: all df = 2,16; all $p > 0.05$) for deposit feeder, grazer, or carnivore FFG, but there were significant differences among sub-bayous for filter feeders (ANOVA: $F_{2,16} = 6.91$; $p = 0.011$) with Bayou Cumbest > North Rigolets (Sidak: $p = 0.012$; Table 19.3). For omnivores, only the constructed reef FFG species richness was greater than natural reef richness (ANOVA: $F_{1,16} = 5.30$; $p = 0.042$). However, in the 22-month assemblages on both natural and constructed reefs in Bayou Cumbest and Crooked Bayou, deposit feeders were the most abundant feeding group (31.98%–50.99%) followed by filter feeders (19.63%–40.91% composition), omnivores (14.20%–26.29%), and reduced percentages of carnivores and grazers (Figure 19.2).

Table 19.2   The Mean Density (# m⁻²) of Each Species Sampled for all Reefs by Reef Type (Constructed [C] vs. Natural [N]) and Their Assigned Functional Groups at the End of the 22-Month Study

Taxa	FFG	Bayou Cumbest		Crooked Bayou		North Rigolets	
		C	N	C	N	C	N
Nematoda	O	15.6	0	0	11.1	171.2	35.6
Platyhelminthes	C	4.4	4.4	0	0	31.1	8.9
Nemertea	C	0	2.2	0	0	2.2	2.2
Dorvilleidae	DF	0	13.3	0	0	2.2	2.2
Glyceridae	C	0	22.2	3.3	15.6	0	42.2
Nereidae	DF	322.4	170.6	166.7	66.7	213.4	166.7
Phyllodocidae	C	0	8.9	0	0	8.9	2.2
Pilargidae	C	0	0	0	0	0	2.2
Capitellidae	DF	24.5	75.6	0	11.1	24.4	166.7
Eunicidae	C	4.4	4.4	0	0	26.7	4.4
Ampharetidae	DF	8.9	0	10.0	0	0	0
Cirratulidae	DF	2.2	0	0	0	0	0
Sabellidae	FF	0	0	0	6.7	0	2.2
Spionidae	DF	4010.9	435.8	4395.5	1316.2	4.4	42.2
Syllidae	C	913.8	233.4	50.0	173.1	8.9	15.6
Assiminea succinea	G	33.3	6.7	6.7	2.2	2.2	0
Cerithidea pliculosa	G	239.9	106.7	170.1	60.0	86.7	57.8
Opisthobranchia	G	4.5	2.2	0	2.2	2.2	0
Amygdalum papyrium	FF	0	0	0	2.2	0	0
Geukensia granosissima	FF	8.9	0	13.3	2.2	15.6	2.2
Ischadium recurvum	FF	415.8	46.7	396.9	0	0	4.4
Crassostrea virginica	FF	1776.4	220.1	2027.7	66.7	738.1	233.4
Balanus spp.	FF	171.2	22.2	0	0	0	2.2
Hargeria rapax	DF	233.4	177.9	20.0	68.9	2.2	17.8
Gammarus mucronatus	O	0	0	30.1	13.3	0	0
Melita spp. (*nitida* complex)	O	4010.9	417.9	4395.5	387.1	4.4	11.8
Grandidierella bonneroides	FF	193.4	64.5	4395.5	387.1	4.4	11.8
Apocorophium louisianum	FF	882.7	389.1	60.0	95.6	8.9	60.0
Apocorophium lacustre	FF	115.6	80.0	0	11.1	0	2.2
Tallidtrid amphipods	DF	0	11.1	263.5	20.0	0	0
Palaemonetes pugio	DF	33.3	6.7	6.7	2.2	2.2	0
Alpheus heterochaelis	C	11.1	4.4	0	0	4.4	2.2
Uca spp.	DF	0	0	0	0	8.9	0
Callinectes sapidus	C	0	2.2	3.3	0	0	2.2
Panopeus simpsoni	C	411.5	31.1	193.4	137.8	133.4	8.9
Panopeid juveniles	C	478.2	106.7	186.8	46.7	86.7	62.0
Pachygrapsus transversus	O	0	0	3.3	0	0	0
Petrolisthes armatus	C	4.4	2.2	0	2.2	2.2	15.6
Megalopae	C	140.1	33.3	50.0	2.2	0	0
Chironomid	O	2.2	4.4	0	2.2	0	0
Ctenogobius boleosoma	C	0	2.2	0	0	0	2.2
Hypsoblennius ionthas	C	6.7	0	0	0	0	0
Gobiesox strumosus	C	2.2	0	0	0	0	0
Opsanus beta	C	13.3	0	0	0	0	2.2

Note: Each species was categorized as a deposit feeder (DF), grazer (G), filter feeder (FF), carnivore (C), or omnivore (O).

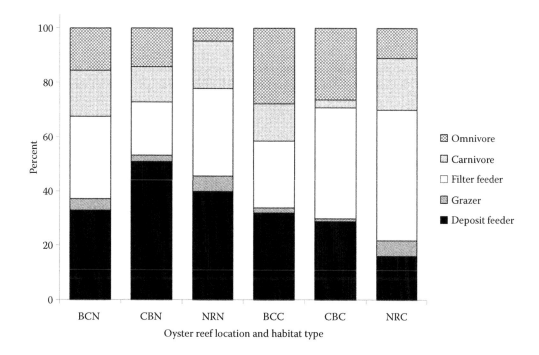

Figure 19.2 Comparison of composition of mean FFG density (# m⁻²) after 22 months in natural and constructed intertidal reefs of Bayou Cumbest (BCN, BCC), Crooked Bayou (CBN, CBC), and North Rigolets (NRN, NRC). N = natural reef, C = constructed reefs. Oyster reef location and habitat type are listed as BCN = Bayou Cumbest, natural reefs, and so on.

Table 19.3 Comparison of FFG Species Richness (Species m⁻²) in Constructed (C) versus Natural (N) Reefs by Sub-bayou after 22 Months

Bayou	Habitat Comparison	Deposit Feeder	Grazer	Filter Feeder	Carnivore	Omnivore
Bayou Cumbest (BC)	C:N	nd	nd	nd	nd	C > N
Crooked Bayou (CB)	C:N	nd	nd	nd	nd	C > N
North Rigolets (NR)	C:N	nd	nd	nd	nd	C > N
Bayou comparison		nd	nd	BC > NR $p = 0.01$	nd	nd

Note: nd, No difference.

The reef assemblages of both habitat types in North Rigolets were numerically dominated by filter feeders (32.09%–48.06% composition), followed by deposit feeders (16.15%–39.87%) and carnivores (17.44%–19.08%), followed by reduced percentages of omnivores and grazers (Figure 19.2).

19.4 DISCUSSION

The primary consideration of efforts to restore oyster reefs as living shoreline should be the long-term self-sustainability of the entire community and not just the oysters themselves, as a major goal of restoration is to develop a functional habitat type where one did not exist or to rehabilitate a degraded habitat type (Simenstad et al. 2006). Much of oyster reef restoration success focuses on the harvested component, but a number of recent papers have addressed the importance of abundances of crab, fish, and other shellfish as metrics of success (Beck et al. 2011; Carroll et al. 2015; Humphreys et al. 2011a; La Peyre et al. 2014a; Luckenbach et al. 2005; Walters and Coen 2006). While these are a few of the multiple possible metrics of restored reef success, and these metrics are typically goal oriented, the primary goal of this study was to assess constructed inter-tidal fringe oyster reef community success in the GBNERR compared to natural fringe reefs using species richness and density, assemblage composition, FFG analysis, and stable isotopic signatures as important metrics. Our results demonstrate that intertidal fringe oyster reefs can be constructed to an extent such that some vital ecosystem services (e.g., density and erosion capabilities, etc.) are at least equivalent to adjacent patchy natural reefs within 22 months (Stricklin et al. 2010). The greater structural complexity made immediately available in constructed reefs provided a heterogeneous habitat that created habitat refuge openings (*sensu* Humphreys et al. 2011b) and attracted fauna. As a result, the constructed intertidal reefs at the end of the 22-month period had similar or increased species richness relative to the natural reefs and higher density of at least 18 major sedentary reef species (polychaetes, arthropods, and mollusks) compared to adjacent natural reefs regardless of where in the landscape they were located. Moreover, constructed reefs had higher densities of all developmental stages of oysters as well (Dillon et al. 2015). Furthermore, the higher density of larval and juvenile crabs on constructed versus natural reefs indicates that the constructed reef functions as a nursery. Similar comparative studies of intertidal reefs in Florida (Boudreaux et al. 2006; Glancy et al. 2003; Tolley and Volety 2005) and Louisiana (La Peyre et al. 2014a; Plunket and La Peyre 2005) also found that more physically complex habitat supported a greater richness and density of fauna compared to degraded or otherwise less complex habitat. However, Carroll et al. (2015) recently noted the role of landscape setting on predation pressure associated with intertidal oyster reefs. The collection of similar species suggests trophic structures that are similar not only between reef types but also among sub-bayous in the GBNERR after 22 months, as determined with stable carbon and nitrogen isotopes at the same study sites (Dillon et al. 2015).

Partitioning organisms into FFGs allows ecosystem function(s) to be evaluated at different locations or periods that may or may not have the same species present; however, it is unclear whether organisms categorized in the same FFG do indeed serve the same ecosystem functions or utilize the same resources. To evaluate whether this is the case at the GBNERR sites (where numerous organisms were collected from three sites; see Dillon et al. 2015), we have grouped organisms from intertidal patchy oyster reefs into the four FFGs presented herein (no grazers were analyzed) and compared their $\delta^{13}C$ and $\delta^{15}N$ values to determine if the stable isotope values follow the functional categorizations. Obviously, some FFGs, such as carnivores and omnivores, would be expected to envelop various taxa with different stable isotope signatures since feeding strate-gies, trophic position, and preferred prey (among other things) are likely to vary. For example, *P. simpsoni* and *O. beta* are both classified as carnivores and have similar $\delta^{13}C$ values; however, their trophic positions revealed by $\delta^{15}N$ values were distinct (Figure 19.3a through c). Interestingly, the amphipods and crabs from GBNERR that were categorized as omnivores (*Melita*, *Gammarus*, and *Pachygrapsus transversus*) all had $\delta^{13}C$ and $\delta^{15}N$ values that were similar to one another, indicating similar diets.

Organisms in other FFGs such as filter feeders and deposit feeders may be expected to have similar isotopic values to other organisms within their FFG because of their less selective feeding strategies. When one compares the similar isotopic values of the Eastern oyster to the southern ribbed mussel (*Geukensia granosissima*) and hooked mussel (*I. recurvum*), it is clear that they are using similar prey items; however, other smaller filter feeders such as *Grandidierella* (Amphipoda) and feather duster worms (family Sabellidae) have more enriched $\delta^{13}C$ values, indicating that their diets are quite different from their bivalve counterparts. The green porcelain crab (*Petrolisthes armatus*) is categorized as a filter feeder, yet its $\delta^{13}C$ value is intermediate between the bivalves and small filter feeders noted above. These results show that not all filter feeders are utilizing the same dietary items, likely because of their different feeding morphologies and strategies. Oysters use cilia

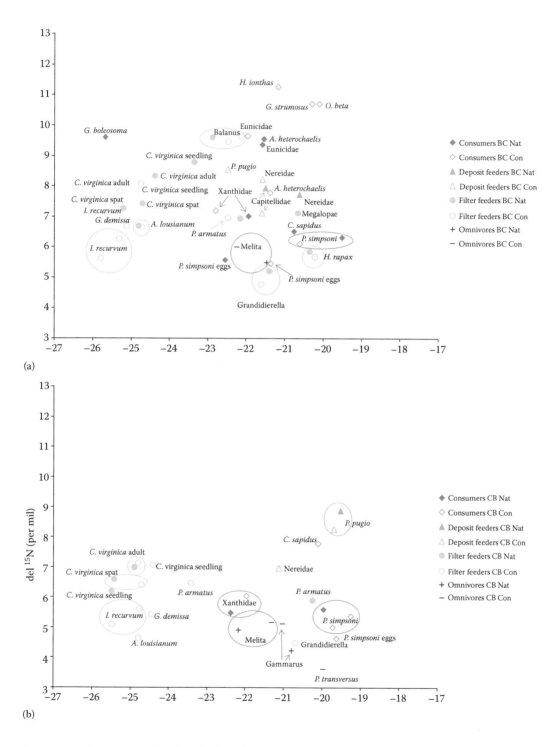

Figure 19.3 Comparison of δ13C and δ15N values of functional group categories after 22 months in natural (Nat) and constructed (Con) intertidal reefs of (a) Bayou Cumbest (BC), (b) Crooked Bayou (CB). Circles and ovals represent replicates for each species or group within a panel and are color coded to the legend. *(Continued)*

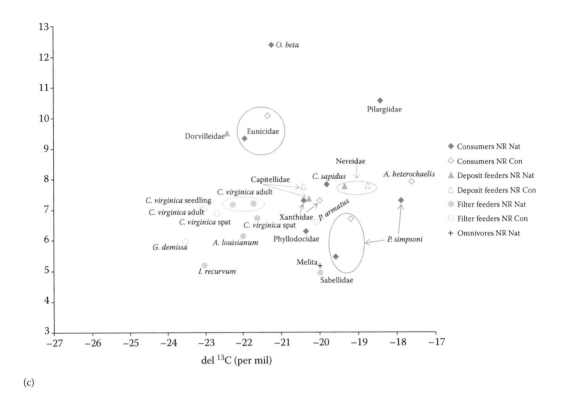

(c)

Figure 19.3 (Continued) Comparison of δ¹³C and δ¹⁵N values of functional group categories after 22 months in natural (Nat) and constructed (Con) intertidal reefs of (c) North Rigolets (NR). Circles and ovals represent replicates for each species or group within a panel and are color coded to the legend.

to push water over their gills, which capture bulk particles in the gill mucus that are then passed to the mouth, while feather duster worms protrude their feeding appendages into the water column and some selection is made at the mouth as to which particles are ingested (Cake 1983; Fauchald and Jumars 1979; Pagliosa 2005; Stanley and Sellers 1986). These two filter feeding strategies appear to select for different particle types with different carbon isotope values, effectively partitioning the available suspended organic particles. Like the omnivores, deposit feeders from these sites were much less versatile in their feeding based on their stable isotope values.

Analysis of FFG patterns showed that several ecosystem services appear enhanced on constructed as compared to natural intertidal reefs. There was an abundance of filter feeders, carnivores, and deposit feeders, with all FFGs present on both natural and constructed reefs. The relative importance of filter feeders on all reef types indicates a potential for increased water filtration and subsequent increased energy transfer from the plankton to the benthos as well as seston removal (Grizzle et al. 2008; La Peyre et al. 2014a,b; Newell 2004; Zu Ermgassen et al. 2013). Constructed oyster reefs appear to facilitate an overall energy transfer to higher trophic levels as deposit feeders consume pseudofeces and feces from filter feeders, and carnivores prey on deposit feeders (Dillon et al. 2015). By attracting and sustaining fauna of lower trophic levels for long periods, the constructed reefs are supporting important prey (e.g., polychaetes, small mollusks, amphipods, xanthid crab juveniles; Dittel et al. 1996) for resident (e.g., gobies, blennies, skilletfish, and toadfish) and transient fishes (e.g., spotted seatrout, spot, Atlantic croaker, and silversides; Breitburg 1999; Kingsley-Smith et al. 2012), as well as larger invertebrates (e.g., mud crabs; Day and Lawton 1988), with xanthid crabs being successful mesopredators on oyster refs (Carroll et al. 2015). Constructed oyster reefs increase fish habitat, provide foraging grounds, and increase the complexity of the trophic structure (Dillon et al. 2015; Hadley et al. 2010; Plunket and La Peyre 2005), thus building sustainable oyster reef assemblages that contribute to living shorelines.

Highly complex environments are common in marine ecosystems. Biogenic reefs are some of the most complex habitats with rich macrofauna assemblages using the reef structure for a variety of reasons. By the end of our study, the elevated faunal richness and density, the presence of all FFGs on both types of reefs, and corresponding stable isotope patterns suggest that the constructed reefs were functionally equivalent to other patchy natural reefs (Dillon et al. 2015). The elevated density of species found on constructed reefs demonstrates that these reefs are capable of surpassing the natural reefs' capacity to provide the same functions and ecosystem services regardless of location within the estuary. Constructed intertidal oyster reefs appear to enhance the number of benthic–pelagic functions of the estuarine ecosystem of the GBNERR, as illustrated by the high richness, density of lower trophic level organisms, and trophic networks (Dillon et al. 2015).

19.5 CONCLUSION

Our results suggest that 22-month-old constructed small, fringe intertidal oyster reefs benefit the GBNERR retrograding deltaic ecosystem as quantified with multiple ecosystem services: (1) although large variation exists in marsh edge growth, the marsh edge adjacent to constructed reefs was less eroded (mean = 0.043 m) than edges adjacent to natural reefs (mean = 0.728 m) regardless of bayou (see our earlier work in Stricklin et al. 2010); (2) constructed reefs developed higher oyster density, similar or greater pooled species richness, and higher phyletic abundance than their natural counterparts (food web structure between paired constructed and natural reefs was not significantly different from their natural counterparts; Dillon et al. 2015); and (3) constructed reefs exhibited greater mean infauna and sedentary macrofauna richness and density than nearby natural reefs. Only a significant habitat effect (ANOSIM: $R = 0.424$, $p = 0.005$) was noted, with assemblages of constructed reefs being greater in density than those of natural reefs. SIMPER analysis indicated that replicate community structures of constructed reefs were more similar to each other (similarity = 52.79%) than those in natural reefs (39.92%), supporting the spread in the NMDS ordination patterns and the associated stable isotope values. Finally, FFG analysis indicated that mean density of deposit feeders, grazers, and carnivores did not differ among sub-bayou or habitat type after the 22-month period, but filter feeder mean density was greater in Bayou Cumbest than in North Rigolets. Omnivore mean density was greater on constructed reefs compared to natural reefs in all bayou sites.

While the intertidal oyster reefs we constructed provide valuable habitat and ecosystem services, this project also required a commitment of resources to prepare cultch material and deploy it to the field sites, a valuable aspect of living shoreline opportunities. We used volunteer labor over several days to bag cultch to make it manageable for transport and deployment using shallow draft boats to access the sites and place cultch material. Volunteer engagement was critical to project success. Projects that engage volunteers in hands-on activities provide opportunities for outreach and education, and they instill a sense of ownership and appreciation for the natural environment. Thus, construction of living shorelines like oyster reefs in combination with other coastal habitats (e.g., seagrass beds and salt marsh) can provide living shorelines that are healthy, sustainable, and resilient to future disturbances and actively involve society in these endeavors.

Overall, this study demonstrated that intertidal oyster reef restoration/enhancement has both short-term and long-term benefits for estuarine ecosystems as noted above and that this type of living shoreline approach compared to shoreline hardening (e.g., bulkheads and riprap) appears to be a valuable method for supplementing/restoring reef habitat and its associated fauna and ecosystem services. Additionally, given the reality of continued urbanization in the coastal zone, this more natural approach not only buffers against continued shoreline erosion but also further provides a number of other ecosystem services, all of which have clear management implications to better maintain, enhance, and restore sustainable estuarine ecosystems.

ACKNOWLEDGMENTS

We thank the Coastal Program of The Nature Conservancy in Alabama (N. Vickey and M.A. Lott) and the GBNERR (D. Ruple and M. Woodrey) for partial financial support of this project. Alex G. Stricklin, J.D. Lopez, P. Grammer, E. Lang, and J. McIlwain supported logistics and laboratory and field work. Drs. R. Heard and J. McClelland discussed FFG categorization by species and Drs. R. Heard and B. Thoma assisted with taxonomic correctness; their assistance is greatly appreciated. This project was conducted under USM-IACUC # 06071301.

REFERENCES

Abeels, H.A., A.N. Loh, and A.K. Volety. 2012. Trophic transfer and habitat use of oyster *Crassostrea virginica* reefs in southwest Florida, identified by stable isotope analysis. *Marine Ecology Progress Series* 462: 125–142.

Abele, L.G. and W. Kim. 1986. *An Illustrated Guide to the Marine Decapod Crustaceans of Florida.* Florida Department of Environmental Protection, Technical Series 8(1) part 2: 327–760.

Alves, A.S., A. Caetano, J.L. Costa, M.J. Costa, and J.C. Marques. 2015. Estuarine intertidal meiofauna and nematode communities as indicator of ecosystem's recovery following mitigation measures. *Ecological Indicators* 54: 184–196.

Angelini, C., A.H. Altieri, B.R. Silliman, and M.D. Bertness. 2011. Interactions among foundation species and their consequences for community organization, biodiversity, and conservation. *BioScience* 61: 782–789.

Beck, M.W., R.D. Brumbaugh, L. Airoldi et al. 2011. Oyster reefs at risk and recommendations for conservation, restoration and management. *BioScience* 61: 107–116.

Borsje, B.W., B.K. VanWesenbeeck, F. Dekker et al. 2011. How ecological engineering can serve in coastal protection? *Ecological Engineering* 37: 115–135.

Boudreaux, M.L., J.L. Stiner, and L.J. Walters. 2006. Biodiversity of sessile and motile macrofauna on intertidal oyster reefs in Mosquito Lagoon, Florida. *Journal of Shellfish Research* 25: 1079–1089.

Bousfield, E.J. and P.M. Hoover. 1997. The amphipod superfamily Corophioidea on the Pacific coast of North America. Part V. Family Corophiidae: Corophiinae, new subfamily, systematics and distributional ecology. *Amphipacifica* II(3): 67–139.

Bracken, M.E.S., B.E. Bracken, and L. Rogers-Bennett. 2007. Species diversity and foundation species: Potential indicators of fisheries yields and marine ecosystem functioning. *CalCOFI Report* 48: 82–91.

Breitburg, D.L. 1999. Are three-dimensional structure and healthy oyster populations the keys to an ecologically interesting and important fish community? In: Luckenbach, M.W., R. Mann, and J.A. Wesson (Eds.), *Oyster Reef Habitat Restoration: A Synopsis and Synthesis of Approaches*, pp. 239–250. Gloucester Point, VA: Virginia Institute of Marine Science.

Bruno, J.F., J.J. Stachowicz, and M.D. Bertness. 2003. Inclusion of facilitation into ecological theory. *Trends in Ecology and Evolution* 18: 119–125.

Brusca, R.C. and G.J. Brusca. 2003. *Invertebrates.* 2nd ed., Sunderland, MA: Sinauer Associates, Inc.

Cake, Jr., E.W. 1983. Habitat suitability index models: Gulf of Mexico American Oyster. Washington D.C.: U.S. Department of the Interior, Fish and Wildlife Service, FWS/OBS-82/10.57.

Carroll, J.M., J.P. Marion, and C.M. Finelli. 2015. A field test of the effects of mesopredators and landscape setting on juvenile oyster, *Crassostrea virginica*, consumption on intertidal reefs. *Marine Biology* 162: 993–1003.

Clarke, K.R. 1993. Non-parametric multivariate analyses of changes in community structure. *Australian Journal of Ecology* 18: 117–143.

Clarke, K.R. and R.N. Gorley. 2006. *PRIMER v6: User manual/tutorial*, Plymouth, UK: PRIMER-E.

Coen, L.D., R.D. Brumbaugh, D. Bushek et al. 2007. Ecosystem services related to oyster restoration. *Marine Ecology Progress Series* 341: 303–307.

Currin, C.A., P.C. Delano, and L.M. Valdes-Weaver. 2007. Utilization of a citizen monitoring protocol to assess the structure and function of natural and stabilized fringing salt marshes in North Carolina. *Wetlands Ecology and Management* 16: 97–118.

Day, E.A. and P. Lawton. 1988. Mud crab (Crustacea: Brachyura: Xanthidae) substrate preference and activity. *Journal of Shellfish Research* 7: 421–426.

Dillon, K.S., M.S. Peterson, and C.A. May. 2015. Functional equivalency of constructed and natural intertidal Eastern oyster reef habitats in a northern Gulf of Mexico estuary. *Marine Ecology Progress Series* 528: 187–203.

Dittel, A., C.E. Epifanio, and C. Natunewicz. 1996. Predation on mud crab megalopae, *Panopeus herbstii* H. Milne Edwards: Effects of habitat complexity, predator species and postlarval densities. *Journal of Experimental Marine Biology and Ecology* 198: 191–202.

Ennis, B., M.S. Peterson, and T.P. Strange. 2014. Modeling of inundation characteristics of a microtidal salt marsh, Grand Bay National Research Reserve, Mississippi. *Journal of Coastal Research* 30(3): 635–646.

Fauchald, K. and P.A. Jumars. 1979. The diet of works: A study of polychaete feeding guilds. *Oceanography and Marine Biology: Annual Reviews* 17: 193–284.

Foster, J.M., B.P. Thoma, and R.W. Heard. 2004. Range extensions and review of the Caprellid amphipods (Crustacea: Amphipods: Caprellidae) from the shallow, coastal waters from the Suwannee River, FL to Port Aransas, TX, with an illustrated key. *Gulf Research Reports* 16(2): 161–175.

Gedan, K.B., L. Kellogg, and D.L. Breitburg. 2014. Accounting for multiple foundation species in oyster reef restoration benefits. *Restoration Ecology* 22(4): 517–524.

Georgia DNR (Georgia Department of Natural Resources). 2013. Living Shorelines along the Georgia Coast: A Summary Report of the First Living Shoreline Projects in Georgia. Coastal Resources Division, Brunswick, GA. 43 pp. plus appendix.

Gittman, R.K., C.H. Peterson, C.A. Currin, F.J. Fodrie, M.F. Piehler, and J.F. Bruno. 2016. Living shorelines can enhance the nursery role of threatened estuarine habitats. *Ecological Applications* 26: 249–263. (http://dx.doi.org/10.1890/14-0716.1).

Glancy, T.P., T.K. Fraser, C.E. Cichra, and W.J. Lindberg. 2003. Comparative patterns of occupancy by decapod crustaceans in seagrass, oyster, and marsh-edge habitats in a northeast Gulf of Mexico estuary. *Estuaries* 26: 1291–1301.

Grabowski, J.H., R.D. Brumbaugh, E.F. Conrad et al. 2012. Economic valuation of ecosystem services provided by oyster reefs. *Bioscience* 52: 900–909.

Grizzle, R.E., J.K. Greene, and L.D. Coen. 2008. Seston removal by natural and constructed intertidal eastern oyster (*Crassostrea virginica*) reefs: A comparison with previous laboratory studies, and the value of *in situ* methods. *Estuaries and Coasts* 31: 1208–1220.

Hadley, N.H., M. Hodges, D.H. Wilber, and L.D. Coen. 2010. Evaluating intertidal oyster reef development in South Carolina using associated faunal indicators. *Restoration Ecology* 18: 691–701.

Harding, J.M. and R. Mann. 1999. Fish species richness in relation to restored oyster reefs, Piankatank River, Virginia. *Bulletin of Marine Science* 65: 289–300.

Heard, R.W. 1982. *Guide to Common Tidal Marsh Invertebrates of the Northeastern Gulf of Mexico*. Ocean Springs, MS: Mississippi–Alabama Sea Grant College, MASGC-79-004.

Hoese, H.D. and R.H. Moore. 1998. *Fishes of the Gulf of Mexico. Texas, Louisiana and Adjacent Waters*. 2nd ed., College Station, TX: Texas A&M University Press.

Humphreys, A.T., M. La Peyre, M.E. Kimball, and L.P. Rozas. 2011a. Testing the effect of habitat structure and complexity on nekton assemblages using experimental oyster reefs. *Journal of Experimental Marine Biology and Ecology* 409: 172–179.

Humphreys, A.T., M. La Peyre, and G.A. Decossas. 2011b. The effect of structural complexity, prey density, and "predator-free space" on prey survivorship at created oyster reef mesocosms. *PLoS ONE* 6(12): e28339.

Jones, C.G., J.H. Lawton, and M. Shachak. 1994. Organisms as ecosystems engineers. *Oikos* 69: 373–386.

Kingsley-Smith, P.R., R.E. Joyce, S.A. Arnott, W.A. Roumillat, C.J. NcDonough, and M.J.M. Reichert. 2012. Habitat use of intertidal eastern oyster (*Crassostrea virginica*) reefs by nekton in South Carolina estuaries. *Journal of Shellfish Research* 31(4): 1009–1021.

Kürten, N., I. Frutos, U. Struck et al. 2013. Trophdynamics and functional feeding groups of North Sea fauna: A combined stable isotope and fatty acid approach. *Biogeochemistry* 113: 189–212.

La Peyre, M.K., A.T. Humphries, S.M. Casas, and J.F. La Payre. 2014a. Temporal variation in development of ecosystem services from oyster reef restoration. *Ecological Engineering* 63: 34–44.

La Peyre, M.K., J. Furlong, L.A. Brown, B.P. Piazza, and K. Brown. 2014b. Oyster reef restoration in the northern Gulf of Mexico: Extent, methods and outcomes. *Ocean & Coastal Management* 89: 20–28.

LeCroy, S.E. 2000a. An Illustrated Identification Guide to the Nearshore Marine and Estuarine Amphipoda of Florida, Volume 1—Families Gammaridae, Hadiidae, Isaeidae, Melitidae, and Oedicerotidae, 1–195. Tallahassee, FL: Florida Department of Environmental Protection.

LeCroy, S.E. 2000b. An Illustrated Identification Guide to the Nearshore Marine and Estuarine Gammaridean Amphipoda of Florida, Volume 2—Families Ampeliscidae, Amphilochidae, Ampithoidae, Aoridae, Argissidae, and Haustoriidae, 196–410. Tallahassee, FL: Florida Department of Environmental Protection.

LeCroy, S.E. 2004. An Illustrated Identification Guide to the Nearshore Marine and Estuarine Gammaridean Amphipoda of Florida, Volume 3—Families Bateidae, Biancolinidae, Cheluridae, Colomastigidae, Corophiidae, Cyproideidae, and Dexaminidae, 411–501.

LeCroy, S.E. 2007. An Illustrated Identification Guide to the Nearshore Marine and Estuarine Gammaridean Amphipoda of Florida, Volume 4—Families Anamixidae, Eusiridae, Hyalellidae, Hyalidae, Iphimediidae, Ischyroceridae, Kysianassidae, Megaluropidae, and Melphippidae, 503–612. Tallahassee, FL: Florida Department of Environmental Protection.

Lenihan, H.S., C.H. Peterson, J.E. Byers, J.H. Grabowski, G.W. Thayer, and D.R. Colby. 2001. Cascading of habitat degradation: Oyster reefs invaded by refugee fishes escaping stress. *Ecological Applications* 11: 764–782.

Luckenbach, M.W., L.D. Coen, P.G. Ross, Jr., and J.A. Stephen. 2005. Oyster reef habitat restoration: Relationships between oyster abundance and community development based on two studies in Virginia and South Carolina. *Journal of Coastal Research* SI40:64–78.

Meyer, D.L., E.C. Townsend, and G.W. Thayer. 1997. Stabilization and erosion control value of oyster cultch for intertidal marsh. *Restoration Ecology* 5: 93–99.

Meyer, D.L. and E.C. Townsend. 2000. Faunal utilization of created intertidal eastern oyster (*Crassostrea virginica*) reefs in the southeastern United States. *Estuaries* 23: 34–45.

Micheli, F. and C.H. Peterson. 1999. Estuarine vegetated habitats as corridors for predator movements. *Conservation Biology* 13: 869–881.

Munsch, S.H., J.R. Cordell, and J.D. Toft. 2015. Effects of shoreline engineering on shallow subtidal fish and crab communities in an urban estuary: A comparison of armored shorelines and nourished beaches. *Ecological Engineering* 81: 312–320.

Needles, L.A., S.E. Lester, R. Ambrose et al. 2015. Managing bay and estuarine ecosystems for multiple services. *Estuaries and Coasts* 38 (Supplemental 1):S35–S48.

Newell, R.I.E. 2004. Ecosystem influences of natural and cultivated populations of suspension-feeding bivalve mollusks: A review. *Journal of Shellfish Research* 23: 51–61.

Pagliosa, P.R. 2005. Another diet of worms: The applicability of polychaete feeding guilds as a useful conceptual framework and biological variable. *Marine Ecology* 26: 246–254.

Pennak, R.W. 1978. *Freshwater Invertebrates of the United States*, 2nd ed., Chapter 35: Diptera (Flies, Mosquitoes, Midges). New York: John Wiley & Sons.

Peterson, C.H., J.H. Grabowski, and S.P. Powers. 2003. Estimated enhancement of fish production resulting from restoring oyster reef habitat: Quantitative valuation. *Marine Ecology Progress Series* 264: 251–256.

Peterson, M.S., G.L. Waggy, and M.S. Woodrey (Eds.). 2007. *Grand Bay National Estuarine Research Reserve: An Ecological Characterization*. Moss Point, MS: Grand Bay National Estuarine Research Reserve.

Piazza, B.P., P.D. Banks, and M.K. La Peyre. 2005. The potential for created oyster shell reefs as a sustainable shoreline protection strategy in Louisiana. *Restoration Ecology* 13(3): 499–506.

Plunket, J. and M.K. La Peyre. 2005. Oyster beds as fish and macroinvertebrate habitat in Barataria Bay, Louisiana. *Bulletin of Marine Science* 77: 155–164.

Ranasinghe, J.A., S.B. Wiesberg, D.M. Dauer, L.C. Schaffner, R.J. Diaz, and J.B. Frithsen. 1993. Chesapeake Bay benthic community restoration goals. Annapolis, MD: United States Environmental Protection Agency Chesapeake Bay Program. CBP/TRS, 107/94.

Rodney, W.S. and K.T. Paynter. 2006. Comparisons of macrofaunal assemblages on restored and non-restored oyster reefs in mesohaline regions of Chesapeake Bay in Maryland. *Journal of Experimental Marine Biology and Ecology* 335: 39–51.

Sanger, D., A. Blair, G. DiDonato et al. 2015. Impacts of coastal development on the ecology of tidal creek ecosystems of the US southeast including consequences to humans. *Estuaries and Coasts* 38(1): 49–66.

Sarver, S.K., M.C. Landrum, and D.W. Foltz. 1992. Genetics and taxonomy of ribbed mussels (*Geukensia* spp.). *Marine Biology* 113: 385–390.

Scyphers, S.B., S.P. Powers, and K.L. Heck Jr. 2015. Ecological value of submerged breakwaters for habitat enhancement on a residential scale. *Environmental Management* 55: 383–391.

Scyphers, S.B., S.P. Powers, K.L. Heck, Jr., and D. Byron. 2011. Oyster reefs as natural breakwaters mitigate shoreline loss and facilitate fisheries. *PLoS ONE* 6: 1–12.

Seim, H.E., B. Kjerfve, and J.E. Sneed. 1987. Tides of Mississippi Sound and the adjacent continental shelf. *Estuarine, Coastal, and Shelf Science* 25: 143–156.

Simenstad, C., D.J. Reed, and M. Ford. 2006. When is restoration not? Incorporating landscape-scale processes to restore self-sustaining ecosystems in coastal wetland restoration. *Ecological Engineering* 26(1): 27–39.

Stanley, J.G. and M.A. Sellers. 1986. Species profiles: Life histories and environmental requirements of coastal fishes and invertebrates (Gulf of Mexico)—American oyster. U.S. Army Corps of Engineers, Vicksburg, MS: U.S. Fish Wildlife Service Biological Report 82(11.64), TR EL-82-4.

Stork, N. 2010. Re-assessing current extinction rates. *Biodiversity and Conservation* 19: 357–371.

Stricklin, A.G., M.S. Peterson, J.D. Lopez, C.A. May, C.F. Mohrman, and M.S. Woodrey. 2010. Do small, patchy constructed intertidal oyster reefs reduce salt marsh erosion as well as natural reefs? *Gulf and Caribbean Research* 22: 21–27.

Stunz, G.W., T.J. Minello, and L.P. Rozas. 2010. Relative value of oyster reef as habitat for estuarine nekton in Galveston Bay, Texas. *Marine Ecology Progress Series* 406: 147–159.

Temmerman, S., P. Meire, T.J. Bouma, P.M.J. Herman, T. Ysebaert, and H.J. De Vriend. 2013. Ecosystem-based coastal defense in the face of global change. *Nature* 504: 79–83.

Tolley, S.G. and A.K. Volety. 2005. The role of oysters in habitat use of oyster reefs by resident fishes and decapod crustaceans. *Journal of Shellfish Research* 24: 1007–1012.

Verdell-Cubedo, D., M. Torralva, A. Andreu-Soler, and F.J. Oiva-Paterna. 2012. Effects of shoreline urban modification on habitat structure and fish community in littoral areas of a Mediterranean coastal lagoon (Mar Menor, Spain). *Wetlands* 32: 631–641.

Walters, K. and L.D. Coen. 2006. A comparison of statistical approaches to analyzing community convergence between natural and constructed oyster reefs. *Journal of Experimental Marine Biology and Ecology* 330: 81–95.

Wieland, R.G. 2007. Habitat types and associated ecological communities of the Grand Bay National Estuarine Research Reserve. In: Peterson, M.S., G.L. Waggy, and M.S. Woodrey (Eds.), *Grand Bay National Estuarine Research Reserve: An Ecological Characterization*, pp. 103–174. Moss Point, MS: Grand Bay National Estuarine Research Reserve.

Williams, A.B. 1983. The mud crab, *Penopeus herbstii*, S. L. partition into six species (Decapoda: Xanthidae). *Fishery Bulletin* 81: 863–882.

Worm, B., E.B. Barbier, N. Beaumont et al. 2006. Impacts of biodiversity loss on ocean ecosystem services. *Science* 314: 787–790.

Yakovis, E.L., A.V. Artemieva, N.N. Shunatova, and M.A. Varfolomeeva. 2008. Multiple foundation species shape benthic habitat islands. *Oecologia* 155: 785–795.

Zu Ermgassen, P.S.E., M.D. Spalding, R. Grizzle, and B.D. Brumbaugh. 2013. Quantifying the loss of a marine ecosystem service: Filtration by the Eastern oyster in US estuaries. *Estuaries and Coasts* 36: 36–43.

CHAPTER 20

Ecosystem Services Provided by Shoreline Reefs in the Gulf of Mexico
An Experimental Assessment Using Live Oysters

Kenneth L. Heck, Jr., Just Cebrian, Sean P. Powers, Nate Geraldi,
Rochelle Plutchak, Dorothy Byron, and Kelly Major

CONTENTS

20.1 INTRODUCTION

In addition to playing their traditional role as producers of a valuable shellfish resource (Breitburg 1999; Coen and Luckenbach 2000; Coen et al. 1999), oyster reefs are increasingly recognized for the many valuable ecosystem services they provide. As discussed by Lenihan and Peterson (1998) and Dame et al. (2000), some of the most important services provided by oyster reefs are shoreline protection, the provision of three-dimensional structures that shelter diverse assemblages of invertebrates and small fishes (Bahr and Lanier 1981; Breitburg 1999; Coen et al. 1999; Harding and Mann 1999; Lehnert and Allen 2002; Lenihan et al. 2001; Zimmerman et al. 1989), and the filtration of large amounts of particulate material from the water column, with potential large-scale effects on phytoplankton assemblages, nutrient dynamics, and sediment biogeochemistry (Dame 1993, 1996; Newell 1988). Nearly 30 years ago, Newell (1988) suggested that intense oyster harvesting in Chesapeake Bay could have increased the Bay's susceptibility to the harmful effects of eutrophication because of the reduction in the abundance of suspension feeders. The formulation of this hypothesis was preceded by evidence that large populations of clams could greatly reduce phytoplankton abundance and thereby increase water clarity in San Francisco Bay (Cloern 1982; Officer et al. 1982) and the Potomac River (Cohen et al. 1984). If similar reductions in algal abundance occurred when oyster densities were restored to former levels, improved water clarity in estuarine systems could lead to enhanced productivity by benthic plants and a system-wide shift in the balance from pelagic to benthic productivity (Baird and Ulanowicz 1989; Newell 1988; Newell et al. 2002).

Despite the increasing recognition of the ecological benefits of oyster reefs, only a few empirical studies quantifying these benefits have been conducted and published. Several investigators have used models (Grabowski and Peterson 2007; Pomeroy et al. 2006; Ulanowicz and Tuttle 1992 [but see responses by Cerco and Noel 2007; Coen et al. 2007; Newell et al. 2007]) or extrapolations from localized, short-term studies (Dame et al. 1984; Porter et al. 2004) to predict the potential effects of oyster removal or enhancement on coastal ecosystems. To date, however, we are aware of only two experimental field tests of the effects of oyster removal or addition on the structure and function of coastal ecosystems. The first was a removal experiment carried out in intertidal reefs in South Carolina marsh complexes (Allen et al. 2007; Dame et al. 2000, 2002). There, the removal of oysters was not clearly related to either changes in water clarity or use of the creek by mobile invertebrates and fishes (Allen et al. 2007; Dame et al. 2000, 2002). The second was an addition experiment carried out in intertidal reefs in North Carolina salt marshes (Cressman et al. 2003; Nelson et al. 2004), where although there were some significant increases in water clarity at the oyster addition study sites in North Carolina, as shown by reduced amounts of suspended solids and chl-a, those differences were inconsistent and not of the magnitude expected.

Although South and North Carolina marshes are representative of the southeastern Atlantic Coast, they are markedly different from other locations where oyster reefs also thrive, such as the microtidal northern Gulf of Mexico. Thus, additional experiments in which the effects of oyster additions or removals are compared with unaltered controls are needed in other locations to assess the many ecosystem services that shoreline oyster reefs are believed to provide. Such studies will be useful in developing the predictive ability necessary to justify the high cost of many shoreline oyster reef restoration programs.

Toward this end, we quantified the most important ecosystem services believed to be provided by nearshore oyster reefs. The central hypothesis of our study was that there would be measurable increases in water clarity, benthic algae, infaunal abundance, and the abundance of juvenile fishes and invertebrates in areas where healthy oyster reefs were restored to pre-harvest levels. To test this hypothesis and quantify the spatial scale at which these effects could be detected, we experimentally introduced oyster reefs to a number of oyster-poor marsh creeks and compared critical ecosystem processes between those creeks and control creeks without oyster additions. We also quantified the effects of oyster reef addition on the dynamics of nutrients and organic matter within the water-column and sediments, including nutrient export from the creeks, and carried out a detailed examination of the effects of oyster reef additions on fish community structure. Some of these results have been previously reported in detail elsewhere (Geraldi et al. 2009; Plutchak 2008; Plutchak et al. 2010) and here we only summarize them. However, they are important in allowing us to present a comprehensive overview of the entire suite of variables measured in our investigation of the ecosystem-level changes that occurred after construction of the reefs.

20.2 METHODS

20.2.1 Study Sites

In the northern Gulf of Mexico, owing to the small tidal range, most large oyster reefs are subtidal (Kilgen and Dugas 1989). In shallow water, oysters often occur as islands or peninsulas that are exposed at extremely low tides. These oyster formations may be oriented either parallel or perpendicular to current flow and often occur near or just seaward of the mouths of marsh creeks. The summary of northern Gulf reefs by Kilgen and Dugas (1989) contains a thorough description of reef development in the microtidal northern Gulf, with diagrams of the different types of subtidal and intertidal reefs and where they are typically found.

To quantify the ecosystem services provided by shallow intertidal oyster reefs in coastal Alabama, we employed a relatively simple replicated Before–After, Control–Impact, Paired (BACIP) design (cf. Osenberg and Schmitt 1996; Underwood 1997). Based on maps, geo-referenced aerial photography (provided by the Mobile Bay National Estuary Program), and a field survey, six small tidal creeks were chosen as study locations along Dauphin Island Bay (Alabama, USA [N 30.255; W 88.109]; Figure 20.1). Tidal creeks were paired, based on physical similarities (i.e., size, depth, orientation, proximity, shoreline vegetation, and natural oyster density), with one paired creek receiving the treatment (oyster reef added) and the other serving as the control (no oyster reef added). Tidal creek surface area, length, and mouth width were determined from the geo-referenced aerial photos using ArcGIS at a scale of 1:2000, which is the smallest scale at which features in the photos were discernible.

Sparse populations of natural oysters were present in all creeks, and to ensure that these natural populations were similar among pairs, surveys of 10% of each creek were conducted in the fall of 2004. Using Hawth's Analysis Tools extension for ESRI ArcGIS 9 (Beyer 2004), random dots were generated with latitude and longitude coordinates so that 10% of the creek would be sampled if each dot represented 1 m^2. During the field survey, the position of each dot was found using a handheld GPS and a 1-m^2 quadrat was then placed on the sediment surface so that the position of the dot would be centered within the quadrat. If quadrats happened to overlap, they were moved until the quadrats were adjacent to each other and new coordinates were recorded. All live oysters visible to the naked eye were then enumerated and returned to their original location, and the total abundances in paired creeks were compared with a two-tailed, paired t test.

20.2.2 Oyster Reef Restoration

One tidal creek from each pair was randomly chosen for oyster reef restoration (DIF 2, DIF 3 and LDI 1; Figure 20.1). All creeks contained scattered clumps of oysters and were judged to be suitable sites for the emplacement of living oyster reefs. Tidal creek area, maximum depth at high tide, and literature values for oyster filtration rates were used to determine the reef area required for the oysters in restored reefs to filter a volume of water equal to the entire volume of water within the creek every 12 h (Table 20.1). We used a 12-h period because we reasoned that the creeks would fill in roughly 12 h and empty roughly 12 h later in accord with the diurnal tidal regime in the study area. Adult oysters were assumed to filter 0.0045m^3 of water per individual per hour (Newell 1988).

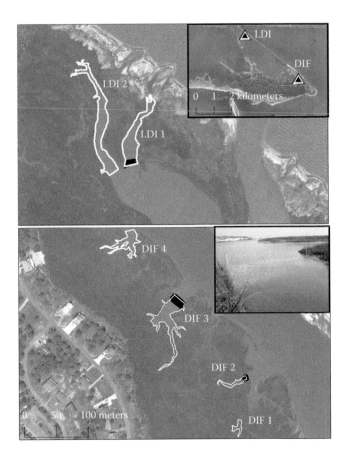

Figure 20.1 Aerial photographs of the six tidal creeks around the east end of Dauphin Island, Alabama (top right inset) that were studied. The upper photo shows the two paired sites on Little Dauphin Island and the lower photo shows the four tidal creeks on Dauphin Island. The sites were paired based on similarity in physical parameters (LDI 1 and 2, DIF 1 and 2, and DIF 3 and 4) and are shown within the white lines. Treatment creeks were randomly chosen and the reef areas are indicated by black polygons. The lower right inset shows one of the reefs.

A target density of 150 adult oysters m^{-2} was used, which is the estimated mean historical density for oyster reefs in the harvested areas of Mobile Bay (cf. May 1971), and should represent an appropriate density for measuring the effects attributed to healthy oyster reefs. From initial calculations, it was determined that approximately 10% of the creek area should receive oysters (Figure 20.1; Table 20.1). Reef sizes were calculated to be 43 m^2 for DIF 2, 207 m^2 for DIF 3, and 161 m^2 for LDI 1. Reefs were placed at and across the mouth of the tidal creek and consisted of an oyster shell pad of approximately 10 cm in height that was covered with live oysters and associated organisms such as polychaetes, mud crabs, barnacles, and mussels. Shell pads were put in place by a contractor (J & W Marine Inc., Bayou la Batre, Alabama) on February 21–24, 2005. Live oysters and associated fauna were tonged by local oystermen from Little Dauphin Island Bay and placed on top of the shell pad on February 28, 2005 to March 9, 2005.

20.2.3 Monitoring

Within each tidal creek, sampling stations were established along a transect perpendicular to the reef, originating 2 m in front of (i.e., toward the mouth of the creek) and extending 4 m behind the

Table 20.1 The Physical Parameters of the Six Tidal Creeks

Site	Treatment	Interpolated Creek Area (m²)	Creek Mouth Width (m)	Depth at Low Tide (m)	Depth at High Tide (m)	Estimated Maximum Volume of Creek (m³)		Estimated Velocity (cm/s)	Calculated Area of Reef (m²)	Live Oysters Added
						Minimum	Maximum			
DIF 1	Control	147	9.14	0.6	1.1	88.18	162	0.06	20	
DIF 2	Treatment	368	8.77	0.65	1.2	239.48	442	0.15	43	8187
DIF 3	Treatment	1752	19.69	0.5	1.05	876.41	1840	0.31	207	34,082
DIF 4	Control	844	11.82	0.4	0.85	337.61	717	0.25	86	
LDI 1	Treatment	1585	27.99	0.6	1.2	951.28	1903	0.20	161	35,232
LDI 2	Control	2116	15.79	0.6	1.2	1270	2540	0.47	314	

Note: Areas were calculated using aerial photographs in ArcGIS. Parameters were used to calculate the size of the reef and the number of oysters that were added to the treatment creeks assuming there were 150 oysters m^{-2}.

reef. Five sampling stations were established along each transect at +2 m, +0.5 m, −0.5 m, −2.0 m, and −4.0 m (+ indicates stations in front of [seaward] and − indicates stations behind [landward of] the reef). Thus, all monitoring of response variables (with the exception of fishes and large mobile macroinvertebrates reported on by Geraldi et al. 2009) was made in the immediate vicinity of the reefs or the creek mouths (in the case of controls). At each station, the following response variables were measured: water column and sediment chl-*a* concentrations and density of infaunal invertebrates. Sampling for physical data (salinity, temperature, dissolved oxygen, and photosynthetically active radiation) was conducted within a meter of the reefs. Salinity, temperature, and dissolved oxygen were measured using a handheld YSI-85 dissolved oxygen and conductivity meter. Photosynthetically active radiation was measured with two spherical sensors connected to a LI-COR 1400 datalogger, which recorded the 5-s averaged value for each sensor. One sensor was placed on the water's surface and the other was placed in 0.5-m increments starting just beneath the water's surface until the sensor reached the sediment. We typically allowed 2 weeks before the LICOR data loggers were downloaded and the data retrieved.

To observe initial (i.e., pretreatment) environmental conditions, response variables were sampled monthly in all creeks for nearly 10 months before construction of experimental reefs. To investigate the effects that oyster reefs had on response variables, monthly sampling continued for 18 months after reef placement.

To ensure that the desired density of 150 live adult oysters m^{-2} was maintained in treatment creeks, biannual sampling on the restored reefs began immediately after construction. Reefs were sampled at low tide by haphazardly tossing a 0.25-m^2 quadrat on the reef surface and removing by hand all oysters and associated fauna above the shell pad from within the quadrat. Six quadrats were collected at DIF 2 and 24 at DIF 3 and LDI 1, for a total sampled area of approximately 3.5% of the reef at DIF 2, 2.9% at DIF 3, and 3.7% at LDI. The removed shell was counted, measured, and placed into four categories: juvenile oysters (less than 3 cm shell height [SH]), adult oysters (greater than 3 cm SH), dead oysters, and mussels. After being counted, all living oysters, shells, and mussels were returned to their original location.

To document how chl-*a* concentration was affected by oyster reefs on a small, localized scale (i.e., directly above each reef), we conducted a small-scale sampling on April 14, 2006. Samples were taken with a device consisting of four 100-cc syringes attached equidistantly along a wooden base. This sampling device, described by Judge et al. (1993), allowed four water samples to be taken simultaneously, thus avoiding any influences of short-term turbidity changes (e.g., the suspension of sediment that might have occurred while accessing the site). Duplicate water samples were taken at the surface of the oyster reef (0 cm) and at 8, 15, and 22.5 cm directly above the reef in each treatment creek, or in the same relative position in control creeks. Upon return to the laboratory 40 mL of water from each depth were filtered onto a Whatman 25-mm glass microfiber filter (GF/F) and frozen at −80°C until analysis. Chlorophyll *a* was extracted from the filters using approximately 10 mL of a 2:3 mixture of dimethyl sulfoxide:90% acetone (Shoaf and Lium 1976), respectively. The chl-*a* content (µg L^{-1}) was fluorometrically (Turner Designs TD-700) determined using the Welschmeyer method, which is designed to be minimally sensitive to chlorophyll *b* and chlorophyll degradation products (Welschmeyer 1994).

20.2.3.1 Microalgae

At each station within each creek, we examined sediment (microphytobenthos) and water-column (phytoplankton) chl-*a* concentrations monthly, as indicators of algal abundance. One liter of water was collected at approximately mid-water depth at each station and 100 mL from each 1-L sample was filtered onto a Pall 47-mm glass microfiber filter (GF/F). Filters were frozen at −80°C until analysis. Additionally, samples from the top 1 cm of sediment were haphazardly collected for chl-*a* analysis where and when water samples were collected, using a modified plastic 60-mL syringe (2.5 cm diameter). The total area of sediment collected by the syringe was approximately 4.9 cm^2. Data from these sampling efforts have been previously reported on by Plutchak et al. (2010) and here we only summarize them.

20.2.3.2 Infauna

The abundance of benthic macrofaunal invertebrates was estimated by a monthly core sample (7.6 cm ID) of infauna at all stations within each tidal creek. Macrofauna was retained on a 0.5-mm mesh and frozen until samples could be processed. Samples were stained with rose bengal and macrofauna sorted from any retained sediment or detritus and identified to the ordinal or class level. One percent of the entire sample set was checked by a second person for consistency in identification. Samples were considered accurate/consistent for both the sorting and identification when they were in 95% agreement, and all samples fell within this range. After identification, the ash-free dry weight (AFDW) of each taxon was determined using a muffle furnace. Densities of benthic macrofauna were expressed per unit area for ease of comparison with previous studies.

20.2.3.3 Fish and Mobile Invertebrates

Demersal fishes and crustaceans present in control and treatment creeks were sampled approximately once a month using seines and block nets. Sampling was conducted during the day, and time of collection was chosen haphazardly so that samples were taken at all tidal heights, except extreme low tides. Each creek was sampled both in front of, and behind, the reef. To maintain a balanced design, control creeks were also seined twice as if an oyster reef existed between the two areas sampled.

Data from both seining and gill net collections have been published by Geraldi et al. (2009) and, as with data on algae, we only summarize their findings here.

20.3 STATISTICAL ANALYSES

20.3.1 Microalgae

Chlorophyll a measurements were initially analyzed by separate two-way analysis of variance (ANOVA) performed on each creek to determine if there were station differences (i.e., with distance from the reef) across dates (Plutchak et al. 2010). If station differences were not significant ($p < 0.05$), values for all stations within a creek were averaged and used for a creek-integrated BACIP analysis.

BACIP values were calculated by subtracting the measured value of a parameter (e.g., water column chl-a) taken at a control creek (a creek without restored oyster reefs) from the value taken at its paired treatment creek (a creek with restored oyster reefs). To identify trends, we looked at whether the impact (= treament) minus control (I – C) values became more positive or negative after the addition of reefs. When values increased in treatment creeks, the I – C values became more positive. When values in the treatment creeks decreased, the I – C values for the pair became more negative. When I – C values after reef deployment were significantly different from those before reef placement, a significant reef effect was indicated.

Although creeks were paired according to physical similarities, the three creek pairs differed from one another in size, location, and some environmental characteristics. Therefore, creek pairs were not treated as replicates but were independently analyzed, effectively removing variability owing to location and making changes owing to oyster reef emplacement more easily identifiable. Creek pairs are referenced as pair 1 (DIF 1 and DIF 2), pair 2 (DIF 3 and DIF 4), and pair 3 (LDI 1 and LDI 2). Two-tailed t tests were conducted between I – C values before and after oyster deployment to determine if there were differences between the measurements obtained before and after reef construction. Data were analyzed using Minitab 15 statistical software, and the threshold of significance was set at $p < 0.05$.

20.3.2 Infauna

Infaunal abundance and biomass data were analyzed by calculating the BACIP values for each paired station (+2 m, +0.5 m, −0.5 m, −2.0 m, and −4.0 m) and each major taxonomic group, as well as for the total number of organisms collected.

To determine if there were significant differences before and after oyster reef restoration, two-tailed t tests were conducted on the calculated BACI values for each station in each of the three creek pairs. Each pair was independently analyzed (as above) and the threshold of significance was set at $p < 0.05$. Data were analyzed using SPSS 12.0 statistical software.

20.3.3 Fish and Mobile Invertebrates

Seine data were analyzed using a BACI intervention analysis (Hewitt et al. 2001) because sampling protocol was different compared to other response variables (cf. Geraldi et al. 2009). For each dependent variable, the difference between the treatment and control creek of each pair was calculated for each sampling date after the "in front" and "behind" reef seine data were averaged. A two-way general linear model (GLM) was run with date (random factor) nested in before–after the addition of the reef (fixed factor) as independent variables and the difference between the treatment and control creek for any given parameter as the dependent variable. GLMs were also run for the abundance and biomass of the five most abundant species.

Although gillnet samples were not taken before construction of the reefs, the effect of the addition of oyster reefs on large transient species caught in the gillnets was quantified using similar analyses. Two-way GLMs were run with date as a random factor and reef (presence/absence) as a fixed factor (cf. Geraldi et al. 2009).

20.4 RESULTS

20.4.1 Study Site Characteristics

Tidal creeks were paired based on creek similarities and proximity, with volume ranging from 162 to 2540 m^3 and depth ranging from 0.85 to 1.2 m (Table 20.1). Temperature ranged between 13.4°C and 35.9°C following a typical seasonal pattern with high temperatures recorded in late spring to early fall and low temperatures during winter months. Salinity ranged from 2.9 to 29 and was highly variable in these shallow creeks.

Surveys of the creeks before reef construction showed that the natural density of oysters within the entire creek area was low, only ranging between 0.59 and 5.58 individuals m^{-2}. Paired t tests for each tidal creek pair revealed no significant difference between any of the tidal creek pairs (all p values >0.05). Densities of oysters on the restored reefs were sampled three times—immediately after reef construction in April 2005, 6 months after reef construction in September 2005, and 1 year after reef construction in February 2006—to verify that the density remained above the target of 150 oysters m^{-2}. Densities of adult oysters (>30 mm in length) were highest immediately after reef construction, ranging from 253 to 446 oysters m^{-2} (Figure 20.2); subsequent sampling indicated that adult oyster density declined at all sites, with ranges of 191 to 354 oysters m^{-2} in September 2005 and 176 to 205

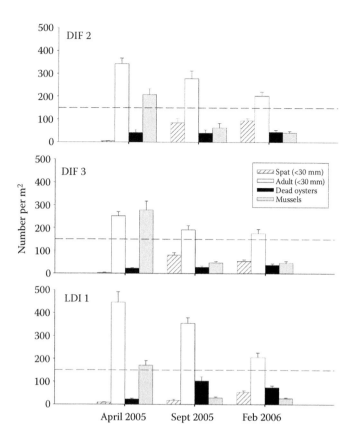

Figure 20.2 Mean number (±SE) of live oysters (spat and adults), dead oysters, and mussels determined from sampling 20% of the reef using 0.25-m² quadrats. The dashed horizontal line indicates the target density of 150 oysters m⁻², which theoretically would filter the entire creek every 12 h.

oysters m⁻² in February 2006. Although the number of oysters at each of the restored reefs declined over time, the density remained above the target density of 150 oysters m⁻². In addition, spat settlement began soon after reef construction and continued throughout the study period (Figure 20.2).

Ten days before reef construction, average chl-*a* concentrations at the permanent monitoring stations were not significantly different between treatment and control creeks (cf. 4.14 ± 0.37 and 4.52 ± 0.38 μg L⁻¹, respectively). However, when measured during midday approximately 1 year after reef construction, the localized water column chl-*a* concentrations directly above the reef, or in similar areas in the control sites, were significantly different between treatment and control creeks (*p* value <0.0001), but varied at marginally significant levels (*p* = 0.053) with increasing distance above the reef (Figure 20.3). The mean chl-*a* concentration directly above the reef in treatment creeks was 2.46 ± 0.24 μg L⁻¹ compared to 7.62 ± 1.32 μg L⁻¹ in control creeks.

20.4.2 Algal Biomass

Phytoplankton chl-*a* concentration tended to be highest in spring and lowest during winter months (Figure 20.4). However, the single highest mean value for chl-*a* was measured in September 2005 (32.2 μg L⁻¹) after the landfall of Hurricane Katrina, while the lowest was recorded just 2 months later in November 2005 (1.5 μg L⁻¹). Trends in microphytobenthic chl-*a* were characterized by high values in spring, and low values in fall/early winter with the highest and lowest mean

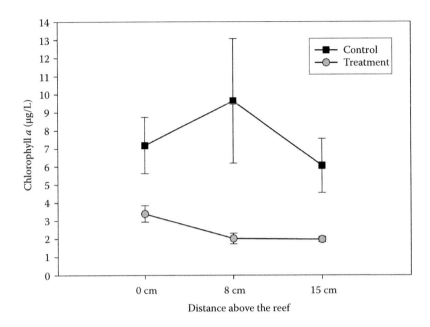

Figure 20.3 Chlorophyll *a* concentration (µg L⁻¹) ± standard error estimated from water samples taken directly above the restored oyster reef or area where a reef would have been restored in the control creeks. (After Plutchak, R., K. Major, J. Cebrian, C. D. Foster, M.-E. C. Miller, A. Anton, K. L. Sheehan, K. L. Heck, Jr. and S. P. Powers. 2010. Impacts of oyster reef restoration on primary productivity and nutrient dynamics in tidal creeks of the north central Gulf of Mexico. *Estuaries and Coasts* 33: 1355–1364.)

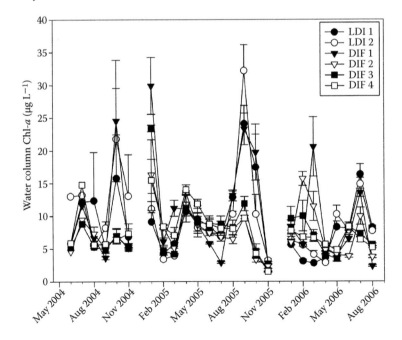

Figure 20.4 Monthly means of water-column chl-*a* for all creeks sampled from June 2004 to August 2006. Bars denote standard errors. (After Plutchak, R., K. Major, J. Cebrian, C. D. Foster, M.-E. C. Miller, A. Anton, K. L. Sheehan, K. L. Heck, Jr. and S. P. Powers. 2010. Impacts of oyster reef restoration on primary productivity and nutrient dynamics in tidal creeks of the north central Gulf of Mexico. *Estuaries and Coasts* 33: 1355–1364.)

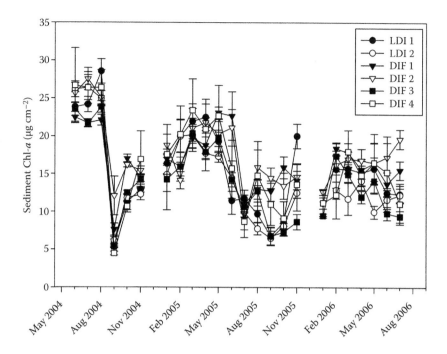

Figure 20.5 Monthly means for sediment chl-*a* sampled from all creeks. Bars denote standard errors. (After Plutchak, R., K. Major, J. Cebrian, C. D. Foster, M.-E. C. Miller, A. Anton, K. L. Sheehan, K. L. Heck, Jr. and S. P. Powers. 2010. Impacts of oyster reef restoration on primary productivity and nutrient dynamics in tidal creeks of the north central Gulf of Mexico. *Estuaries and Coasts* 33: 1355–1364.)

chl-*a* concentrations obtained in August 2004 (28.5 µg cm^{-2}) and September 2004 (4.5 µg cm^{-2}), respectively, immediately before and after Hurricane Ivan (Figure 20.5).

There were no obvious trends in water column chl-*a* between any of the creek pairs (Figure 20.4). Similarly, sediment chl-*a* showed no overall pattern (Figure 20.5). However, a significant ($p = 0.027$) effect was identified for creek pair 1, where the treatment creek exhibited lower chl-*a* concentrations compared to that of the control after construction of the oyster reef.

20.4.3 Infauna

Infaunal assemblages were dominated by polychaetes, which made up nearly 96% of all individuals collected, although other invertebrate taxa such as amphipods, and some epifaunal species of isopods, gastropods, bivalves, small shrimp, and crabs were occasionally found. Polychaete abundances ranged from 227 to 21,136 individuals m^{-2}; biomass ranged from 0.02 to 7.68 g m^{-2} AFDW.

Although creeks were paired based on physical similarities (size, depth, area, etc.), there were large infaunal differences before reef construction between treatment and control creeks, with treatment creeks having greater numbers of organisms than control creeks. This is illustrated by the representative results for the −0.5-m station shown in Figure 20.6. These differences were reduced after a highly active 2004 hurricane season, which brought BACI values to near zero, indicating that the treatment and control creeks were similar. This trend was not observed the following year (2005), which was also a highly active hurricane season along the Northern Gulf Coast, and infaunal abundances varied substantially among pairs with no consistent pattern (Figure 20.6).

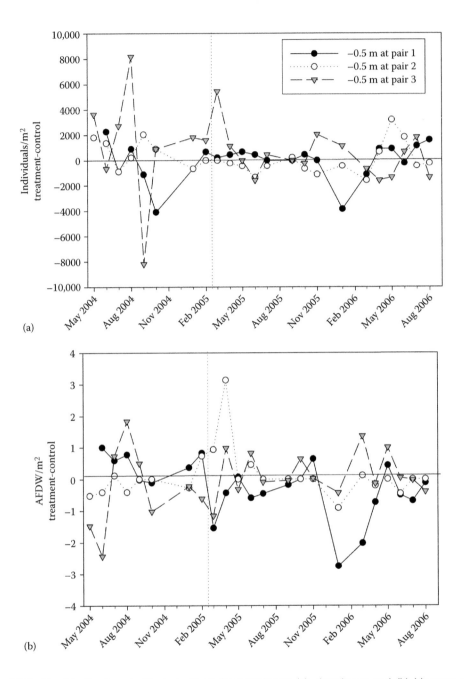

Figure 20.6 Monthly Treatment—Control values for polychaete (a) abundance and (b) biomass at the −0.5-m station for each creek pair separately. Dashed line indicates time of reef deployment. Pair 1 = DIF 1 and DIF 2, pair 2 = DIF 3 and DIF 4, and pair 3 = LDI 1 and LDI 2.

20.4.4 Fish and Mobile Macroinvertebrates

20.4.4.1 Seine Results

As reported by Geraldi et al. (2009), biomass in seine hauls was dominated by blue crab, grass shrimp, striped mullet, brown shrimp, spot, and pinfish, and abundance by grass shrimp, which made up of more than 48% of the individuals caught. Both biomass and abundance were highly variable, with biomass ranging from 0.01 to 38.40 g m^{-2} and abundance ranging from 0.08 to 115.19 individuals m^{-2}.

Two-way GLMs of date and reef addition showed no significant effect of date on the biomass or abundance of fish, demersal fish, sciaenids, or crustaceans (Figure 20.7a). Reef presence was significant for demersal fish abundance (Figure 20.7b; $p < 0.01$), but not for the abundance or biomass of other faunal groups, although all groups were more abundant after reef addition.

20.4.4.2 Gillnet Results

Geraldi et al. (2009) found that biomass in gillnet collections ranged from 0 to 814 g m^{-2} and abundance ranged from 0 to 2.06 individuals m^{-2}. Together, red drum (20%), blue crab (12%), hardhead catfish (12%), southern flounder (11%), silver perch (9%), and speckled trout (9%) made up 73% of the biomass, while silver perch (22%), blue crab (15%), croaker (12%), hardhead catfish (10%), and speckled trout (8%) were most abundant.

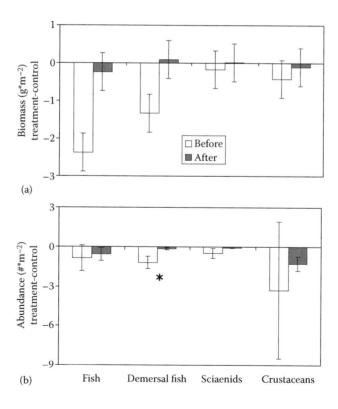

Figure 20.7 Mean differences (±1SE) in biomass (a) and abundance (b) of taxa collected by seine between treatment and control creeks, both before and after reef construction (*$p > 0.05$). (After Geraldi, N., S. P. Powers, K. L. Heck, Jr. and J. Cebrian. 2009. Can habitat restoration be redundant? Response of mobile fishes and crustaceans to oyster reef restoration in marsh tidal creeks. *Marine Ecology Progress Series* 389: 171–180.)

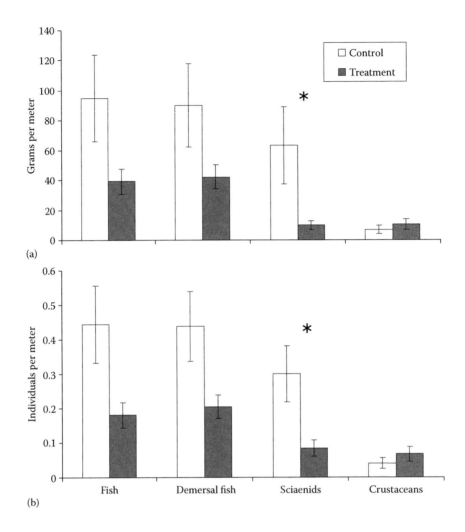

Figure 20.8 Mean differences (±1SE) in biomass (a) and abundance (b) of taxa collected by gillnet between treatment and control creeks (*$p < 0.05$). (After Geraldi, N., S. P. Powers, K. L. Heck, Jr. and J. Cebrian. 2009. Can habitat restoration be redundant? Response of mobile fishes and crustaceans to oyster reef restoration in marsh tidal creeks. *Marine Ecology Progress Series* 389: 171–180.)

Two-way ANOVAs of fish, demersal fish, sciaenids, and crustaceans found significantly fewer sciaenids in reef than control creeks (Figure 20.8a). All other fish groups had, on average, less biomass and fewer individuals, although not significantly so, in reef than control creeks.

Two-way ANOVAs of biomass and abundance of the five most important species caught in gill nets (southern flounder, silver perch, blue crabs, croaker, and catfish) found only southern flounder to significantly increase in the presence of reefs, while creeks with reefs had significantly reduced biomass and abundance of silver perch.

20.5 DISCUSSION

Restoration of oyster reefs within the tidal creeks was successful: populations of adult oysters remained above the targeted 150 oysters m^{-2} for 2 years post-construction; the abundance of juvenile oysters increased; and the number of dead adults remained constant, and relatively low, as the study progressed. Despite the expectation that the addition of oysters at the densities used in this study would reduce phytoplankton biomass in the water column of the creeks with reefs, there was no significant impact on water column chl-*a* in any of the three creek pairs (Figure 20.4; Plutchak et al. 2010). Instead, the only observable effects of the reefs on algal biomass were at small distances directly above the reefs (Figure 20.3; Plutchak et al. 2010). Given that there were no large changes in water clarity that would enhance the abundance and production of microphytobenthos, and the fact that these primary producers did not substantially respond to reef emplacement, it was not surprising that benthic infaunal macroinvertebrates showed no large reef-associated responses (Figure 20.6). Somewhat more unexpectedly, Geraldi et al. (2009) reported few significant reef effects on the abundance or biomass of large, mobile macroinvertebrates or finfish (Figures 20.7 and 20.8). Below, we discuss several potential reasons for the absence of consistent effects of reef construction on many of our response variables.

20.5.1 Lack of Effects on Algal Biomass and Infaunal Abundance

There are a number of non–mutually exclusive factors that may explain the noticeably absent "oyster reef impact" on water column and sediment chl-*a*, infaunal invertebrates, and fish and macrocrustaceans, and we consider them in turn. First, the creeks we studied were completely surrounded by *Juncus roemerianus* (black needlerush). Undoubtedly, runoff from marshes flooded by high tides (especially during summer or particularly heavy rain events) contributed nutrients and suspended solids to the creeks (cf. Mallin et al. 2002). Runoff may have stimulated algal production above that, which is possible at ambient nutrient levels, thereby partially overcoming the ability of oysters to reduce water column chl-*a* concentrations. Perhaps, although less likely, oyster fecal deposits containing plankton-derived particulate nitrogen could have been rapidly remineralized to produce ammonium that, in turn, led to increased phytoplankton growth. This type of feedback loop would obscure the impacts of filtration except when in proximity to the reef.

Depending on temperature, adult oysters can filter around 9.5–19.9 L of water per hour (0.0095–0.0199 m^3 h^{-1}), and at temperatures above 28°C, filtration rate rapidly rises (Korringa 1952). During our study, water temperatures were predominantly above 28°C from May to September each year, indicating that filtration rates were likely very high during these months. While one can question our initial assumption that oysters filtered water at a rate of 0.0045 m^3 h^{-1}, since this rate is based on results obtained at 25°C (Pomeroy et al. 2006), the average water temperature over the study duration was, according to records from the monitoring station at Dauphin Island, Alabama (http://www.mobilebaynep.com/mondata/mainmenu.cfm), approximately 23°C. Thus, it seems unlikely that the deviations in filtration as a result of temperature could have resulted in large errors in our *a priori* calculations.

Alternatively, since reefs were only built to a height of 0.5 m, they were not always in contact with the entire water column, particularly in the summer when south winds resulted in water levels that were elevated above those predicted. In addition, our reefs were located at the mouth of the tidal creeks and incoming and outgoing tidal waters interacted with the reef during a small portion of the tidal cycle. Although our measurements did show a significant reduction in chl-*a* at small distances above the reef, measurements made at distances only 0.50 m from the reef were beyond the "zone of impact" and outside the range at which oysters had a significant impact on water clarity.

We also believe that, for much of the time, oysters in treatment creeks may never have come in contact with substantial portions of the water column because of stratification. Our initial assumption was that the shallowness of the water column (mean depth <1 m) would ensure complete mixing; however, Riisgard et al. (2007) reported that similarly shallow water columns in Denmark remained stratified unless winds blew consistently at velocities greater than 8 m s^{-1}. Thus, filter feeders in the shallow fjord studied by Riisgard et al. (2007) often did not come in contact with much of the water column and their effective filtration ability was lower when compared to other rates reported in the literature. Only when wind velocities were above the 8 m s⁻¹ threshold did mixing occur and substantial filtration of the water column take place. Since wind velocities are usually low in the north-central Gulf of Mexico (monthly means always <8 m s⁻¹, Zhao and Chen 2008), it is quite possible that only a small portion of the lower water column regularly was available for filtration by oysters on the reefs.

We also had the opportunity to examine the impact of tropical storms on the creeks. A dramatic decrease in sediment chl-*a* concentration was observed after the landfall of Hurricane Ivan on September 16, 2004 (Cebrian et al. 2007). This catastrophic storm resulted in atypically low chl-*a* concentrations during September 2004. During the following year, the Gulf Coast again experienced several tropical storms, which likely contributed to the unusually low chl-*a* concentrations observed throughout this study. Indeed, substantial drops in sediment chl-*a* concentration were observed after Tropical Storm Arlene (June 11, 2005), Hurricane Cindy (July 6, 2005), Hurricane Dennis (July 10, 2005), and Hurricane Katrina (August 29, 2005), all of which made landfall near our study sites. Frequent storms repeatedly scoured and disturbed the sediments, resulting in an entire summer of depressed algal biomass (Cebrian et al. 2007). Thus, storm events likely masked any impacts of oyster emplacements on the microphytobenthos. Similarly, storms could have homogenized the sediments, thereby reducing the overall impacts of the treatment reefs on infauna. For example, the infaunal differences that were evident before Hurricane Ivan subsequently disappeared after the storm (Figure 20.6).

Storms were also likely responsible for the high water column chl-*a* concentrations noted in post–storm sampling events, and these events were probably a result of large short-term contributions of suspended microphytobenthic chl-*a* to the water column. Moreover, storm impacts on the water column were short lived, as the creeks were flushed by tides and particulate matter settled from suspension.

Because of Hurricane Ivan, pre-reef data in 2004 were likely lower than they otherwise would have been, and it appeared that Ivan and subsequent storms of 2005 overwhelmed any effects of reef construction on chl-*a* concentrations. However, the central Gulf of Mexico has historically experienced substantial hurricane activity, and organisms in the Gulf have evolved with frequent tropical storm/natural disturbance. Thus, we might have expected to observe strong reef effects, if they exist, even during periods of elevated tropical storm activity. Furthermore, the summer of 2006 was free of tropical storm activity; thus, if a reef effect was present, we would expect to see it during this time of low disturbance.

As noted previously, the lack of obvious reef effects on macrofaunal invertebrates seems to follow directly from the paucity of significant effects on benthic algae. This is because the deposition of reef-associated fecal material or the improved water clarity attributed to oysters should have stimulated microphytobenthic algal production that would, in turn, lead to increased numbers of deposit-feeding invertebrates. Since the presence of reefs had little effect on benthic algae, there was similarly little effect on macroinvertebrates in treatment creeks.

20.5.2 Small Effects on Fish and Mobile Macroinvertebrates

Demersal fish were the only species that significantly increased in treatment creeks, even though their abundance was still greater in control creeks after the addition of reefs (Figure 20.7; Geraldi et al. 2009). For those groups that did not show significant effects, our conclusions are likely robust because of the BACI-type design (Peterson et al. 2003). Although we had only three paired sites, and our ability to detect significant impacts could have been increased with more replicates, this would not likely have affected the general nature of our conclusions, given the high *p* values obtained in a number of our tests.

The few significant increases in abundance of fishes after reef addition reported by Geraldi et al. (2009) is at odds with the current paradigm of fish enhancement by oyster reefs (Coen and

Luckenbach 2000; Lehnert and Allen 2002; Lenihan et al. 2001; Peterson et al. 2003). Although there are no other investigations of the effect of the addition of oyster reefs in semi-enclosed systems such as ours, several other studies in open systems did not find that oyster reefs significantly increased the abundance of nonresident species (Allen et al. 2007; Dame et al. 2002; Grabowski et al. 2005; Harding and Mann 1999, 2001; Luckenbach et al. 2005).

Our findings also agree with Peterson et al. (2003), who reported that many of the species caught in this study were not augmented by the presence of oyster reefs. For example, the most abundant demersal species caught by seine—blue crab, brown shrimp, darter goby, pinfish, spot, and white shrimp—were not found by Peterson et al. (2003) to be enhanced by reef restoration. Four species caught by gillnet—Atlantic croaker, blue crab, red drum, and speckled trout—were also found by Peterson et al. (2003) to be unaffected by the presence of oyster reefs. However, southern flounder was enhanced by oyster reef in both our study and those reviewed by Peterson et al. (2003).

Grabowski et al. (2005) found increased juvenile fish on oyster reefs surrounded by mud flats. They did not, however, find similar enhancement on oyster reefs adjacent to salt marsh or seagrass, which they attributed to the functional redundancy of adjacent "nursery" habitats such as marshes and seagrass meadows. Others have noted the importance of the surrounding landscape and the synergistic effects that adjacent "nursery habitats" may have on fishes (Peterson and Lipcius 2003). We believe that functional redundancy of biogenic structure in the form of marsh plants likely caused our finding that oyster reefs did not significantly increase mobile fish and macrocrustaceans. That is, biogenic structure was not a limiting factor at our study sites because of the abundant salt marsh creek habitat that existed before reef construction (Geraldi et al. 2009).

20.6 CONCLUSIONS

While there were small-scale increases in water clarity, and scattered, inconsistent treatment effects, we found few persistent changes in our response variables between treatment and control creeks after the construction of oyster reefs. The absence of dramatic effects was likely caused by a combination of factors, including lack of vertical mixing, inputs of nutrients from adjacent marshes, suspension of microphytobenthos, abundance of adjacent "nursery" habitat, and intense tropical storm activity. It is also certainly possible that reef impacts might occur at times beyond the 18-month post–reef monitoring we carried out. Clearly, monitoring for longer periods, perhaps on the order of the generation time of the longest-lived organisms being studied (cf. Dayton and Tegner 1984), is desirable. However, the time encompassed by our study was sufficient to have detected large impacts of the reefs if they existed, since most of the study taxa (e.g., algae and infaunal invertebrates) have fast turnover rates and would have gone through several generations during our study period.

We find it noteworthy that our results are qualitatively similar to those of the other experimental studies that have investigated the ecosystem services provided by oyster reefs. For example, a study of intertidal reefs in North Carolina salt marshes (Cressman et al. 2003; Nelson et al. 2004) found some significant increases in water clarity, as shown by reduced amounts of suspended solids and chl-*a*, although the differences between sites with and without oyster additions were inconsistent and small. Similarly, the removal of oysters from marsh creeks in South Carolina was not clearly related to changes in water clarity or use of the creeks by mobile invertebrates and fishes (Allen et al. 2007; Dame et al. 2000, 2002).

Thus, there is little existing evidence of large ecosystem-wide effects of oyster enhancement or removal in studies done to date. Regardless of the reasons, we did not find substantial changes in water clarity and biological activity in the treatment creeks. In addition, our results suggest that if a major goal of oyster reef enhancement is to directly benefit mobile fishes and macroinvertebrates, surrounding landscapes can play a critical role. In closing, we emphasize that our results should not be interpreted as to meaning that restoration of nearshore oyster reefs should not be carried out, especially since our reef restoration was successful, and because functional oyster reefs offer many benefits such as enhanced benthic–pelagic coupling (Baird and Ulanowicz 1989) and denitrification (Newell et al. 2002). They do mean, however, that the types, magnitudes, and spatial extent of changes in ecosystem services that should be expected after reef restoration need to be reevaluated.

ACKNOWLEDGMENTS

We thank Cissie Davis, Kevan Gregalis, Drew Foster, Lesley Baggett, Mairi Miller, Josh Goff, Andrea Anton, and Kate Sheehan for help in carrying out this project. Support was provided by NOAA via the University of South Alabama Oyster Reef Restoration Program and the Dauphin Island Sea Lab (DISL).

REFERENCES

Allen, D.M., S.S. Haertel-Borer, B.J. Milan, D. Bushek and R.F. Dame. 2007. Geomorphological determinants of nekton use of intertidal salt marsh creeks. *Marine Ecology Progress Series* 329: 57–71.

Bahr, L.M. and W.P. Lanier. 1981. The ecology of intertidal oyster reefs of the south Atlantic coast: A community profile. U.S. Fish and Wildlife Service, Office of Biological Service. Washington, D.C., FWS/OBS-81/15. 105 pp.

Baird, D. and R.E. Ulanowicz. 1989. The seasonal dynamics of the Chesapeake Bay ecosystem. *Ecological Monographs* 59: 329–364.

Beyer, H. L. 2004. Hawth's Analysis Tools for ArcGIS. Available at http://www.spatialecology.com/htools.

Breitburg, D. C. 1999. Are three-dimensional structure and healthy oyster populations the keys to an ecologically interesting and important fish community? In: Luckenbach M.W., R. Mann, J.A. Wesson (Eds.), *Oyster Reef Habitat Restoration: A Synopsis and Synthesis of Approaches*, pp. 239–250. Virginia Inst. Mar. Sci. Press, Gloucester Point, VA.

Cebrian, J., C.D. Foster, R. Plutchak, K.L. Sheehan, M.-E.C. Miller, A. Anton, K. Major, K.L. Heck, Jr. and S.P. Powers. 2007. The impact of Hurricane Ivan on the primary productivity and metabolism of marsh tidal creeks. *Aquatic Ecology* 42: 391–404.

Cerco, C.F. and M.R. Noel. 2007. Can oyster restoration reverse cultural eutrophication in Chesapeake Bay? *Estuaries and Coasts* 30: 1–13.

Cloern, J.E. 1982. Does the benthos control phytoplankton biomass in south San Francisco Bay? *Marine Ecology Progress Series* 9: 191–202.

Coen, L.D., R.D. Brumbaugh, D. Bushek, R. Grizzle, M.W. Luckenback, M.H. Posey, S.P. Powers and S.G. Tolley. 2007. Ecosystem services related to oyster restoration. *Marine Ecology Progress Series* 341: 303–307.

Coen, L.D. and M.W. Luckenbach. 2000. Developing success criteria and goals for evaluating oyster reef restoration: Ecological functioning or resource exploitation? *Ecological Engineering* 15: 323–343.

Coen, L.D., M.W. Luckenbach and D.L. Breitburg. 1999. The role of oyster reefs as essential fish habitat: A review of current knowledge and some new perspectives. In: Benaka, L.R. (Ed.), *Fish Habitat: Essential Fish Habitat and Restoration*. American Fisheries Society Symposium 22: 438–454.

Cohen, R.R.H., P.V. Dresler, E.J.P. Philips and R.L. Cory. 1984. The effect of the Asiatic clam, *Corbicula fluminea*, on phytoplankton of the Potomac River, Maryland. *Limnology and Oceanography* 29: 170–180.

Cressman, K.A., M.H. Posey, M.A. Mallin, L.A. Leonard and T.D. Alphin. 2003. Effects of oyster reefs on water quality in a tidal creek estuary. *Journal of Shellfish Research* 22: 753–762.

Dame, R.F. 1993. *Bivalve Filter Feeders and Coastal and Estuarine Ecosystem Processes*. Springer-Verlag, Heidelberg. 579 pp.

Dame, R.F. 1996. *Ecology of Marine Bivalves: An Ecosystem Approach*. CRC Press, Boca Raton, Florida. 254 pp.

Dame, R., D. Bushek, D. Allen, D. Edwards, L. Gregory, A. Lewitus, S. Crawford, E. Koeppler, C. Corbett, B. Kjerfve and T. Prins. 2000. The experimental analysis of tidal creeks dominated by oyster reefs: The premanipulation year. *Journal of Shellfish Research* 19: 361–369.

Dame R., D. Bushek, D. Allen, A. Lewitus, D. Edwards, E. Koeplfler and L. Gregory. 2002. Ecosystem response to bivalve density reduction: Management implications. *Aquatic Ecology* 36: 51–65.

Dame, R.F., R.G. Zingmark and E. Haskins. 1984. Oyster reefs as processors of estuarine materials. *Journal of Experimental Marine Biology and Ecology* 164: 147–159.

Dayton, P.K. and M.J. Tegner. 1984. The importance of scale in ecology: A kelp forest example with terrestrial analogs. In: Price, P. W., C. N. Slobodchikoff and W. S. Gaud (Eds.), *A New Ecology: Novel Approaches to Interactive Systems*, pp. 457–481. Wiley, New York.

Geraldi, N., S.P. Powers, K.L. Heck, Jr. and J. Cebrian. 2009. Can habitat restoration be redundant? Response of mobile fishes and crustaceans to oyster reef restoration in marsh tidal creeks. *Marine Ecology Progress Series* 389: 171–180.

Grabowski, J.H., A.R. Hughes, D.L. Kimbro and M.A. Dolan. 2005. How habitat setting influence restored oyster reef communities. *Ecology* 86: 1926–1935.

Grabowski, J.H. and C.H. Peterson. 2007. Restoring oyster reefs to recover ecosystem services. In: Cuddington, K., J.E. Byers, W.G. Wilson, and A. Hastings (Eds.), *Ecosystem Engineers*. Elsevier Academic Press, Burlington, MA, pp. 281–298.

Harding, J.M. and R. Mann. 1999. Fish species richness in relation to restored oyster reefs, Piankatank River, Virginia. *Bulletin of Marine Science* 65: 289–300.

Harding, J.M. and R. Mann. 2001. Oyster reefs as fish habitat: Opportunistic use of restored reefs by transient fishes. *Journal of Shellfish Research* 20: 951–959.

Hewitt, J.E., S.E. Thrush and V.J. Cummings. 2001. Assessing environmental impacts: Effects of spatial and temporal variability at likely impact scales. *Ecological Applications* 11: 1503–1516.

Judge, M.L., L.D. Coen and K.L. Heck, Jr. 1993. Does *Mercenaria mercenaria* encounter elevated food levels in seagrass beds? Results from a novel technique to collect suspended food resources. *Marine Ecology Progress Series* 92: 141–150.

Kilgen, R.H. and R.J. Dugas. 1989. *The ecology of oyster reefs of the northern Gulf of Mexico: An open file report*. US Department of the Interior, Fish and Wildlife Service, Research and Development, National Wetlands Research Center.

Korringa, P. 1952. Recent advances in oyster biology. *Quarterly Review of Biology* 27: 266–308.

Lehnert, R.L. and D.M. Allen. 2002. Nekton use of subtidal oyster shell habitat in a southeastern U.S. estuary. *Estuaries* 24: 1015–1024.

Lenihan, H.S. and C.H. Peterson. 1998. How habitat disturbance through fishery disturbance enhances effects of hypoxia on oyster reefs. *Ecological Applications* 8: 128–149.

Lenihan, H.S., C.H. Peterson, J.E. Byers, J.H. Grabowski, G.W. Thayer and D. Colby. 2001. Cascading of habitat degradation: Oyster reefs invaded by refugee fishes escaping stress. *Ecological Applications* 11: 746–782.

Luckenbach, M.W., L.D. Coen, P.G. Ross and J.A. Stephen. 2005. Oyster reef habitat restoration: Relationships between oyster abundance and community development based on two studies in Virginia and South Carolina. *Journal of Coastal Research* 40: 64–78.

Mallin, M.A., S.H. Ensign, T.L. Wheeler and D.B. Mayes. 2002. Pollutant removal efficacy of three wet detention ponds. *Journal of Environmental Quality* 31: 654–660.

May, E.B. 1971. A survey of the oyster and shell resources of Alabama. Alabama Marine Resources Bulletin No. 4.

Nelson, K.A., L.A. Leonard, M.H. Posey, T.D. Alphin and M.A. Mallin. 2004. Using transplanted oyster (*Crassostrea virginica*) beds to improve water quality in small tidal creeks: A pilot study. *Journal of Experimental Marine Biology and Ecology* 298: 347–368.

Newell, R.I.E. 1988. Ecological changes in the Chesapeake Bay: Are they the result of overharvesting the American oyster, *Crassostrea virginica*? In: Lynch, M.P. and E.C. Krome (Eds.), *Understanding the Estuary: Advances in Chesapeake Bay Research*, pp. 536–546. Chesapeake Bay Research Consortium, Publ. 129 CBP/TRS 24/88, Gloucester Point, VA.

Newell, R.I.E., J.C. Cornwell and M.S. Owens. 2002. Influence of simulated bivalve biodeposition and microphytobenthos on sediment nitrogen dynamics: A laboratory study. *Limnology and Oceanography* 47: 1367–1379.

Newell, R.I.E, W.M. Kemp, J.D. Hagy III, C.F. Cerco, J.M. Testa and W.R. Boynton. 2007. Top-down control of phytoplankton by oysters in Chesapeake Bay, USA: Comment on Pomeroy et al. (2006). *Marine Ecology Progress Series* 341: 293–298.

Officer, C.B., T.J. Smayda and R. Mann. 1982. Benthic filter feeding: A natural eutrophication control. *Marine Ecology Progress Series* 9: 203–210.

Osenberg, C.W. and R.J. Schmitt. 1996. Detecting ecological impacts caused by human activities. In: Schmitt, R.J. and C.W. Osenberg (Eds.), *Detecting Ecological Impacts: Concepts and Applications in Coastal Habitats*, pp. 3–16. Academic Press, NY.

Peterson, C.H., J.H. Grabowski and S.P. Powers. 2003. Estimated enhancement of fish production resulting from restoring oyster reef habitat: Quantitative valuation. *Marine Ecology Progress Series* 264: 249–264.

Peterson, C.H. and R.N. Lipcius. 2003. Conceptual progress towards predicting quantitative ecosystem benefits of ecological restorations. *Marine Ecology Progress Series* 264: 297–307.

Plutchak, R. 2008. The impact of oyster reef restoration on the nutrient dynamics and primary productivity in tidal marsh creeks of the northcentral Gulf of Mexico. M.S. Thesis, University of South Alabama. Mobile, AL. 74 pp.

Plutchak, R., K. Major, J. Cebrian, C.D. Foster, M.-E.C. Miller, A. Anton, K.L. Sheehan, K.L. Heck, Jr. and S.P. Powers. 2010. Impacts of oyster reef restoration on primary productivity and nutrient dynamics in tidal creeks of the north central Gulf of Mexico. *Estuaries and Coasts* 33: 1355–1364.

Pomeroy, L.R., C.F. E'Elia and L.C. Shaffner. 2006. Limits to top-down control of phytoplankton by oysters in Chesapeake Bay. *Marine Ecology Progress Series* 325: 301–309.

Porter, E.T., J.C. Cornwell and L.P. Sanford. 2004. Effect of oysters *Crassostrea virginica* and bottom shear velocity on benthic–pelagic coupling and estuarine water quality. *Marine Ecology Progress Series* 271: 61–75.

Riisgard, H.U., J. Lassen, M. Kortegaard, L.F. Moller, M. Friedrichs, M.H. Fensen and P.S. Larsen. 2007. Interplay between filter-feeding zoobenthos and hydrodynanics in the shallow Odense Fjord (Denmark)—Earlier and recent studies, perspectives and modeling. *Estuarine, Coastal and Shelf Science* 75: 281–295.

Shoaf, W.T. and B.W. Lium. 1976. Improved extraction of chlorophyll *a* and *b* from algae using dimethyl sulfoxide. *Limnology and Oceanography* 21: 926–928.

Ulanowicz, R.E. and J.H. Tuttle. 1992. The trophic consequences of oyster stock rehabilitation in Chesapeake Bay. *Estuaries* 15: 298–306.

Underwood, A.J. 1997. *Experiments in ecology: Their logical design and interpretation using analysis of variance*. Cambridge University Press.

Welschmeyer, N.A. 1994. Fluorometric analysis of chlorophyll *a* in the presence of chlorophyll *b* and pheopigments. *Limnology and Oceanography* 39: 1985–1992.

Zhao, H. and Q. Chen. 2008. Characteristics of extreme meteorological forcing and water level in Mobile Bay, Alabama. *Estuaries and Coasts* 31: 704–718.

Zimmerman, R., T. Minello, T. Baumer and M. Castiglione. 1989. Oyster reef as habitat for estuarine macrofauna. Technical Memorandum. NMFS-SEFC-249, National Oceanic and Atmospheric Administration, Galveston, TX.

CHAPTER 21

Benches, Beaches, and Bumps
How Habitat Monitoring and Experimental Science Can Inform Urban Seawall Design

Jeffery R. Cordell, Jason D. Toft, Stuart H. Munsch, and Maureen Goff

CONTENTS

21.1 BACKGROUND

On February 8, 2001, Washington State experienced one of its largest recorded earthquakes, measuring 6.8 on the moment magnitude scale and lasting approximately 45 s. In Washington's largest city, Seattle, a 100-ft-long by 10-ft-wide section of a surface street adjacent to the shoreline settled, raising concerns about the condition of the city's seawall. Further assessment of the 71-year-old structure showed it to be in worse shape than expected; in addition to the effects of aging, the wooden foundation of the seawall was deteriorating because of the activities of wood-boring invertebrates. Over the next 10 years, the seawall continued to deteriorate despite regular maintenance, and it became evident that it needed to be replaced. Replacement of the seawall presented an opportunity to explore designs for improving the habitat conditions of the structure and its surroundings, and the city formed an interdepartmental team to focus on habitat enhancements that could be incorporated into the seawall design. A series of meetings were conducted, which included scientists from the University of Washington, local environmental consultants, and city engineers and environmental planners. These meetings resulted in habitat enhancement concepts developed and employed in Seattle's new seawall. In this chapter, we review habitat impacts of seawalls and how these impacts can be mitigated and present a case study of Seattle's seawall habitat improvements.

21.2 ECOLOGICAL EFFECTS OF SEAWALLS

Marine shorelines around the world are being transformed as the demand for infrastructure increases, and shoreline alteration is projected to accelerate as populations along coastal areas grow and as the threat of sea level rise increases (Bulleri and Chapman 2010; Chapman and Underwood 2011; Neumann et al. 2015). Replacing natural beaches with armoring structures is one of the most common methods to provide for human use and protect infrastructure. Vertical seawalls represent an extreme example of this: of shoreline armoring techniques, seawalls are the least complex, typically built of smooth vertical concrete slabs that often transform complex habitats into relatively featureless surfaces (Chapman and Underwood 2011). The lack of habitat complexity in seawalls means that they typically support only some of the organisms that occur in natural rocky intertidal habitats, resulting in altered recruitment, survival, densities, fecundity, and species interactions (Bulleri and Chapman 2010; Chapman and Bulleri 2003; Chapman and Underwood 2011; Klein et al. 2011). Seawalls have lower diversity, supporting fewer mobile species than rocky shores, but biological effects are also manifested in changes of density, size, and reproductive capability of the organisms (Bulleri and Chapman 2004; Chapman 2003). Thus, the ecology of armored waterfronts is fundamentally different from natural shorelines.

The ecology of intertidal areas is driven by the physical environment, which is modified by shoreline armoring. Important differences between seawalls and natural rocky intertidal shorelines include habitat heterogeneity and complexity (e.g., slope, roughness, crevices, and overhangs) that tend to be lacking on seawalls. Slope and shade impact intertidal communities by affecting recruitment, thermal stress, desiccation, and survival (Blockley and Chapman 2008; Helmuth and Hofmann 2001; Menconi et al. 1999; Wethey 1984). Surface roughness (small-scale variations in the height of a surface) and crevices are important habitat features, especially as refuges from physical disturbance for invertebrates such as mussels, chitons, limpets, and snails (Bergeron and Bourget 1986; Faller-Fritsch and Emson 1986; McKindsey and Bourget 2001; Menconi et al. 1999; Moreira et al. 2007). Because they are flat, seawalls also have less space, a major limiting resource in rocky intertidal habitats, than natural hard substrata (Little and Kitching 1996; Raffaelli and Hawkins 1996). Limited space on seawalls creates more abrupt vertical zonation than naturally craggy, sloped rocky shorelines and may make organisms more vulnerable to increased competition and predation (Iveša et al. 2010; Klein et al. 2011).

There are several ways in which seawalls and other armoring structures can be constructed to minimize long-term impacts on aquatic ecosystems including supplementing structures (e.g., adding boulders, beach nourishment), decreasing footprint of the structure, not repairing or replacing structures as they degrade, or removing them altogether (see review in Nordstrom 2014). However, in highly urbanized settings, such techniques are often inconsistent with retaining the shoreline protection function of the armoring. When shoreline protection is essential, complexity can be added to armoring structures in the form of natural materials or quasi-natural structures, or features such as beaches can be engineered into the shoreline to break up the armored landscape. Ecological enhancements can occur at larger scales, such as introducing slopes, terraces, or tidepools (Chapman and Blockley 2009, this chapter) or adding engineered beaches (e.g., Toft et al. 2013), or at smaller scales such as changing the complexity or chemical composition of armoring surfaces (e.g., Coombes et al. 2015; Firth 2014; Perkol-Finkel and Sella 2014). Recent studies have shown that employing these engineered enhancements can significantly affect the local ecology (Browne and Chapman 2011; Chapman and Underwood 2011; Dugan et al. 2011). The ability to accurately measure (for existing projects) or predict (for proposed projects) the potential ecological benefits of a particular enhancement is important for assessing the success of a project and for comparing alternatives.

21.3 CASE STUDY: SEATTLE'S SEAWALL

The case study that we present in this chapter is located in Puget Sound, an estuarine fjord in Washington State, USA, where shorelines are primarily glacial sediment beaches, embayments, and deltas, including mudflats and tidal marshes, with rocky coasts limited mostly to the northern and seaward margins of the Sound (Shipman 2008). Beaches in Puget Sound have been modified by the addition of hard substrata through armoring, intertidal fills, seawalls, groins and jetties, and overwater structures (Shipman 2008). In particular, natural shorelines have declined in urban bays, with 68% modified in King County, the most populous county in the state where Elliott Bay and the City of Seattle are located (WDNR 2001). Mixed gravel-cobble beaches that originally characterized Elliott Bay with low- to high-bank bluffs to landward and low tide terraces to seaward were filled in the early 1900s and are now armored with seawalls. Currently Seattle's central waterfront has more than 3 km of seawall and numerous piers built adjacent to deep water, resulting in very little remaining shallow sloping intertidal beach (Figure 21.1) (WDNR 2001). As such, the heavily armored margin of Elliott Bay represents an extreme example of an urban-affected shoreline. Along less-affected Puget Sound shorelines, successful habitat enhancement designs have included complete removal of seawalls and restoration of the beach terrace (Toft et al. 2014), as well as incorporating soft shoreline armoring techniques and placement of logs, vegetation, and sediments (Johannessen et al. 2014). While these types of enhancements were not possible owing to infrastructure constraints on Seattle's vertical seawall, we describe other potentially viable alternatives to improving the seawall's habitat in this chapter.

Much of the driving force for habitat improvement in Puget Sound is the regional importance of Pacific salmon and their presence as juveniles in shallow nearshore waters. Chinook salmon (*Oncorhynchus tshawytscha*) are of particular concern because they are listed as threatened under the Endangered Species Act. Although a nonnatural environment of overwater piers and other structures that create extensive shading dominate Seattle's waterfront, juvenile pink (*Oncorhynchus gorbuscha*), chum (*Oncorhynchus keta*), and Chinook salmon are very abundant there during their

Figure 21.1 Downtown waterfront of Seattle, Washington, showing a segment of the seawall and pedestrian sidewalk in the center, a pier on the left, and another overwater structure (the Seattle Aquarium) on the right.

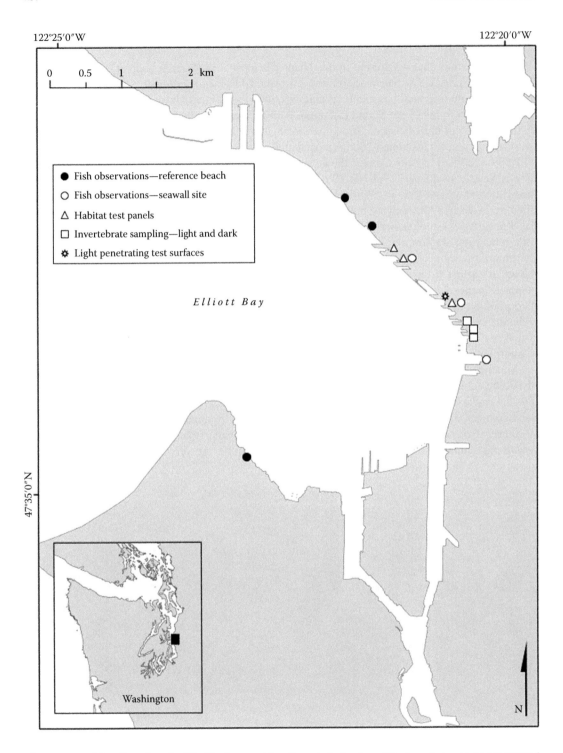

Figure 21.2 Elliott Bay, Seattle, Washington, and location of study sites. Filled symbols represent created beaches; open symbols represent seawall sites—see text for more detailed descriptions.

spring seaward migration. During this time, their proclivity for shallow water and shoreline orientation means that they are closely associated with the seawall and other overwater structures (Munsch et al. 2014; Toft et al. 2007). Also at this time, salmon feed extensively on terrestrial and aquatic invertebrates including various types of insects, harpacticoid copepods, and gammarid amphipods (Duffy et al. 2010; Feller and Kaczynski 1975; Healey 1979; Sibert 1979; Toft et al. 2007; Webb 1991). Juvenile salmon are active visual predators, and shade cast by piers reduces feeding intensity (Munsch et al. 2014). In addition, salmon collected near armoring tend to feed less on typical prey items than those collected at man-made beaches (Munsch et al. 2015).

Habitat enhancement options chosen for Seattle's seawall included (1) creation of an artificial beach and placement of intertidal benches and stone-filled marine mattresses designed to create shallow water low gradient habitat; (2) incorporating texture and relief into the seawall face, with the goal of increasing complexity and enhancing production of invertebrates and algae; and (3) placement of light-penetrating surfaces (LPS) in the sidewalk directly overhead of the seawall, intended to provide a light "corridor" such that juvenile salmon can migrate and feed normally, and which may improve productivity under piers. Design of seawall habitat enhancements was significantly informed both by a local history of monitoring urban habitat enhancements and by studies specifically targeted to identify enhancement techniques to improve the vertical face of the seawall itself and to enhance light penetration into shallow water (Figure 21.2). In the remainder of this chapter, we summarize this research and how it contributed to the design and construction of Seattle's seawall.

21.4 RESEARCH INFORMING SEAWALL ENHANCEMENT

21.4.1 Beaches and Benches—Olympic Sculpture Park

Along the shoreline north of Seattle's central waterfront (location of the seawall replacement) is the Seattle Art Museum's Olympic Sculpture Park (second site from the top in Figure 21.2). The Seattle Art Museum and the Trust for Public Land purchased a 7.3-acre former industrial site and developed it into a showcase for outdoor art, and also built habitat enhancements along the shoreline. When the park opened in 2007, it included two major enhancements intended to benefit juvenile salmon: (1) a ~290-m-long habitat bench projecting from the base of the seawall with sediment simulating more natural conditions than those found along most of Seattle's waterfront and (2) a ~100-m-long pocket beach excavated from a stretch of riprap-armored shoreline and surfaced with pebbles and cobbles (Figure 21.3). Monitoring at the site was conducted before enhancement (2005) and after enhancement (2007, 2009, and 2011), with comparisons to adjacent seawall and riprap shorelines. Full methods and biological and physical results through 2009 are reported in Toft et al. (2013). Biological monitoring included snorkel observations of juvenile salmon and other fishes and sampling their potential prey invertebrates using two methods: an epibenthic suction pump that vacuums small aquatic invertebrates from near the bottom, and fallout traps (plastic trays with soapy water placed on the ground) that collect terrestrial insects. Figure 21.4 shows results of this monitoring updated through the end of the study in 2011: both the pocket beach and habitat bench shoreline enhancements benefited juvenile salmon, evidenced by higher salmon densities and feeding rates as well as more abundant and diverse invertebrate assemblages compared to pre-enhancement armored conditions.

21.4.2 "Bumps"—Seawall Surface Study

One of the outcomes of meetings between the City of Seattle and the University of Washington was a study to determine whether or not the vertical face of the seawall could be improved by adding

(a)

(b)

Figure 21.3   Olympic Sculpture Park pocket beach habitat enhancement before (a) and after (b) construction. Enhancement at the site also included a habitat sediment bench (only visible at low tide) at the base of the seawall that can be seen in the distance.

physical complexity—"bumps" in the words of one of our agency colleagues. The study site was the seawall along the central waterfront of Seattle (Figures 21.1 and 21.2). Logistical constraints limited study locations, but three replicate locations that had similar conditions of low freshwater influence, elevation at the toe of the seawall, orientation to sun and waves, and usable length for deploying experiments were found. With a southwest orientation, the seawall and its associated biota were subject to afternoon sun during summer low tides and exposure to wind, waves, and low temperatures in winter.

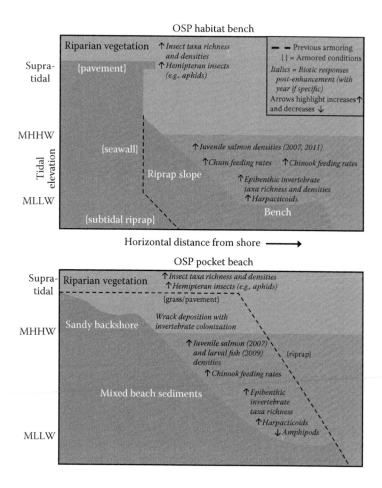

Figure 21.4 Conceptual model of the Olympic Sculpture Park monitoring results based on postrestoration sampling in 2007, 2009, and 2011, compared to prerestoration sampling in 2005 (see Toft et al. 2013 for detailed methods and results through 2009). Arrows indicate statistically significant results compared to prerestored and armored conditions (ANOVA for densities and taxa richness, chi-square for feeding rates, $p < 0.05$ designating significance). Intertidal range is designated by mean higher high water (MHHW) and mean lower low water (MLLW).

With design, fabrication, and deployment help from City of Seattle engineers and contractors, experimental concrete test panels measuring approximately 1.5 m by 2.3 m were placed at each of the three study sites during winter 2007–2008. Panel treatments included three types of larger-scale relief (finned, stepped, and flat) and two surface textures (smooth and cobble) for a total of six treatments (Figure 21.5). At each site, the six treatments were attached in random order to the seawall face with the bottom of each panel at a tidal elevation of 0.0 m MLLW (mean lower low water). Reference (undisturbed original seawall) and control (pressure-washed seawall) sections were also randomly selected at each location when the panels were installed to provide a comparison to existing conditions and as a "time-zero" for evaluating development of test panel assemblages. Biota on the panels were sampled annually in the spring and summer, 2008–2011.

Quadrat sampling was used to quantify biota at three panel elevations (relative to MLLW): (1) "upper" from approximately 1.5 to 2.3 m, (2) "middle" from approximately 0.7 to 1.5 m, and (3) "lower" from approximately 0 to 0.7 m. Three random quadrat locations were chosen at each elevation for a total of nine quadrats per panel. On step and fin panels, subsamples were taken from

Flat smooth   Step smooth   Fin smooth

Flat cobble   Step cobble   Fin cobble

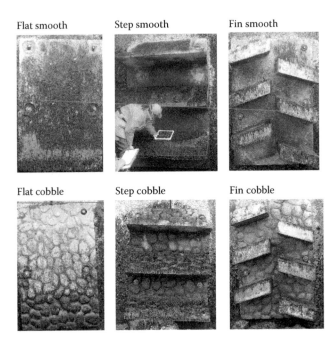

Figure 21.5 Six types of habitat treatment panels as they appeared several months after deployment on Seattle's seawall. Treatments not shown include power-washed and undisturbed seawall sections the same size as the panels (control and reference, respectively).

both vertical and sloped surfaces at each elevation to enable separate analyses of different substratum angles. Areas on the underside of fins and steps were not sampled. Invertebrates and algae were scanned in 25 cm × 25 cm gridded quadrats to identify species and estimate percent cover (Murray et al. 2006). Invertebrates and algae falling within each of 25 regularly spaced grid cells within the quadrats were identified and a percent cover was estimated for the entire quadrat (Dethier et al. 1993). Algae were grouped into major functional groups based on Steneck and Dethier (1994).

During 2008, the first year of the study, proportions of algae and invertebrates on engineered panels and on the control areas were quite different from the reference (Figure 21.6). In early stages of colonization, the panels were almost completely covered with algae (more than 80%) with small patches of bare space. The first algae to appear on the upper and middle elevations of panels consisted mainly of foliose forms dominated by the green algae *Ulva* spp. and the red algae *Porphyra* spp., with secondary contributions by green and brown microalgal biofilms (Figure 21.6). Dense mats of the green filamentous alga *Acrosiphonia coalita* were present at the lower elevations. Invertebrates on the panels were limited to new barnacle recruits and small snails. In contrast, the reference panels had a much lower percent cover of algae. The reference panels also had more bare space and dead algae and barnacles. A year later in June 2009, communities on the test panels had become more similar to those on the reference and controls and remained so through the rest of the study period (Figure 21.6). However, one difference noted at the end of the study in June 2011 was that assemblages on the flat panels, controls, and references were dominated by only one or two organism types, whereas those on the step and fin panels were distributed more evenly into the various functional groups. This was further demonstrated by applying Pielou's evenness measure, a measure of diversity that indicates how evenly the functional groups were distributed among the different treatments, which showed higher values at the step and fin panels (0.79 and 0.91, respectively) compared to the flat, control, and reference treatments (0.58, 0.65, and 0.57, respectively).

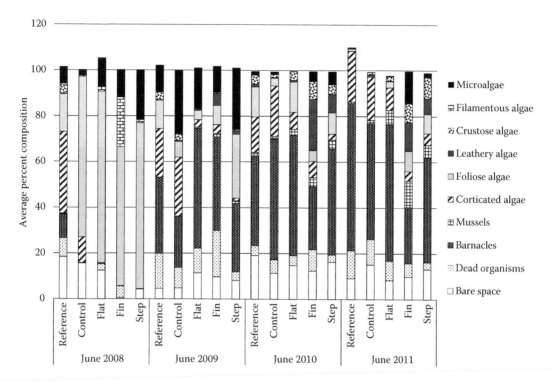

Figure 21.6 Percent cover of organism functional groups by main relief treatments (e.g., flat, fin, step) on seawall test panels over the study period. Cover exceeded 100% in cases where one type of organism overlaid another (e.g., algae on top of barnacles).

At the end of the second year, panels with the cobble surface treatment had significantly greater coverage of *Mytilus* mussels at all three elevations, indicating better recruitment to this surface texture. At this point, presence of larger relief step and fin features did not have a significant effect on mussels. During the last 2 years of the study, the cobble surface effect on mussel cover had disappeared and they showed increased abundances on all of the panel types compared to the control and references, with highest abundances on the step and fin panels (Figure 21.7b). From the second year of the study on, the rockweed *Fucus distichus* had consistently higher abundances on steps and fins than on vertical surfaces (flat panels, reference, and control) (Figure 21.7a). The higher abundances of both *Mytilus* and *Fucus* on cobble or step and fin panels were considered to be a beneficial contribution to the seawall habitat because both are globally considered to be ecosystem engineers encouraging rich faunal communities by creating habitat, altering sediment fluxes and nutrients, and providing food and refuge (Koivisto and Westerbom 2010 and references cited therein).

21.4.3 Pier and Light Effects Studies

The shade cast by piers in Elliott Bay has large effects on juvenile salmon behavior and the invertebrates that the salmon feed on. Juvenile salmon in Elliott Bay rarely occur or feed in shaded areas under piers and often aggregate directly adjacent to the shade (Munsch et al. 2014). These effects are particularly evident at high tides when light penetration under piers is limited because of higher water levels. Juvenile salmon, especially Chinook and chum salmon, feed on a variety of arthropod invertebrates, mainly harpacticoid copepods, gammarid amphipods, and dipteran flies (Duffy et al. 2010; Feller and Kaczynski 1975; Healey 1979; Sibert 1979; Toft et al. 2007; Webb 1991). These taxa are often associated with marine algae and as such are affected by pier shading.

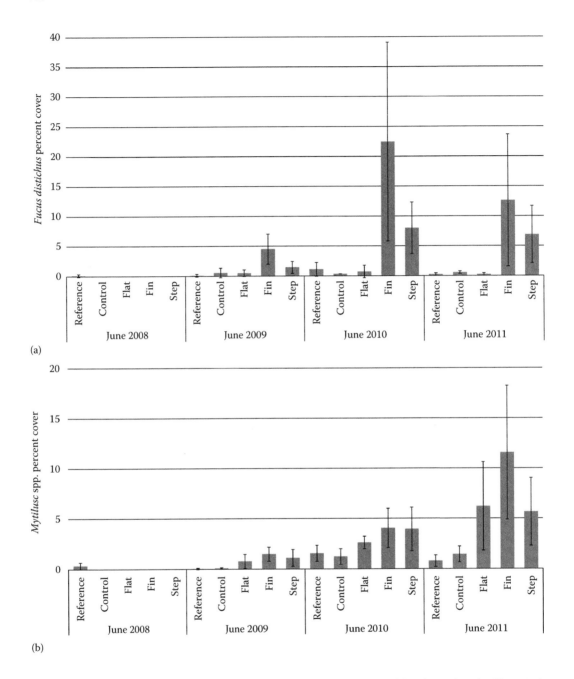

Figure 21.7 Average percent cover of *Fucus distichus* (a) and *Mytilus* spp. (b) in June of each of four study years. Error bars represent 95% confidence intervals.

In 2014, we sampled epibenthic invertebrates in April and June under three piers in Elliott Bay and in three adjacent sunlit areas, using an epibenthic suction pump (Figure 21.2; see Toft et al. 2013 for methods). We used generalized linear models (GLMs) to assess the effect of pier shading on the abundance of these taxa and overall taxa richness. This approach allowed us to account for differences in abundances among piers and between months, in addition to the nonnormal distribution of count data. GLMs indicated a significant ($p < 0.05$) negative effect of pier shading on the abundance

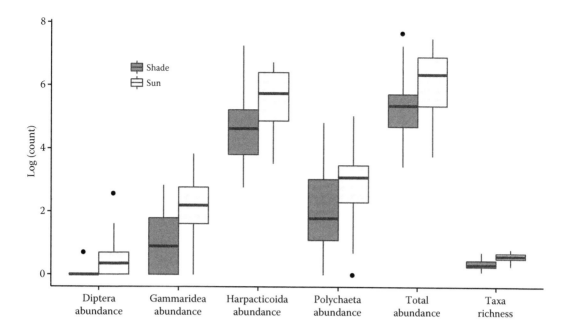

Figure 21.8 Averages for abundances of three invertebrate groups important in juvenile salmon diets, overall invertebrate abundance, and taxa richness combined from May and June 2014, in shaded areas under three piers and sunlit areas adjacent to the piers.

of Diptera, Gammaridea, and Harpacticoida, as well as overall invertebrate abundance and taxa richness (Figure 21.8).

In order to reduce the negative effects of shading under piers and other poorly lit areas, LPS were proposed to illuminate the aquatic habitat along the seawall. In a preconstruction pilot study, three types of LPS were installed into the surface of a pier approximately 5 m from the shoreline: metal grating (3 × 1.5 m), glass panels (3 × 1.5 m), and a solar tube (0.5 m diameter—similar to a residential skylight). The effect of the LPS on light levels was quantified by taking measurements of photosynthetically active radiation (PAR), the portion of the light spectrum from 400 to 700 nm that supports photosynthesis. After the LPS installation in 2013, measurements occurred adjacent to the pier, directly underneath the LPS, and in shaded areas next to the LPS (controls). Light levels were recorded at the surface and at 0.5-m-depth intervals in the depths that juvenile salmon occupy. Visual surveys of fish occurred April–July, along snorkel transects 3 and 10 m from shore under and adjacent to piers (see detailed methods in Munsch et al. 2014). Each observation recorded the species identity and school size of juvenile salmon, the presence or absence of feeding behavior, and the location of the school relative to the shade cast by the pier.

Light measurements before LPS installation indicated much lower PAR in areas underneath the pier compared to away from the pier, and decreasing PAR with greater distance underneath the pier (Table 21.1). Although direct comparisons could not be made among the LPS types because they occurred at different distances from the pier edge (e.g., the grating LPS was closest to the pier edge and thus had the highest PAR) and were of different sizes, the LPS clearly mitigated for some of the light lost. However, light levels under the LPS were substantially less than ambient conditions.

Fish surveys before installation of the LPS recorded both juvenile pink and Chinook salmon, while those after LPS installation recorded only juvenile Chinook salmon (juvenile pink salmon in this region are present every other year). Chinook salmon were more evenly distributed compared to the prior (non-LPS) year, including 10 observations (of 25 total) in shaded areas underneath the pier (Figure 21.9). On one occasion, Chinook salmon were observed feeding in the shaded areas under

Table 21.1   **Photosynthetically Active Radiation (PAR; μ mol⁻² s⁻¹) Observed above the Surface of the Water in Ambient Conditions, under Three Types of LPS Installed on a Pier, and LPS Controls (Dark Areas Adjacent to Each LPS) Averaged from Four Different Days**

Treatment	Mean PAR
Ambient	1028.2
Grating	142.8
Grating control	43.8
Glass blocks	41.3
Glass blocks control	12.6
Solar tube	11.0
Solar tube control	4.9

Note: Treatments are listed in order of increasing distance under the pier.

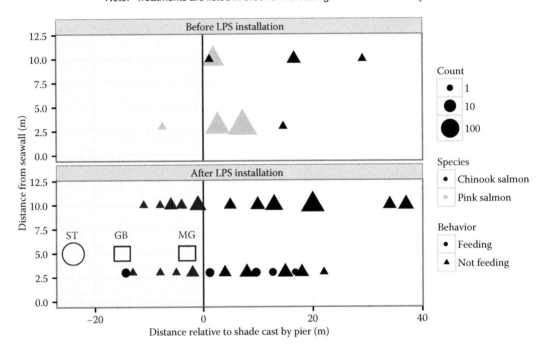

Figure 21.9   Locations, behavior, and school size of juvenile salmon relative to shade cast by Pier 62/63 before and after LPS installations of a solar tube (ST), glass blocks (GB), and metal grating (MG). Shaded area on the graph corresponds to the shaded area under the pier.

LPS. Of the five schools of Chinook salmon that were observed under piers in Elliott Bay the year prior, none were feeding (Munsch et al. 2014). Although this preliminary study was unreplicated and there were uneven number of salmon observations before and after LPS installation, the results suggested that LPS may have encouraged juvenile salmon use of the under-pier habitat.

21.5 IMPLEMENTATION

Ultimately, enhancements derived from the findings of the research we have described were included in the seawall design, adding texture and relief, intertidal benches, and a corridor near shore illuminated by glass panels (Figure 21.10). However, some compromises were made to meet

(a)   (b)

Figure 21.10 Habitat enhancements in the new construction of Seattle's seawall: light-penetrating surfaces (LPS) in the sidewalk (a), intertidal benches, and seawall relief and texturing under LPS (b). The sheet pile retaining wall on the water side was removed postconstruction. (Right photo credit: Seattle Department of Transportation.)

the needs of the various stakeholders such as the Seattle Department of Transportation, state and federal resource agencies, the Army Corps of Engineers, and local Native American tribes. Among the concerns that needed to be addressed in the design were preservation of existing businesses and associated transportation, minimizing interference with navigation and other in-water activities, and preservation and enhancement of habitat important to juvenile salmon. Technical panels had proposed intertidal benches of 30 or 60 m width, but the final design included benches of 4 m width or less that would not encroach on navigation and commerce needs. Also, because of engineering requirements, intertidal benches along much of the waterfront, especially under piers, are constructed of confined fill (Figure 21.10)—stacked mesh mattresses containing quarry rock rather than the finer sediment that was shown to be productive at the Olympic Sculpture Park. Real estate for constructing pocket beaches such as at the Olympic Sculpture Park was largely unavailable along Seattle's central waterfront, but a similar beach adjacent to Seattle's ferry terminal will be constructed as part of the seawall habitat enhancements. During the design phase, it became evident that in order to maintain preexisting pedestrian space, the sidewalk would be cantilevered 10–15 ft over the water in many places, shading much of the area intended for use by juvenile salmon. At this point, the concept of providing light in the sidewalk surface via glass panels was included in the final design and is being implemented in the construction, not only in piers but also in the sidewalk (Figure 21.10).

21.6 FUTURE CONSIDERATIONS

Construction of Seattle's seawall and its habitat enhancements represent a unique experiment on an unprecedented scale. Upon completion, the project will replace approximately 7000 linear feet of relatively featureless vertical seawall with a wall containing texture and relief and having a uniform elevation shallow bench at its base. The nearshore will consist of a corridor illuminated by LPS designed to provide juvenile salmon with the ability to migrate and feed more naturally than they would in a shaded environment—local media has characterized the project as a "highway for fish" (http://www.seattletimes.com/seattle-news/environment/seattles-new-seawall-also-a-highway-for-fish/).

The potential ecological benefit of enhancing habitat complexity of seawalls and other marine structures is receiving increased attention (Airoldi et al. 2005; Bulleri and Chapman 2010; Chapman and Blockley 2009; Chapman and Bulleri 2003; Dafforn et al. 2015; Davis et al. 2002; Dyson and Yocum 2015; Glasby and Connell 1999). The use of ecological criteria in seawall design may mitigate some negative impacts, while still serving societal needs for erosion protection and infrastructure support (Bulleri and Chapman 2010). In Sydney Harbor, Australia, seawall enhancements such as small tide pools resulted in significant increases of algal and sessile invertebrate species diversity, especially at higher elevations of the intertidal zone (Browne and Chapman 2011; Chapman and Blockley 2009). There, features such as crevices on seawalls can provide important microhabitats and slope along urbanized shorelines, which can increase distribution, cover, and types of sessile and mobile invertebrates (Chapman 2003; Chapman and Underwood 2011; Moreira et al. 2006, 2007). Similarly, the research we presented in this chapter indicates that benches, beaches, and "bumps" added to Seattle's seawall can have beneficial effects. We note, however, that this scale of manipulation and testing of seawall habitat has been rare, and it is important to continue to test the generality of the conclusions reached from Sydney Harbor and Elliott Bay before comprehensive guidelines for enhancing seawalls in urban waters can be developed with confidence.

The uniqueness of Seattle's seawall enhancements also means that there are several unknowns with regard to whether or not they will function as intended. For example, LPS will provide only a fraction of the light that was available before the seawall was moved back, and this may not be enough for fish to orient themselves or feed normally, or for productivity of invertebrates to reach levels beneficial to the fish. It is difficult to predict how juvenile salmon will react because although much is known about the effect of photoperiod (day length) on salmonid behavior and growth, comparatively little is known about the effects (and minimum requirements) of light intensity (Boeuf and Le Bail 1999). Other questions, such as whether or not confined fill marine mattresses are as productive as benches constructed with finer sediments, are also posed by this project. It is often difficult to find postproject support for monitoring habitat enhancement projects to acquire sound scientific knowledge about restoration or enhancement efficacy, but in this case, Seattle's Department of Transportation has developed a 10-year monitoring and adaptive management plan designed to address the unknowns (SDOT 2013). The plan includes postconstruction monitoring of all the habitat enhancements using methods consistent with or identical to those used in preconstruction studies (Table 21.2). The nearshore habitat enhancement objectives for Seattle's seawall are to improve salmon migration along the shoreline and provide enhanced habitat for salmon and other fish, and accordingly the following hypotheses will be tested by the monitoring plan (SDOT 2013):

- Hypothesis 1—Salmon Migration. Creation of an intertidal migratory corridor results in improved salmon migration along the seawall shoreline.
- Hypothesis 2—Physical. The created habitat benches and beach are dynamically stable shoreline features up to the 10-year storm and wave design event, as measured by cross sections and sediment gradation.
- Hypothesis 3—Nearshore Ecosystem Biota. Habitat enhancement features provide improved habitat for salmon and other fish, as measured by fish, epibenthic and benthic invertebrates, insect, algae, and riparian vegetation assemblages.

Habitat improvements along extremely urbanized shorelines are obviously constrained by human uses and the infrastructure needs of a city. Enhancements along Seattle's seawall will be extensive given the status quo of a monotypic seawall. Other habitat improvements were feasible, but were limited in implementation owing to engineering or concessions to other needs of the shoreline such

Table 21.2 Postconstruction Monitoring for the Seattle Seawall

Monitoring Type	Years Conducted (after Construction)	Monitoring Elements
Illumination	Years 1, 2, and 3	Light levels above/below cantilevered sidewalks with LPS, reference open, reference shaded
Corridor physical characteristics	Years 1, 3, 5, and 10	Bench, beach, and marine mattress physical characteristics (slope, substrate sizing, erosion)
Invertebrate colonization of benches and beach	Years 1, 3, and 5	Epibenthic invertebrate colonization on benches and benthic invertebrates at beach
Salmon presence and behavior	Years 1, 2, 3, 5, and 10	Salmon behavior and presence at benches, beach, cantilevered sidewalks with LPS, and piers

Source: Modified from Seattle Department of Transportation (SDOT). 2013. *Elliott Bay Seawall Project Post-Construction Monitoring and Adaptive Management Plan.*

as business and transportation. Other enhancements that could be considered along urban seawalls in the future include plantings of vegetation close to the water's edge or overhanging the water if possible (e.g., if on a seawall or pier). Not only would this provide a natural aesthetic component, but it would potentially provide input of insect prey to juvenile salmon. Removing rather than mitigating for artificial shading caused by piers and cantilevered sidewalks should be a primary goal. LPS should be used only where decreasing the shade footprint is not possible owing to engineering or infrastructure requirements. However, not all enhancements need conflict with urban waterfront needs. For example, constructed beaches provide enhanced habitat for juvenile salmon as well as public parks for people (e.g., Toft et al. 2013).

Several questions remain on how this type of urban seawall enhancement fits in to the broader spectrum of living shorelines and restoration techniques. What is the best-case scenario that we can expect from enhancements along extremely urbanized shorelines? How should managers weigh the costs versus benefits? Should ecosystem services be monetarily or culturally appraised to inform economic design of waterfronts? In what cases does this type of seawall enhancement apply, for example, only to seawalls that extend into deeper water and are necessary to protect urban infrastructure, or also to smaller and less extensive seawalls? Can these techniques be combined with other soft armoring and living shoreline techniques? Should biodiversity enhancement and other multifunctional goals (e.g., pollution mitigation) be incorporated into policy as Dafforn et al. (2015) suggest? It is clear that there are limitations to what can be restored with seawall enhancements. For example, nearshore processes that depend on a healthy upper intertidal beach terrace such as deposition and breakdown of beach wrack by invertebrates (Heerhartz et al. 2014), natural sediment input from backshore sources, and beach spawning by forage fish will likely continue to be compromised by the presence of seawalls. Answers to these questions will depend on postconstruction data from projects such as Seattle's, which will inform future designs worldwide.

ACKNOWLEDGMENTS

Funding for the research described in this chapter was provided by the Seattle Department of Transportation (SDOT), a grant from King Conservation District to Seattle Public Utilities, Washington Sea Grant, the Estuary and Salmon Restoration Program, and the Washington Department of Fish and Wildlife. SDOT also provided for fabrication and deployment of the experimental habitat panels. S.H.M. was supported by a National Science Foundation Graduate Research Fellowship. We are particularly indebted to Elizabeth Armbrust, Michael Caputo, Erin Morgan, and other members of the University of Washington's Wetland Ecosystem Team, who spent many hours in the field and laboratory.

REFERENCES

Airoldi, L., M. Abbiati, M.W. Beck, S.J. Hawkins, P.R. Jonsson, D. Martin, P.S. Moschella, A. Sundelof, R.C. Thompson, and P. Aberg. 2005. An ecological perspective on the deployment and design of low-crested and other hard coastal defence structures. *Coastal Engineering* 52: 1073–1087.

Bergeron, P.E. and E. Bourget. 1986. Shore topography and spatial partitioning of crevice refuges by sessile epibenthos in an ice disturbed environment. *Marine Ecology Progress Series* 28: 129–145.

Blockley, D. J. and M. G. Chapman. 2008. Exposure of seawalls to waves within an urban estuary: Effects on intertidal assemblages. *Austral Ecology* 33: 168–183.

Boeuf, G. and P.Y. Le Bail. 1999. Does light have an influence on fish growth? *Aquaculture* 177: 129–152.

Browne, M.A. and M.G. Chapman. 2011. Ecologically informed engineering reduces loss of intertidal biodiversity on artificial shorelines. *Environmental Science and Technology* 45: 8204–8207.

Bulleri, F. and M.G. Chapman. 2004. Intertidal assemblages on artificial and natural habitats in marinas on the north-west coast of Italy. *Marine Biology* 145: 381–391.

Bulleri, F. and M.G. Chapman. 2010. The introduction of coastal infrastructure as a driver of change in marine environments. *Journal of Applied Ecology* 47: 26–35.

Chapman, M.G. 2003. Paucity of mobile species on constructed seawalls: Effects of urbanization on biodiversity. *Marine Ecology Progress Series* 264: 21–29.

Chapman, M.G. and D.J. Blockley. 2009. Engineering novel habitats on urban infrastructure to increase intertidal biodiversity. *Oecologia* 161: 625–635.

Chapman, M.G. and F. Bulleri. 2003. Intertidal seawalls—New features of landscape in intertidal environments. *Landscape and Urban Planning* 62: 159–172.

Chapman, M.G. and A.J. Underwood. 2011. Evaluation of ecological engineering of "armoured" shorelines to improve their value as habitat. *Journal of Experimental Marine Biology and Ecology* 400: 302–313.

Coombes, M.A., E.C. La Marca, L.A. Naylor, and R. C. Thompson. 2015. Getting into the groove: Opportunities to enhance the ecological value of hard coastal infrastructure using fine-scale surface textures. *Ecological Engineering* 77: 314–323.

Dafforn, K.A., T.M. Glasby, L. Airoldi, N.K. Rivero, M. Mayer-Pinto, and E.L. Johnston. 2015. Marine urbanization: An ecological framework for designing multifunctional artificial structures. *Frontiers in Ecology and the Environment* 13: 82–90.

Davis, J., L. Levin, and S. Walther. 2002. Artificial armored shorelines: Sites for open-coast species in a southern California bay. *Marine Biology* 140(6): 1249–1262.

Dethier, M.N., E.S. Graham, S. Cohen, and L.M. Tear. 1993. Visual versus random-point percent cover estimations: 'Objective' is not always better. *Marine Ecology Progress Series* 96: 93–100.

Duffy, E.J., D.A. Beauchamp, R.M. Sweeting, R.J. Beamish, and J.S. Brennan. 2010. Ontogenetic diet shifts of juvenile Chinook salmon in nearshore and offshore habitats of Puget Sound. *Transactions of the American Fisheries Society* 139: 803–823.

Dugan, J.E., L. Airoldi, M.G. Chapman, S.J. Walker, and T. Schlacher. 2011. 8.02-Estuarine and coastal structures: environmental effects, a focus on shore and nearshore structures. In: Luckenbach, M.W., R. Mann, and J.A. Wesson (Eds.), *Oyster Reef Habitat Restoration: A Synopsis and Synthesis of Approaches*, pp. 239–250. Waltham: Academic Press.

Dyson, K. and K. Yocom. 2015. Ecological design for urban waterfronts. *Urban Ecosystems* 18: 189–208.

Faller-Fritsch, R.J. and R.H. Emson. 1986. Causes and patterns of mortality in *Littorina rudis* (Maton) in relation to intraspecific variation: A review. In: Peterson, M.S., G.L. Waggy, and M.S. Woodrey (Eds.), *Grand Bay National Estuarine Research Reserve: An Ecological Characterization*, pp. 103–174. New York: Columbia University Press.

Feller, R.J. and V.W. Kaczynski. 1975. Size selective predation by juvenile chum salmon (*Oncorhynchus keta*) on epibenthic prey in Puget Sound. *Journal of the Fisheries Board of Canada* 32: 1419–1429.

Firth, L.B., R.C. Thompson, K. Bohn, M. Abbiati, L. Airoldi, T.J. Bouma, F. Bozzeda, V.U. Ceccherelli, M.A. Colangelo, A. Evans, and F. Ferrario. 2014. Between a rock and a hard place: Environmental and engineering considerations when designing coastal defence structures. *Coastal Engineering* 87: 122–135.

Glasby, T. and S. Connell. 1999. Urban structures as marine habitats. *Ambio* 28: 595–598.

Healey, M.C. 1979. Detritus and juvenile salmon production in the Nanaimo Estuary: I. Production and feeding rates of juvenile chum salmon (*Oncorhynchus keta*). *Journal of the Fisheries Board of Canada* 36: 488–496.

Heerhartz, S.M., M.N. Dethier, J.D. Toft, J.R. Cordell, and A.S. Ogston. 2014. Effects of shoreline armoring on beach wrack subsidies to the nearshore ecotone in an estuarine fjord. *Estuaries and Coasts* 37: 1256–1268.

Helmuth, B.S. and G.E. Hofmann. 2001. Microhabitats, thermal heterogeneity, and patterns of physiological stress in the rocky intertidal zone. *The Biological Bulletin* 201: 374–384.

Iveša, L., M.G. Chapman, A.J. Underwood, and R.J. Murphy. 2010. Differential patterns of distribution of limpets on intertidal seawalls: Experimental investigation of the roles of recruitment, survival and competition. *Marine Ecology Progress Series* 407: 55–69.

Johannessen, J., A. MacLennan, A. Blue, J. Waggoner, S. Williams, W. Gerstel, R. Barnard, R. Carman, and H. Shipman. 2014. *Marine Shoreline Design Guidelines*. Washington Department of Fish and Wildlife, Olympia, Washington.

Klein, J.C, A.J. Underwood, and M.G. Chapman. 2011. Urban structures provide new insights into interactions among grazers and habitat. *Ecological Applications* 21: 427–438.

Koivisto, M.E. and M. Westerbom. 2010. Habitat structure and complexity as determinants of biodiversity in blue mussel beds on sublittoral rocky shores. *Marine Biology* 157: 1463–1474.

Little, C. and J.A. Kitching. 1996. *The Biology of Rocky Shores*. Oxford and New York: Oxford University Press.

McKindsey, C.W. and E. Bourget. 2001. Diversity of a northern rocky intertidal community: The influence of body size and succession. *Ecology* 82: 3462–3478.

Menconi, M., L. Benedetti-Cecchi, and F. Cinelli. 1999. Spatial and temporal variability in the distribution of algae and invertebrates on rocky shores in the northwest Mediterranean. *Journal of Experimental Marine Biology and Ecology* 233: 1–23.

Moreira, J., M.G. Chapman, and A.J. Underwood. 2006. Seawalls do not sustain viable populations of limpets. *Marine Ecology Progress Series* 322: 179–188.

Moreira, J., M.G. Chapman, and A.J. Underwood. 2007. Maintenance of chitons on seawalls using crevices on sandstone blocks as habitat in Sydney Harbour, Australia. *Journal of Experimental Marine Biology and Ecology* 347: 134–143.

Munsch, S.H., J.R. Cordell, and J.D. Toft. 2015. Effects of seawall armoring on juvenile Pacific salmon diets in an urban estuarine embayment. *Marine Ecology Progress Series* 535: 213–229.

Munsch, S.H., J.R. Cordell, J.D. Toft, and E.E. Morgan. 2014. Effects of seawalls and piers on fish assemblages and juvenile salmon feeding behavior. *North American Journal of Fisheries Management* 34: 814–827.

Murray, S.N., R.F. Ambrose, and M.N. Dethier. 2006. *Monitoring Rocky Shores*. Berkeley: University of California Press.

Nordstrom, K.F. 2014. Living with shore protection structures: A review. *Estuarine, Coastal and Shelf Science* 150: 11–23.

Neumann, B., A.T. Vafeidis, J. Zimmermann, and R.J. Nicholls. 2015. Future coastal population growth and exposure to sea-level rise and coastal flooding—A global assessment. *PLoS ONE* 10(3): e0118571.

Perkol-Finkel, S. and I. Sella. 2014. Ecologically active concrete for coastal and marine infrastructure: Innovative matrices and designs. In *Proceeding of the 10th ICE Conference: From Sea to Shore—Meeting the Challenges of the Sea. ICE Publishing, 18–20 September 2013-Edinburgh, UK*, 1139–1150.

Raffaelli, D.G. and S. J. Hawkins. 1996. *Intertidal Ecology*. London and New York: Chapman and Hall.

Seattle Department of Transportation (SDOT). 2013. *Elliott Bay Seawall Project Post-Construction Monitoring and Adaptive Management Plan*.

Shipman, H. 2008. *A Geomorphic Classification of Puget Sound Nearshore Landforms*. Washington Department of Ecology.

Sibert, J.R. 1979. Detritus and juvenile salmon production in the Nanaimo estuary: II. Meiofauna available as food to juvenile chum salmon (*Oncorhynchus keta*). *Journal of the Fisheries Board of Canada* 36: 497–503.

Steneck, R.S. and M.N. Dethier. 1994. A functional group approach to the structure of algal-dominated communities. *Oikos* 69: 476–498.

Toft, J.D., J.R. Cordell, and E.A. Armbrust. 2014. Shoreline armoring impacts and beach restoration effectiveness vary with elevation. *Northwest Science* 88: 367–375.

Toft, J.D., J.R. Cordell, C.A. Simenstad, and L.A. Stamatiou. 2007. Fish distribution, abundance, and behavior along city shoreline types in Puget Sound. *North American Journal of Fisheries Management* 27: 465–480.

Toft, J.D., A.S. Ogston, S.M. Heerhartz, J.R. Cordell, and E.E. Flemer. 2013. Ecological response and physical stability of habitat enhancements along an urban armored shoreline. *Ecological Engineering* 57: 97–108.

WDNR. 2001. *The Washington State ShoreZone Inventory*. Nearshore Habitat Program, Washington State Department of Natural Resources, Olympia, WA.

Webb, D.G. 1991. Effect of predation by juvenile Pacific salmon on marine harpacticoid copepods. I. Comparisons of patterns of copepod mortality with patterns of salmon consumption. *Marine Ecology Progress Series* 72: 25–36.

Wethey, D.S. 1984. Sun and shade mediate competition in the barnacles *Chthamalus* and *Semibalanus*: A field experiment. *The Biological Bulletin* 167: 176–185.

CHAPTER 22

The Ecological Impacts of Reengineering Artificial Shorelines
The State of the Science

Mark Anthony Browne and M.G. Chapman

CONTENTS

22.1 PROBLEMS OF EXISTING SHORELINES

Globally, 70% of countries with coastlines have their largest urban areas built on the coast or on "reclaimed" land (Harris 2012; McGranahan et al. 2007). To protect these urban areas from waves and erosion, many coastlines have been armored by building seawalls, revetments, bulkheads, and groynes. Seawalls, revetments, and bulkheads are physical structures built parallel to the shore to reduce the rate at which land is eroded, while groins are structures built perpendicular to the shore, which cause sediment in the water to accumulate on one side and to be removed from the other side (see review by Nordstrom 2014 and references therein). These structures are getting taller and longer because of increasing urbanization, rising sea levels, and more stormy weather and cover between 10% and 60% of the main coastline of some countries (i.e., China, Huang et al. 2015; United States, Gittman et al. 2015). This has led to the modification of natural habitats and their replacement by artificial habitats, causing diverse ecological impacts, so there have been a number of recent attempts at reengineering armored shorelines to increase their ecological value. Here, we review the evidence about the ecological impacts of reengineering seawalls, bulkheads, revetments, and groynes on artificial shorelines.

22.2 REPORTED ECOLOGICAL EFFECTS OF INTERTIDAL ARTIFICIAL SHORELINES

Armoring shorelines affect intertidal organisms, populations, and assemblages. Here, we discuss our understanding of the nature of such impacts (i.e., the level of biological organization affected, the techniques used, whether distributions or abundances of species or ecological processes are affected), the quality of the data obtained (i.e., anecdotal vs. structured sampling, sampling linked to experimentation), and the scale (i.e., temporal and spatial) of any impacts. We have focused our attention on seawalls and similar habitats, particularly in Australia and the UK, with some attention to work in the United States because that is where much of the experimental work has focused. Thus, much of the research described regard populations of invertebrates. We have also largely confined our discussion to impacts on rocky habitats as the effect of artificial habitats on intertidal and subtidal sediments has been recently thoroughly reviewed by Dugan et al. (2011).

First, data are needed to determine whether there are different populations and assemblages on seawalls or other armored shores compared to natural shores and, once these patterns have been established, research is needed to determine the features of the artificial habitats and the ecological process(es) affected (e.g., recruitment, growth, survival, competition) that cause the observed patterns of impact(s).

The quality of evidence is important because anecdotal evidence of any impacts, unsupported by quantitative data, cannot guide appropriate managerial action aimed at ameliorating such purported impacts. Similarly, data from one place or one time cannot, by themselves, provide evidence about what might be appropriate actions elsewhere. Ecological knowledge advances from well-structured sampling to identify spatial and temporal variability in ecological patterns, followed by experimental tests to evaluate the processes leading to the observed patterns (Underwood et al. 2000). Applying a similar approach to mitigating effects of armored shorelines is particularly important given how expensive such structures are to build and repair. With quantitative data (preferably supported by experimental evidence where possible), managers can alter features of infrastructure or attempt to control biological features of the assemblages or populations occupying them, to maintain key processes and reduce any impacts. The process of ecological engineering symbiotically combines engineering and ecological techniques to redesign and rebuild structures (Chapman and Underwood 2011) and associated natural assemblages in the ecosystem so that they benefit populations of humans and wildlife. This cannot be successful, however, without a detailed ecological understanding of the impacts.

A number of types of ecological impacts have commonly been associated with armored shorelines: (i) changes to population structure of individual species, their behavior, or interactions; (ii) reduced native biodiversity; (iii) changes to biodiversity in adjacent habitats'; and (iv) increases in nonindigenous species.

22.2.1 Changes to Populations of Individual Species and Their Interactions

Most research on the effects of seawalls and similar structures on the behavior of individual species has focused on limpets because they are often numerous on artificial structures, they are important interactors in intertidal assemblages, their ecology in natural habitats has been well documented, and they are relatively easy to manipulate experimentally in both natural and artificial habitats. The limpet *Siphonaria denticulata* is extremely common on seawalls in Sydney Harbor, which would indicate that the habitats are suitable for this species. Nevertheless, the individuals are smaller and have been shown to produce smaller egg masses and eggs (Moreira et al. 2006). This raises questions about the long-term local persistence of the species, if seawalls replace a considerable amount of their native habitat.

The choice of material used to build armored shorelines is also important. Surveys by Moreira (2006) showed differences in the abundances of different species of limpets between sandstone and concrete seawalls; for example, sandstone seawalls had large densities of *Siphoniaria denticulata* and smaller densities of *Patelloida latistrigata* than were found on concrete seawalls. Subsequent manipulative experiments suggested that differences in recruitment and competition among the species were largely responsible for some of the patterns observed (Iveša et al. 2010).

Similar work linking surveys with manipulative experiments have shown that the greater cover of oysters on seawalls provide refuge for the small grazer *P. latistrigata* from the larger grazer *Cellana tremoserica* (Klein et al. 2011), in a similar way that large barnacles provide refuge for *P. latistrigata* from competition on natural shores (Creese 1982). Oysters similarly provide habitat for whelks on seawalls, with increased survival and growth on walls where there is a thick crust of oysters (Jackson et al. 2008).

As well as potentially affecting interactions among species, as described above, artificial shorelines may affect the ways in which animals forage and disperse. Thus, the large limpet, *Cellana tramoserica*, tends to disperse further and show less tendency to home when foraging on seawalls compared to natural shores (Bulleri et al. 2004). Intertidal chitons also showed increased dispersal on seawalls when the crevices in which they shelter were filled in, showing the importance of the complexity of the surface of a seawall to its suitability of habitat for some intertidal species (Martins et al. 2010; Moreira et al. 2007).

22.2.2 Reduced Native Biodiversity on the Shoreline Itself

The biodiversity of an area can be defined by the biological variability (e.g., phenotypic, genetic) among (i.e., number of species) and within each species (subtaxonomic diversity). In Australia, a combination of structured surveys has been linked with manipulative experiments to identify reductions in native biodiversity on armored shores and some of the processes that underlie these changes. Chapman (2003) showed that assemblages on seawalls lack 50% of the mobile species (common grazing snails, predatory whelks, starfish, sea urchins, sipunculids, limpets, and chitons) and that most of the rare species were found on the rocky shores. Extensive surveys by Bulleri et al. (2005) similarly showed that seawalls lacked many mobile gastropods (*Nerita atramentosa*, *Onchidella patelloides*) and common brown algae (*Hormosira banksii*, *Sargassum* sp.). These changes to biodiversity vary according to whether the seawall is shaded or made of sandstone or concrete. For instance, surveys have shown that there are more mobile grazers and algae on sun-exposed walls and more sessile animals on shaded walls (Blockley 2007; Blockley and Chapman 2006). Experiments by Blockley and Chapman (2006) showed that these differences were attributed to differences in recruitment.

Outside of Australia, there has been less work combining observations of impacts from surveys with manipulative experiments designed to test specific hypotheses about processes underlying these impacts. In an extensive study in Europe, Moschella et al. (2005) compared intertidal assemblages on offshore groynes to natural rocky shores and showed similar patterns, although patterns were very different from country to country. They also demonstrated that provision of microhabitats in the form of pits could increase diversity locally. In the United Kingdom, Attrill et al. (1999) sampled the Thames Estuary and showed more species in areas with more vegetation. Similarly, Hoggart et al. (2012) surveyed sites along the Thames, above and below the mean high tide and showed more species on larger brick walls (including algae lower on the shore) and fewer species on concrete walls.

Experiments are needed to determine what causes differences such as these, because various alternative models could explain such patterns, for example, the features of the brick walls facilitate recruitment of algae and invertebrates independently, increased recruitment of algae onto the brick walls facilitates survival of the insects, recruitment is similar on the two surfaces, but crevices in the brick walls reduce subsequent mortality. Understanding these processes will make it more cost-effective to reengineer such habitats, for example, decide whether concrete walls should be replaced by bricks, whether it would be cheaper to add microhabitats to concrete structures.

Other surveys have similarly shown considerable differences in assemblages of species among artificial shores of different structures and when compared to natural shores, for example, on riprap compared to natural shores (North America, Davis et al. 2002; Morley et al. 2012; Pister 2009; South America, Aguilera et al. 2014; United Kingdom, Russell 2000). In very few studies, however, the description of the patterns has been combined with manipulative experiments to try to understand the processes underlying the reported differences.

Building artificial shores may also increase or decrease the movement of biological material (Heerhartz et al. 2014, 2015), organisms (Bulleri et al. 2004; Moreira et al. 2007), gametes, and genes among populations on shorelines by altering ecological (competition, facilitation, predation, and production of viable offspring; e.g., Iveša et al. 2010; Klein et al. 2011; Moreira 2006), physiological (growth, survival; e.g., Iveša et al. 2010; Klein et al. 2011), and hydrological (i.e., how water is circulated through advection and diffusion; Denny et al. 2003; Gentile and Landò 2007; Neelamani and Sandhya 2005) processes that might affect the ability of organisms to disperse their larvae. Although armoring shorelines has been shown to reduce the deposition of wrack among shores (Heerhartz et al. 2014, 2015) and cause grazing limpets (Bulleri et al. 2004) and chitons (Moreira et al. 2007) to disperse further when foraging, less is known about the movement of larvae, gametes, and genes among populations of organisms on separate natural and artificial shores. This connectivity can be measured directly or indirectly using a range of chemical and genetic material from organisms collected as part of surveys and experiments (see review by Cowen and Sponaugle 2009).

Genetic techniques have identified disconnections among populations of organisms on natural and artificial shores. Fauvelot et al. (2009) used structured sampling to show that limpets (*Patella caerula*) living on artificial shores were genetically less diverse than limpets living on rocky shores, although experiments are needed to determine the hydrological and ecophysiological processes that cause limpets on seawalls to have fewer genotypes. Two explanatory models could explain the observed patterns. First, the limpets on rocky shores and seawalls may be poorly connected (in space and time), causing larvae of the diverse genotypes of limpets found on rocky shores to arrive and settle on seawalls in smaller numbers. Second, the larvae of the less diverse genotypes of limpets found on seawalls may be more viable, allowing them to survive and reproduce at greater rates after settlement. It is therefore clear that genetic methods alone are unable to provide robust evidence about the ability of species to persist in landscapes fragmented by the activities of humans (see review by Lowe and Allendorf 2010) because to test these models requires experiments to estimate the dispersal distances (e.g., using "transgenerational" chemicals tags; Pecl et al. 2010) and rates of survival of this species among natural and artificial shores. Together with modeling, such experiments would allow research to determine which populations of limpets are connected and how that varies in space and time (Cowen and Sponaugle 2009), the ecological consequences for the structure and dynamics of populations and the processes responsible (see review by Hughes et al. 2008). This type of research is needed to substantiate unsupported claims that a very large and recent seawall is facilitating the spread of native and nonnative species between North and South China (Dong et al. 2016).

22.2.3 Changes to Biodiversity in Adjacent Habitats

Armored shorelines, for example, groynes, seawalls, and bulkheads and marinas to provide shelter for boats, are frequently built adjacent to sedimentary habitats, to reduce erosion and movement of sand. The most common adjacent habitats are, therefore, subtidal sediments and intertidal beaches. The impact of adding hard structures to intertidal sedimentary habitats is dealt with in detail in Dugan et al. (2011) but includes direct loss of the sedimentary habitat itself (Dugan et al. 2008), changes to coastal vegetation (Hughes and Paramor 2004, but see Bozek and Burdick 2005, who showed evidence that seawalls were not affecting saltmarsh assemblages), rates of accumulation of marine wrack (Sobocinski et al. 2010), and changes to benthic assemblages (Dugan et al. 2008), among other impacts (Dugan et al. 2011). Impacts can be smaller or absent if the modifications are higher on the shore, rather than toward low shore levels (Jaramillo et al. 2002). Impacts to subtidal areas include changes in circulation and residence time of water, attenuation of wave action, and accumulation of pollutants, all of which can affect subtidal assemblages (Airoldi et al. 2005a). Because of limited space, this chapter has, however, largely focused on intertidal habitats, particularly rocky shores and seawalls. A combination of well-designed structured sampling and experiments have shown that armoring shorelines can cause cascading ecological impacts by reducing the deposition of wrack, which results in a reduction of the assemblages on the shore (Heerhartz et al. 2014, 2015).

There have been fewer studies on effects of armoring on adjacent rocky habitat, although surveys by Goodsell et al. (2007) and Goodsell (2009) showed that many rocky shore species were less abundant on rocky shores that were abutted by one or more artificial seawalls and on those that abutted other natural habitats, such as mangroves. Thus, assemblages differed among rocky shores in Sydney Harbor according to the proximity and amount of adjacent artificial shoreline.

Degradation of old stone seawalls can create adjacent patches of boulders from the falling stones. Boulder fields are important intertidal habitats in many areas, supporting rare marine invertebrates (Chapman 2005), but Chapman (2006) showed that, in Sydney Harbor, patches of boulders created by blocks from seawalls did not support the rare specialist species found in natural boulder fields. This is surprising given the ease with which artificial boulder fields are colonized (Chapman 2012), and it may be that further research shows that boulders created by degrading artificial shorelines should be left *in situ* because they can support a more diverse biota.

Changes to organisms in adjacent habitats can potentially have flow-on effects on assemblages living on the armored shores themselves. For example, there has been considerable research showing changes to fish assemblages in waters adjacent to armored shorelines compared to natural shores (Able et al. 1998, 2013; Munsch et al. 2014; Peterson et al. 2000), although there have been very few experimental tests of the features of the artificial structures that influence fish assemblages (Coleman and Connell 2001). Because many fish feed on organisms living on rocky reefs, they can also prey on species living on the artificial shorelines, potentially changing the structure of these assemblages. There is, however, a paucity of experimental studies on such effects (but see Cordell et al. 2016 and references therein).

22.2.4 Increases in Nonindigenous Species

Much of our understanding about increases in nonindigenous species on artificial structures comes from subtidal assemblages, with less known about intertidal assemblages. Sampling in a South African estuary shows that addition of concrete seawalls can increase the abundance and spread of the invasive bioengineering polychaete, *Ficopomomatus enigmaticus* (McQuaid and Griffiths 2014), altering both the biomass and diversity of associated species compared to those found in natural habitats. Seawalls in Sydney Harbor support large populations of the mussel, *Mytilus galloprovincialis*, which is invasive in many parts of the world, although its status in Australia has been questioned (Beu 2004). Like other bioengineering species, clumps of mussels support a diverse biota. These assemblages differ among mussels occupying natural or artificial structures (People 2006). In addition, when mussels overgrow and replace natural algal habitat, the associated assemblage changes to smaller densities and sizes of animals (Chapman et al. 2005).

In some parts of the world, the spread of invasive species on intertidal artificial structures is very extensive, for example, the mussel *M. galloprovincialis* and the alga *Caulerpa fragile* on groynes off the coast of Italy (Airoldi et al. 2005b; Bacchiocchi and Airoldi 2003; Bulleri and Airoldi 2005). Intertidal species can be found living on artificial surfaces in countries, such as Belgium, which has no natural rocky shore. Many of these are not generally considered invasive species, such as species of littorinid snails, but can be quite widespread (Johannesson and Wamoes 1990). The mechanism of invasion and spread is not understood for marine species that have direct-development (animals that develop directly into juveniles that are smaller versions of adults), but if direct-developers can invade an area, a small number of individuals can rapidly establish viable populations (Johannesson 1988).

22.2.5 Effects of Armored Shores on Populations versus Individuals

The abundances, persistence, and rates of growth of populations are controlled by the rates at which individuals grow, survive, and reproduce. Because individuals can be injured or removed from populations without ecological impacts (i.e., cases where biological effects do not always translate to ecological impacts; Browne et al. 2015), it is important to link any effects of artificial shores shown to affect individuals to their populations through modeling before broad-scale ecological impacts can be shown. This would allow one to integrate spatiotemporal changes in, for example, the size and distribution of ages of a collection of organisms to assess how their populations are affected by interactions with other organisms inhabiting armored shores, features of the artificial shore, and the surrounding environment.

Population models are useful because they provide a means to assess (a) whether or not a population is declining or increasing, (b) the cause of the decline or increase, (c) which stage of the life cycle needs managerial action, and (d) the probable outcome for the population (Caswell 2001). These models have, however, not been used for organisms affected by artificial shorelines, but they would allow one to determine whether managerial actions achieve their aims of increasing the size and rate of growth of populations of native species and slow the growth of populations of nonnative species so that they can be exterminated or their invasion be halted.

Models are based on mathematical equations (Forbes and Calow 1999), matrices (Caswell 1989), and integrals (Easterling et al. 2000) that incorporate scientific understanding about how a number of abiotic and biotic processes influence the speed at which individual organisms grow, survive, and reproduce. In the case of artificial shores, more complicated models may be appropriate because populations consist of smaller groups, some of which inhabit fragments of natural shores and others fragments of artificial shores (Goodsell et al. 2007), with potentially different ecological processes affecting abundances, sizes, and so on, on each (Moreira et al. 2007). Depending on their life history, they may be linked, but this is unlikely as adult populations and recruits are seldom spatially linked for species with planktonic development—supply-side ecology (e.g., Caffey 1985; Underwood and Fairweather 1989).

Although these models provide a means to link the individual to the population, the strength of this linkage will depend on the amount of detail about how features of the engineered shores and the wider landscape affect recruitment, survival, and reproduction among individuals in the population. Estimates of these parameters need to be critically evaluated in terms of (i) their quality (and hence reliability), (ii) the sizes of the samples, (iii) the precision and accuracy of estimates, and (iv) whether the estimates used apply to the population that is being modeled (Caswell 2001). Data from rigorous well-designed surveys and experiments are recognized as the most valuable in determining the relative importance of different biotic (e.g., competition, facilitation, predation) and abiotic (e.g., wave-exposure, shade, temperature) agents in regulating a population (Caswell 2001; Hastings 2013). For many organisms, the impacts of artificial shores can only be assessed if we understand their life history (i.e., direct-developer, lecithotrophic, planktotrophic) and how success of reproduction is affected by alterations to the artificial shore and its ecological assemblage. This would allow sensitivity analyses to make predictions about how and when to try to assemble viable populations of native species and eradicate populations of nonnative species on artificial shores. This has yet to be done for species that appear to be affected by extensive shoreline alteration.

22.3 REENGINEERING ARMORED SHORELINES TO IMPROVE THEIR VALUE AS HABITAT FOR BIOTA

There are mainly three general ways in which armored shorelines have been altered in attempts to ameliorate their undesirable ecological impacts. These are removal of the structures, either alone or with replacement of natural habitats (Nordstrom 2014; Nordstrom and Jackson 2013), leaving the structures in place but adding natural habitat to adjacent areas, or modifying the structures themselves, either when they are being built or repaired or while they remain *in situ* (as discussed by Chapman and Underwood 2011).

22.3.1 Removal of Seawalls and Replacement with Natural Habitats

At a large scale, old seawalls and other defensive structures have been totally removed to allow the tide to penetrate further inland, in order to increase extent of marshes and other such habitats (Borsje et al. 2011). This is commonly referred to as managed realignment. At a smaller scale, walls can be removed from shores where they are no longer needed. This is more common for walls at the tops of beaches, above the high tide level (Dugan et al. 2011). As part of a beach restoration project in Washington, for example, a seawall was removed from the top of a beach and the sedimentary fauna compared to that on a nearby beach. Fauna was reduced adjacent to the wall compared to the same shore level on the natural beach and recovery had not occurred 16 months after removal (Toft et al. 2014).

In sheltered areas, walls can be removed and replaced by marshes (Davis et al. 2006), or hybrid designs that incorporate vegetation into or in conjunction with the artificial shorelines can be developed (Bilkovic and Mitchell 2013). These can reduce the effects of wave action on a shore and provide additional habitat, although there are limited data on how successful such manipulations are in restoring natural levels of diversity (Chapman and Underwood 2011). Such modifications are often referred to as hybrid structures—combinations of living and nonliving material (Smith 2006). This contrasts with so-called Living Shorelines, which describe situations where walls are removed and replaced by natural shoreline vegetation, with the expectation that the natural material will have a similar buffering effect against wave action (Davis et al. 2006). This can only be considered in relatively sheltered conditions and the topic is covered in detail elsewhere in this volume.

22.3.2 Modifications to Habitat Adjacent to Seawalls

Some modifications are aimed solely at protection of the armored shorelines, without any ecological intentions. For example, it has been suggested that building low breakwaters offshore can reduce the effects of waves on seawalls, thus protecting their structure (Bettington and Cox 1997). This action was prompted solely to protect the structure, not to change its value as ecological habitat, although offshore artificial habitats built solely for shoreline protection can support a diverse biota (Green et al. 2012).

Adding additional structures offshore can, however, have important unintended ecological effects, especially if they are placed on sediments (Barros et al. 2001) and provide novel hard surfaces, generally for subtidal species (Davis et al. 2006). Patches of rocks placed offshore from walls can reduce wave action and, it is assumed, damage to seawalls in areas where such patches are stable and can persist (see also the use of mussel and oyster beds to similar effect; Borsje et al. 2011). Providing additional habitat such as this may be a cheap way of augmenting local species diversity, especially if a mosaic of different types of habitat are used, assuming increased diversity of habitat is reflected in increased diversity of biota (Davis et al. 2006). The addition of such structures without clear understanding of what are the original impacts of the armored shoreline and without clear understanding of what causes such impacts if they exist, can, however, lead to extensive managerial activities and wasted money to fix a problem that does not exist, or which is caused by factors other than the lack of habitat, as shown by an extensive review of similar forms of restoration in rivers (Palmer et al. 2010).

Similarly, shading sun-exposed seawalls by the addition of small areas of wharves that shade the surface may enhance the diversity of intertidal species living on the wall because different suites of species tend to occupy shaded or sun-exposed areas (Blockley 2007; Blockley and Chapman 2006). Large areas of shade caused by large piers, and so on can, however, reduce the abundance and diversity of fish in the water under the pier compared to areas that are not shaded (Able et al. 1998, 2013; Munsch et al. 2014). There are likely to be similar effects on the intertidal species living on the walls and piers under extensive shaded areas, although this has not been measured. Therefore, one must take care in attempting to mitigate any negative effects of one form of engineered shoreline by adding additional forms of engineering without a clear understanding of what additional impacts may be likely.

22.3.3 Reengineering Seawalls to Add Additional Habitat

Reengineering seawalls has largely focused on adding complexity or specific structures to their surfaces, generally in an attempt to provide additional habitat for those species that are absent from many walls, but also to deflect damage caused by wave action (Neelamani and Sandhya 2005). At the smallest scale, small pits or crevices are added to the surface of the wall, as has also been done on low-crested offshore structures off European shores (Moschella et al. 2005). Small holes and pits are commonly used as habitat during low-tide on natural rocky shores (e.g., Borsje et al. 2011; Chapman 1994; Martins et al. 2010).

Martins et al. (2010) showed increased recruitment of the exploited limpet *Patella candei* onto seawalls that had additional drilled pits than into sites on featureless seawalls. Adult limpets also used slightly larger pits as habitat during low tide (see also Borsje et al. 2011; Firth et al. 2014b). Similarly, Moreira et al. (2007) showed increased densities of chitons in the crevices between blocks of sandstone on seawalls in Sydney Harbor compared to the faces of the blocks. The chitons also dispersed further when these crevices were filled in, a managerial practice that is common where seawalls are built of blocks of stone, for both structural and aesthetic reasons (M.G.C., personal observation). Chapman and Underwood (2011) showed increases in diversity in areas where the mortar between adjoining blocks was indented to create small crevices compared to areas where the mortar was flush with the surface of the blocks, but most crevices were occupied by sessile animals and algae and not the larger mobile animals that appear to be most deleteriously affected by converting natural shores to walls (Chapman 2003). Simply changing this simple practice, if feasible, could therefore improve habitat for many intertidal organisms, but, unfortunately, these small habitats can be completely occupied or overgrown as time goes on and the walls are colonized by larger organisms and more dense assemblages, thus negating their original ameliorative effects (Borsje et al. 2011; Chapman and Underwood 2011; Firth et al. 2014a,b).

Some of the larger modifications to seawalls themselves have involved adding specifically engineered habitats to retain water during low tide and mimic some aspects of rock pools. Thus, Chapman and Blockley (2009) created shaded water-retaining features in seawalls built by facing a concrete wall with building blocks of natural stone, omitting stones at three different shore levels and adding a lip at the front of these cavities to retain water during low tide. Compared to the face of the wall itself, there was increased diversity of foliose algae and sessile and mobile animals, especially higher on the shore and many species generally confined to low shore levels increased their range upshore. With the reduced areas of intertidal habitat created by converting sloping natural shores to vertical walls, increasing the amount of useable intertidal habitat could be very important to diversity of intertidal species. These habitats did not, however, ameliorate the negative effects of the wall on many mobile species, either because a longer time is needed for them to recruit into these habitats, or because the habitats were not suitable for all species. Nevertheless, there were more species in these constructed pools than in nearby natural pools, perhaps because they were shaded, thus providing novel habitat, or perhaps because of the absence of competitors or predators.

This concept was extended by Browne and Chapman (2011, 2014) who used custom-made concrete pots, akin to flower pots (Figure 22.1), which can be attached to seawalls using metal brackets. Their advantage is that, in contrast to the example described above, they can be attached to any existing seawall and do not shade organisms, but the disadvantage is that they are very heavy to manipulate and less durable because they are susceptible to damage from strong waves. Although this pilot study was only short term (7 months deployment of the pots), they showed an increase in the number of species occupying the site of 64%. Work is continuing in this promising concept, with both longer-term studies and tests of impacts of these structures on fish in the waters adjacent to seawalls. Recent work in a location in Wales has added rock pools to riprap using drilling techniques used previously by intertidal ecologists (Underwood and Skilleter 1996) to create rock pools on natural shores. Through this work, Evans et al. (2015) and Firth et al. (2014a) added novel species to riprap with drilled pools supporting larger numbers of species than adjacent rocky surfaces. These assemblages were, however, different to those found in natural pools on adjacent rocky shores, with drilled pools lacking coralline algae and containing novel species of ephemeral algae and sessile polychaetes and barnacles.

There have also been attempts to increase the amount of intertidal area by building walls with less slope, or with stepped horizontal and vertical surfaces, often using unconsolidated boulders to increase habitat complexity (Chapman and Underwood 2011). Although this seems an intuitively obvious way to increase the diversity of habitats and, hence, of species occupying them, results of experimental tests to date are not convincing, possibly because these structures can only replace seawalls in very sheltered areas where intertidal diversity may be naturally small (Chapman and Underwood 2011). Clearly, more research is needed in this area in naturally more diverse sites.

Figure 22.1 Flower pots creating novel habitat on seawalls in Sydney Harbor. (Photo taken by Gemma Deavin.)

22.4 DEVELOPING APPROPRIATE HYPOTHESES TO TEST EFFECTS OF ALTERING ARMORED SHORELINES

It is crucial that we advance our understanding of the ecological impacts of armored shorelines and the potential for minimizing such impacts by ecologically engineering the structures themselves or the biota living on them. Clear thought has, however, to go into the hypotheses that are being tested by any data collected. Although this may seem obvious, it is clear from many of the published studies that the hypotheses have not been always clearly thought out. As a consequence, it is often not clear what the data are, in fact, testing. All ecologists doing research on such an important topic need to pay careful attention to hypotheses and data, considering the fact that their results can be used by managers to change managerial practices and that this is likely to be a very expensive exercise.

Here, we highlight a few of the more common discrepancies between aims of the research, the hypotheses, and the data collected to evaluate the success of the aims. These are not all of the problems that can be found, but are presented to highlight the need for more rigorous thoughts and practices.

22.4.1 Hypotheses about Univariate Variables

Univariate variables might include predictions about increases in adult sizes of a species, increases in the population density, or numbers of species, in response to alterations of habitat in association with armored shorelines. Influenced by a long history of successfully using statistical procedures rigorously to advance our understanding of ecological patterns and processes, especially in experimental ecology (Underwood 1997), data to test hypotheses about such univariate measures are frequently analyzed using analyses of variance including techniques such as general linear models. Often, however, the hypothesis requires something quite different.

As an example, if one adds a novel habitat, such as artificially created rock pools to a seawall, breakwater, or similar habitat, in order to provide habitat that can be occupied by species that do not live on the existing structure, the hypothesis clearly requires one to know (i) what are the species on the original structure and (ii) what additional species are found in created habitat, for example, pools (Browne and Chapman 2011; Tywan Breakwater in Firth et al. 2014b). Using analyses of variance to compare the mean number of species in samples of the two habitats (Firth et al. 2014b) does not answer that question. Finding more species in pools will show that at least some species in pools were not sampled on the surrounds, but if the number of samples is small relative to the area of the original structure, the rarer species may not be sampled in the original habitat, even when found there, but may accumulate in pools in greater numbers. In contrast, pools could contain the same mean number of species per sample as the surrounding area, or even fewer, but the hypothesis would be retained if these were a different set of species than lived on the surrounds. Yet, analysis of variance on the mean number of species per sample would indicate no effect of pools. Such a hypothesis can, however, be tested with simple data, showing the number of species sampled on the original structure and the number of additional species found in pools but not on the original structure (Chapman and Blockley 2009).

Analyzing the mean number of species per sample of habitat is a test of species density (the number of species per area), not a test of the number of species in each habitat. There may be appropriate models that would lead to hypotheses regarding species density, but the hypothesis that the novel habitat would add additional species to the original structure is not one. Consider a scenario with six species, four of which can live in pools and on the surrounding area, but two of which do not recruit to, or cannot survive in pools. Those that share the habitat are attracted to pools and, therefore, more likely to be found in pools. Sampling could show that pools have a mean of 3.75 species, as illustrated in Figure 22.2a. Because of reduced densities on the surrounding area, samples

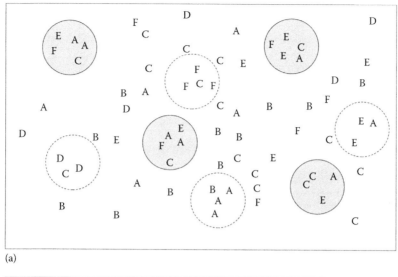

(a)

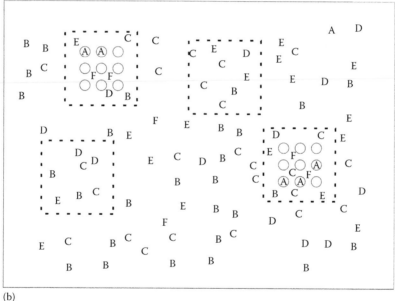

(b)

Figure 22.2 (a) Consider sampling four pools (solid lines) and four areas of background habitat (dotted lines). The pools contain more species per sample (mean 3.75) than the background area (mean 2), suggesting greater number of species in pools. There are, however, only four species on all the pools and six in the four samples of the background. The test of mean number of species per sample is not appropriate for a hypothesis that pools add additional species to the original structure. (b) Consider sampling two areas with added pits (circles) and two areas of background habitat. The background samples have four species in the two samples and the areas with pits have six. Species A is confined to pits and species F is confined to the area between the pits, suggesting that adding pits supports a greater number of species. If the pits have attracted species A and F from the surrounds, however, they would be present if the pits had not been created, and this contrast is inadequate. One also needs to contrast this habitat with the pits to another area where no pits have been added to identify whether species A and F do live in the area if there are no pits at all. The contrast between areas with and without pits needs to encompass one or more spatial scales that can distinguish between attraction and increased production of species. *(Continued)*

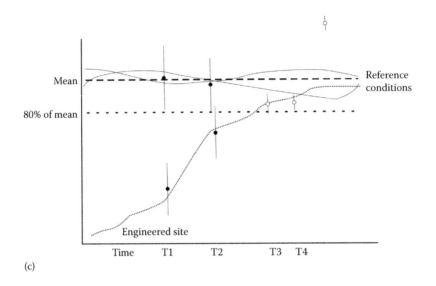

(c)

Figure 22.2 (Continued) (c) Consider a test of difference between the mean values (with error bars) of a univariate measure taken from an ecologically engineered site (dotted line) and the mean of two reference conditions (heavy dashed line). At Time 1, there is a difference between the means, but at Time 2, there is no significant difference because of the large error bars (imprecise sampling). One would conclude using a test of difference that the engineered site had converged with the reference sites at Time 2. Consider a test of bioequivalence that requires that the mean value in the engineered site must exceed 80% of the mean of the reference sites. With large error bars, this is unlikely to ever occur. With more precise sampling, that is, smaller error bars, there is still overlap with the 80% value at Time 3, indicating that the engineered sites still have not converged with the reference sites. It is only at Time 4 that one could conclude that engineering had achieved its target of 80% of the mean values in the reference sites.

could have a mean of two species per equivalent area, suggesting that pools support more species. If, however, all pools contain the same subset of species that use both habitats (species A, C, E, and F in Figure 22.2a), but samples on the surrounding area contain all species (A–F, Figure 22.2a), it is clear that pools support fewer species than the original habitat. The hypothesis that pools increase the number of species requires data on the number and identity of species per habitat, not the mean number per replicate.

Comparisons can become more complex if the dispersions of individual species vary between habitats. It is a common pattern for many intertidal animals during low tide that a species can use two different habitats; however, because it can be very clustered in the original exposed habitat, it may not be sampled in the other habitat. This problem becomes more acute if the area sampled (generally a few quadrats) is small relative to the extent of the habitat (a seawall or breakwater). If the species occurs at all in the created habitat (e.g., pools), it will always be sampled when pools are sampled completely (Browne and Chapman 2011; Firth et al. 2014b). It is important to realize that many comparisons are between complete censuses of the created habitat and samples of the original habitat. Ideally, one needs to compare species area curves (graphs showing how the area of a sampled habitat relates to the number of species found within that area) to compensate for the difference in area of the two habitats if the hypothesis is about the habitats as a whole and not the number of species per unit area of each habitat.

It is more straightforward to evaluate the model that novel microhabitats added to an artificial structure will increase the number of species on the structure by predicting that areas of original habitat to which novel microhabitats are added (e.g., the addition of pits to the surface, Firth et al. 2014b;

Martins et al. 2010) would support more species than areas of the same artificial habitat without additional microhabitats. Similarly, one can add different types of habitat to a wall (e.g., Kirribilli, in Chapman and Underwood 2011), with the hypothesis that one type of structure would support more species than the other. Such hypotheses can be tested by comparing replicate samples of both sets of habitat, that is, areas of the original structure with the added microhabitats and areas without, with the prediction that the former will contain more species than the latter. As long as provision is made for potentially sampling increased surface area in patches with additional microhabitat (which may be substantial when anything larger than small pits are added) and ensuring that the areas of the wall are not spatially confounded, analysis of mean number of species per patch should successfully test the hypothesis.

There can be a problem if the areas without the added habitat are in proximity to those with the habitat, such that the latter can affect species in the former. For example, in Figure 22.2b, species A are only found in pits and species F are only found in the areas among the pits, whereas the other species occur in all areas; thus, there are six species in areas with pits and four elsewhere. If, however, species A and F have been attracted to the novel habitat from the surrounding area, where they would be found, albeit at smaller densities, were the pits not present, then the hypothesis would be supported, but the model that the addition of pits increases species diversity will still be incorrect.

When one is comparing a restored situation, for example, the habitat after removal of a seawall (Toft 2005) to a control area, where a wall is left intact, then analysis of variance is appropriate to test the hypothesis that the former area will have larger densities or diversities than the latter. If, in contrast, the comparison is between the restored area and an area that never had a seawall, the reference conditions, then analysis of variance is not appropriate. One is not testing for a difference, but a similarity and tests of bioequivalence are appropriate (Chapman 1999a). These require *a priori* decisions about what level of similarity would be adequate to conclude that success has been achieved (Figure 22.2c).

In many cases of restoration, recovery is very slow. When it is likely that responses to the engineered habitat have not yet occurred, or there is no information about the rate at which changes should occur, it may be more useful to compare trajectories of change (Simenstad and Thom 1996), rather than the values at any single point in time. Then, one could determine whether the rate and direction of change in species living on engineered habitat are converging with those in natural habitat.

22.4.2 Hypotheses about Multivariate Variables

If the aim is to change an assemblage in an engineered habitat to resemble that in a natural habitat, the hypothesis is by nature multivariate. It requires comparison of the suite of species and their relevant abundances. Reducing the data to univariate measures, such as species number, that is, the mean number of species per sample (often incorrectly referred to as species richness, i.e., the total number of species over all samples), cannot test such a hypothesis. For example, an engineered habitat with 10 species of algae and two of barnacles cannot be considered equivalent to a natural area with three species of algae, three of barnacles, two of sponges, and two of ascidians, although the species richness and, possibly, species numbers, are equivalent. Multivariate analyses, which consider the identity and abundances of each species are necessary to show the ways in which assemblages differ.

Unfortunately, such analyses are also influenced by dispersions of the different species among replicates and there are no tests of bioequivalence of multivariate assemblages currently easily available. In fact, it is difficult to develop the types of sophisticated hypotheses that are necessary to evaluate the importance of such activities as engineering armored shorelines. Most hypotheses are still evaluated by tests for differences between assemblages. At the assemblage level, we still need to put more formal thought into what the aim of the exercise is, for example, to provide habitat for rare species, for a certain percentage of the species, to develop viable reproductive populations of certain types of species. Hopefully, if we become more careful and articulate about what we are hoping to achieve by such engineering works, we will stimulate the development of new analytical techniques that can address such complex problems; for example, we have obtained 80% or more of the required abundance of at least 60% of the species, with natural spatial and temporal patterns of variance. At this stage, unfortunately, questions such as these, despite their importance, are best answered by a series of different univariate hypotheses and tests.

22.5 CONCLUSIONS

Humans are now intrinsic agents of landscape evolution with decisions on how, when, and where to place, modify, or remove artificial shores depending on inputs from ecologists, engineers, and policy makers (Nordstrom 2014). Identifying uncertainty is a critical part of ecology (e.g., Underwood 1997), engineering (Komar and McDougall 1988; Petroski 1985, 1994), ecological engineering (Chapman and Underwood 2011), and management (Underwood 1995) because misplaced certainty can be considered the greatest form of ignorance (Goodman 1992). A useful strategy to reduce uncertainties in the understanding and management of the ecological impacts caused by artificial shorelines is to use explicit hypotheses, structured sampling, where possible linked to manipulative experiments and population modeling. For many scientists, engineers, and policy makers, this may require specific changes in the way they think and interact.

A major advantage of ecologists and engineers working together has been that where ecological impacts have been shown and there has been the possibility of field experiments to determine whether artificial shores can be reengineered, for example, by adding cages (Espinosa et al. 2006), crevices (Chapman and Underwood 2011), holes (Martins et al. 2010), and pools (Browne and Chapman 2011, 2014; Chapman and Blockley 2009; Evans et al. 2015; Firth et al. 2014a), there has been evidence for reduction in the impacts. Nevertheless, such research is still small scale, spatially scattered, and temporally confounded (at the same time in a few sites, or at different times in different sites; Chapman and Underwood 2011). They also do not entirely eliminate the impacts of artificial shores on native biodiversity, to some extent because the data and analyses do not allow rigorous tests of their effects. In addition, most research to date has been on measures of ecological structure, with little to no experimental work on biological (e.g., rates of feeding, survival, growth by organisms) and ecological (e.g., rates of recruitment, competition, predation, population growth) functions that sustain biodiversity and societal services (e.g., maintenance of water quality, biogeochemical cycling, production of food). Unfortunately, the linkages between biological and ecological functions and societal services are unclear (Browne et al. 2015). It is therefore not at all clear what functions and services most shoreline armoring affect, or what suite (or subset) of species are necessary to restore such functions and services. This is concerning given that coastal armor is getting taller and longer through accelerating urbanization and climatic change.

An important issue that is timely for discussion among policy makers, the public, managers, and ecologists is what changes to the ecology of intertidal shorelines are people willing to accept, given growing coastal urbanization and the need to protect important infrastructure. For commercial fish species that have been shown to be affected by shoreline development (Toft et al. 2014), spending time and effort to improve the habitat is likely to be perceived as important by a wider public. For single species such as these, there is adequate ecological knowledge about the requirements of different ages of focus fish species and that detailed modifications to shorelines can be proposed; these can be experimentally tested and then managers (with public support) can make modifications accordingly and test their effectiveness. In the context of adaptive management (Thom 2000), such modifications need be relatively cheap and flexible as a single recipe is unlikely to be appropriate everywhere.

With respect to more conservation-oriented objectives, the role of reengineered shorelines is even less clear. We are clearly entering a world of novel or emerging ecosystems (Hobbs et al. 2006). Impacts are clearly more intense on land than in the sea, and compared to pollution, over-fishing, and other marine impacts, changes to shorelines may not be considered unduly important. There is, however, evidence that such changes are associated with local loss of diversity, homogenization of biota, and spread of invasive species. The costs of trying to halt or reverse these trends are potentially very large and the public are not usually willing to pay much for conservation of noncharismatic or utilitarian species of habitats (Martín-López et al. 2007). Hence, to date, experiments have been scattered and small scale, depending on the support of local agencies in individual countries. Issues revolve around what biodiversity really is, how important it is to retain all species if the goal is to maintain certain ecological functions (e.g., rates at which assemblages filter water), do rare cryptic species really matter, and do we know enough to protect them anyway (Chapman 1999b). How much are we prepared to accept a "zoo" in our back garden (so to speak), when we know that it can never again be a natural habitat? Although scientists and managers can assist in directing such questions, the decision ultimately is up to the public and policy makers about how much the habitats are valued and what it would cost to try to maintain them in a more natural state.

There is thus a major challenge to develop understanding of the processes and techniques that alter artificial shores and assemblages. This will require detailed data on the impacts, clear hypotheses about what ecological engineering might achieve, well-designed experiments, with relevant control and reference locations to test such hypotheses, robust data to test such hypotheses, and, almost certainly, the development of novel analytical procedures to deal with such data to test such hypotheses, which will almost certainly be very complex. We still have a long way to go before we can evaluate the relative benefits of retrospective eco-engineering against the inevitable costs, but it is a direction in which we need to progress, given the continued deterioration of our coastlines.

REFERENCES

Able, K.W., J.P. Manderson, and A.L. Studholme. 1998. The distribution of shallow water juvenile fishes in an urban estuary: The effects of manmade structures in the lower Hudson River. *Estuaries* 21: 731–744.

Able, K.W., T.M. Grothues, and I.M. Kemp. 2013. Fine-scale distribution of pelagic fishes relative to a large urban pier. *Marine Ecology Progress Series* 476: 185–198.

Aguilera, M.A., B.R. Broitman, and M. Thiel. 2014. Spatial variability in community composition on a granite breakwater versus natural rocky shores: Lack of microhabitats suppresses intertidal biodiversity. *Marine Pollution Bulletin* 87: 257–268.

Airoldi, L., M. Abbiati, M.W. Beck, S.J. Hawkins, P.R. Jonsson, D. Martin, P.S. Moschella, A. Sundelöf, R.C. Thompson, and P. Åberg. 2005a. An ecological perspective on the deployment and design of lowcrested and other hard coastal defence structures. *Coastal Engineering* 52: 1073–1087.

Airoldi, L., F. Bacchiocchi, C. Cagliola, F. Bulleri, and M. Abbiati. 2005b. Impact of recreational harvesting on assemblages in artificial rocky habitats. *Marine Ecology Progress Series* 299: 55–66.

Attrill, M.J., D.T. Bilton, A.A. Rowden, S.D. Rundle, and R.M. Thomas. 1999. The impact of encroachment and bankside development on the habitat complexity and supralittoral invertebrate communities of the Thames Estuary foreshore. *Aquatic Conservation: Marine and Freshwater Ecosystems* 9: 237–247.

Bacchiocchi, F. and L. Airoldi. 2003. Distribution and dynamics of epibiota on hard structures for coastal protection. *Estuarine, Coastal and Shelf Science* 56: 1157–1166.

Barros, F., A.J. Underwood, and M. Lindegarth. 2001. The influence of rocky reefs on structure of benthic macrofauna in nearby soft-sediment. *Estuarine, Coastal and Shelf Science* 52: 191–199.

Beu, A.G. 2004. Marine mollusca of oxygen isotope stages of the last 2 million years in New Zealand. Part 1. Revised generic positions and recognition of warm-water and cool-water migrants. *Journal of the Royal Society of New Zealand* 34: 111–265.

Bettington, S.H. and R.J. Cox. 1997. Low reflectance structures for small harbours (double skirt breakwaters with a perforated front skirt). In: *Pacific Coasts and Ports '97*. pp. 179–183. Centre for Advanced Engineering, University of Canterbury, Christchurch, NZ.

Bilkovic, D. and Mitchell, M. 2013. Ecological tradeoffs of stabilized salt marshes as a shoreline protection strategy: Effects of artificial structures on macrobenthic assemblages. *Ecological Engineering* 61: 469–481.

Blockley, D.J. 2007. Effect of wharves on intertidal assemblages on seawalls in Sydney Harbour, Australia. *Marine Environmental Research* 63: 409–427.

Blockley, D.J. and Chapman, M.G. 2006. Recruitment determines differences between assemblages on shaded or unshaded seawalls. *Marine Ecology Progress Series* 327: 27–36.

Borsje, B., B.K. van Wesenbeeck, F. Dekker, P. Paalvast, T.J. Boumab, M.M. van Katwijk, and M.B. de Vries. 2011. How ecological engineering can serve in coastal protection. *Ecological Engineering* 37: 113–122.

Bozek, C.M. and D.M. Burdick. 2005. Impacts of seawalls on saltmarsh plant communities in the Great Bay Estuary, New Hampshire USA. *Wetlands Ecology and Management* 13: 553–568.

Browne, M.A. and M.G. Chapman. 2011. Ecologically informed engineering reduces loss of intertidal biodiversity on artificial shores. *Environmental Science and Technology* 45: 8204–8207.

Browne, M.A. and M.G. Chapman. 2014. Mitigating against the loss of species by adding artificial intertidal pools to existing seawalls. *Marine Ecology Progress Series* 497: 119–129.

Browne, M.A., A.J. Underwood, M.G. Chapman, R. Williams, R.C. Thompson, and J.A. van Franeker. 2015. Linking effects of anthropogenic debris to ecological impacts. *Proceedings Royal Society Series B* 282: 20142929.

Bulleri, F. and L. Airoldi. 2005. Artificial marine structures facilitate the spread of a non-indigenous green alga, *Codium fragile* ssp. *tomentosoides*, in the north Adriatic sea. *Journal of Applied Ecology* 42: 1063–1072.

Bulleri, F., M.G. Chapman, and A.J. Underwood. 2004. Patterns of movement of the limpet *Cellana tramoserica* on rocky shores and retaining seawalls. *Marine Ecology Progress Series* 281: 121–129.

Bulleri, F., M.G. Chapman, and A.J. Underwood. 2005. Intertidal assemblages on seawalls and rocky shores in Sydney Harbour (Australia). *Austral Ecology* 30: 655–667.

Caffey, H.J. 1985. Spatial and temporal variation in settlement and recruitment of intertidal barnacles. *Ecological Monographs* 55: 313–332.

Caswell, H. 1989. The analysis of life table response experiments. I. Decomposition of treatment effects on population growth rate. *Ecological Modelling* 46: 221–237.

Caswell, H. 2001. *Matrix population models: Construction, analysis, and interpretation*, 2nd ed. Sinauer Associates, Sunderland, MA, USA.

Chapman, M.G. 1994. Small-scale patterns of distribution and size-structure of the intertidal littorinid, *Littorina unifasciata* (Gastropoda: Littorinidae) in New South Wales. *Australian Journal of Marine and Freshwater Research* 45: 635–642.

Chapman, M.G. 1999a. Improving sampling designs for measuring restoration in aquatic habitats. *Journal of Aquatic Ecosystem Stress and Recovery* 6: 235–251.

Chapman, M.G. 1999b. Are there adequate data to assess how well theories of rarity apply to marine invertebrates? *Biodiversity and Conservation* 8: 1295–1318.

Chapman, M.G. 2003. Paucity of mobile species on constructed seawalls: Effects of urbanization on biodiversity. *Marine Ecology Progress Series* 264: 21–29.

Chapman, M.G. 2005. Molluscs and echinoderms under boulders: Tests of generality of patterns of occurrence. *Journal of Experimental Marine Biology and Ecology* 325: 65–68.

Chapman, M.G. 2006. Intertidal seawalls as habitats for molluscs. *Journal of Molluscan Studies* 72: 247–257.

Chapman, M.G. 2012. Restoring intertidal boulder-fields as habitat for "specialist" and "generalist" animals. *Restoration Ecology* 20: 277–285.

Chapman, M.G. and D. Blockley. 2009. Engineering novel habitats on urban infrastructure to increase intertidal biodiversity. *Oecologia* 161: 625–635.

Chapman, M.G., J. People, and D. Blockley. 2005. Intertidal assemblages associated with natural Corallina turf and invasive mussel beds. *Biodiversity and Conservation* 14: 1761–1776.

Chapman, M.G. and A.J. Underwood. 2011. Evaluation of ecological engineering of "armoured" shorelines to improve their value as habitat. *Journal of Experimental Marine Biology and Ecology* 400: 302–311.

Coleman, M.A. and S.D. Connell. 2001. Weak effects of epibiota on the abundances of fishes associated with pier pilings in Sydney Harbour. *Environmental Biology of Fishes* 61: 231–239.

Cordell, J.R., J.D. Toft, S.H. Munsch, and M. Goff. 2016. Benches, beaches, and bumps: How habitat monitoring and experimental science can inform urban seawall design. In: Bilkovic, D.M., M. Mitchell, J. Toft, M. La Peyre (Eds.), *Living Shorelines: The Science and Management of Nature-Based Coastal Protection*. CRC Press, Boca Raton, USA.

Cowen, R.K. and S. Sponaugle. 2009. Larval dispersal and marine population connectivity. *Annual Review of Marine Science* 1: 443–466.

Creese, R.G. 1982. Distribution and abundance of the acmaeid limpet, *Patelloida latistrigata*, and its interaction with barnacles. *Oecologia* 52: 85–96.

Davis, J.L.D., L.A. Levin, and S.M. Walther. 2002. Artificial armored shorelines: Sites for open-coast species in a southern California Bay. *Marine Biology* 140: 1249–1262.

Davis, J.L.D., R.L. Takacs, and R. Schnabel. 2006. Evaluating ecological impacts of living shorelines and shoreline habitat elements: An example from the upper western Chesapeake Bay. In: Erdle, S.Y., J.L.D. Davis, and K.G. Sellner. (Eds.), *Proceedings of the 2006 Living Shoreline Summit*, pp. 55–61. CRC Publication No. 08-164, VA, USA.

Denny, M.W., L.P. Miller, M.D. Stokes, L.J.H. Hunt, and B.S.T. Helmuth. 2003. Extreme water velocities: Topographical amplification of wave-induced flow in the surf zone of rocky shores. *Limnology and Oceanography* 48: 1–8.

Dong, Y.-W., X.-W. Huang, W. Wang, Y. Li, and J. Wang. 2016. The marine 'great wall' of China: Localand broad-scale ecological impacts of coastal infrastructure on intertidal macrobenthic communities. *Diversity and Distributions* 1–14.

Dugan, J.E., L. Airoldi, M.G. Chapman, S. Walker, and T. Schlacher. 2011. Estuarine and coastal structures: Environmental effects. A focus on shore and nearshore structures. In: Elliott, M. and J. Dugan (Eds.), *Treatise on Estuarine and Coastal Science*. Elsevier Press, NY, USA.

Dugan, J.E., D.M. Hubbard, I.F. Rodil, and D. Revell. 2008. Ecological effects of coastal armoring on sandy beaches. *Marine Ecology* 29: 160–170.

Easterling, M.R., S. Ellner, and P. Dixon. 2000. Size-specific sensitivity: Applying a new structured population model. *Ecology* 81: 694–708.

Espinosa, F., J.M. Guerra-Garcia, D. Fa, and J.C. García-Gómez. 2006. Effects of competition on an endangered limpet *Patella ferruginea* (Gastropoda: Patellidae): Implications for conservation. *Journal of Experimental Marine Biology and Ecology* 330: 482–492.

Evans, A.J., L.B. Firth, S.J. Hawkins, E.S. Morris, H. Goudge, and Moore, P.J. 2015. Drill-cored rock pools: An effective method of ecological enhancement on artificial structures. *Marine and Freshwater Research* 67: 123–130.

Fauvelot, C., F. Bertozzi, F. Costantini, L. Airoldi, and M. Abbiati. 2009. Lower genetic diversity in the limpet *Patella caerulea* on urban coastal structures compared to natural rocky habitats. *Marine Biology* 156: 2313–2323.

Firth, L.B., M. Schofield, F.J. White, M.W. Skov, and S.J. Hawkins. 2014a. Biodiversity in intertidal rock pools: Informing engineering criteria for artificial habitat enhancement in the built environment. *Marine Environmental Research* 102: 122–130.

Firth, L.B., R.C. Thompson, K. Bohn, M. Abbiati, L. Airoldi, T.J. Bouma, F. Bozzeda, V.U. Ceccherelli, M.A. Colangelo, A. Evans, F. Ferrario, M.E. Hanley, H. Hinz, S.P.G. Hoggart, J.E. Jackson, P. Moore, E.H. Morgan, S. Perkol-Finkel, M.W. Skov, E.M. Strain, J. van Belzan, and S.J. Hawkins. 2014b. Between a rock and a hard place: Environmental and engineering considerations when designing coastal defence structures. *Coastal Engineering* 87: 122–135.

Forbes, V.E. and P. Calow. 1999. Is the per capita rate of increase a good measure of population-level effects in ecotoxicology? *Environmental Toxicology and Chemistry* 18: 1544–1556.

Gentile, R. and L.R. Landò. 2007. Statistical behaviour of directional bound long waves. *International Journal of Offshore and Polar Engineering* 16: 183–194.

Gittman, R.K., F.J. Fodrie, A.M. Popowich, D.A. Keller, J.F. Bruno, C.A. Currin, C.H. Peterson, and M.F. Piehler. 2015. Engineering away our natural defenses: An analysis of shoreline hardening in the US. *Frontiers in Ecology and the Environment* 13: 301–307.

Goodman, G.T. 1992. Introduction. In: Dooge, J.C.I., G.T. Goodman, J.W.M. Rivière, J. Marton-Lefèvre, and T. O'Riordan (Eds.), *An Agenda of Science for Environment and Development into the 21st Century*, p. 71. Cambridge University Press, Cambridge, UK.

Goodsell, P.J. 2009. Diversity in fragments of artificial and natural marine habitats. *Marine Ecology Progress Series* 384: 23–31.

Goodsell, P.J., M.G. Chapman, and A.J. Underwood. 2007. Differences between biota in anthropogenically fragmented and in naturally patchy habitats. *Marine Ecology Progress Series* 351: 15–23.

Green, D.S., M.G. Chapman, and D.J. Blockley. 2012. Ecological consequences of the type of rock used in the construction of artificial boulder-fields. *Ecological Engineering* 46: 1–10.

Harris, P.G. 2012. Environmental policy and sustainable development in China: Hong Kong in global context. Policy Press, Bristol, UK.

Hastings, A. 2013. *Population biology: Concepts and models*. Springer Verlag, NY, USA.

Heerhartz, S.M., M.N. Dethier, J.D. Toft, J.R. Cordell, and A.S. Ogston. 2014. Effects of shoreline armoring on beach wrack subsidies to the nearshore ecotone in an estuarine fjord. *Estuaries and Coasts* 37: 1256–1268.

Heerhartz, S.M., J.D. Toft, J.R. Cordell, M.N. Dethier, and A.S. Ogston. 2015. Shoreline armoring in an estuary constrains wrack-associated invertebrate communities. *Estuaries and Coasts* 39: 171–188.

Hobbs, R.J., S. Arico, J. Aronson, J.S. Baron, P. Bridgewater, V.A. Cramer, P.R. Epstein, J.E. Ewel, C.A. Klink, A.E. Lugo, D. Norton, D. Ojima, D.M. Richardson, E.W. Sanderson, F. Valladeres, M. Vila, R. Zamora, and M. Zobel. 2006. Novel ecosystems: Theoretical and management aspects of the new ecological world order. *Global Ecology and Biogeography* 15: 1–7.

Hoggart, S.P.G., R.A. Francis, and M.A. Chadwick. 2012. Macroinvertebrate richness on flood defence walls of the tidal River Thames. *Urban Ecosystems* 15: 327–346.

Huang, X.W., W. Wang, and Y.W. Dong. 2015. Complex ecology of China's seawall. *Science* 347: 1079–1080.

Hughes, A.R., B.D. Inouye, M.T.J. Johnson, N. Underwood, and M. Vellend. 2008. Ecological consequences of genetic diversity. *Ecology Letters* 11: 609–623.

Hughes, R.G. and O.A.L. Paramor. 2004. On the loss of saltmarshes in south-east England and methods for their restoration. *Journal of Applied Ecology* 41: 440–448.

Iveša, L., M.G. Chapman, A.J. Underwood, and R.J. Murphy, R.J. 2010. Differential patterns of distribution of limpets on intertidal seawalls: Experimental investigation of the roles of recruitment, survival and competition. *Marine Ecology Progress Series* 407: 405–469.

Jackson, A.C., M.G. Chapman, and A.J. Underwood. 2008. Ecological interactions in the provision of habitat by urban development: Whelks and engineering by oysters on artificial seawalls. *Austral Ecology* 33: 307–316.

Jaramillo, E., H. Contreras, and A. Bollinger. 2002. Beach and faunal response to the construction of a seawall in a sandy beach in South Central Chile. *Journal of Coastal Research* 18: 523–529.

Johannesson, K. 1988. The paradox of Rockall: Why is a brooding gastropod (*Littorina saxatilis*) more widespread than one having a planktonic larval dispersal stage (*L. littorea*)? *Marine Biology* 99: 507–513.

Johannesson, K. and T. Warmoes. 1990. Rapid colonization of Belgian breakwaters by the direct developer, *Littorina saxatilis* (Olivi) (Prosobranchia, Mollusca). *Hydrobiologia* 193: 99–108.

Klein, J.C., A.J. Underwood, and M.G. Chapman. 2011. Urban structures provide new insights into interactions among grazers and habitat. *Ecological Applications* 21: 427–438.

Komar, P.D. and W.G. McDougal. 1988. Coastal erosion and engineering structures: The Oregon experience. *Journal of Coastal Research* 4: 77–92.

Lowe, W.H. and F.W. Allendorf. 2010. What can genetics tell us about population connectivity? *Molecular Ecology* 19: 3038–3051.

Martín-López, B., C. Montes, and J. Benayas, J. 2007. The non-economic motives behind the willingness to pay for biodiversity conservation. *Biological Conservation* 139: 67–82.

Martins, G.M., R.C. Thompson, A.I. Neto, S.J. Hawkins, and S.R. Jenkins. 2010. Enhancing stocks of the exploited limpet *Patella candei* d'Orbigny via modifications in coastal engineering. *Biological Conservation* 143: 203–211.

McGranahan, G., D. Balk, and B. Anderson. 2007. The rising tide: Assessing the risks of climate change and human settlements in low elevation coastal zones. *Environment and Urbanization* 19: 17–37.

McQuaid, K.A. and C.L. Griffiths. 2014. Alien reef-building polychaete drives long-term changes in invertebrate biomass and diversity in a small, urban estuary. *Estuarine, Coastal and Shelf Science* 138: 101–106.

Moschella, P.S., M. Abbiati, P. Åberg, L. Airoldi, J.M. Anderson, F. Bacchiocchi, F. Bulleri, G.E. Dinesen, M. Frost, E. Gacia, L. Granhag, P.R. Jonsson, M.P. Satta, A. Sundelof, R.C. Thompson, and S.J. Hawkins. 2005. Low-crested coastal defence structures as artificial habitats for marine life: Using ecological criteria in design. *Coastal Engineering* 52: 1053–1071.

Moreira, J. 2006. Patterns of occurrence of grazing molluscs on sandstone and concrete seawalls in Sydney Harbour (Australia). *Molluscan Research* 26: 51–60.

Moreira, J., M.G. Chapman, and A.J. Underwood. 2006. Seawalls do not sustain viable populations of limpets. *Marine Ecology Progress Series* 322: 179–188.

Moreira, J., M.G. Chapman, and A.J. Underwood. 2007. Maintenance of chitons on seawalls using crevices on sandstone blocks as habitat in Sydney Harbour, Australia. *Journal of Experimental Marine Biology and Ecology* 347: 134–143.

Morley, S.A., J.D. Toft, and K.M. Hanson. 2012. Ecological effects of shoreline armoring on intertidal habitats of a Puget Sound urban estuary. *Estuaries and Coasts* 35: 774–784.

Munsch, S.B., J.R. Cordell, J.D. Toft, and E. Morgan. 2014. Effects of seawalls and piers on fish assemblages and juvenile salmon feeding behavior. *North American Journal of Fisheries Management* 34: 814–827.

Neelamani, S. and N. Sandhya. 2005. Surface roughness effect of vertical and sloped seawalls in incident random wave fields. *Ocean Engineering* 32: 395–416.

Nordstrom, K.F. 2014. Living with shore protection structures: A review. *Estuarine, Coastal and Shelf Science* 150: 11–23.

Nordstrom, K.F. and N.L. Jackson. 2013. Removing shore protection structures to facilitate migration of landforms and habitats on the bayside of a barrier spit. *Geomorphology* 199: 179–191.

Palmer, M.A., H.L. Menninger, and E. Bernhardt. 2010. River restoration, habitat heterogeneity and biodiversity: A failure of theory or practice? *Freshwater Biology* 55: 205–222.

Pecl, G.T., Z.A. Doubleday, L. Danyushevsky, S. Gilbert, and N.A. Moltschaniwskyj. 2010. Transgenerational marking of cephalopods with an enriched barium isotope: A promising tool for empirically estimating post-hatching movement and population connectivity. *ICES Journal of Marine Science* 67: 1372–1380.

People, J. 2006. Mussel beds on different types of structures support different macroinvertebrate assemblages. *Austral Ecology* 31: 271–281.

Peterson, M.S., B.H. Comyns, J.R. Hendon, P.J. Bond, and G.A. Duff. 2000. Habitat use by early life-history stages of fishes and crustaceans along a changing estuarine landscape: Differences between natural and altered shoreline sites. *Wetlands Ecology and Management* 8: 209–219.

Petroski, H. 1985. *To engineer is human: The role of failure in successful design.* St. Martin's Press, NY, USA.

Petroski, H. 1994. *Design paradigms: Case histories of error and judgment in engineering.* Cambridge University Press, Cambridge, UK.

Pister, B. 2009. Urban marine ecology in southern California: The ability of riprap structures to serve as rocky intertidal habitat. *Marine Biology* 156: 861–873.

Russell, G. 2000. The algal vegetation of coastal defences: A case study from NW England. *Botanical Journal of Scotland* 52: 31–42.

Simenstad, C.A. and R.M. Thom. 1996. Functional equivalency trajectories of the restored Gog-Le-Hi-Te estuarine wetland. *Ecological Applications* 6: 38–56.

Smith, K.M. 2006. Integrating habitat and shoreline dynamics into living shoreline applications. In: Erdle, S.Y., J.L.D. Davis, and K.G. Sellner (Eds.), *Management, Policy, Science, and Engineering of Nonstructural Erosion Control in the Chesapeake Bay,* pp. 9–11. CRC Publ. No. 08-164, Chesapeake Bay.

Sobocinski, K.I., J.R. Cordell, and C.A. Simenstad. 2010. Effects of shoreline modifications on supralittoral macroinvertebrate fauna on Puget Sound, Washington beaches. *Estuaries and Coasts* 33: 699–711.

Thom, R.M. 2000. Adaptive management of coastal ecosystem restoration projects. *Ecological Engineering* 15: 365–372.

Toft, J. 2005. Benthic macroinvertebrate monitoring of Seahurst Park 2004: Pre-construction of seawall and removal. University of Washington School of Aquatic and Fishery Sciences, Washington, USA.

Toft, J.D., J.R. Cordell, and E.A. Armbrust, E.A. 2014. Shoreline armoring impacts and beach restoration effectiveness vary with elevation. *Northwest Science* 88: 367–375.

Underwood, A.J. 1995. Ecological research and (and research into) environmental management. *Ecological Applications* 5: 232–247.

Underwood, A.J. 1997. *Experiments in Ecology. Their Logical Design and Interpretation Using Analysis of Variance.* Cambridge University Press, Cambridge, UK.

Underwood, A.J., M.G. Chapman, and S.D. Connell. 2000. Observations in ecology: You cannot make progress on processes without understanding the patterns. *Journal of Experimental Marine Biology and Ecology* 250: 97–115.

Underwood A.J. and P.G. Fairweather. 1989. Supply-side ecology and benthic marine ecology. *Trends in Ecology and Evolution* 4: 16–20.

Underwood, A.J. and G.A. Skilleter. 1996. Effects of patch-size on the structure of assemblages in rockpools. *Journal of Experimental Marine Biology and Ecology* 197: 63–90.

PART V

Summary and Future Guidance

CHAPTER 23

Gaps in Knowledge
Information We Still Need to Know about Living Shoreline Erosion Control

Jana Davis

CONTENTS

23.1 INTRODUCTION

The concept of using natural habitat elements to protect shorelines from erosion, instead of or in addition to hard shoreline armor, has advanced significantly since the 1970s, when one could argue that natural habitat elements were first intentionally used for this purpose (e.g., Garbisch and Garbisch 1994). At that time, the community knew less about the physics of these systems and the impact of erosive forces on natural habitat elements, especially "softer" natural habitat elements like vegetation. If a marsh once existed along an eroding shoreline, but no longer did, how long could a rebuilt marsh at that site last? Did the answer depend on why the marsh was lost in the first place? What if a marsh never existed at the site, but land managers wanted to use vegetation instead of "harder" substrates?

The community still does not have the answers to all of these questions, but advances in both the ecological and physical sciences of shoreline erosion control have narrowed our gaps in knowledge. In addition, work on the social side—after all, erosion "problems" exist predominantly when infrastructure or some other resource valued by humans is threatened (Pilkey et al. 1992)—has allowed advances in understanding how humans make decisions about shoreline erosion control.

Still, several major gaps in knowledge contribute to debates about when to use shoreline erosion control practices at all, when to use natural features in erosion control, how many natural features to use, whether natural features are enough to protect shorelines from erosion, and the impact of using shoreline erosion control features on natural resources at the site in question and on neighboring shorelines. These debates exist due to fundamental differences of philosophy about humans' interaction with their environment, but also because certain information is missing.

These gaps negatively affect several key constituencies, such as managers and designers, but they provide an opportunity for others, such as scientists. Regulators are faced with decisions about approving individual project requests, often relying on assumptions about performance and impact on resources rather than data. Designers, engineers, and contractors in the shoreline erosion control community are using the best available information, but most agree that improved design guidelines would help them. Two constituencies can benefit from the needs of all of these communities: These gaps in knowledge can help guide scientists toward key research topics and help them focus their research efforts on producing a product for resource managers and policy makers. In addition, research funders can use these gaps to help them target resources to questions that need answers (Strayer and Findlay, Chapter 16, this book).

23.2 WHEN IS IT APPROPRIATE TO USE SHORELINE EROSION CONTROL PRACTICES AT ALL?

One of the biggest debates in the management community has less to do with the inner workings of living shorelines, but remains at the big picture level: How much erosion is acceptable at a site? How much erosion is natural? When should shorelines be "buttoned up," whether with natural habitat features or hard shoreline armor? Much of this debate centers around philosophical views, but certain gaps in both natural and social scientific information could help refine the discussion.

Shoreline retreat, or erosion, is a natural process. On the scale of tens of thousands and hundreds of thousands of years, sea level naturally rises and falls as a result of processes such as polar ice cap formation and melt and thermal expansion, and land naturally lifts and subsides in response to processes such as glacial rebound. Over the scale of millions of years (though meaningfully in the intertidal zone, sometimes shorter [Plafker 1965]), land naturally lifts because of tectonic processes, countered by constant erosion. Along estuarine shorelines, sediment is constantly carved by rains from the tops of banks of shorelines (top-down erosion) and constantly undercut from bottom of banks by streams, rivers, estuaries, and the ocean (bottom-up erosion).

These natural processes are linked to the ecology of many species. Input of sediments, especially large-grained sediments, in certain quantities supports submerged aquatic vegetation (SAV) beds, wetlands, unvegetated beach habitat important for species like horseshoe crabs and terrapin, and dynamic sand spits and other similar features that protect low energy coves, which, in turn, can be important habitat for seabirds and other wildlife (e.g., Kirwan and Megonigal 2013; Palinkas and Koch 2012; Patrick et al. 2014).

While humans may prefer to see their waterfront property stay in place, sediment is naturally constantly moving from one place to another, alternately exposed by sea level drop and submerged with sea level rise at longer time scales and linked to healthy ecological processes on shorter time scales. Should, therefore, one human be permitted to lock that sediment in place? Does the question of degree of anthropogenic portion of the cause of erosion determine what that human is permitted to do?

One gap in knowledge that weighs into this debate is quantifying how much erosion at a site is attributed to anthropogenic forces. One might have a hard time arguing for stopping a natural process, but if a large component of erosion is anthropogenic, does that provide a higher ground on which to stand philosophically? Several components of erosion can be anthropogenically driven: Vessel wakes in some areas can dominate the wave energy regime and lead to erosion (Houser 2011; McConchie and Toleman 2003; Walters et al., Chapter 12, this book). Engineers, when designing a site, often struggle with factoring in boat wake energy when calculating energy attributed to fetch and other forces. Waterway managers often ignore impacts of boat wakes on shoreline erosion when designating speed limits.

Also debated is the component of sea level change in the current era because of anthropogenic components of climate change. On the East Coast of the United States, for example, as New England rebounds from glacial pressure, the lower Chesapeake continues to sink—a natural process (Hammar-Klose and Thieler 2001). However, sea level is rising as a result of polar ice melt and warmer ocean temperatures as well, which most scientists agree is driven by anthropogenic forces (Nicholls 2004). Along the Gulf Coast of the United States, land subsidence has been linked to oil and gas drilling, also anthropogenic (Morton et al. 2006).

Humans have also been modifying the sediment budget, preventing accretion to sites that may be eroding but would have been balanced by natural accretion. Dams prevent sediment from entering the system through rivers, shoreline armor prevents sediment eroding from one shoreline segment from accreting to another, and dredging may draw sediment away from shorelines (Pilkey et al. 1992).

Gap #1: Quantifying amount of erosion due to anthropogenic forces (e.g., sea level rise due to the anthropogenic component of climate change, vessel wakes, and disruption of accretion forces such as damming of rivers and inlets).

In some areas, careful inventory of the tidal shoreline has been made, and those tools are used extensively by managers and landowners to make decisions about whether erosion control practices should be used at a site (e.g., Berman et al. 2000). If for each given region, managers knew how much sediment was in motion because of anthropogenic forces, the debate about whether to install erosion control practices that prevent both natural and unnatural erosion would be limited to the philosophical realm, as the scientific component of the question would be answered.

Gap #2: Quantifying regional sediment budgets in order to know how much sediment is transported due to both anthropogenic and natural forces.

Though most agree that coastal erosion is a natural process, some managers suggest erosion control techniques should be a tool to combat water pollution impairments (Beck et al., Chapter 14, this book). Before delving into the debate about whether curtailing a natural source of sediment and nutrient loads should provide "credit," we first need more data on the magnitude of potential reductions from such a management practice.

Gap #3: Quantifying water quality benefits (nutrient and sediment load reductions) of erosion control projects.

Those scientists and managers who disagree with use of such a tool argue that there has been a failure to distinguish among sediment sources or erosion processes. Instead, in the models, incentives are provided for reducing sediment loads no matter the sediment quality or from what process, natural or anthropogenic, the sediment originated (Linker et al. 2002).

Eroding coastal shorelines can contain both large- and small-grained sediment, the latter reducing visibility, light penetration, and potentially growth of such resources as seagrass beds. Therefore, while inputs of some sediment types and levels may benefit seagrasses, others may harm its growth. Should erosion control projects be credited with reducing sediment "pollution" in impaired water bodies? Understanding the trade-offs between ecological processes that depend on some level of erosion and those that may be harmed remains a challenge in the community.

Shoreline erosion control projects may have impacts on neighboring shorelines (e.g., possible sediment starvation) or on living resources that depend on some level of sediment input from erosive processes (e.g., seagrass). What are the trade-offs between risks and benefits of reducing erosion, and does it depend on site or region characteristics?

Gap #4: Understanding trade-offs—Information about ecological and physical impacts of disrupting the sediment budget by "buttoning up" some percentage of the shoreline in a region.

23.3 WHEN SHOULD NATURAL HABITAT FEATURES BE USED IN SHORELINE EROSION CONTROL?

Assuming consensus is reached that erosion cannot be tolerated at a site or that reducing erosion has a net benefit for some other reason as discussed above (e.g., sediment reduction goals for impaired water bodies), the coastal community still does not always agree on when it is appropriate to incorporate natural habitat features into erosion control projects. Much of this lack of agreement is based on uncertainty about the erosion control performance of natural habitat features like vegetation, oyster shell or oyster reefs, and woody debris relative to hard shoreline armor like riprap rock, seawalls, or bulkheads. Can certain natural habitat elements, especially soft habitat elements like vegetation, be sustainably included at very high energy sites (e.g., Currin et al., Chapter 11, this book)? Can "softer" features be successful erosion control elements at sites in which no degree of shoreline loss can be tolerated, like shipyards, marinas, harbors, or other heavily developed areas in which infrastructure is very close to or in the intertidal zone (Figure 23.1).

Some argue that vegetated living shoreline projects will not be successful at high-energy sites or in areas at which no movement of sediment along the shoreline can be tolerated. While some have begun to address this issue by assessing structural integrity and quantifying movement of sediment within, from, and to living shoreline sites (Currin et al. 2010; Manis 2013; Scyphers et al. 2011; Toft et al. 2013), many questions remain. To fully test whether certain design features, like vegetation, can withstand wind and wave energy at sites with long fetches or can be successful in urbanized sites, additional research on effectiveness and design is needed (Cordell et al. Chapter 21, this book; Hall et al., Chapter 13, this book; Strayer and Findlay, Chapter 16, this book).

Gap #5: Research into amount of sediment movement at vegetated living shorelines versus shoreline armor at all energy regimes, but especially at high energy sites where some managers doubt soft elements can be used effectively.

Gap #6: Research into the design of living shoreline projects at highly developed sites, and whether it is possible to design or retrofit built infrastructure to tolerate the degree of sediment movement that occurs within living shorelines.

Figure 23.1   Use of a living shoreline to protect infrastructure in proximity to the shoreline, leaving little room for tolerating migration or movement of the shoreline erosion control project or the land it is protecting.

Gap #7: Social science research: Understanding landowner tolerance for movement of sediments within, from, and to shoreline sites (Arkema et al., Chapter 2, this book; Cordell et al., Chapter 21, this book; Strayer and Findlay, Chapter 16, this book).

23.4   WHAT FEATURES ARE CONSIDERED NATURAL?

Notwithstanding debate about "success" of living shorelines in combating erosion control, a philosophical debate within the living shoreline community is underway about the focus on vegetation. Especially in the United States along the Gulf Coast and the Atlantic Coast from the mid-Atlantic south where estuarine shorelines tend to be "soft," a living shoreline is generally not considered "living" unless there is a significant vegetative component. In Maryland, USA, for example, a living shoreline permit from the resource agency requires 85% vegetation with a minimum 50–50 split between *Spartina alterniflora* and *Spartina patens.*

However, should it always be appropriate to include vegetation in erosion control projects at sites in which vegetation has not been present in recent geologic history? If a site has not been vegetated, but instead has been (in recent history) a beach environment, can vegetation still be considered a "natural" feature? In the Chesapeake Bay region, living shoreline projects in which the predominant "natural" feature is *Spartina* marsh have been constructed at three general types of sites:

a. Sites that have or at one point had eroding *Spartina* marshes (Figure 23.2)
b. Eroding sites that could be characterized as riparian habitat in which the "natural" condition would be fallen trees (and therefore coarse woody debris habitat) and cut banks leading down to intertidal mudflat without significant *Spartina* spp. vegetation (Figure 23.3)
c. High-energy sites in which the shoreline type is generally unvegetated beach habitat, used by species such as terrapin, horseshoe crabs, and others (Figure 23.4)

Figure 23.2 Example of a vegetated living shoreline used at a site that had prior wetland vegetation. (a) Before living shoreline installation. (b) After living shoreline installation.

Some resource managers have criticized the use of *Spartina*-based living shorelines in scenarios b (Figure 23.3) and c (Figure 23.4) below, indicating that these types of sites should never have had such vegetation and that to artificially construct a *Spartina*-based living shoreline at such a site is actually quite unnatural. Instead, such practitioners feel that the definition of living shorelines, "the use of as many natural habitat features as possible to protect shorelines from erosion," should drive designers to consider what should be "natural" at the site in question. If the natural characteristic of the eroding site is unvegetated beach habitat, shouldn't a living shoreline have unvegetated beach habitat as its dominant feature, and provide as much access to beach habitat for fauna as possible? If the site is an eroding bank with downed trees and significant coarse woody debris in a medium-energy environment, shouldn't the living shoreline aim to incorporate significant coarse woody debris rather than *Spartina* spp. vegetation?

(a)

(b)

Figure 23.3 Example of a vegetated living shoreline used at a site that had little prior wetland vegetation before construction. (a) Before installation of the project. (b) After installation.

Other resource managers have argued that because most of our regions have lost more wetland than any other habitat, it is not inappropriate to build wetlands where none existed in the past. They argue that wetlands should not only be restored, they should be created, and the community should accept any opportunity to do so, even in areas in which wetlands did not historically exist. Loss of coastal wetland acres is extreme, with a large portion of wetlands lost in many regions. Some estimate that half of global wetlands have been lost to date (Boesch et al. 1994; Kearney et al. 2002; Zedler and Kercher 2005), getting more attention than loss of riparian habitat along estuarine coasts or unvegetated beach habitat. Because vegetated habitats are highly valuable, and generally harbor greater densities of macrofauna and infauna than either coarse woody debris habitats or unvegetated sediment (Davis et al. 2008; Heck and Thoman 1984; Kneib 1997), created wetlands are valuable, even more so when they serve an erosion protection function as well.

Figure 23.4 Example of a vegetated living shoreline used at a site that had no prior wetland vegetation before construction but was instead an unvegetated beach habitat. (a) Before project installation. (b) After project installation.

While, as above, this debate is not just steeped in science, but has philosophical underpinnings that prevent one "right" answer, additional information about the ecological effects of vegetation will contribute to a clearer debate. It is assumed that vegetation added to a high-energy environment, where only beach existed before, will lead to increased densities of most species and will not serve as a detriment to key species. This assumption should be tested. Should it be the case that vegetated living shorelines qualify as habitat loss for, for example, horseshoe crabs, permitting agencies may decide to change requirements for living shoreline permits.

Gap #8: Assessment of the ecological effects of adding vegetation to a shoreline with no history of vegetation.

23.5 IF NONNATURAL FEATURES ARE INCLUDED IN A LIVING SHORELINE DESIGN, DOES A PRACTICE LOSE ITS "LIVING SHORELINE" LABEL?

Living shoreline purists feel that no nonnatural structural habitat components, like rocks in coastal Louisiana where none exist naturally, should ever be included. In the 1970s in the Chesapeake Bay, the initiators of living shorelines felt the same way, but found that a marsh installed where a marsh once eroded tended to face the same fate: erosion. As a result, the "hybrid" living shoreline was developed.

In a hybrid living shoreline, a nonnative hard substance is used in conjunction with a natural habitat feature to provide longer-term stability. The hard substance can be used to protect the shoreward edge or some other component from wave energy (Figure 23.5). At present, there is no tested

(a)

(b)

Figure 23.5 Example of (a) sills (perpendicular to shore), (b) segmented sills (parallel to shore). *(Continued)*

(c)

Figure 23.5 (Continued)   Example of (c) continuous sill with a window (parallel to shore).

guidance on how these hard structures should be designed, and what expense to the natural feature they pose (Strayer and Findlay, Chapter 16, this book).

A segment of the living shoreline community argues for keeping the stone used as minimal as possible, with the lowest, smallest sill structures possible, and the smallest rock size. But why? Is it an aesthetic argument? (But is that a valid basis for a position?) A cost argument? (But what if the landowner is willing to pay the difference?) Is the hypothesis that the ecological function of the living shoreline will be impaired when more stone is used? While some studies show that sills can have an altered invertebrate community compared to soft sediment (Bilkovic and Mitchell, Chapter 15, this book), others comparing hybrid living shorelines to natural marsh control sites have actually found greater diversity and density of organisms in hybrid living shoreline sites, perhaps due to the offering of a diversity of microhabitats (Davis et al. 2008). So is rock bad?

Living shoreline purists point to sites in which a large quantity of very large rocks that take up half the project area are used, leaving little room for the wetland component. They question whether the project is really a living shoreline at all or simply a revetment with a bit of green in it, and have accused such builders of greenwashing. Defenders of these designs will say that the large amount of rock was used because the landowners could not tolerate any degree of movement of the shoreline and could not take the risk of using less rock. While purists might disagree with its use, in most jurisdictions, landowners currently have the right to install such features according to current law, whether they fit in with regional sediment budgets or are consistent with natural geologic processes, offend our good senses, or not.

The reason this debate rages is that the community still does not know the ecological impact of including nonnatural structural features, such as rock sills or rock groins or wood retaining structures, in living shoreline designs. Is degree of rock use correlated with reduced ecological value? Does the presence of a rock sill in a living shoreline disrupt regional energy budgets as much as hard shoreline armor, and are therefore just as detrimental from the geomorphologic perspective?

Gap #9: Ecological studies comparing hybrid living shorelines to "pure" living shorelines in which no nonnatural hard structure is used as protection insurance.

23.6 IMPACTS TO ADJACENT CROSS-SHORE RESOURCES—THE OFFSHORE SUBTIDAL ZONE AND UPLAND RIPARIAN ZONE

Because living shorelines built along soft shorelines are generally of a more gentle slope than the alternatives (seawalls, revetments, bulkheads), by definition they are going to take up a larger surface area when constructed. Often, living shorelines with a wetland base either will encroach into the subtidal zone, requiring fill atop subtidal bottom, increasing its elevation, or will encroach up into the riparian zone, cutting into the bank, which could affect trees or other vegetation on the shoreline.

Not surprisingly, this process has engaged, sometimes enraged, but always interested those responsible for resources in those two realms. Those who are charged with protection of subtidal bottom claim that fill by definition harms the subtidal zone's resources. Those charged with protecting riparian resources, such as tree buffers, have expressed concern that trees are removed to make way for another habitat type, even if it is valuable wetland. Questions about the trade-offs have begun to be addressed but still persist (Bilkovic and Mitchell 2013), and often management of these resources is not fully integrated into one regulatory or management process. Often, each resource (subtidal, intertidal, and upland) is managed by its own resource agency, and the process is not integrated into a holistic view, which can confuse landowners (Du Bois, Chapter 6, this book). Those who work in the intertidal realm maintain that of all three habitat types, loss of wetland resources has been greatest in recent centuries, and in fact as a result of wetland loss, amount of shallow unvegetated subtidal bottom has actually increased. Therefore, sacrifice of subtidal habitat to create new wetlands is justified.

Gap #10: Understanding subtidal/wetland trade-offs—Quantifying what subtidal resources are lost when intertidal wetland habitat is created requiring fill.

Gap #11: Understanding riparian/wetland trade-offs—Quantifying the benefits of wetland creation that requires some loss of riparian habitat relative to the costs.

23.7 DEBATES ABOUT OPTIMAL DESIGN: HOW TO MAXIMIZE ECOLOGICAL PERFORMANCE WITHOUT SACRIFICING STRUCTURAL PERFORMANCE, AND HOW TO MAXIMIZE STRUCTURAL PERFORMANCE WITHOUT SACRIFICING ECOLOGICAL PERFORMANCE?

The impetus to using living shorelines is to increase habitat value. Recent studies now show that living shorelines, whether hybrid or not, provide greater habitat value than bulkheads or revetments (Bilkovic and Mitchell 2013; Chapman and Underwood 2011; Currin et al. 2010; Davis et al., 2008; Toft et al. 2013). However, many debates still exist about how best to design living shorelines.

Some of these questions affect the permitting process and force design changes despite the fact that data do not yet exist to support them. Some designers create sills 4 ft above mean lower low water at sites with a fetch of less than 5 miles, and debates ensue in the field between designers and permitters about whether the sill is too high. Some designers propose sills only 1 ft above mean lower low water at sites with a fetch of 10 miles, and landowners question whether this level provides enough protection for their properties. Designs can be as much art as science. Each of these design debates represents an information gap that could be addressed by the scientific community (Bilkovic and Mitchell, Chapter 15, this book).

Part of the problem is landowners and managers often fail to define "success," that is, clear goals either ecologically or physically. Is there an aesthetic argument for the landowner against additional rock? What is the tolerance level for any movement/shifting/loss of sediment within/from the erosion control project or from the shoreline it is intended to protect (Strayer and Findlay, Chapter 16, this book)? Is any movement considered a failure? What if the created erosion control project shifts significantly, but the upland area remains protected? Is that a success or failure? Is 100% vegetation cover of the living shoreline required, or is some degree of vegetation loss acceptable, especially if ecological function remains high (or higher), measured by density or diversity of, for example, fish species? Each project is going to have different success criteria for the different parties involved (landowner, permitter, contractor).

23.7.1 Need for Design Standards

The nonstructural/living shoreline erosion control community currently does not have design standards—rates of failure in suites of different energy regimes, substrate types, bathymetry, topography, and other characteristics (Walker et al. 2011). The best way to determine these is experimentally through comparative installations varying only one feature at a time (e.g., building projects in the same energy regimes and all other factors equal with intertidal slopes ranging from 4:1 to 20:1). Given landowners' understandable unwillingness to experiment given the risks, both modeling and descriptive studies can help arrive at ideal and "safe" (see definition of success issue above) levels of these factors given site conditions.

Gap #12: Design standards providing guidelines for structural component material, size, height, shape, and placement; intertidal slope; and other design features based on site characteristics such as energy regime, substrate, bathymetry, and topography (Rella et al., Chapter 5, this book). Physical studies need to be conducted to examine risk of failure with different design characteristics.

23.7.2 Understanding Ecological and Structural Ramifications of Windows

Many hybrid living shoreline sites are designed to use a sill to protect the toe of a vegetated living shoreline project. The toe may be constructed of stone, oyster reef, oyster bags, or other material. Many resource managers encourage or require such sills to contain "windows" or offsets or gaps. These windows are intended—assumed—to allow motile macrofauna access to the vegetated area from offshore at high tide and prevent trapping behind the sill as the tide ebbs. However, designers are sometimes resistant to windows because they add expense (both time and materials) depending on how they are built. It has not yet been tested quantitatively whether projects with windows provide "better" habitat than those without (Bilkovic and Mitchell, Chapter 15, this book).

Gap #13: Research to determine whether continuous sills do indeed prevent access or trap motile fauna, and whether windows increase species density, diversity, richness, and otherwise provide better habitat.

Designers are sometimes uneasy about including windows because they can leave the vegetated platform exposed to wave energy. It has not yet been tested quantitatively whether there is a difference in erosion protection between continuous sills and sills with windows at sites with the same characteristics.

Gap #14: Research to determine whether installation of windows, intended for improved ecological value, compromises the structural value of the erosion control projects.

23.7.3 Perpendicular versus Parallel Structure Features—Groins versus Sills

Some designers of hybrid living shorelines use groins (stone features perpendicular to shore) instead of sills or breakwaters (stone features installed parallel to shore) to provide a structural component to a vegetated living shoreline (Figure 23.5). The premise is that groin projects use less rock and, hence, are less expensive, and also provide more access for fauna to the vegetated intertidal edge, but neither the ecological nor physical value has been compared among design options.

Gap #15: As with Gap #12, studies are needed to evaluate the importance of a vegetation–water interface devoid of rock to evaluate whether groin projects are ecologically more valuable.

Gap #16: Design guidance on when or why to use sills or breakwaters versus groins. The structural performance of the two design types needs to be compared in like environments.

23.7.4 Value of Oyster Reef and Other Natural Materials as a Structural Sill Component

In certain areas, especially the Gulf Coast of the United States, healthy intertidal oyster reefs thrive, and many designers prefer to use oyster shell, reef balls, oyster reef, or other oyster structures as sills rather than rock (e.g., Bilkovic and Mitchell, Chapter 15, this book; La Peyre et al., Chapter 18, this book; Peterson et al., Chapter 19, this book). In fact, some consider living shorelines that use stone as "green-washed." In other areas that may be far from quarries, local, "native" timber is used instead of stone to construct a sill. However, natural substrates like oyster and timber have their challenges. Timber will biodegrade, and oyster reef can have survivorship challenges. Are living shorelines that use oyster or timber as a structural element "better" than those using rock, and are the costs worth the benefits? Do living shorelines with oyster reef, oyster shell, or timber toe protection provide the same degree of erosion protection for the same duration as those with rock as toe protection? Do living shorelines that use oyster as a structural element support greater diversity and density of species than those with rock, to make any trade-offs in erosion control value worth it?

Gap #17: Research on the structural value of oysters and timber in living shorelines to determine whether they are as effective as stone in hybrid living shorelines.

Gap #18: Research on the ecological impacts of oysters and timber used for structural purposes in living shorelines to determine whether there is ecological value in their use.

23.7.5 Value of Installing Habitat Features Not Necessary for Erosion Control Value

Some in the living shoreline community have noted the lack of variability or habitat heterogeneity in project designs and have questioned whether features can be engineered or installed (at minimal net cost) that have no impact either positively or negatively on structural performance but that can improve or even maximize habitat benefit (Arkema et al., Chapter 2, this book; Bilkovic and Mitchell, Chapter 15, this book; Browne and Chapman, Chapter 22, this book). One example is the inclusion of tree trunks and branches, especially those that might have been removed to construct the site, in the design to introduce a coarse woody debris habitat (Figure 23.6). Another example in sites with wide intertidal platforms is the inclusion of pools, depressions, or simply variability in intertidal heights (Figure 23.7). A third example comes from estuaries in which hard substrate is the dominant natural shoreline type: the inclusion of horizontal areas with tidepools or rugose features to provide diversity of habitat, even features like flowerpots, within seawalls (Browne and Chapman 2011; Browne and Chapman, Chapter 22, this book; Chapman and Underwood 2011; Cordell et al., Chapter 21, this book; Dugan et al. 2011; Piazza et al. 2005).

Gap #19: Analyzing the costs (financial, structural, and possible ecological) and benefits (ecological value) of introducing habitat-enhancing variability, such as coarse woody debris, oysters not intended to have an erosion protection function, intertidal pools, or other variations in bathymetry/topography.

Figure 23.6   Example of coarse woody debris designed into a living shoreline site for hypothesized habitat benefit.

Figure 23.7   Example of pools designed into living shoreline projects for a hypothesized habitat heterogeneity benefit.

23.7.6 Impacts of Shading

Many living shoreline designers installing living shorelines in heavily buffered areas have suggested that shading from riparian vegetation is an issue and recommend removing trees to improve survivorship of wetland vegetation. Tree removal has been controversial, as it assumes that the net wetland benefits outweigh the net benefits offered from trees (trees absorb runoff, provide habitat, and serve a multitude of other benefits). While research on the various subtidal, intertidal, and upland trade-offs can help with this debate (see Gap #3), quantifying survivorship of wetland plants in various shading scenarios is needed.

Gap #20: An analysis of the degree of shade from shoreline trees that prevent wetland establishment and growth.

23.8 UNDERSTANDING EXPECTATIONS ABOUT SUSTAINABILITY: WE KNOW ARMOR DOES NOT LAST FOREVER. DO LIVING SHORELINES FARE BETTER OR WORSE?

Nothing lasts forever. Or can wetlands created through living shorelines last forever? We do know that an unprotected shoreline is likely to change over time, erode, or possibly accrete. Between 50% (Dolan et al. 1990) and 75% (Atlantic), 77% (Gulf), and 83% (Pacific) (data calculated from Figure 1 of Pilkey and Thieler 1992) of all coastline in the United States is eroding, with only 17%–25% stable or accreting. Shoreline erosion control is meant to reduce erosion for some period, but it is generally assumed that such projects do not provide erosion protection for an unlimited period. Despite the prevalence of requirements to install living shorelines and the activity in the management community surrounding living shorelines, few studies have actually compared the longevity of living shorelines to armor alternatives, or rather the duration of time they prevent erosion of the upland area intended to be protected (Arkema et al., Chapter 2, this book). Few also have compared maintenance costs of living shorelines and armor, and assuming that not all projects are fully maintained in perpetuity, what happens to (and what are the ecological effects of) the non-native materials in either hybrid living shorelines or armor after the projects have degraded and their materials no longer serve their intended protection function (Browne and Chapman, Chapter 22, this book)?

Gap #21: Studies on the sustainability of living shoreline projects compared to armor: Understanding how long do living shorelines themselves last compared to armor, and over what period do they protect the intended upland target compared to armor (Arkema et al., Chapter 2, this book).

While long-term sustainability is the most important component of the temporal question, storm damage is a short-term interest in the public sector. Erosion and therefore expenses to repair shorelines tend to occur episodically (Hall et al. 2002; Ralston and Geyer 2009). Much debate but few studies exist about how well "natural" shoreline erosion control projects fare against storm wind and wave energy compared to armor such as seawalls, bulkheads, and revetments (e.g., Wowk and Yoskowitz, Chapter 4, this book).

Gap #22: Quantified comparisons of effects of storm damage on living shoreline versus armor in multiple regions considering (a) persistence of the shoreline erosion control project itself and (b) fate of the upland property it is intended to protect.

Some have argued that when designed properly in site conditions that will support them, vegetated living shorelines can actually be sustainable indefinitely, because as sea level rises (and presumably falls over longer time scales), the wetland can migrate upland both due to accretion in the wetland platform and migration up the bank as sea level rises, just as natural wetlands can (Arkema et al., Chapter 2, this book; Currin et al. 2010; Currin et al., Chapter 11, this book; Morris et al. 2002).

Gap #23: Understanding how living shorelines' sustainability is affected by sea level rise: Determining whether wetlands installed as a component of living shoreline projects can keep up with sea level rise by migrating shoreward and accreting (Arkema et al., Chapter 2, this book; Bilkovic and Mitchell, Chapter 15, this book; Currin et al., Chapter 11, this book).

23.9 HOW CAN USE OF LIVING SHORELINES INSTEAD OF ARMOR BE EXPANDED?

The majority of estuarine shoreline is privately owned. While government can help, and governments can make choices for public land, private landowners will have to be engaged in the conversation, especially as shoreline protection installed on one property can have impacts on adjacent properties.

Ideally, shoreline management decisions would be made on a watershed scale, considering areas along a shoreline more prone to erosion, areas that contain finer-grained sediment that serves as a net detriment to water quality if permitted to erode, and areas that if protected would not negatively affect other natural resources. However, shoreline is generally not owned or managed on a watershed scale. Instead, in most jurisdictions, landowners have some rights to protect their individual properties, without consideration of larger watershed processes or cumulative impacts (Strayer and Findlay, Chapter 16, this book).

Landowners can either be required to install living shorelines if they choose to reduce erosion affecting their waterfront property (regulation) or encouraged to do so (voluntary measures). There is a key advantage to relying more heavily on voluntary measures, including maintenance and sustainability. If landowners want to use living shorelines, they are more likely to properly care for them and to install them more quickly. While some work has been done by management agencies to understand landowner willingness and factors that affect landowner decisions, like impact of living shorelines or shoreline erosion control in general on property value, few have been published in the scientific literature for others to use (Arkema et al., Chapter 2, this book and Strayer and Findlay, Chapter 16, this book, but see Landry and Hindsley 2008).

Gap #24: Understanding attitudes toward living shorelines, including perceptions of barriers and benefits of installing them (Strayer and Findlay, Chapter 16, this book). To obtain this information, public opinion surveys on barriers, benefits, and willingness to install living shorelines must be conducted.

Gap #25: Once better understanding of the perceived barriers and benefits of living shorelines is obtained, marketing campaigns containing appropriate messages must be developed (Esteves and Williams, Chapter 9, this book).

Presumably, one of the key barriers to landowners installing living shorelines is uncertainty about cost. Considering life span, maintenance to the erosion control project, and maintenance to the upland area being protected (risk of non-protection), what is the annual cost of a living shoreline versus armor (Strayer and Findlay, Chapter 16, this book)?

Gap #26: An economic analysis of the life cycle costs of living shoreline erosion control relative to structural erosion control.

In addition to voluntary actions, in some regions, laws requiring living shorelines have been passed or considered (Currin et al. 2010; Pace 2011). These laws have faced some opposition. Some argue that the business community will be affected, and that marine contractors will have to invest in additional equipment or additional training that adds cost and reduces their profit margin. However, we do not have data about whether it is more difficult or more expensive for contractors to build living shorelines than bulkheads or revetments.

Gap #27: An analysis of the impact of living shoreline requirements on marine contractors: Understanding degree of difficulty and expense to build living shorelines relative to armor.

If difficulties for contractors are perceived, but not real, training programs for contractors can help them enter the living shoreline field. If difficulties for contractors are real, programs to help them overcome obstacles can be developed.

Gap #28: Training programs for marine contractors to help them design and construct living shorelines.

In some cases, regulatory agencies struggle, both within and between agencies, with conflicting policies or perceived conflict between policies. In some states with living shoreline laws requiring use of nonstructural or hybrid projects instead of armor when conditions permit, agencies tasked with managing adjacent resources—subtidal resources in the offshore direction and riparian buffer resources in the upland direction—have expressed concern over living shoreline projects, even those that meet the shoreline law requirements. For example, in Maryland, living shorelines proposed or required in some areas (and supported by the programmatic arm of the National Oceanic and Atmospheric Administration [NOAA]) have met with opposition from the NOAA's regulatory arm, which has commenting authority on the Joint Federal–State Permit Process. The concern is based on the grounds that using fill to increase elevation in the subtidal zone to construct the intertidal platform for the living shoreline can interfere with subtidal resources, potential SAV habitat, or other essential fish habitat. NOAA commenters have instead suggested construction of the intertidal platform for the project in the other direction, reducing elevation of the upland area to intertidal elevations. However, building the intertidal platform in the upland direction has drawn concern from the Critical Area Commission and county building permit reviewers, tasked with managing riparian tree buffer resources. These entities have expressed concern that reducing elevation of a section of the upland will affect trees, even requiring tree removal, to make way for the intertidal platform, and is not consistent with the Critical Area Law.

Many landowners do not expect to receive conflicting information from different resource agencies, and certainly not from agents within the same agency. Most landowners likely do not fully understand the number of agencies involved in a regulatory action (Pace, Chapter 3, this book) and may assume that one government agent speaks for all. For this reason, in addition to the ecological reasons, individual policies that tangentially address a secondary resource, especially if they might provide conflicting direction, should be integrated with policies that primarily address the secondary resource. A holistic approach to the system, from offshore to upland, is needed from an ecological perspective, given that each resource interacts physically or biologically with the others, and from a practical perspective, to aid in permitting and management decisions.

Gap #29: Integration of laws, regulations, and policies that currently treat individual resources of shoreline habitat, subtidal resources, and upland riparian resources separately.

23.10 MISSING TOOLS

Many of the gaps in knowledge discussed above are not unique to a region or sector and could be more efficiently filled with the development of tools that would allow use or sharing of information across the region. For example, many of these gaps require rigorous scientific studies to address them, and because of the inherent variability in the system(s) (Strayer and Findlay, Chapter 16, this book), they need to be designed with large sample sizes. Our power to test these hypotheses would benefit by being able to better compare studies, and even combine data collected by more than one science team in a meta-analysis (Wowk and Yoskowitz, Chapter 4, this book). To do so, however, data collection protocols would have to be consistent.

Gap #30: Consistent protocols for data collection on biological and physical parameters of living shorelines and control sites (Rella et al., Chapter 5, this book).

While experimental studies are often the best way to isolate individual questions and test hypotheses, in the case of questions about issues like restoration techniques, such experimental design is likely to be cost-prohibitive. However, with careful project selection and use of controls, descriptive studies of existing projects can be very valuable in addressing some of the scientific and design questions described above. Researchers often struggle with site selection for studies, and often miss ideal projects to include a descriptive study design because of lack of knowledge of the project inventory. Efforts to establish a national project repository in the United States have been initiated. For example, the American Society of Civil Engineers has (at the time of this publication) a living shorelines database (www.mycopri.org//livingshorelines) that contains information on the environmental site conditions and physical and ecological parameters of designs. Practitioners and landowners could be more strongly encouraged to enter their projects in it.

Gap #31: A database of projects, site conditions, project characteristics, and project performance.

Finally, the community should consider a way of ensuring that information gleaned in one sector can be quickly transferred to another sector for which the gap in question is most debilitating. In many of the gaps discussed above, the scientific community is best poised to obtain the missing information. At such exchanges, scientific knowledge can be synthesized, translated to actionable steps, and then transferred to regulatory/practitioner communities. In turn, new gaps experienced by the regulatory/practitioner communities can be shared with the scientific community to inform research activity.

Gap #32: Forums in which new information and information needs can be shared back and forth between sectors of the shoreline management community.

23.11 WHAT DO RESOURCE MANAGERS, SCIENTISTS, PRACTITIONERS, AND OTHERS IN THE COMMUNITY CONSIDER AS GAPS IN KNOWLEDGE?

Many of these gaps in knowledge were discussed at a Living Shoreline Summit held in Maryland in 2013, attended by 233 individuals that were identified as members of the scientific community, program management community, designer/builder sector, and regulatory/policy maker communities (Maryland Department of Natural Resources 2013). A survey exercise was conducted in which the sectors were asked to develop their top priorities of gaps. A total of 29 gaps were identified across the four sectors. Then, the full attendance was asked to vote for their top two gaps from the list of 29 generated by each sector, with the idea that other sectors may have proposed important gaps not initially considered.

23.11.1 Initial Brainstorming within Sectors

Comparing the top gaps identified by each sector indicates opportunity for one community to meet the needs of another. When initially asked, scientists indicated that development of design criteria and issues pertaining to monitoring and assessment were most important. In this case, one of the other four sectors (designer/builder sector) can help with the first need, development of design criteria. The second need rests squarely within their own research community.

When initially brainstorming, program managers indicated that cost–benefit analyses were the biggest gap, a gap that could be filled by either the social science sector or their own community. In contrast, the biggest gap identified initially by the designer/builder community rested within another of the four sectors: the regulatory and policy realm (permitting challenges). The regulatory and policy community listed two biggest gaps, a self-acknowledgment of gaps in the regulatory realm (planning shoreline projects at a watershed scale, conflicting regulations, interaction with permittees) as well as gaps in the research realm on ecosystem trade-offs.

These initial brainstormed gaps show how increased communication among sectors can help groups identify where they can help other sectors. The research community needs help from the designer/builder community on design criteria. The designer/builder community needs assistance from the regulatory/policy community to fill gaps in permitting. The regulatory/policy community needs input from the research community on ecosystem trade-offs.

23.11.2 Ranking the Full Range of Gaps by Sector

After each sector identified key gaps, all individuals were permitted to cast two votes for their top two gaps (Figure 23.8). As part of the exercise, each individual could consider any of the 29 gaps proposed, even if not by his or her own sector. A total of 276 votes were cast. Considerable overlap

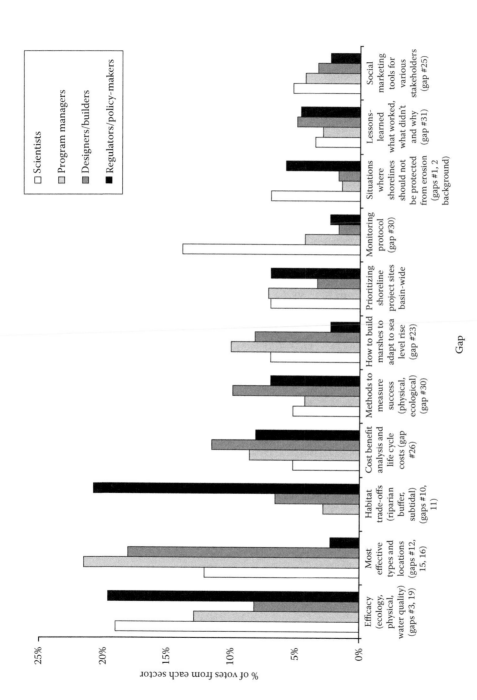

Figure 23.8 Gaps in knowledge identified by various sectors at the 2013 Mid-Atlantic Living Shoreline Summit (Maryland Department of Natural Resources 2013). Each gap is labeled on the x axis as worded by the Summit participants. Corresponding gaps in this chapter, where applicable, are listed in parentheses.

was observed among the top five gaps prioritized by the members of each sector, with some notable exceptions. The regulatory/policy community was the only one to rank ecological trade-offs (Gap #10 and #11 of this chapter) in its top five priority gaps. Such a high ranking by this group indicates that this community recognizes the conflicting policies that often govern subtidal, intertidal, and riparian resources. The research community was the only one to highly rank need for consistent monitoring protocols. All four sectors were in agreement on certain gaps: all highly ranked gaps in understanding of efficacy (long term and short term) in ecological resources, physical performance, and water quality.

23.12 SUMMARY

One of the goals of this publication is to outline what the community has learned about living shorelines, especially over the last 40 years. However, a key goal is to bring to light gaps in knowledge in the realm of living shoreline erosion control. This information is intended to be useful for natural, physical, and social scientists as they aim to produce research that can help users of the information such as regulators, managers, designers, and more. The information may also be helpful for others in a position to produce both information and tools, such as resource managers. The next step is to put a plan in place to fill these gaps. Scientists can play a role in many of them. Managers can play a role in others. Funders can play a role in making resources available for both of those constituencies' ability to fill the gaps.

REFERENCES

Arkema, K.K., S.B. Scyphers, and C. Shepard. In press. Living shorelines for people and nature. In *Living Shorelines: The Science and Management of Nature-Based Coastal Protection*, CRC Press.

Beck, A., R.M. Chambers, M. Mitchell, and D.M. Bilkovic. In press. Evaluation of living shoreline marshes as a tool for reducing nitrogen pollution in coastal systems. In *Living Shorelines: The Science and Management of Nature-Based Coastal Protection*, CRC Press.

Berman, M.R., H. Berquist, S. Dewing, J. Glover, C.H. Hershner, T. Rudnicky, D.E. Schatt, and K. Skunda. 2000. Mathews County Shoreline Situation Report, Special Report in Applied Marine Science and Ocean Engineering No. 364, Comprehensive Coastal Inventory Program, Virginia Institute of Marine Science, College of William and Mary, Virginia.

Bilkovic, D.M. and M. Mitchell. In press. Designing living shoreline salt marsh ecosystems to promote coastal resilience. In *Living Shorelines: The Science and Management of Nature-Based Coastal Protection*, CRC Press.

Bilkovic, D.M. and M.M. Mitchell. 2013. Ecological tradeoffs of stabilized salt marshes as a shoreline protection strategy: Effects of artificial structures on macrobenthic assemblages. *Ecological Engineering* 61: 469–481.

Boesch, D.F., M.N. Josselyn, A.J. Mehta, J.T. Morris, W.K. Nuttle, C.A. Simenstad, and D. J. Swift. 1994. Scientific assessment of coastal wetland loss, restoration and management in Louisiana. *Journal of Coastal Research*, i-103.

Browne, M.A. and M.G. Chapman. 2011. Ecologically informed engineering reduces loss of intertidal biodiversity on artificial shorelines. *Environmental Science and Technology* 45: 8204–8207.

Browne, M.A. and M.G. Chapman. In press. The ecological impacts of reengineering artificial shorelines: The state of the science. In *Living Shorelines: The Science and Management of Nature-Based Coastal Protection*, CRC Press.

Chapman, M.G. and A.J. Underwood. 2011. Evaluation of ecological engineering of "armoured" shorelines to improve their value as habitat. *Journal of Experimental Marine Biology and Ecology* 400: 302–313.

Cordell, J.R., J. Toft, S. Munsch, and M. Goff. In press. Benches, beaches, and bumps: How habitat monitoring and experimental science can inform urban seawall design. In *Living Shorelines: The Science and Management of Nature-Based Coastal Protection*, CRC Press.

Currin, C.A., J. Davis, and A. Malhotra. In press. Response of salt marshes to wave energy provides guidance for successful living shoreline implementation. In *Living Shorelines: The Science and Management of Nature-Based Coastal Protection*, CRC Press.

Currin, C.A., W.S. Chappell, and A. Deaton. 2010. Developing alternative shoreline armoring strategies: The living shoreline approach in North Carolina. Puget Sound Shorelines and the Impacts of Armoring—Proceedings of a State of the Science Workshop: 91–102.

Davis, J.L.D., R. Schnabel, and R. Takacs. 2008. Evaluating ecological impacts of living shorelines and shoreline habitat elements: An example from the upper western Chesapeake Bay. In: S. Erdle, J.L.D. Davis, and K.G. Sellner (Eds.), *Management, Policy, Science and Engineering of Nonstructural Erosion Control in the Chesapeake Bay: Proceedings of the 2006 Living Shoreline Summit*, CRC Publ. No. 08-164.

Dolan, R., S. Trossbach, and M. Buckley. 1990. New shoreline erosion data for the Mid-Atlantic Coast. *Journal of Coastal Research* 6: 471–477.

Du Bois, K. In press. Overcoming barriers to living shoreline use and success: Lessons from Southeastern Virginia's coastal plain. In *Living Shorelines: The Science and Management of Nature-Based Coastal Protection*, CRC Press.

Dugan J.E., L. Airoldi, M.G. Chapman, S.J. Walker, and T. Schlacher. 2011. Estuarine and coastal structures: Environmental effects, a focus on shore and nearshore structures. In: Wolanski, E. and McLusky, D.S. (Eds.), *Treatise on Estuarine and Coastal Science*, Vol. 8, pp. 17–41. Waltham: Academic Press.

Esteves, L.S. and J.J. Williams. In press. Managed realignment in Europe: A synthesis of methods, achievements, and challenges. In *Living Shorelines: The Science and Management of Nature-Based Coastal Protection*, CRC Press.

Garbisch, E.W. and J.L. Garbisch. 1994. Control of upland bank erosion through tidal marsh construction on restored shores: Application in the Maryland portion of Chesapeake. *Environmental Management* 18: 677–691.

Hall, J.W., I.C. Meadowcroft, E.M. Lee, and P.H. van Gelder. 2002. Stochastic simulation of episodic soft coastal cliff recession. *Coastal Engineering* 46: 159–174.

Hall, S.G., R. Beine, M. Campbell, T. Ortego, and J. Risinger. In press. Growing living shorelines and ecological services via coastal bioengineering. In *Living Shorelines: The Science and Management of Nature-Based Coastal Protection*, CRC Press.

Hammar-Klose, E.S. and E.R. Thieler. 2001. Coastal Vulnerability to Sea-Level Rise: A Preliminary Database for the U.S. Atlantic, Pacific and Gulf of Mexico Coasts. U.S. Geological Survey Digital Data Series—68.

Heck, K.L. and T.A. Thoman. 1984. The nursery role of seagrass meadows in the upper and lower reaches of the Chesapeake Bay. *Estuaries* 7: 70–92.

Houser, C. 2011. Sediment resuspension by vessel-generated waves along the Savannah River, Georgia. *Journal of Waterway, Port, Coastal, and Ocean Engineering* 137: 246–257.

Kearney, M.S., A.S. Rogers, J.R. Townshend, E. Rizzo, D. Stutzer, D., J. Stevenson, and K. Sundborg. 2002. Landsat imagery shows decline of coastal marshes in Chesapeake and Delaware Bays. *EOS, Transactions American Geophysical Union* 83: 173–178.

Kirwan, M.L. and J.P. Megonigal. 2013. Tidal wetland stability in the face of human impacts and sea-level rise. *Nature* 504: 53–60.

Kneib, R.T. 1997. The role of tidal marshes in the ecology of estuarine nekton. In: H. Barnes, A.D. Ansell, R.N. Gibson, M. Barnes (Eds.). *Oceanography and Marine Biology*, Vol. 35, pp. 163–220.

La Peyre, M., L. Schwarting Miller, S. Miller, and E. Melancon. In press. Comparison of oyster populations, shoreline protection service, and site characteristics at seven created fringing reefs in Louisiana: Key parameters and responses to consider. In *Living Shorelines: The Science and Management of Nature-Based Coastal Protection*, CRC Press.

Landry, C. and P. Hindsley. 2008. Willingness to pay for risk reduction and amenities: Applications of the hedonic price method in the coastal zone. In: S. Erdle, J.L.D. Davis, and K.G. Sellner (Eds.), *Management, Policy, Science and Engineering of Nonstructural Erosion Control in the Chesapeake Bay: Proceedings of the 2006 Living Shoreline Summit*, CRC Publ. No. 08-164.

Linker, L.C., G.W. Shenk, P. Wang, K.J. Hopkins, and S. Pokharel. 2002. A short history of Chesapeake Bay modeling and the next generation of watershed and estuarine models. *Proceedings of the Water Environment Federation* 2002: 569–582.

Manis, J.E. 2013. *Assessing the Effectiveness of Living Shoreline Restoration and Quantifying Wave Attenuation in Mosquito Lagoon, Florida* (Master's Thesis, University of Central Florida Orlando, Florida). 73 pp.

Maryland Department of Natural Resources. 2013. Proceedings of the 2013 Mid-Atlantic Living Shorelines Summit. http://dnr.maryland.gov/ccs/pdfs/ls/2013summit/SummitProceedings.pdf.

McConchie, J.A. and I.E.J. Toleman. 2003. Boat wakes as a cause of riverbank erosion: A case study from the Waikato River, New Zealand. *Journal of Hydrology. New Zealand* 42: 163–179.

Morris, J.T., P.V. Sundareshwar, C.T. Nietch, B. Kjerfve, and D.R. Cahoon. 2002. Responses of coastal wetlands to rising sea level. *Ecology* 83: 2869–2877.

Morton, R.A., J.C. Bernier, and J.A. Barras. 2006. Evidence of regional subsidence and associated interior wetland loss induced by hydrocarbon production, Gulf Coast region, USA. *Environmental Geology* 50: 261–274.

Nicholls, R.J. 2004. Coastal flooding and wetland loss in the 21st century: Changes under the SRES climate and socio-economic scenarios. *Global Environmental Change* 14: 69–86.

Pace, N. In press. Permitting a living shoreline: A look at the legal framework governing living shoreline projects at the federal, state, and local level. In *Living Shorelines: The Science and Management of Nature-Based Coastal Protection*, CRC Press.

Pace, N.L. 2011. Wetlands or seawalls? Adapting shoreline regulation to address sea level rise and wetland preservation in the Gulf of Mexico. *Journal of Land Use & Environmental Law* 26: 327–363.

Palinkas, C.M. and E.W. Koch, 2012. Sediment accumulation rates and submersed aquatic vegetation (SAV) distributions in the mesohaline Chesapeake Bay, USA. *Estuaries and Coasts* 35: 1416–1431.

Patrick, C.J., D.E. Weller, X. Li, and M. Ryder. 2014. Effects of shoreline alteration and other stressors on submerged aquatic vegetation in subestuaries of Chesapeake Bay and the mid-Atlantic Coastal Bays. *Estuaries and Coasts* 37: 1516–1531.

Peterson, M.S., K.S., Dillon, and C.A. May. In press. Species richness and functional feeding group patterns in small, patchy, natural and constructed intertidal fringe oyster reefs. In *Living Shorelines: The Science and Management of Nature-Based Coastal Protection*, CRC Press.

Piazza, B.P., P.D. Banks, and M.K. La Peyre. 2005. The potential for created oyster shell reefs as a sustainable shoreline protection strategy in Louisiana. *Restoration Ecology* 13: 499–506.

Pilkey, O.H. and E.R. Thieler. 1992. Erosion of the United States Shoreline. Quaternary Coasts of the United States: Muine and Lacusvine Systems. SEPM Special Publication No. 48: 3–7.

Pilkey, O.H., W.J. Neal, and D.M. Bush. 1992. Coastal zones and estuaries. In: Pilkey, O.H., and Thieler, E.R. (Eds.), *Coastal Erosion*. Society of Economic Paleontologists and Mineralogists.

Plafker, G. 1965. Tectonic deformation associated with the 1964 Alaska earthquake. The earthquake of 27 March 1964 resulted in observable crustal deformation of unprecedented areal extent. *Science* 148: 1675–1687.

Ralston, D.K. and W.R. Geyer. 2009. Episodic and long-term sediment transport capacity in the Hudson River estuary. *Estuaries and Coasts* 32: 1130–1151.

Rella, A., J. Miller, and E. Hauser. In press. An overview of the living shorelines initiative in New York and New Jersey. In *Living Shorelines: The Science and Management of Nature-Based Coastal Protection*, CRC Press.

Scyphers S.B., S.P. Powers, K.L. Heck Jr., and D. Byron. 2011. Oyster reefs as natural breakwaters mitigate shoreline loss and facilitate fisheries. *PLoS ONE* 6(8): e22396. DOI:10.1371/journal.pone.0022396.

Strayer, D.L. and S.E.G. Findlay. In press. Ecological performance of Hudson River shore zones: What we know and what we need to know. In *Living Shorelines: The Science and Management of Nature-Based Coastal Protection*, CRC Press.

Toft, J.D., A.S. Ogston, S.M. Heerhartz, J.R. Cordell, and E.E. Flemer. 2013. Ecological response and physical stability of habitat enhancements along an urban armored shoreline. *Ecological Engineering* 57: 97–108.

Walker, R., B. Bendell, and L. Wallendorf, L. 2011 Defining engineering guidance for living shoreline projects. *Coastal Engineering Practice* 2011: 1064–1077.

Walters, L., M. Donnelly, P. Sacks, D. Campbell. In press. Lessons learned from living shoreline stabilization in popular tourist areas: Boat wakes, volunteer support, and protecting historic structures. In *Living Shorelines: The Science and Management of Nature-Based Coastal Protection*, CRC Press.

Wowk, K.M. and D. Yoskowitz. In press. Socioeconomic and policy considerations of living shorelines—US context. In *Living Shorelines: The Science and Management of Nature-Based Coastal Protection*, CRC Press.

Zedler, J.B. and S. Kercher. 2005. Wetland resources: Status, trends, ecosystem services, and restorability. *Annual Review Environment and Resources* 30: 39–74.

CHAPTER 24

A Synthesis of Living Shoreline Perspectives

Jason D. Toft, Donna Marie Bilkovic, Molly M. Mitchell, and Megan K. La Peyre

CONTENTS

24.1 INTRODUCTION

The main goal of this summary chapter is to synthesize author perspectives across the contributed chapters, make recommendations on the correct usage of the term living shorelines, and offer guidance for planning in the future. Nature-based approaches are being applied globally, as signified by the breadth of geographic coverage in this book. The author's institutions and locations of study span the East, Gulf, and West Coasts of the United States, including the states of Massachusetts, New York, New Jersey, Maryland, Virginia, North Carolina, Florida, Alabama, Mississippi, Louisiana, Texas, California, Washington, and several national perspectives, including Hawaii; British Columbia in Canada; the Netherlands, as well as perspectives across Europe also including Belgium, Denmark, France, Germany, Spain, and the United Kingdom; Sydney Harbor in Australia; and Belize. Living shoreline techniques are very diverse and practices can vary by region, salinity and tidal regime, and degrees of natural and artificial components. Techniques covered in this book include restoring oyster reefs, eelgrass, and mangroves, planting marshes with and without supportive sills (e.g., stone, oyster shell bags, coir logs), incorporating structures such as logs and reef balls, nourishing beaches and dunes with sediment, engineering habitat features into seawalls, and managed realignment. All of these can have a variety of components, such as permitting, land acquisition, design, and monitoring. However, given the diverse representation, there are some shared commonalities that can help inform and direct shoreline management moving forward.

24.2 SYNTHESIS OF AUTHOR PERSPECTIVES

To synthesize author perspectives and identify commonalities among regions, we posed the following three focal areas to organize some collective lines of thought:

1. Lessons learned from the practice of shoreline restoration/conservation.
2. Longevity and stability of projects in the near and long term with considerations for climate change and human development.
3. What is the path forward? Research needs, strategies for working across different disciplines, training options, and future opportunities.

For each of these three focal areas, we synthesized author perspectives and highlighted the most recurring concepts. Identified in Table 24.1 are the top three perspectives for each focal area that had overlap and consistency across chapters and regions. The top three lessons learned focused on understanding the environmental setting, developing interdisciplinary approaches, and emphasizing the application of natural components in project design. These lessons learned cover the realm of natural history, recognize the strength of collaborations across specialties, and highlight the use of natural components over artificial components. By applying these main perspectives to future projects, we can build upon the past and update with new information. The top three perspectives in longevity and stability focused on coastal squeeze and adaptation to sea level rise, the ability of natural components to be self-sustaining, and degrees of resiliency to wind and waves compared to storm surge. These perspectives address issues of scale both in the size of projects and in the degree of climate forces, and again highlight the resiliency of natural components over artificial components. It will be imperative to address these issues as they will likely only continue to increase in their degree of importance. The top three perspectives regarding the path forward focused on outreach and education, long-term monitoring, and improved permitting. These perspectives highlight

Table 24.1   The Top Three Author Perspectives in Three Focal Areas

Focal Area	Top 3 Perspectives
Lessons learned	Understanding of the environmental setting can help fit the design of a project to the local conditions (e.g., wave energy, sediment processes, tidal regime, predator and competitor habitat suitability).
	Develop interdisciplinary approaches to encourage project success by collaborating with managers, engineers, and landowners.
	Natural components such as oyster reefs, fringing marshes, and sediment nourishment should be considered first over artificial components such as stone sills.
Longevity and stability	Coastal squeeze will probably have a higher impact on small projects that may not have the space and ability to adapt to climate change, although small projects have proven to be effective in the short term. The potential for landward migration or tidal elevation shift of vegetation will facilitate adaptation.
	Natural components such as oyster reefs and marsh vegetation have the ability to expand and become self-sustaining over time, and therefore can be responsive to sea level rise. Engineered components may need maintenance.
	Living shorelines are resilient to wind and waves, but may not be at the scale to address storm surge even though that is a prime consideration.
Path forward	Keep a broad perspective by promoting education and outreach to a diverse array of stakeholders, and adaptive management strategies at multiple scales.
	Conduct long-term quantitative monitoring of living shoreline projects, integrating research that combines biological and physical processes with input from engineers.
	Improve the process for permitting living shorelines in order to encourage nature-based approaches as opposed to armoring, including providing incentives to landowners.

the broad audience that are involved with living shorelines, the opportunities for learning as sites develop, and the incentives that could assist in navigating the often complicated permitting field. These all require a steady effort, from project development through implementation and measurement of success.

Overall, the findings of the contributed chapters show that nature-based approaches have proven effective, and the synthesis of these focal areas is meant to improve upon the foundation that has been developed. We encourage their reference and utility in the field, and anticipate further refinement as new information becomes available as sites progress through time (Gittman et al. 2016; La Peyre et al. 2014). In addition to these, we recommend incorporating an ecosystem services framework as outlined in the Introduction section by Arkema et al., and emphasize the gaps in knowledge addressed in the Summary section by Davis.

24.3 WHAT'S IN A NAME? RECOMMENDATIONS FOR PROPER USAGE

Following from our chapter in the Introduction section that presented the terminology and defined features of living shorelines, here we place into context the correct usage of living shorelines with other similar terms. The four terms that we will discuss here are (1) restoration, (2) living shorelines, (3) ecological engineering, and (4) novel ecosystems. Although all of these overlap, there are specifics associated with each of them (Figure 24.1). Restoration can simply be defined as "to restore" and implies a return to premodified natural conditions. The literature in this field is vast, with dedicated journals and books. As one example of how living shorelines can fit into the restoration context, we offer a quote from Perring et al. (2015): "Ongoing challenges include setting realistic, socially acceptable goals for restoration under changing environmental conditions, and prioritizing actions in an increasingly space-competitive world." Living shorelines definitely fit into this realm of environmental change with space at a premium and can help fulfill restoration

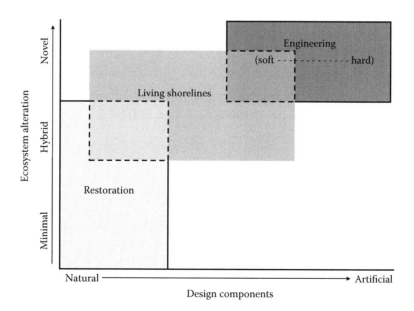

Figure 24.1 Conceptual diagram illustrating the degree of overlap of projects categorized as restoration, living shorelines, and ecological engineering, with a range of ecosystem alteration and design components.

goals. We have discussed the definition of living shorelines in the Introduction chapter, with further context given by the contributed chapters. As a lead in to ecological engineering, the definition of living shorelines by the recent Restore America's Estuaries report (RAE 2015) recognizes the incorporation of engineered elements where appropriate: "Any shoreline management system that is designed to protect or restore natural shoreline ecosystems through the use of natural elements and, if appropriate, manmade elements." Guidance documents often present a spectrum of green (softer) and gray (harder) stabilization techniques (NOAA 2015; SAGE 2015). Manmade elements of course utilize engineering, and definitions of ecological engineering typically include benefits to both humans and the natural environment (Chapman and Underwood 2011; Mitsch 2012; Odum et al. 1963), again linking the two sides of nature and humans this time from the engineering perspective so that infrastructure can have increased habitat benefits. Subcategories of hard and soft engineering designate the degree to which these may fall into the living shorelines realm. Last, the concept of novel ecosystems recognizes that the world is changing, and reverting to historic conditions often may not be feasible along developed waterfronts (Hobbs et al. 2009). Thus, restoration aims to push the system toward historical conditions. This is an important reality to acknowledge in a world where novel ecosystems are becoming common—that restoration is a goal, of which living shorelines and nature-based approaches can hope to reach, but not necessarily fully achieve. This should be viewed positively, at least in situations where full restoration is not possible, as that may be the most dependable solution for waterfronts with a history of development and ongoing human use. As an example phrase using all four of the discussed terms in context: when faced with management of a novel ecosystem, the restoration goals should incorporate aspects of living shorelines, which could include both soft and hard ecological engineering techniques.

Living shorelines should be implemented with the restoration goal at its foundation, given the pressure of alternative needs such as erosion control. Consistency is key, not only for clearly stating goals and objectives, but there are also benefits of building momentum and having national and international practices with similar guidelines (NOAA 2015). One question is whether the living shorelines bandwagon has progressed to the point of housing subdisciplines, similar to the term "restoration." When a certain threshold is reached, terminology moves beyond its simple definition and becomes a baseline as a reference for other developing techniques. This signifies achievement, as it provides an opportunity for new ideas to latch onto, but care must be taken not to misuse its intended purpose (Nordstrom 2014).

24.4 THE CHALLENGE: CROSSING BOUNDARIES AND PLANNING FOR THE FUTURE

At the start of editing this book, we as editors had several questions regarding how unified the living shorelines community was. To what extent are we working together across fields? To what extent are we working together across regions? It is apparent that there are key aspects that are both common and unique across regions and sections of the book (management, physical, biological). Methods that are applicable to areas with oyster reefs may not be applicable to areas without, yet the common goal of maintaining shoreline functions that are resilient to human development and sea level rise are the same. Different regions have specific habitats and species that are targeted for conservation and restoration goals, such as eelgrass habitats and endangered species that are regionally specific. Again, although the specifics often differ, the generalities of providing a living shoreline framework can help manage across systems. Permitting is necessary to implement living shoreline projects, design and construction are necessary to build them, and physical and biological monitoring are necessary to document whether living shoreline projects are stable and provide the target ecological functions. Although it can be difficult to address the differences across regions and specialities, we hope this book has moved the conversation forward to create a more synthesized approach, and given new ideas for practitioners to broaden their application.

One way to move the conversation forward is to increase the presence of virtual living shoreline forums, which can facilitate the flow of information from different regions and practitioners. Throughout the development of this book, we relied on the benefits of remote communication via video and audio Internet services, as many projects do. Promoting organized outreach and communication tools are essential for reaching across boundaries, as travel costs are high and can be daunting with limited time and funding. As modern communication continues to evolve, so must the expectations of how to manage the multiple needs of in-person and remote meetings; agencies can assist by promoting regular outreach activities that seek to reach a diverse audience both regionally and beyond local jurisdictions.

Our expectations for the reference utility of this book are that the breadth of coverage speaks across specialties and brings a common framework to managers, scientists, students, consultants, and others engaged in the design and implementation of living shorelines. Our advice for summaries in the future is to continue tackling the difficult questions that are dynamic through time, such as the following:

1. How can living shoreline projects help mitigate the impacts of sea level rise and adapt with climate change?
2. When do we know if we're making a difference at local to regional scales?
3. Are there "thresholds" in the amount of armoring and living shorelines that can be used to help plan the functional efficiency of shoreline ecosystems?
4. Are there specific cases and sites that should be prioritized for living shorelines, apart from availability of land? And how realistic are these due to issues of land acquisition and permitting?

Nature is constantly changing in our human-dominated world, and it is up to us to be adaptive and plan for the future (Bilkovic et al. 2016). By laying the framework of nature-based approaches and building upon past successes and failures, we can hope to maintain natural features in our shorelines while faced with the formidable climate change scenarios that are down the road (Arkema et al. 2013; Duarte et al. 2013; Popkin 2015). Innovative approaches are needed to address the complexities involved (Sutton-Grier et al. 2015), which can only be accomplished with continued experimentation, especially along urban shorelines (Toft et al. 2013). We thank the many contributors and reviewers of this book and look forward to the continued application and development of their ideas in the future.

REFERENCES

Arkema, K.K., G. Guannel, G. Verutes, S.A. Wood, A. Guerry, M. Ruckelshaus, P. Kareiva, M. Lacayo, and J.M. Silver. 2013. Coastal habitats shield people and property from sea-level rise and storms. *Nature Climate Change* 3:1–6.

Bilkovic, D.M., M. Mitchell, P. Mason, and K. Duhring. 2016. The role of living shorelines as estuarine habitat conservation strategies. *Coastal Management* 44: 161–174.

Chapman, M.G. and A.J. Underwood. 2011. Evaluation of ecological engineering of "armoured" shorelines to improve their value as habitat. *Journal of Experimental Marine Biology and Ecology* 400: 302–313.

Duarte, C.M., I.J. Losada, I.E. Hendriks, I. Mazarrasa, and N. Marba. 2013. The role of coastal plant communities for climate change mitigation and adaptation. *Nature Climate Change* 3: 961–968.

Gittman, R.K., C.H. Peterson, C.A. Currin, F.J. Fodrie, M.F. Piehler, and J.F. Bruno. 2016. Living shorelines can enhance the nursery role of threatened estuarine habitats. *Ecological Applications* 26: 249–263.

Hobbs, R.J., E. Higgs, and J.A. Harris. 2009. Novel ecosystems: Implications for conservation and restoration. *Trends in Ecology and Evolution* 24: 599–605.

La Peyre, M.K., A.T. Humphries, S.M. Casas, and J.F. La Perye. 2014. Temporal variation in development of ecosystem services from oyster reef restoration. *Ecological Engineering* 63: 34–44.

Mitsch, W.J. 2012. What is ecological engineering? *Ecological Engineering* 45: 5–12.

National Oceanic and Atmospheric Administration (NOAA). 2015. Guidance for considering the use of living shorelines. Technical report. 36 pp.

Nordstrom, K.F. 2014. Living with shore protection structures: A review. *Estuarine, Coastal, and Shelf Science* 150: 11–23.

Odum, H.T., W.L. Siler, R.J. Beyers, and N. Armstrong. 1963. Experiments with engineering of marine ecosystems. *Publications of the Institute of Marine Science, University of Texas* 9: 374–403.

Perring, M.P., R.J. Standish, J.N. Price, M.D. Craig, T.E. Erickson, K.X. Ruthrof, A.S. Whiteley, L.E. Valentine, and R.J. Hobbs. 2015. Advances in restoration ecology: Rising to the challenges of the coming decades. *Ecosphere* 6(8): 131. http://dx.doi.org/10.1890/ES15-00121.1

Popkin, G. 2015. Breaking the waves. *Science* 350: 756–759.

Restore America's Estuaries (RAE). 2015. *Living Shorelines: From Barriers to Opportunities*. Arlington, VA.

Systems Approach to Geomorphic Engineering (SAGE). 2015. Natural and structural measures for shoreline stabilization brochure.

Sutton-Grier, A.E., K. Wowk, and H. Bamford. 2015. Future of our coasts: The potential for natural and hybrid infrastructure to enhance the resilience of our coastal communities, economies and ecosystems. *Environmental Science and Policy* 51: 137–148.

Toft, J.D., A.S. Osgon, S.M. Heerhartz, J.R. Cordell, and E.E. Flemer. 2013. Ecological response and physical stability of habitat enhancements along an urban armored shoreline. *Ecological Engineering* 57: 97–108.

Index

Printed and bound by CPI Group (UK) Ltd, Croydon, CR0 4YY

24/10/2024

01778290-0018